W9-ABZ-706

PRIMATE BIOGEOGRAPHY

DEVELOPMENTS IN PRIMATOLOGY: PROGRESS AND PROSPECTS

Series Editor:
Russell H. Tuttle
University of Chicago, Chicago, Illinois

This peer-reviewed book series melds the facts of organic diversity with the continuity of the evolutionary process. The volumes in this series exemplify the diversity of theoretical perspectives and methodological approaches currently employed by primatologists and physical anthropologists. Specific coverage includes: primate behavior in natural habitats and captive settings; primate ecology and conservation; functional morphology and developmental biology of primates; primate systematics; genetic and phenotypic differences among living primates; and paleoprimatology.

ALL APES GREAT AND SMALL

Volume I: African Apes
Edited by Birute' M.F. Galdikas, Nancy Erickson Briggs, Lori K. Sheeran, Gary L. Shapiro and Jane Goodall

THE GUENONS: DIVERSITY AND ADAPTATION IN AFRICAN MONKEYS

Edited by mary E. Glenn and Marina Cords

ANIMAL MINDS, HUMAN BODIES

By W. A. Hillix and Duane Rumbaugh

COMPARATIVE VERTEBRATE COGNITION

Edited by Lesley J. Rogers and Gisela Kaplan

ANTHROPOID ORIGINS: NEW VISIONS

Edited by Callum F. Ross and Richard F. Kay

MODERN MORPHOMETRICS IN PHYSICAL ANTHROPOLOGY

Edited by Dennis E. Slice

BEHAVIORAL FLEXIBILITY IN PRIMATES: CAUSES AND CONSEQUENCES

By Clara B. Jones

NURSERY REARING OF NONHUMAN PRIMATES IN THE 21ST CENTURY

Edited by Gene P. Sackett, Gerald C. Ruppenthal and Kate Elias

NEW PERSPECTIVES IN THE STUDY OF MESOAMERICAN PRIMATES: DISTRIBUTION, ECOLOGY, BEHAVIOR, AND CONSERVATION

Edited by Alejandro Estrada, Paul A. Garber, Mary S. M. Pavelka, and LeAndra Luecke

HUMAN ORIGINS AND ENVIRONMENTAL BACKGROUNDS

Edited by Hidemi Ishida, Martin Pickford, Naomichi Ogihara and Masato Nakatsukasa

PRIMATE BIOGEOGRAPHY

Edited by Shawn M. Lehman and John G. Fleagle

PRIMATE BIOGEOGRAPHY
Progress and Prospects

Shawn M. Lehman
John G. Fleagle

With 72 Figures

 Springer

BOWLING GREEN STATE
UNIVERSITY LIBRARY

Shawn M. Lehman
Dept. of Anthropology
University of Toronto
Toronto, ON M5S3G3
Canada
slehman@chass.utoronto.edu

John G. Fleagle
Dept. of Anatomical Sciences
Health Sciences Center
Stony Brook University
Stony Brook, New York 11794
jfleagle@notes.cc.sunysb.edu

Library of Congress Control Number: 2006921645

ISBN-10: 0-387-29871-1   e-ISBN 0-387-31710-4
ISBN-13: 978-0387-29871-9

Printed on acid-free paper.

© 2006 Springer Science+Business Media, LLC
All rights reserved. This work may not be translated or copied in whole or in part without the written
permission of the publisher (Springer Science+Business Media, LLC, 233 Spring Street, New York,
NY 10013, USA), except for brief excerpts in connection with reviews or scholarly analysis. Use in
connection with any form of information storage and retrieval, electronic adaptation, computer software,
or by similar or dissimilar methodology now known or hereafter developed is forbidden.
The use in this publication of trade names, trademarks, service marks, and similar terms, even if they
are not identified as such, is not to be taken as an expression of opinion as to whether or not they are
subject to proprietary rights.

Printed in the United States of America.   (TB/MVY)

9 8 7 6 5 4 3 2 1

springer.com

CONTENTS

PREFACE

Biogeography, the study of the distribution of organisms over the surface of the earth, plays a central role in our understanding of virtually all aspects of the biology of primates and other animals. Biogeography is critical for determining systematics and mechanisms of speciation and for evaluating population genetics and demography. The distribution of primates relative to aspects of climate and habitat, including altitude, forest type, and food availability, form the basis for our understanding of ecological and behavioral adaptations. Likewise, the biogeography of primates in the past is a major component of our understanding of their evolutionary history. Despite the importance of biogeography in our understanding of primate evolution and biology, and the broad representation of research on this subject in journals, field guides, and edited volumes on different regions of the world, Primate Biogeography has never before been the focus of a single volume.

Our goal in bringing together these papers on the biogeography of primates, past and present, is to provide an introduction to Primate Biogeography as a discipline, to draw attention to the many factors that may influence the distribution of primates, and to demonstrate the wide range of approaches that are available to understanding the distribution of this order. In many ways, primates are an ideal subject for studies of biogeography. The systematics of the group is well documented and is the subject of constant, ongoing revision. Compared with many other groups of mammals, primates are relatively large, colorful, noisy, mostly diurnal, and relatively easy to locate and identify in the field. Unlike birds or fishes, their movement patterns are relatively easy to document. Moreover, in recent years, the genetics of many primate species have been widely studied and published. In addition, primates are relatively abundant in the fossil record of most continents and the broad outlines of their evolutionary history are well established. Thus the raw data for studies of primate biogeography are more abundant and available than for virtually any other group of mammals.

Except for the introductory chapter and the final section, which provide historical and evolutionary overviews of primate biogeography, we have organized the volume by major geographical regions. This organization reflects the fact that living primates are largely restricted to the tropics and that the faunas of the Neotropics, Africa, Madagascar, and Asia, respectively, are unique to each region. This organization also highlights the diversity of approaches that can be used to reconstruct the biogeography of the primates in a single part of the world. Each of the sections begins with a short introduction to the major geographical features of the region and a summary of the primates in the area.

Chapter 1, "Primate Biogeography: A Review", provides a broad review of primate biogeography in the context of its development as a discipline. It provides an introduction to the many approaches used in the study of biogeography and a context for subsequent chapters. The biogeography of primates in Central and South America are the subject of Chapters 2, 3, and 4. Each of these chapters demonstrates a different approach to the subject including, statistical analysis of distribution patterns, studies of genetic similarities, and multivariate analysis of field surveys. African primates are the subject of Chapters 5, 6, and 7. Again, each of the chapters reflects a different approach to understanding the biogeography of primates and its many aspects, including genetics, comparative ecology, and comparative anatomy. Chapters 8, 9, and 10, are devoted to the primates of Madagascar, perhaps the most unique fauna of primates anywhere on earth. The authors of these chapters provide insights based on an overview of field research, detailed studies of molecular systematics, and a broad consideration of environmental variables. Chapters 11 and 12 discuss factors that have influenced the distribution of primates in two different regions of Asia—mainland Southeast Asia and the Sunda Shelf. The authors emphasize the roles of geological history and human activity in generating current patterns of primate distribution. Chapters 13, 14, 15, and 16 extend our perspective into the evolutionary past and look at primate biogeography and phylogeny over the past 55 million years.

This project has been possible only with the help of many people. Some of the contributions were initially presented in a symposium sponsored by The American Association of Physical Anthropologists. All of the papers benefited greatly from the time and insights offered by numerous reviewers. Jason Kamilar and Luci Betti provided invaluable organization and artistic skills, respectively. At Kluwer/Plenum/Springer we owe a special debt to Andrea Macaluso, Krista Zimmer, and the Series Editor Russ Tuttle.

LIST OF CONTRIBUTORS

K. Christopher Beard, Section of Vertebrate Paleontology, Carnegie Museum of Natural History, 4400 Forbes Avenue, Pittsburgh, PA 15213. Email: beardc@clpgh.org

Jane Philips-Conroy, Department of Anthropology, Washington University, St. Louis, MO 63130 and Department of Anatomy and Neurobiology, Washington University Medical School, St. Louis, MO 63110. Email: baboon@thalamus.wustl.edu

Todd. R. Disotell, Department of Anthropology, New York University, 25 Waverly Place, New York, NY 10003. Email: todd.disotell@nyu.edu

Julie A. Ellsworth, Department of Biology, Truckee Meadows Community College, 7000 Dandini Boulevard, Reno NV 89512-3999. Email: jellsworth@tmcc.edu

John G. Fleagle, Department of Anatomical Sciences, Health Sciences Center, Stony Brook University, Stony Brook, NY 11794-8081. Email: jfleagle@notes.cc.sunysb.edu

Jörg U. Ganzhorn, Department Animal Ecology and Conservation, Biozentrum Grindel, Martin-Luther-King-Platz 3, 20146 Hamburg, Germany. Email: ganzhorn@zoologie.uni-hamburg.de

Christopher C. Gilbert, Interdepartmental Doctoral Program in Anthropological Sciences, Stony Brook University, Stony Brook, NY 11794-4364. Email: cgilbert@ic.sunysb.edu

M. Katherine Gonder, Department of Biology, University of Maryland, College Park, MD 20903. Email: mgonder@umd.edu

Steven M. Goodman, Field Museum of Natural History, 1400 South Lake Shore Drive Chicago, Illinois 60605, and WWF, BP 738, Antananarivo (101), Madagascar. Email: SGoodman@wwf.mg

Colin P. Groves, Australian National University, School of Archaeology and Anthropology, Canberra, A.C.T. 0200, Australia. Email:colin.groves@anu.edu.au

Terry Harrison, Center for the Study of Human Origins, Department of Anthropology, New York University, 25 Waverly Place, New York, NY 10003. Email: terry.harrison@nyu.edu

Kellie L. Heckman, Department of Ecology and Evolutionary Biology, Yale University, P.O. Box 208105, 21 Sachem Street, New Haven, CT 06520. Email: kellie.heckman@yale.edu

Christopher P. Heesy, Department of Anatomy, New York College of Osteopathic Medicine, Old Westbury, New York 11568. Email: cheesy@nyit.edu

Guy A. Hoelzer, University of Nevada Reno, Biology Department, m/s 314, Reno, NV 89557. Email: hoelzer@unr.edu

Jason M. Kamilar, Interdepartmental Doctoral Program in Anthropological Sciences, Stony Brook University, Stony Brook, NY 11794-4364. Email: jason.kamilar@sunysb.edu

John S. Krigbaum, Department of Anthropology, University of Florida, Gainesville, FL 32611-7305. Email: krigbaum@anthro.ufl.edu

Shawn M. Lehman, Department of Anthropology, University of Toronto, 100 St. George Street, Toronto, Ontario, M5S 3G3, CANADA. Email: slehman@chass.utoronto.ca

Jessica Manser, Department of Anthropology, New York University, 25 Waverly Place, New York, NY 10003. Email: JMM6294@nyu.edu

W. Scott McGraw, Department of Anthropology, The Ohio State University, 1680 University Drive, Mansfield, OH 44907. Email: mcgraw.43@osu.edu

Erik Meijaard, Australian National University, School of Archaeology and Anthropology, Canberra, A.C.T. 0200, Australia. Email: emeijaard@tnc.org

Stephan Nash, Conservation International, 1919 M Street NW, Washington D.C. 20036, USA and Department of Anatomical Sciences, Health Sciences Center, Stony Brook University, Stony Brook NY 11794-8081, USA. Email: stephen.nash@sunysb.edu

Patrick M. O'Connor, Department of Biomedical Sciences, Ohio University, Athens, OH 45701. Email: oconnorp@ohiou.edu

W. Prince, Iwokrama International Centre for Rain, Forest Conservation and Development, Georgetown, Guyana. Email: wprince@iwokrama.org

James B. Rossie, Department of Anthropology, Stony Brook University, Stony Brook, NY 11794-4364. Email: jrossie@notes.cc.sunysb.edu

Karen E. Samonds, Biology Department, Mount Holyoke College, 50 College Street, South Hadley, Massachusetts 01075. Email: ksamonds@ic.sunysb.edu

Erik R. Seiffert, Department of Earth Sciences and Museum of Natural History, University of Oxford, Parks Road, Oxford, OX1 3PR, United Kingdom. Email: erik.seiffert@earth.ox.ac.uk

Nancy J. Stevens, Department of Biomedical Sciences, Ohio University College of Osteopathic Medicine, Athens, OH 45701. Email: stevensn@ohio.edu

Robert W. Sussman, Department of Anthropology, Washington University, St. Louis, MO 63130. Email: rwsussma@artsci.wustl.edu

Urs Thalmann, Anthropological Institute and Museum, University of Zürich, Winterthurerstr. 190, 8057 Zürich, Switzerland. Email: uthal@aim.unizh.ch

Anne D. Yoder, Departments of Biology & BAA, Duke University, Box 90338, Durham, NC 27708. Email: anne.yoder@duke.edu

CHAPTER ONE

Biogeography and Primates: A Review

Shawn M. Lehman and John G. Fleagle

ABSTRACT

In this paper, we present an introduction to primate biogeography at a continental level and then review the literature as it pertains to primate studies. Primate species diversity is highest in the Neotropics and Asia. Most primates range into rain/humid forests in Africa, Asia, and the Neotropics. Asia contains the highest total number of primate species (N = 38) that are considered to require conservation attention, followed closely by the Neotropics (N = 33 species). These biogeographic patterns reflect complex phylogenetic, geologic, and ecological processes. The various biogeographic theories and models used to explain these patterns can be organized into several broad categories (1) descriptive studies, (2) comparative-quantitative approaches, (3) refugia theory, (4) phylogenetic approaches, (5) community ecology, and (6) conservation biology. Descriptive models have been derived from distribution data obtained during collecting expeditions. These models focused on geographic variations in species characteristics and barriers to dispersal (e.g., Gloger's Rule, Bergmann's Rule, Allen's Rule, river barrier hypothesis). With the advent of digitized statistical procedures, these barriers became testable biogeographic hypotheses using comparative-quantitative models. Thus, many researchers have noted the importance of rivers as

Shawn M. Lehman • Department of Anthropology, University of Toronto, Toronto, Ontario M5S 3G3 **John G. Fleagle** • Department of Anatomical Sciences, Health Sciences Center, Stony Brook University, Stony Brook, NY 11794-8081

Primate Biogeography, edited by Shawn M. Lehman and John G. Fleagle.
Springer, New York, 2006.

1

geographical subdivisions of populations of a species. Comparative-quantitative models have also involved studies of species-area and distribution-abundance relationships. Generally, larger areas are more species rich and widely distributed primates tend to exist at higher densities. Many researchers have also investigated various ecological correlates (e.g., rainfall, latitude) to patterns of primate species richness. There has been considerable debate regarding the importance of Pleistocene Refugia for understanding the historical biogeography of primates. Phylogenetic or cladistic biogeography focuses on shared derived characters, which can be used to reconstruct biogeographical history. The presence or absence of species within a geographic area has been investigated extensively through studies of community ecology. Similarities between primate communities are most likely if they share a common biogeographic history. Composition of primate communities can also reflect evolutionary niche dynamics. Finally, researchers studying primate conservation biology have synthesized methods from various biogeographic models to understand and predict primate rarity and extinction events. Much of the renewed interest in primate biogeography tends to focus on the spatial and temporal patterns that influence species origins and diversity.

Key Words: Primates, ecological biogeography, historical biogeography, diversity, Neotropics, Africa, Madagascar, Asia.

INTRODUCTION

Biogeography is the study of the distribution and diversity of organisms in space and time (Cox and Moore, 2005). There are two main approaches to producing and testing hypotheses of species distribution and diversity: (1) ecological biogeography and (2) historical biogeography (Lomolino *et al.*, 2005). Ecological biogeography is used to investigate distribution and diversity patterns based on the interactions between an organism and its physical and biotic environment (Huggett, 2004). Historical biogeography determines the series of events that led to the origin, dispersal, and extinction of tropical taxa (Crisci *et al.*, 2003). Using this approach, researchers have explained the biogeography of plants and animals as the result of the appearance of barriers and the disappearance of barriers (Wiley, 1988). The biogeography of many organisms is likely the result of a complex relationship between ecological and historical factors (e.g., Bush, 1994; Tuomisto and Ruokolainen, 1997; Lomolino *et al.*, 2005). In this paper, we present an introduction to primate biogeography at a continental level and then review the literature as it pertains to primate studies in order to put the papers from this volume in a broader historical perspective.

Primate Biogeography at the Continental Level

There are approximately 348 exant primate species in the world (Appendix 1), although this number varies depending on which taxonomy is used (e.g., Rylands et al., 2000; Groves, 2001; Grubb et al., 2003; Brandon-Jones et al., 2004; Isaac et al., 2004). Morever, there have been at least 46 new species discovered or redescribed in the last 10 years (Rylands, 1998; Silva and Noronha, 1998; van Roosmalen, 1998; Kobayashi and Langguth, 1999; Rasoloarison et al., 2000; Thalmann and Geissmann, 2000; van Roosmalen et al., 2000; Rylands et al., 2002; van Roosmalen et al., 2002; Mayor et al., 2004; Jones et al., 2005). Extant primates are found almost exclusively in one of the following four tropical regions: Neotropics (Central and South America), Africa, Madagascar, and southern and eastern Asia. In an effort to obtain a broad overview of primate ecology in a biogeographical perspective we have summarized broad patterns of primate ecology of living primates by continent (Table 1). In our overall semi-quantitative review of primate adaptations and biogeography, we have relied heavily on secondary sources (e.g., Rowe, 1996) because they provide a breadth of data reduced to a common format.

Species diversity is highest in the Neotropics and Asia. At higher taxonomic levels, the most genera are located in Africa whereas the most families are found in Madagascar and the Neotropics. The high taxonomic diversity for Madagascar is remarkable because it is considerably smaller in area (587,040 km^2) than any of the other regions (Reed and Fleagle, 1995) and only 10–20% of the original forest cover remains in this country (Green and Sussman, 1990; Du Puy

Table 1. Primate species, genera, and family diversity in four main biogeographic regions

Region	Species	Genera	Families
Neotropics	116	18	5
Africa	83	21	4
Madagascar	59	14	5
Asia	90	16	4
Total	348	69	18

and Moat, 1998). There are also extremely high levels of endemicity (81%–100%) for primates, vascular plants, reptiles, and amphibians in Madagascar (e.g., Ganzhorn *et al.*, 1999; Garbut, 1999; Goodman and Benstead, 2005).

Most primates range into rain/humid forests in Africa, Asia, and the Neotropics (Table 2). Patterns of forest use are somewhat different for lemurs. Of the 48 lemur species for which there are habitat data, 64.0% (N = 31) range into dry forests. Exploitation of woodlands and wooded grasslands is most common among African primates, which allows these animals to range over a wider area than that covered only by forests. Neotropical and African primates use riparian habitats more often than taxa in either Madagascar or Asia. Use of swamp and montane habitats is common among Neotropical and Asian primates. In the Neotropics, numerous primate species, particularly those in the Callitrichidae, exploit secondary/edge habitats.

Fruit is exploited by many primate species in each region, and particularly in the Neotropics where all species studied to date eat at least some fruit (Table 3). Leaves are eaten by many primates in all regions, but are exploited by only a few taxa in the Neotropics. Conversely, a higher proportion of primates exploit gums and tree exudates in the Neotropics. Insects and fauna are eaten commonly by primates in the Neotropics, Africa, and Asia; but infrequently by those in Madagascar. Although few lemurs exploit seeds as food, many species eat flowers.

Primate conservation priorities at the species level differ between regions (Table 4). Asia contains the highest total number of primate species (N = 38) that are considered to require conservation attention, followed closely by the Neotropics (N = 33 species). The Neotropics contain 42.8% (N = 9) of the 21 total primate species that are critically endangered worldwide (*Ateles hybridus*, *Brachyteles hypoxanthus*, *Callicebus barbarabrownae*, *C. coimbrai*, *Cebus xanthosternos*, *Leontopithecus caissara*, *L. chrysopygus*, *Oreonax flavicauda*, and *Saguinus bicolor*). There are six primate species listed as critically endangered in Asia (*Hylobates moloch*, *Macaca pagensis*, *Pongo abelii*, *Rhinopithecus avunculus*, *Trachypithecus delacouri*, and *T. poliocephalus*). There are four critically endangered primates in Madagascar (*Eulemur albocollaris*, *Hapalemur aureus*, *Prolemur simus*, and *Propithecus tattersalli*) and two in Africa (*Piliocolobus rufomitratus* and *P. tephrosceles*). Of the 51 primate species considered to be endangered, 21 are located in Asia and 13 in Africa. The Neotropics contain 33.3% (N = 14) of the 42 primate species listed as vulnerable worldwide.

Table 2. Percentage of primate species that range into eight main habitats in the Neotropics, Africa, Madagascar, and Asia

Region	N species	% total recognized species	Wet/humid forests	Dry/deciduous forests	Woodland forests	Riparian forests	Spiny forests/scrub	Swamp forests	Montane forests	Edge/secondary forests
Neotropics	84	72.4	91.7	22.6	3.6	33.3	3.6	21.4	14.3	44.0
Africa	60	72.3	80.0	43.3	30.0	38.3	10.0	11.7	8.3	31.7
Madagascar	50	84.7	54.0	64.0	8.0	22.0	8.0	2.0	8.0	30.0
Asia	62	68.9	91.9	27.4	1.6	11.3	1.6	19.4	32.3	24.2
Total	256	73.6	81.6	36.7	10.2	27.0	5.5	14.8	16.0	33.6

Table 3. Percentage of primate species that exploit seven main food categories in the Neotropics, Africa, Madagascar, and Asia

Region	N species	% total recognized species	Fruit	Leaves	Exudate	Insects/fauna	Seeds	Flowers	Other
Neotropics	70	60.3	100.0	38.6	38.6	67.1	40.0	38.6	30.0
Africa	55	66.3	81.8	65.5	10.9	67.3	34.5	25.5	10.9
Madagascar	55	93.2	69.1	52.7	16.4	32.7	9.1	56.4	25.5
Asia	52	57.8	96.2	84.6	3.8	59.5	42.3	57.7	46.2
Total	232	66.7	89.8	60.2	19.5	58.8	32.7	45.1	28.8

Table 4. Number of primate species at three levels of conservation risk in the
Neotropics, Africa, Madagascar, and Asia

Region	No. critically endangered	No. endangered	No. vulnerable	Total	% total recognized species
Neotropics	9	10	14	33	28.4
Africa	2	13	7	22	26.5
Madagascar	4	7	10	21	35.6
Asia	6	21	11	38	42.2
Total	21	51	42	114	32.8

The above biogeographic patterns reflect complex phylogenetic, geologic, and ecological processes (Eisenberg, 1979; Terborgh and van Schaik, 1987; Ayres and Clutton-Brock, 1992; Reed and Fleagle, 1995; Fleagle and Reed, 1996; Pastor-Nieto and Williamson, 1998; Wright, 1999; Harcourt, 2000b; Laws and Eeley, 2000; Harcourt *et al.*, 2005). For organizational structure, we have grouped biogeographic theories and models into several broad categories (1) descriptive studies, (2) comparative-quantitative approaches, (3) refugia theory, (4) phylogenetic approaches, (5) community ecology, and (6)conservation biology. Historically, descriptive models have been derived from distribution data obtained during collecting expeditions (e.g., Wallace, 1853; Darwin, 1859; Bates, 1863; Wagner, 1868). These models focus on differences in species distribution and abundance due to barriers to dispersal at the continental level. The comparative and quantitative models enable researchers to narrow the geographic focus to patterns of local species richness. Two primary variables have been extensively examined in these models: species number and some ecological characteristic(s) of the environment (e.g., area, latitude, and rainfall). An especially important series of biogeography theories have emphasized the importance of Pleistocene refugia. The phylogenetic approach to biogeohraphy developed from Hennig's (1966) method of analyzing taxa with respect to shared characters which have been derived from an ancestor common only to themselves. In phylogenetic or cladistic biogeography, shared characters were replaced by shared geographic regions. The resulting area cladograms could then be used to reconstruct the biota of a historic region. Community ecology incorporates the perspective on how species composition and interactions relate to biogeographic processes. Conservation biology synthesizes methods from other biogeographic approaches to understand and predict primate rarity and extinction events at various levels (e.g., species, sites, landscape, regions, continent, and global). From these models, precise predictions of the distribution and diversity of species could be generated and tested.

Descriptive Models

There has been an explosion of research interest on the identification, classification, and study of mammals during the last 150 years (Wallace, 1853; Darwin, 1859; Grandidier, 1875–1921; Hesse *et al.*, 1937; Mayr, 1942; Darlington, 1957; Simpson, 1965; Futuyma, 1998; Groves, 2001). As researchers catalogued and analyzed new species, they began to develop rules and general descriptive models to explain biogeographic processes. For example, geographic differences in climate were used to explain clinal variations in skin pigmentation (Gloger's Rule), body size (Bergmann's Rule), and appendage length (Allen's Rule). Increased research interest into allopatric speciation led researchers to investigate how barriers, such as rivers, caused geographical subdivision of populations of a species. Rivers have long been thought to influence the biogeography of tropical taxa (e.g., Wallace, 1853; Darwin, 1859; Bates, 1863; Wagner, 1868). Wallace (1853) is credited with first proposing that rivers influence the geographic distribution of tropical species:

> During my residence in the Amazon district, I took every opportunity of determining the limits of species, and I soon found that the Amazon, Rio Negro and the Madeira formed the limits beyond which certain species never passed (p. 5).

Observations such as this led to the formulation of the river theory of biogeography. River theory holds that differentiation of tropical biota occurred as the result of populations being split into isolated subpopulations by networks of rivers. The constant processes of erosion and silt deposition cause changes in the course of a tropical river. Forest habitats along the riverbanks are also altered as a river changes course. The combination of meandering rivers and mosaic forests creates habitat heterogeneity, which is associated with increased opportunities to specialize and avoid interspecific competition (Salo *et al.*, 1986; Räsänen *et al.*, 1987). River based explanations have been used by researchers studying the distribution and diversity of birds (e.g., Sick, 1967;

Remsen and Parker, 1983; Caparella, 1992), reptiles (Rodrigues, 1991), and non-volant mammals (Eisenberg, 1981; Eisenberg, 1989). However, recent studies of patterns of genetic population differentiation in rodents (Patton *et al.*, 1994) and frogs (Gascon *et al.*, 1996, 1998) along the Jurua River in Brazil do not support the river barrier hypothesis. Although the population structure for some loci in the rodents and frogs were consistent with differentiation along opposite river banks, the results were due largely to substantial differentiation at one or a few collecting localities. Gascon and co-workers (1998) concluded that patterns of geographic variation in four frog species were the result of the sampling region being a zone of secondary contact.

Rivers and their floodplains have been shown to influence the adaptive radiation and distribution of Malagasy strepsirhines (Martin, 1972; Tattersall, 1982; Meyers *et al.*, 1989; Thalmann and Rakotoarison, 1994; Goodman and Ganzhorn, 2003), New World monkeys (Hershkovitz, 1968; Hershkovitz, 1977; Eisenberg, 1979; Hershkovitz, 1984; Ayres, 1986; Hershkovitz, 1988; Cheverud and Moore, 1990; Froehlich *et al.*, 1991; Ayres and Clutton-Brock, 1992; Peres *et al.*, 1996; Wallace *et al.*, 1996; Peres, 1997; Lehman, 1999; Lehman, 2004), Old World monkeys (Booth, 1958; Grubb, 1990; Colyn and Deleporte, 2002), and apes (Hill, 1969; Gonder *et al.*, 1997). For example, Ayres and Clutton-Brock (1992) conducted a preliminary biogeographic survey of the distribution of Amazonian primates and found that similarity of species across riverbanks was negatively correlated with river discharge, length/discharge, and width. There was also a negative correlation between the distance from the headwaters of the Amazon River and the similarity of primate species between its banks. However, some studies of Old World monkeys do not support the river barrier hypothesis. Colyn (1988) and Oates (1988) reviewed data on the distribution of guenons in western Africa and in Zaire. They concluded that although rivers may somewhat impede gene flow in guenons, there is little evidence that rivers are major barriers to the dispersal of forest monkeys in western and central Africa (but see Colyn and Deleporte, 2002). Similarly, Meijaard and Groves (this volume) emphasize that the effect of specific rivers as biogeographic barriers is influenced by a variety of other factors, including their history.

In the mid-20th century, new data on tropical geology and animal distribution lead to the theory of panbiogeography. Panbiogeography focuses on the coevolution of geographic barriers and biotas (Croizat, 1958, 1976). This theory employs the notion that biotas evolve together with barriers (Cracraft, 1988; Cracraft and Prum, 1988b). Thus, the barrier cannot be older than the disjunction. Darwin (1859) recognized the role of vicariance in evolution when he proposed that: "barriers of any kinds, or obstacles to free migration, are related in a close and important manner to the differences between the productions of various regions" (p. 347). If populations are isolated by vicariance events for extended periods of time, then speciation may occur via allopatry (reviewed in Wiens and Graham, 2005). The basic method is to plot the distributions of organisms on maps and connect the disjunct distribution areas together with lines called tracks. A track is the spatial coordinates of a species or groups of species. If the superimposed tracks of unrelated species overlap, the resulting overlapped lines indicate the presence of ancestral biotas that were fragmented by geologic or climatic change. For example, Croizat (1976) suggested that faunal differences to the east and west of the Andes are due to the uplift of this mountain range.

Comparative-Quantitative Approaches

Quantitative approaches developed since the middle of the mid-20th century have vastly improved our understanding of biogeography. One of the first and best examples of ecological patterns that grew out of analyses of these data was the relationship between species number and area (Rosenzweig, 1995). Species-area relationships predict that there is a positive relationship between the number of species and the size of an area (Preston, 1962; Williams, 1964; MacArthur and Wilson, 1967). This relationship is expressed as the equation:

$$S = CA^z$$

which is usually expressed in the log-transformed form,

$$\log S = \log C + z \log A$$

where S is the number of species, A is the area, z the slope of the line, and C is a constant usually referred to as the intercept. Species-areas relationships have been investigated at various biogeographic levels in primates (Reed and Fleagle, 1995; Jones, 1997; Bates et al., 1998; Eeley and Laws, 1999; Harcourt, 1999; Laws and Eeley, 2000; Lomolino, 2000; Biedermann, 2003; Lehman, 2004; Harcourt and Doherty, 2005). For example, Reed and Fleagle

(1995) documented a high correlation ($R^2 = 0.87$) between the number of primate species and the area of rain forest for major continents (South America, Africa, and SE Asia) and large islands (Madagascar, Borneo, Sumatra, and Java).

Increased understanding of species-area relationships and the role of behavior and diet in determining an animal's ability to persist in habitats of varying size led to the ecological specialization hypothesis (Hanski, 1982; Brown, 1984; Hanski et al., 1993; Hanski and Gyllenberg, 1997; Irschick et al., 2005). According to this hypothesis (Brown, 1984), species that exploit a wide range of resources (generalists) are both locally common (high density) and widely distributed, whereas species that exploit a narrow range of resources (specialists) have a limited distribution and tend to be locally uncommon (low density). Studies of ecological specialization in primates have provided conflicting results (Arita et al., 1990; Jones, 1997; Eeley and Foley, 1999; Peres and Janson, 1999; Harcourt et al., 2002; Harcourt, 2004; Lehman, 2004; Harcourt et al., 2005). At the global level, Wright and Jernvall (1999) found a "remarkably linear" relationship between the geographic range of primates and habitat breadth, but not dietary breadth. Conversely, Harcourt et al. (2002) found that dietary breadth was the only trait to covary with rarity in primate genera. Finally, Eely and Foley (1999) documented positive relationships between species range size and both habitat breadth ($r = 0.851$) and dietary breadth ($r = 0.634$) in African anthropoid primates. Recent studies have revealed the need to refine methods used to test species-area relationships and associated models, such as ecological specialization (Vazquez and Simberloff, 2002; Fernandez and Vrba, 2005a; Irschick et al., 2005). For example, dietary niche breadth is often measured by summing the total number of food categories (fruit, leaves, flowers, insects, etc.) exploited by a species (Eeley and Foley, 1999; Wright and Jernvall, 1999; Harcourt et al., 2002). It is important to note that this dietary categorization does not discriminate between dietary type breadth (number of food categories exploited) and dietary species diversity (number of plant species exploited). For example, a hypothetical primate species could be a dietary type specialist if it exploits only two food categories (e.g., fruits and leaves) but a dietary species generalist if it exploits hundreds of plant species within each of these two food types. Irschick et al. (2005) argued that specialization should be measured using data on resource availability and exploitation, and that researchers should integrate phylogenetic data into their models.

Ecological gradients also influence the biogeography of many tropical organisms. Many abiotic and biotic factors form a gradient within the environment (Hutchinson, 1957). Although some species are eurytopic (ecologically tolerant) and others are stenotopic (ecologically intolerant), each can survive within only a certain environmental range (range of optimum). This range is bounded at both ends of the gradient by zones of physiological stress, which are areas where a species finds it increasingly difficult to survive. Thus, a variety of environmental gradients (e.g., temperature, humidity, and latitude as well as geological features) may influence primate biogeography (Stevens and O'Conner, this Volume; Kamilar, this volume).

Correlates between rainfall and primate diversity have been investigated at the continental level (Reed and Fleagle, 1995; Cowlishaw and Hacker, 1997; Kay et al., 1997; Peres and Janson, 1999). Reed and Fleagle (1995) found a high correlation between species diversity and mean annual rainfall for Africa ($R^2 = 0.75$), Madagascar ($R^2 = 0.70$), and South America ($R^2 = 0.67$). They concluded that although more data are needed on specific abiotic and biotic factors, primate diversity at the global and continental levels is highly correlated with geography and climate. In another example, Kay et al. (1997) found that primate species richness in South America exhibits a unimodal relationship with rainfall; peaking at ca. 2500 mm and then declining. They then reanalyzed Reed and Fleagle's (1995) data for Asia and found similar results. Kay and co-workers (1997) concluded that in areas with very high rainfall, soil leaching depletes nutrient levels and cloud cover reduces the light available for solar radiation for plants. Thus, plant productivity and primate species richness actually decline in areas of highest rainfall. Peres and Janson (1999) conducted a zoogeographical review of primate species distribution and environmental factors at 185 forest sites in the Neotropics. Their data did not support the hypothesis that primate richness is correlated with rainfall. Instead, they suggested that this relationship holds only in deciduous (dry) closed canopy forests where precipitation may be a limiting abiotic factor. In evergreen rain forests, where rainfall is not a limiting factor, precipitation is not a major determinant of primate richness in the Neotropics. Many researchers have cited geographic variation in rainfall as the proximate factor influencing lemur evolutionary ecology (Albrecht et al., 1990; Godfrey et al., 1990; Albrecht and Miller, 1993; Ravosa et al., 1993, 1995; Wright, 1999; Ganzhorn, 2002; Godfrey et al., 2004; Lehman et al., 2005). For example, resource seasonality may apply to some extant Indriidae (*Indri, Avahi,* and *Propithecus*) in that the largest *Propithecus* are found in the eastern humid forests with progressively smaller forms being found in the dry forests of western, northern, and southern Madagascar (Albrecht *et al.*, 1990; Ravosa *et al.*, 1993, 1995). Seasonal fluctuations in rainfall are more pronounced and the length of the dry season tends to be longer in dry forests compared to humid forests (Ganzhorn, 1994; Ganzhorn *et al.*, 1997; Ganzhorn, 2002). Based on these biogeographic data, low annual rainfall and a long dry season should produce strong selective pressures for larger adult body size in indriids (Ravosa *et al.*, 1995). However, Lehman *et al.*, (2005) investigated ecogeographic size variations in sifakas and found a positive rather than negative correlation between body size and rainfall. This positive relationship may reflect reduced leaf and fruit quality due to nutrient leaching from soils in areas of high rainfall in Madagascar.

Latitudinal gradients have been suggested to influence primate richness and diversity (Cowlishaw and Hacker, 1997; Gaston *et al.*, 1998; Peres and Janson, 1999; Harcourt, 2000b; Böhm and Mayhew, 2005). The mechanisms controlling latitudinal variation in species richness and range size are poorly understood (for review, see Gaston *et al.*, 1998 and Willig, 2003). Brown (1984) argued that range size decreases in areas of high species richness because of increased levels of interspecific competition. Conversely, Stevens (1989) suggested that greater ecological flexibility of high-latitude species enables them to exist in ephemeral populations at lower altitudes. Many researchers have linked patterns of species richness to the combined effects of latitude and rainfall (Schall and Pianka, 1978; Stevens, 1989; Pagel *et al.*, 1991; Ruggiero, 1994; Cowlishaw and Hacker, 1997; Kay *et al.*, 1997; Pastor-Nieto and Williamson, 1998; Conroy *et al.*, 1999; Harcourt, 2000b; Harcourt and Schwartz, 2001; Harcourt *et al.*, 2002; Fernandez and Vrba, 2005b). Regions close to the equator exhibit increased habitat heterogeneity and rainfall, which tend to result in more niches and higher mammalian species richness (Emmons, 1999). However, in a recent study, Böhm and Mayhew (2005) used historical biogeography techniques to investigate patterns of species richness for primates in Africa and Asia. They found that these patterns result from the passage of time since colonization and rates of cladogenesis rather than latitude. Cowlishaw and Hacker (1997) tested Rapoport's rule, that latitudinal ranges of species become progressively smaller toward the equator, using the distribution and diversity of 64 species of African primates. Although latitude only influenced the geographical range of species south of the equator, rainfall was a better predictor of the geographic range of African primates north of the equator and south of the equator. Peres and

Janson (1999) reviewed the effect of latitudinal gradients on primate species richness in the Neotropics. They found that within latitudinal belts the number of primate species found in Central and South America was very weakly correlated with latitude.

Dispersal biogeography developed from attempts to correlate present day distribution patterns with dispersal of ancestral species (i.e., historical biogeography). Dispersal biogeography holds that species move from a center of origin and undergo jump dispersal across pre-existing barriers to outlying areas (Cox and Moore, 2005). Understanding the distribution of fossils is essential because the oldest fossils are presumed to be located near the center of origin. This model requires dispersal to occur after the development of isolating barriers (Gaston, 1994). These barriers are often polarized, allowing migration in only one direction (Por, 1978). Polarization is due to ecological conditions, such as species richness and composition on either side of the barrier(s). Colonizing individuals may become isolated for such an extended period of time that they undergo speciation. Dispersalism relies on biotic factors, such as differential abilities of some species to colonize an outlying area (Myers and Giller, 1988).

Dispersal biogeography has been used to explain the distribution and diversity of primates in eastern Venezuela, Guyana, Suriname, and French Guiana (Eisenberg, 1989; Norconk et al., 1997). For example, Eisenberg (1989) proposed dispersal of primates into Venezuela and Guyana via two routes: (1) from the SW through western Amazonia (Brazil and Colombia); and (2) across the Andes bordering Venezuela and Colombia. Norconk et al. (1997) elaborated further on this theory by suggesting that widespread tropical savannas, rivers, and mountain ranges represent contemporary barriers to the dispersal of primates in Guyana. There have been many criticisms of the dispersalist approach (Craw and Weston, 1984). Dispersal explanations for species distribution often constitute untestable hypotheses that do not provide a general framework for the analyses of multiple taxa. Thus, *ad hoc* explanations for the disjunct distribution of one taxon cannot be applied to other taxon or taxa.

Refugia Theory

Increased understanding of the historical biogeography of tropical flora and fauna led to the formation of the refuge hypothesis (Mayr and O'Hara, 1986;

Brown, 1987; Prance, 1987). Analyses of several groups of South American plants and animals showed overlapping areas of endemism as well as hybridization zones located between these areas (Vanzolini and Williams, 1970; Haffer, 1982). Haffer (1969) and Vanzolini and Williams (1970) hypothesized that varying humid and arid conditions since the Quaternary period resulted in speciation and subspeciation among tropical organisms. Forest areas contracted whereas savannas expanded during arid periods. In humid periods, the forest refuges re-expanded and joined. Some animal populations that became isolated in the restricted forest areas differentiated at the species or subspecies level before geographical overlap was reestablished with other isolated populations (Haffer, 1982). The resulting species then colonized new habitats following expansion of forest biota. This theory has four assumptions: (1) allopatry is required for geographic differentiation; (2) allopatry leads to differentiation; (3) differentiation takes many thousands of generations; and (4) differentiating characters are selectively neutral (Prance, 1987). The refuge theory has been used to model species diversity in numerous taxa and biogeographic regions (Kingdon, 1971; Diamond and Hamilton, 1980; Prance, 1987; Avise and Walker, 1998). However, the refuge hypothesis has been criticized by many researchers (Endler, 1982; Colinvaux, 1987; Cracraft and Prum, 1988b; Bush et al., 1990; Bush, 1994; Colinvaux et al., 1996; Knapp and Mallet, 2003; Bridle et al., 2004). Colinvaux (1987) and Bush (1994) reviewed data on the paleoecological record in the Amazon basin of South America. They concluded that glacial cooling and reduced atmospheric CO_2 caused disturbance of refuge areas. Thus, it was proposed that refugia were areas of maximal disturbance rather than areas of minimal disturbance. Furthermore, the refuge theory has, at times, been supported using biased or inadequate patterns of endemism and character change across geographic areas (Mayr and O'Hara, 1986; Prance, 1987; Gentry, 1989; Nelson et al., 1990). Researchers have found that species level diversification for many tropical organisms occurred before the Pleistocene (Endler, 1977; Heyer and Maxson, 1982; Cracraft, 1988; Cracraft and Prum, 1988a; Bush et al., 1990).

There is considerable debate regarding the influence of forest refugia on primate biogeography (e.g., Kinzey and Gentry, 1979; Kinzey, 1982; Froehlich et al., 1991; Evans et al., 2003). For example, Kinzey and Gentry (1979) suggested that the distribution of dusky titi monkeys (*Callicebus moloch*) and collared titi monkeys (*Callicebus torquatus*) are the result of these taxa being restricted to different forest refugia during the Pleistocene and that they consequently developed species-specific adaptations to flora and fauna associated with different soils that have persisted. However, the habitat differences have been questioned (Defler, 1994). Researchers conducting genetic studies of chimpanzees (*Pan troglodytes*) have also questioned the role of Pleistocene refugia in primate biogeography in Africa (Morin *et al.*, 1994; Goldberg, 1996; Gonder *et al.*, 1997). Morin *et al.* (1994) documented that populations of chimpanzees exchanged genes across large geographic regions regardless of forest refugia. Collins and Dubach (2000) found similar results for *Ateles*, in that most speciation events predated the Pleistocene in the Neotropics, and Disotell and Raaum (2002) suggest dates for many guenon taxa in the Miocene, well before Pleistocene climatic fluctuations.

Phylogenetic Approaches

Phylogenetic or cladistic biogeography focuses on shared derived characters which can be used to reconstruct biogeographical history (Brooks, 1990; Hovenkamp, 1997; Humphries and Parenti, 1999). Phylogeography uses results of molecular systematics to infer biogeography (Avise, 2000). Brooks (1990) suggests two reasons why a species lives where it lives: (1) it may live in an area because its ancestor lived in that area and the descendant evolved there; or (2) it may have evolved elsewhere and dispersed into the area where it now resides. If the first case holds true, then the history of the species should coincide with the history of the area (association by descent). In the second case, there should no relationship between species history and area history (association by colonization). Thus, areas that have been connected most recently share more species and characters in common than those areas that have been separated for longer periods of time.

Cladistic biogeography has been used extensively in studies of living and fossil primates (Froehlich *et al.*, 1991; Albrecht and Miller, 1993; Da Silva and Oren, 1996; Goldberg and Ruvolo, 1997; Grubb, 1999; Ron, 2000; Jensen-Seaman and Kidd, 2001; Cortés-Ortiz *et al.*, 2003), and is well-represented in this collection of articles (Ellsworth and Hoelzer, this volume; Gonder and Disotell, this volume; McGraw and Fleagle, this volume; Yoder and

Heckman, this volume; Heesy *et al.*, this volume; Beard, this volume; Rossies and Seiffert, this volume). Phylogenetic studies of primate biogeography have employed many types of data. Froehlich *et al.* (1991) analyzed 76 cranio-dental measurements on 284 spider monkeys. They concluded that the craniodental morphology of spider monkeys is the result of a complex relationship among dispersal from Pleistocene refugia, adaptation to non-flooded forest near seasonal swamp forest, and isolation by rivers and habitat barriers. However, other investigations of the distribution of Amazonian primates with cladistic methods revealed that diversity patterns do not match those predicted by the refugia model (Da Silva and Oren, 1996; Ron, 2000). Instead, there was consistent support for rivers acting as barriers to dispersal that ultimately led to allopatric speciation. In another example, Grub (1999) theorized that speciation that occurred due to cladogenesis required more than one vacariance event. Thus, the evolutionary history of large-bodied primates may have been in response to a series of vicariance events in Africa. If this theory finds support from, for example, molecular data, then researchers must consider determining environmental conditions before, during, and after a sequence of variance events (Hovenkamp, 1997). Researchers have often looked at only one vicariance event when studying the evolutionary biology of primates (e.g., Brandon-Jones, 1996; Medeiros *et al.*, 1997; Cropp *et al.*, 1999). Moreover, researchers investigating colonization abilities in extant taxa tend to have utilized a constant dispersal rate or distance (e.g., Zagt *et al.*, 1997; Losos *et al.*, 1998; Berggren *et al.*, 2002). Grubb (1999) hypothesized that expansion and contraction of African biomes led to changes in dispersal rates for many species. Finally, Grubb (1999) theorized that increased forest fragmentation actually leads to heightened geographical variation in species. He suggested that primate data support this model in that the number of taxa within zoogeographical primate species seem to be significantly positively correlated with total range.

Community Ecology

The presence or absence of species within a geographic area has been investigated extensively through studies of community ecology (Gee and Giller, 1987; Schoener, 1988; Wiens, 1989; Findley, 1993; Ricklefs and Schluter, 1993;

Thiollay, 1994; Pugesek *et al.*, 2002). Species may rely on each other, or one upon another, for a variety of things (e.g., food, shelter, predator detection, and parasite protection). Thus, the biogeography of one species may be positively influenced by the distribution and density of another species (Huston, 1996). In other cases, competitive exclusion may occur whereby the presence of one species prevents one or more species from occupying an area (Lotka, 1925; Volterra, 1926; Gause, 1934). This phenomenon can occur naturally or as the result of native species being displaced by an invader (Connell, 1961; Silander and Antonovics, 1982).

There have been numerous biogeographic studies of primate community structure (Fleagle and Mittermeier, 1980; Bourliere, 1985; Soini, 1986; Waser, 1986; Terborgh and van Schaik, 1987; Peres, 1988; Ganzhorn, 1992; Peres, 1993a, b; Fleagle and Reed, 1996; Ganzhorn, 1997; Godfrey *et al.*, 1997; Peres, 1997; Tutin *et al.*, 1997; Julliot and Simmen, 1998; Cowlishaw and Dunbar, 1999; Fleagle *et al.*, 1999; Fleagle and Reed, 1999; Janson and Chapman, 1999; Peres and Janson, 1999; Reed, 1999; Lehman, 2000; Peres and Dolman, 2000; Ganzhorn and Eisenbeiss, 2001; Fleagle and Reed, 2004; Haugaasen and Peres, 2005). Similarities between primate communities are most likely if they share a common biogeographic history (Fleagle and Reed, 1996; Ganzhorn, 1998). Composition of primate communities can also reflect evolutionary niche dynamics (e.g., Webb *et al.*, 2002; Des devises *et al.*, 2003; Wiens and Graham, 2005). For example, the presence of species in a primate community, such as those in eastern Madagascar, can represent assemblages of functional groups of omnivores, frugivores, and folivores (Ganzhorn, 1997). Species entering a community following extinctions or climatic changes seem to fill adaptive or functional gaps. These cycles of adding new species continue until each functional group is represented in a community.

Many researchers have noted that the collections of species in relatively depauperate communities are not random subsets of larger assemblages. Rather they are often ordered, nested subsets of species from species-rich sites (Darlington, 1957; Patterson, 1987). Different communities of a faunal area are considered to be nested if each species in community A, which has few species, is also represented in the larger, more species-rich community B. The size and isolation of the habitat plays a critical role in determining nestedness (Yiming *et al.*, 1998). A large habitat will tend to contain more species than, for example, three small habitats of the same total size. Nestedness is thought to be due to three mechanisms: (1) differential colonization abilities of species, (2) nested distribution of habitats, and (3) differential extinction of species associated with reduced habitat area (Patterson and Atmar, 1986; Boecklen, 1997). For example, Ganzhorn (1998) documented that species-poor communities tend to represent nested subsets of species-rich communities in Madagascar and Lehman (this volume) provides a similar analysis of the primates of Guyana. However, there have been no other studies of nestedness patterns in primate communities.

Conservation Biology

Although specifics of primate evolutionary ecology are widely debated, there is consensus that primate evolution is closely linked to the use tropical forest habitats (e.g., Cartmill, 1972; Sussman, 1991; Cartmill, 1992; Martin, 1993). Forest-dwelling primates are increasingly threatened by logging, agriculture, and hunting (Cowlishaw and Dunbar, 2000; Chapman and Peres, 2001). Numerous studies have provided insights into how primates respond to habitat disturbances and hunting pressures (Johns and Skorupa, 1987; Mittermeier *et al.*, 1994; Ganzhorn *et al.*, 1996/1997; Ganzhorn, 1997; Chiarello, 1999; Peres, 1999; Lehman and Wright, 2000; Onderdonk and Chapman, 2000; Peres, 2000; Peres and Dolman, 2000; Radispiel and Raveloson, 2001; Laurance *et al.*, 2002; Goodman and Raselimanana, 2003; Marsh, 2003; Sussman *et al.*, 2003; Paciulli, 2004; Johnson *et al.*, 2005; Lehman *et al.*, this volume). For example, frugivorous lemurs may be particularly susceptible to habitat disturbance because there are few fruiting trees in Madagascar (Ganzhorn *et al.*, 1999). Of the fruiting trees available, most tend to produce small crops with long intervals between fruiting periods (Ganzhorn, 1997). In southeast Madagascar, White *et al.* (1995) found that density estimates for *Varecia variegata variegata* dropped from 1.6 individuals/ha in primary forests to 0.01 individuals/ha in disturbed forests. Sympatric *Eulemur fulvus* and *E. rubriventer* were less affected by forest disturbance because of their greater locomotor flexibility for vertical clinging and leaping, and because they exploit smaller fruit trees than *V. v. variegata*. This disparity in the size of feeding trees is important because one of the consequences of fragmentation is a reduction in the number of large trees, particularly near fragment edges (Laurance *et al.*, 1997).

Many researchers have investigated the effects of forest fragmentation on primate biogeography (Jones, 1997; Estrada *et al.*, 1999; Oka *et al.*, 2000; Onderdonk and Chapman, 2000; Passamani and Rylands, 2000; Ganzhorn *et al.*, 2001; Ganzhorn and Eisenbeiss, 2001; Evans *et al.*, 2003; Ganzhorn *et al.*, 2003; Marsh, 2003; Norconk and Grafton, 2003; Sussman *et al.*, 2003; Baranga, 2004; Mbora and Meikle, 2004; Chapman *et al.*, 2005; Harcourt and Doherty, 2005). Larger fragments tend to have more habitats and larger total population limits, which in turn allow them to host more species (Rosenzweig, 1995). However, this positive relationship between fragment size and species richness can mask important ecological information when species are lumped together without regard to variations in rarity, habitat requirements, or range limits (Zanette, 2000). Ideally, species should be categorized according to several ecological variables, which provide more deterministic analyses and predictions. This approach is important because there is not a consistent positive relationship between species richness and fragment size (Matthiae and Stearns, 1981; Onderdonk and Chapman, 2000; Harcourt and Doherty, 2005). For example, Ramanamanjato and Ganzhorn (2001) found that capture rates and population characteristics of *Microcebus rufus* were not affected by fragment size in the littoral forests of southern Madagascar. Harcourt and Doherty (2005) investigated how forest fragmentation influenced primate richness at global, continental, and site scale. They found that primate richness declined with fragment area at all spatial scales, except in Africa. Most importantly for conservation biology, Harcourt and Doherty (2005) noted that estimates of minimum area requirements for primate species tend to exceed the size of most forest fragments. Moreover, a recent edited volume contains data that indicates that there is species-specific or even individual flexibility in how primates respond to forest fragmentation (Marsh, 2003). Clearly, more data are needed to understand why and how some primate species do better than others in terms of their population dynamics in fragmented forest landscapes.

Forest fragmentation causes a dramatic increase in the amount of habitat edge (Lovejoy *et al.*, 1986; Laurance and Yensen, 1991; Chen *et al.*, 1992). Edges are dynamic zones characterized by the penetration, to varying depths and intensities, of abiotic conditions (e.g., wind, temperature, humidity, solar radiation) from the matrix into the forest interior (Chen *et al.*, 1992; Malcolm, 1994). The penetration of abiotic factors into the forest interior results in changes to vegetation structure, microclimate, and food resources

(Lovejoy *et al.*, 1986; Laurance and Yensen, 1991; Malcolm, 1994; Murcia, 1995; Laurance *et al.*, 1997; Woodroffe and Ginsberg, 1998; Fagan *et al.*, 1999; Cadenasso and Pickett, 2001). For example, trees in forest edges are prone to higher rates of canopy-gap formation, damage, and mortality because of microclimatic changes and increased wind turbulence (Laurance *et al.*, 1997; Laurance, 2000). When the total area of a forest fragment decreases linearly, the relative amount of interior forest decreases more rapidly than forest edge (e.g., Laurance and Yensen, 1991; Murcia, 1995; Zheng and Chen, 2000). Thus, smaller fragments will contain a relatively higher proportion of edge to interior forest than larger fragments. Although there have been numerous studies of the influence of forest edges on tropical taxa (Murcia, 1995; Ries *et al.*, 2004; Harper *et al.*, 2005), there are relatively few studies of edge effects on primates (e.g., Norconk and Grafton, 2003; Lehman *et al.*, 2006a; Lehman *et al.*, 2006b). Increased use of secondary forests/edges in Neotropical primates is driven largely by habitat selection in the speciose Callitrichidae (*Cebuella*, *Mico*, *Callithrix*, *Saguinus*, *Leontopithecus*, and *Callimico*). Callitrichidae may exploit these habitats because they contain an abundance of insect prey (Rylands and de Faria, 1993). Conversely, use of edge habitats may be an artifact of the number of studies conducted in these habitats versus those on conspecifics in natural habitats (Rylands, 1996). Edge effects are particularly relevant to lemurs. Madagascaran forests are highly fragmented and, therefore, may be prone to extreme edge effects (Green and Sussman, 1990; Du Puy and Moat, 1998; Lehtinen *et al.*, 2003; Watson *et al.*, 2004). Lehman *et al.* (2006a) found that density estimates for *Avahi laniger* and *Microcebus rufus* were higher in edge habitats in SE Madagascar. Clinal variations in food quality rather than abundance represents a possible covariate to the distribution and abundance of *A. laniger*. Specifically, Ganzhorn (1995) documented higher protein concentration in leaves near forest edges. Thus, the quality of leaves may be highest near forest edges which results in higher *A. laniger* densities in these habitats. The density and distribution of *M. rufus* and their food trees were positively correlated. Preference for edge habitats can have significant negative impacts on primate conservation. For example, there is evidence for increased hunting pressures by humans in edge habitats, which may place edge tolerant lemurs at greater risk for species extirpations (Lehman, in press).

Conservation biologists have applied biogeographic models to questions on rarity and extinction patterns in primates (Arita *et al.*, 1990; Jones, 1997;

Jernvall and Wright, 1998; Mittermeier *et al.*, 1998; Harcourt, 1999; Mittermeier *et al.*, 1999; Wright, 1999; Myers *et al.*, 2000; Harcourt and Schwartz, 2001; Harcourt *et al.*, 2002; Ratsimbazafy, 2002; Dehgan, 2003; Harcourt, 2004; Whittaker *et al.*, 2005). Harcourt and Schwartz (2001) investigated what biological traits distinguish taxa susceptible to extinction from less susceptible taxa among primates in SE Asia. They found that traits associated with extinction risk appear to be large body mass, low density, large annual home range, and low maximum latitude. Expected traits that did not correlate with susceptibility were low interbirth interval, high percent frugivory, high group mass, low altitudinal range, and small geographic range. Jernvall and Wright (1998) sought to answer this question by analyzing the ecological characteristics of extant primates in various categories of endangerment of extinction. They used these data to predict the ecological integrity of communities in the future, assuming extinctions proceed according to current rankings of endangerment. The most severe change in ecological range is projected to happen in Madagascar, while Africa has less severe, but ecologically specific extinctions. Loss in the ecological range of Asian primates is severe but only a little more severe than would be expected based on the decline in Asian primate species richness. South American extinctions affect taxonomic more than ecological aspects of diversity. Despite advances made in applying biogeographic models to primate conservation, we have few longitudinal data on correlates to species rarity (Coppeto and Harcourt, 2005). For example, Chapman *et al.* (2005) analyzed primate survey data collected over 28 years in Kibale, Uganda. They found that primate recovery in logged areas was either slow or did not occur for some species. Thus, future biogeographic studies should investigate how forest fragmentation, habitat loss, and edge effects operate synergistically to influence the survival and extinction patterns of primates.

Biodiversity hotspots are used by some conservation biologists to assign conservation priorities when a lack of resources requires maximization of thediversity of biological features (Prendergast *et al.*, 1993; Pressey *et al.*, 1996; Mittermeier *et al.*, 1998; Mittermeier *et al.*, 1999; Myers *et al.*, 2000; Hamilton *et al.*, 2001; Meijaard and Nijman, 2003; Watson *et al.*, 2004). Although definitions of hotspots vary widely, they are typically defined as geographic areas characterized by high numbers of rare, endemic species (Myers, 1988; Vane-Wright *et al.*, 1991). Application of hotspot methodology indicates that 34 biogeographic regions, which comprise only 2.3% of the Earth' surface, contain approximately 75% of the world's most threatened mammals, birds, and amphibians (Mittermeier *et al.*, 2005).

There has been considerable debate regarding the practical value of assessing conservation priorities based solely on unweighted indices of species biodiversity (e.g., Prendergast *et al.*, 1993; Harcourt, 2000a; Brummitt and Lughadha, 2003). The biodiversity hotspot approach assumes that each endemic species has equal weight or value in terms of conservation priorities. Numerous theoretical and empirical studies have revealed that phylogenetic relationships among taxa are also an important measure for conservation biology (May, 1990; Vane-Wright *et al.*, 1991; Williams *et al.*, 1991; Faith, 1992a, b, 1993, 1994a, c, 1996; Croizer, 1997; Heard and Mooers, 2000; Owens and Bennett, 2000; Faith, 2002). For conservation purposes, these relationships can be measured as indices of phylogenetic diversity. Phylogenetic diversity of a species can be measured either as the inverse proportion of the relative number and closeness of its phylogenetic relatives (Vane-Wright *et al.*, 1991) or by summing the lengths of all those phylogenetic branches spanned by a data set (Faith, 1994b). For example, a novel application of hotspot and phylogenetic diversity methods revealed that lemurs represent the world's highest conservation priority for primates (Sechrest *et al.*, 2002). Furthermore, Lehman (in press) found that the phylogenetic component of lemur diversity is greatest for *Daubentonia madagascariensis, Allocebus trichotis, Lepilemur septentrionalis, Indri indri*, and *Mirza coquereli*. It is unfortunate that many of these high-priority lemur taxa are also amongst the least-studied of all primates.

Numerous studies have been conducted on the behavioral ecology of a few species at well-established sites, with relatively little attention paid to determining the geographic distribution for each species (Scott *et al.*, 2002). Despite a lack of data on the distribution of many tropical mammals, range maps are often produced in articles and books. Ultimately, distribution limits represent hypotheses that must be tested with fieldwork (MacArthur, 1972). As such, many researchers have investigated methods for determining the geographical range of species (e.g., Fortin *et al.*, 1996; Lidicker, 1999; Peterson, 2001; Bauer and Peterson, 2005). Range limits for some species are abrupt and can be demarcated by a barrier to dispersal (Caparella, 1992). However, many species exhibit a clinal decrease in their distribution, with no observable barrier to dispersal (Terborgh, 1971). Fortin *et al.* (2005) reviewed methods for quantifying distribution patterns and suggested that the following questions need to be addressed for many species: (1) how large are geographic ranges?; (2) how can geographic range boundaries be identified?; (3) are range boundaries gradual or sharp transitions?; (4) are the shapes of species' boundaries jagged or smooth?; (5) how much variation in the use of the landscape is found within range boundaries?; (6) are there internal boundaries?; and (7) is the range fragmented? Furthermore, it is important to realize that range limits for a species are not static and tend to change through time. Many primate species have experienced a drastic reduction in their geographic distribution within the last 2,000 years (e.g., Jungers *et al.*, 1995; Godfrey *et al.*, 1997; Godfrey *et al.*, 1999; Harcourt and Schwartz, 2001; Miller *et al.*, 2004; Lehman *et al.*, 2006c). For example, skeletal remains of *Indri indri* have been recovered from sites thousands of kilometers past the range limits of extant conspecifics (Jungers *et al.*, 1995). Range contractions in mammals are often the direct result of human disturbance (Channell and Lomolino, 2000), although global warming can also alter habitat structure and primate distributions (Jungers *et al.*, 1995; Dunbar, 1998). The question arises as to the long-term consequences of range contraction on population dynamics of primates (Cowlishaw, 1999).

Current Issues

Primate biogeography is entering a period of intense research and synthesis. Much of this interest tends to focus on documenting the spatial patterns of species in the world today and changes that can be reconstructed from records of the past in an effort to identify past trends and predict patterns for the future (e.g., Wright and Jernvall, 1999). Hoever our ability to reconstruct the processes that drive primate biogeography depend heavily on our understanding of several basic, but poorly aspects of primate biology:

1. **The dispersal abilities of individual primate taxa and the factors that influence these abilities.**

   All primate taxa (except possibly humans) are surrounded by areas where the species cannot maintain a population because of different physical conditions or a scarcity of required resources (Fortin *et al.*, 1996; Legendre *et al.*, 2002; Fagan *et al.*, 2003; Bauer and Peterson, 2005; Stevens and O'Connor, this volume). These dispersal barriers may be related to a variety of topographic and ecological conditions. For example, species can be limited in their distribution by topographic barriers. Thus, mountain ranges act as efficient barriers because their elevation can present conditions too cold for most primates that have adapted to the warmer conditions found at lower elevations. The ultimate barrier to primate dispersal is the physiology of primate species, which tends to be adapted to a limited range of environmental conditions. Researchers tend to invoke the multidimensional niche concept (MNC) when discussing dispersal patterns and limitations of a species. The MNC is a theoretical explanation of how different environmental factors limit abundance and distribution (Hutchinson, 1957). Because each species has a range of tolerances and preferences along every niche axis (habitat, diet, rainfall, etc.), a species can only occur in those areas where niche axes are within ranges of tolerance. Population growth rates are highest where the greatest number of niche axes is closest to most optimal conditions (Brown, 1995). Thus, there is great interest in determining how the physical limitations of a primate species and environmental gradients interact to form historical and ecological patterns of dispersal.

2. **Reconstructing the history and influence of disturbances upon primate taxa.**

   Reconstructive studies place our understanding of distribution patterns of primate species, and the habitats they range into, in a temporal context. In biogeography, a disturbance is any ecological or human-related process that disrupts the structure and/or composition of a habitat type. The effects of the disturbance can be either temporary or permanent. It is informative to divide disturbances into two classes: (1) those that influence the structure of a ecosystem, and (2) those that affect primate community structure within a habitat, region, and/or ecosystem. For example, primatologists tend to focus on how anthropogenic disturbances (e.g., fire, logging, and human land-clearing activities) influence primate community structure. Disturbances can also influence extinction and extirpation patterns of certain species. Although anthropogenic disturbances are often cited in discussions of primate biogeography and conservation, the effects of natural disturbances (e.g., flooding regimes, tree falls) on primate community structure have rarely been studied.

APPENDIX 1. Distribution, habitat use, diet, and conservation ratings for 348 species of extant primates ("+" indicates use of that habitat or food category by a species)

Species[1]	Region	Habitat use[2]								Diet[3]							Threat[4]
		WHF	DDF	WF	RF	SSF	SF	MF	ESF	FR	LE	EX	I/F	SE	FL	OT	
Allenopithecus nigroviridis	Africa	+					+			+			+		+		
Arctocebus calabarensis	Africa	+								+			+				
Cercocebus atys	Africa	+	+							+			+	+			
Cercocebus galeritus	Africa	+	+		+					+	−			+			
Cercocebus sanjei	Africa	+	+														
Cercocebus torquatus	Africa	+	+							+	+		+		+		
Cercopithecus albogularis	Africa	+	+	+					+	+	+		+				
Cercopithecus ascanius	Africa	+	+						+	+	+		+		+		
Cercopithecus campbelli	Africa	+	+						+	+	+		+	+	+		
Cercopithecus diana	Africa	+			−				−	+	+						EN
Cercopithecus doggetti	Africa	+	+						+	+	+		+				
Cercopithecus erythrogaster	Africa	+								+	+		+				EN
Cercopithecus erythrotis	Africa	+								+	+		+				VU
Cercopithecus hamlyni	Africa	+							+	+	+						
Cercopithecus kandti	Africa	+															
Cercopithecus lhoesti	Africa	+							+	+	+		+				
Cercopithecus lowei	Africa	+															
Cercopithecus mitis	Africa	+	+		+			+	+	+	+		+				
Cercopithecus mona	Africa	+	+		+			+	+	+	+		+	+			
Cercopithecus neglectus	Africa	+	+		+			+	+	+	+		+				
Cercopithecus nictitans	Africa	+	+						+	+	+		+				
Cercopithecus petaurista	Africa	+	+			+			+	+	+		+		+		
Cercopithecus pogonias	Africa	+	+			+				+	+		+				EN
Cercopithecus preussi	Africa						+										
Cercopithecus roloway	Africa	+								+	+		+				VU
Cercopithecus solatus	Africa	+					+										EN
Cercopithecus sclateri	Africa	+	+		+												

(cont.)

APPENDIX 1. (*Continued*)

Species[1]	Region	Habitat use[2]								Diet[3]							Threat[4]
		WHF	DDF	WF	RF	SSF	SF	MF	ESF	FR	LE	EX	I/F	SE	FL	OT	
Chlorocebus aethiops	Africa	+	+						+	+	+		+	+			
Chlorocebus cynosuros	Africa									+	+		+	+			
Chlorocebus djamdjamensis	Africa																
Chlorocebus pygerythrus	Africa																
Chlorocebus sabaeus	Africa																
Chlorocebus tantalus	Africa																
Colobus angolensis	Africa	+							+	+	+			+			
Colobus guereza	Africa	+	+	+					+	+	+						
Colobus polykomos	Africa	+			+						+			+			
Colobus satanas	Africa	+	+		+				+	+	+			+	+		VU
Colobus vellerosus	Africa			+	+		+			+			+	+	+		VU
Erythrocebus patas	Africa															+	
Euoticus pallidus	Africa	+								+							
Galago alleni	Africa	+			+					+			+				
Galago cameronensis	Africa																
Galago cocos	Africa																
Galago demidoff	Africa	+	+		+					+		+	+				
Galago gallarum	Africa		+	+	+					+			+	+			
Galago granti	Africa	+															
Galago matschiei	Africa				+							+	+				
Galago moholi	Africa	+		+	+					+		+	+				
Galago nyasae	Africa																
Galago orinus	Africa																
Galago rondoensis	Africa											+					EN
Galago senegalensis	Africa		+	+	+								+				
Galago thomasi	Africa					+							+				
Galago zanzibaricus	Africa	+								+							
Galagoides granti	Africa										+						
Gorilla beringei	Africa	+						+			+					+	EN

Biogeography and Primates: A Review

Species	Region																Status
Gorilla gorilla	Africa	+		+	+		+	+				+				+	EN
Lophocebus albigena	Africa			+	+											+	
Macaca sylvanus	Africa				+										+	+	VU
Mandrillus leucophaeus	Africa				+											+	EN
Mandrillus sphinx	Africa				+											+	VU
Miopithecus talapoin	Africa															+	
Otolemur crassicaudatus	Africa					+				+	+		+			+	
Otolemur garnettii	Africa									+	+		+			+	
Otolemur monteiri	Africa															+	
Pan paniscus	Africa	+		+	+		+	+					+			+	EN
Pan troglodytes	Africa			+	+		+	+	+		+	+	+	+		+	EN
Papio anubis	Africa			+	+	+	+	+	+							+	
Papio cynocephalus	Africa								+		+	+	+	+		+	
Papio hamadryas	Africa				+		+	+	+		+	+	+	+		+	
Papio papio	Africa																
Papio ursinus	Africa						+	+	+		+	+	+	+		+	
Perodicticus potto	Africa						+	+	+	+	+	+	+	+	+	+	
Piliocolobus badius	Africa				+		+	+	+		+	+	+	+	+	+	EN
Piliocolobus foai	Africa							+	+		+	+	+	+		+	
Piliocolobus gordonorum	Africa															+	VU
Piliocolobus kirkii	Africa															+	EN
Piliocolobus pennantii	Africa				+		+	+	+	+		+	+	+		+	EN
Piliocolobus preussi	Africa						+	+		+		+					
Piliocolobus rufomitratus	Africa															+	CR
Piliocolobus tephrosceles	Africa															+	CR
Procolobus verus	Africa					+									+	+	
Theropithecus gelada	Africa	+					+	+	+							+	EN
Bunopithecus hoolock	Asia	+	+	+		+	+	+		+						+	EN
Hylobates agilis	Asia	+	+	+		+	+	+		+						+	EN
Hylobates albibarbis	Asia															+	
Hylobates klossii	Asia															+	VU
Hylobates lar	Asia	+	+	+	+	+	+		+							+	EN
Hylobates moloch	Asia	+	+	+		+	+							+		+	CR
Hylobates muelleri	Asia	+	+	+		+	+	+								+	

(cont.)

APPENDIX 1. (Continued)

Species[1]	Region	WHF	DDF	WF	RF	SSF	SF	MF	ESF	FR	LE	EX	I/F	SE	FL	OT	Threat[4]
		Habitat use[2]								Diet[3]							
Hylobates pileatus	Asia	+						+		+	+		+		+	+	VU
Loris lydekkerianus	Asia	+	+												+	+	
Loris tardigradus	Asia	+	+			+		+			+	+			+		EN
Macaca arctoides	Asia	+	+			+				+	+		+	+	+	+	VU
Macaca assamensis	Asia	+						+		+	+		+	+	+	+	VU
Macaca fascicularis	Asia	+					+			+	+		+	+	+	+	
Macaca fuscata	Asia	+	+					+		+	+		+	+	+	+	
Macaca becki	Asia	+															
Macaca leonina	Asia	+	+							+	+		+	+	+	+	VU
Macaca maura	Asia	+								+	+						EN
Macaca mulatta	Asia	+								+							
Macaca nemestrina	Asia	+				+	+			+	+		+	+	+	+	VU
Macaca nigra	Asia	+			+			+		+	+		+		+	+	EN
Macaca nigrescens	Asia	+								+				+	+	+	
Macaca ochreata	Asia	+								+				+	+	+	
Macaca pagensis	Asia									+	+					+	CR
Macaca radiata	Asia	+	+							+	+		+	+	+	+	
Macaca siberu	Asia	+															
Macaca silenus	Asia	+	+							+	+		+	+	+		EN
Macaca sinica	Asia	+	+							+	+		+	+	+	+	VU
Macaca thibetana	Asia	+						+		+	+		+	+	+	+	
Macaca tonkeana	Asia	+								+	+		+	+	+		
Nasalis larvatus	Asia				+		+			+	+			+			EN
Nomascus concolor	Asia																EN
Nomascus gabriellae	Asia																EN
Nomascus hainanus	Asia																VU
Nomascus leucogenys	Asia																
Nomascus siki	Asia																
Nycticebus bengalensis	Asia	+											+				

Species	Region	IUCN Status
Nycticebus coucang	Asia	VU
Nycticebus pygmaeus	Asia	
Pongo pygmaeus	Asia	CR
Pongo abelii	Asia	EN
Presbytis chrysomelas	Asia	
Presbytis comata	Asia	EN
Presbytis femoralis	Asia	
Presbytis frontata	Asia	
Presbytis hosei	Asia	
Presbytis melalophos	Asia	
Presbytis natunae	Asia	
Presbytis potenziani	Asia	VU
Presbytis rubicunda	Asia	
Presbytis siamensis	Asia	
Presbytis thomasi	Asia	
Pygathrix cinerea	Asia	EN
Pygathrix nemaeus	Asia	EN
Pygathrix nigripes	Asia	
Rhinopithecus avunculus	Asia	CR
Rhinopithecus bieti	Asia	EN
Rhinopithecus roxellana	Asia	VU
Semnopithecus ajax	Asia	
Semnopithecus dussumieri	Asia	
Semnopithecus entellus	Asia	
Semnopithecus hector	Asia	
Semnopithecus hypoleucos	Asia	
Semnopithecus priam	Asia	
Semnopithecus schistaceus	Asia	
Simias concolor	Asia	EN
Symphalangus syndactylus	Asia	
Tarsius bancanus	Asia	
Tarsius dentatus	Asia	
Tarsius pelengensis	Asia	
Tarsius pumilus	Asia	

(cont.)

APPENDIX 1. (*Continued*)

Species[1]	Region	Habitat use[2]								Diet[3]							Threat[4]
		WHF	DDF	WF	RF	SSF	SF	MF	ESF	FR	LE	EX	I/F	SE	FL	OT	
Tarsius sangirensis	Asia																
Tarsius syrichta	Asia	+											+				
Tarsius tarsier	Asia																
Trachypithecus auratus	Asia	+							+	+	+		+		+		EN
Trachypithecus barbei	Asia	+	+		+				+	+	+					+	
Trachypithecus cristatus	Asia	+	+			+				+	+						CR
Trachypithecus delacouri	Asia	+	+		+					+	+						
Trachypithecus ebenus	Asia	+	+							+	+				+		EN
Trachypithecus francoisi	Asia	+	+		+					+	+				+		EN
Trachypithecus geei	Asia			+													
Trachypithecus germaini	Asia	+	+		+			+		+	+				+	+	EN
Trachypithecus hatinhensis	Asia																
Trachypithecus johnii	Asia	+	+					+	+	+	+		+		+	+	
Trachypithecus laotum	Asia	+	+							+	+		+	+	+	+	
Trachypithecus obscurus	Asia	+	+	+				+	+	+	+		+	+	+	+	EN
Trachypithecus phayrei	Asia	+								+	+						CR
Trachypithecus pileatus	Asia	+						+	+	+	+	+	+		+	+	EN
Trachypithecus poliocephalus	Asia	+	+		+					+	+				+	+	EN
Trachypithecus shortridgei	Asia					+											
Trachypithecus vetulus	Asia	+	+		+					+	+	+					VU
Allocebus trichotis	Madagascar	+	+	+	+												
Avahi laniger	Madagascar																
Avahi occidentalis	Madagascar									+	+		+	+		+	
Avahi unicolor	Madagascar									+	+						
Cheirogaleus adipicaudatus	Madagascar																
Cheirogaleus crossleyi	Madagascar																
Cheirogaleus major	Madagascar	+	+		+					+	+		+	+	+	+	
Cheirogaleus medius	Madagascar																
Cheirogaleus minusculus	Madagascar															+	

(cont.)

Species	Location	Status
Cheirogaleus ravus	Madagascar	
Cheirogaleus sibreei	Madagascar	
Daubentonia madagascariensis	Madagascar	EN
Eulemur albifrons	Madagascar	
Eulemur albocollaris	Madagascar	CR
Eulemur cinereiceps	Madagascar	
Eulemur collaris	Madagascar	
Eulemur coronatus	Madagascar	VU
Eulemur fulvus	Madagascar	VU
Eulemur macaco	Madagascar	VU
Eulemur mongoz	Madagascar	VU
Eulemur rubriventer	Madagascar	
Eulemur rufus	Madagascar	
Eulemur sanfordi	Madagascar	
Hapalemur alaotrensis	Madagascar	CR
Hapalemur aureus	Madagascar	EN
Hapalemur griseus	Madagascar	VU
Hapalemur occidentalis	Madagascar	VU
Indri indri	Madagascar	
Lemur catta	Madagascar	
Lepilemur dorsalis	Madagascar	
Lepilemur edwardsi	Madagascar	
Lepilemur leucopus	Madagascar	
Lepilemur microdon	Madagascar	
Lepilemur mustelinus	Madagascar	
Lepilemur ruficaudatus	Madagascar	
Lepilemur septentrionalis	Madagascar	VU
Microcebus berthae	Madagascar	
Microcebus griseorufus	Madagascar	EN
Microcebus murinus	Madagascar	
Microcebus myoxinus	Madagascar	EN
Microcebus ravelobensis	Madagascar	
Microcebus rufus	Madagascar	

APPENDIX 1. (*Continued*)

Species[1]	Region	Habitat use[2]								Diet[3]							Threat[4]
		WHF	DDF	WF	RF	SSF	SF	MF	ESF	FR	LE	EX	I/F	SE	FL	OT	
Microcebus sambiranensis	Madagascar		+							+			+				
Microcebus tavaratra	Madagascar		+							+			+				
Mirza coquereli	Madagascar		+		+					+		+	+		+	+	VU
Phaner electromontis	Madagascar																
Phaner furcifer	Madagascar		+		+					+	+	+	+		+	+	
Phaner pallescens	Madagascar																
Phaner parienti	Madagascar																
Prolemur simus	Madagascar	+								+							CR
Propithecus coquereli	Madagascar		+						+	+	+				+		
Propithecus deckenii	Madagascar		+						+	+	+			+	+		
Propithecus diadema	Madagascar	+							+	+	+			+	+		EN
Propithecus edwardsi	Madagascar	+		+					+	+	+				+		
Propithecus perrieri	Madagascar			+						+	+				+		CR
Propithecus tattersalli	Madagascar		+	+						+	+				+		VU
Propithecus verreauxi	Madagascar		+						+	+	+				+		
Varecia rubra	Madagascar	+								+		+			+		
Varecia variegata	Madagascar	+								+		+			+		EN
Alouatta sara	Neotropics	+								+	+						
Alouatta belzebul	Neotropics	+								+	+						
Alouatta caraya	Neotropics	+	+		+	+				+	+				+	+	
Alouatta coibensis	Neotropics	+					+			+	+				+	+	
Alouatta guariba	Neotropics	+							+	+	+				+		
Alouatta niggerima	Neotropics								+								
Alouatta palliata	Neotropics	+	+		+		+		+	+	+				+	+	
Alouatta pigra	Neotropics	+	+		+	+			+	+	+				+	+	EN
Alouatta seniculus	Neotropics	+	+		+	+			+	+	+				+	+	
Aotus azarai	Neotropics	+						+	+	+					+		
Aotus nigriceps	Neotropics									+					+	+	
Aotus nancymaae	Neotropics																

(cont.)

Species	Realm	1	2	3	4	5	6	7	8	9	10	11	12	Status
Aotus trivirgatus	Neotropics		+		+	+	+					+	+	
Aotus hershkovitzi	Neotropics												+	
Aotus lemurinus	Neotropics								+				+	VU
Aotus miconax	Neotropics								+				+	VU
Aotus vociferans	Neotropics												+	
Ateles paniscus	Neotropics				+	+	+			+		+	+	
Ateles belzebuth	Neotropics	+	+			+	+				+		+	VU
Ateles fusciceps	Neotropics	+	+			+	+					+	+	
Ateles geoffroyi	Neotropics	+	+	+		+	+		+	+			+	CR
Ateles hybridus	Neotropics						+		+				+	EN
Ateles marginatus	Neotropics			+	+		+					+	+	EN
Brachyteles arachnoides	Neotropics	+	+	+		+	+						+	CR
Brachyteles hypoxanthus	Neotropics	+	+	+		+	+				+		+	
Cacajao calvus	Neotropics	+	+	+	+	+	+	+		+	+		+	
Cacajao calvus	Neotropics		+	+	+	+	+			+	+		+	
Callicebus brunneus	Neotropics						+				+		+	
Callicebus cinerascens	Neotropics			+			+	+					+	
Callicebus cupreus	Neotropics						+	+					+	
Callicebus donacophilus	Neotropics						+						+	
Callicebus hoffmannsi	Neotropics				+	+	+	+		+	+		+	
Callicebus moloch	Neotropics					+	+			+			+	
Callicebus baptista	Neotropics							+					+	
Callicebus pallescens	Neotropics			+	+		+	+			+		+	
Callicebus torquatus	Neotropics						+						+	
Callicebus medemi	Neotropics												+	
Callicebus barbarabrownae	Neotropics												+	CR
Callicebus coimbrai	Neotropics										+		+	CR
Callicebus melanochir	Neotropics								+				+	VU
Callicebus modestus	Neotropics												+	VU
Callicebus nigrifrons	Neotropics											+	+	
Callicebus oenanthe	Neotropics												+	VU
Callicebus olallae	Neotropics												+	VU
Callicebus ornatus	Neotropics		+			+							+	VU
Callicebus personatus	Neotropics						+	+					+	VU

APPENDIX 1. (*Continued*)

Species[1]	Region	Habitat use[2]								Diet[3]							Threat[4]
		WHF	DDF	WF	RF	SSF	SF	MF	ESF	FR	LE	EX	I/F	SE	FL	OT	
Callimico goeldii	Neotropics	+								+		+	+				
Callithrix acariensis	Neotropics									+	+	+	+				
Callithrix argentata	Neotropics	+	+		+					+	+	+	+				EN
Callithrix aurita	Neotropics	+	+							+	+	+	+				EN
Callithrix flaviceps	Neotropics	+	+						+	+	+	+	+	+			VU
Callithrix geoffroyi	Neotropics	+								+				+			
Callithrix humilis	Neotropics									+		+	+				
Callithrix jacchus	Neotropics	+		+	+	+	+		+	+	+	+	+				
Callithrix kuhlii	Neotropics	+	+		+	+			+	+	+	+	+				
Callithrix manicorensis	Neotropics	+															
Callithrix mauesi	Neotropics	+															
Callithrix melanura	Neotropics									+		+	+				
Callithrix nigriceps	Neotropics	+		+					+	+	+	+	+	+			
Callithrix penicillata	Neotropics	+		+	+				+	+	+	+	+				
Callithrix pygmaea	Neotropics								+								
Callithrix saterei	Neotropics																
Cebus albifrons	Neotropics	+	+		+		+			+		+	+	+		+	
Cebus apella	Neotropics	+	+		+			+		+		+	+	+		+	
Cebus capucinus	Neotropics							+	+		+				+	+	
Cebus libidinosus	Neotropics																
Cebus nigritus	Neotropics				+		+			+			+	+			
Cebus olivaceus	Neotropics	+	+							+							
Cebus robustus	Neotropics	+							+								
Cebus xanthosternos	Neotropics	+								+		+	+	+	+		VU
Chiropotes albinasus	Neotropics	+			+					+		+	+	+	+		CR
Chiropotes satanas	Neotropics	+			+					+		+	+	+	+		EN
Lagothrix brunneus	Neotropics	+															
Lagothrix cana	Neotropics	+															EN
Lagothrix lagotricha	Neotropics	+			+		+		+	+		+	+	+		+	

	Status		Realm
Lagothrix lugens	VU		Neotropics
Lagothrix poeppigii			Neotropics
Leontopithecus caissara	CR		Neotropics
Leontopithecus chrysomelas	EN		Neotropics
Leontopithecus chrysopygus	CR		Neotropics
Leontopithecus rosalia	EN		Neotropics
Mico humeralifer			Neotropics
Mico leucippe			Neotropics
Mico melanurus			Neotropics
Mico intermedius			Neotropics
Mico emiliae			Neotropics
Mico nigriceps			Neotropics
Mico marcai			Neotropics
Mico chrysoleucus			Neotropics
Mico saterei			Neotropics
Oreonax flavicauda	CR		Neotropics
Pithecia aequatorialis			Neotropics
Pithecia albicans			Neotropics
Pithecia irrorata			Neotropics
Pithecia monachus			Neotropics
Pithecia pithecia			Neotropics
Saguinus martinsi			Neotropics
Saguinus pileatus			Neotropics
Saguinus tripartitus			Neotropics
Saguinus nigricollis			Neotropics
Saguinus mystax			Neotropics
Saguinus midas			Neotropics
Saguinus labiatus			Neotropics
Saguinus inustus			Neotropics
Saguinus imperator			Neotropics
Saguinus bicolor	CR		Neotropics
Saguinus fuscicollis			Neotropics
Saguinus geoffroyi			Neotropics
Saguinus graellsi			Neotropics

(*cont.*)

APPENDIX 1. (*Continued*)

Species[1]	Region	Habitat use[2]								Diet[3]							Threat[4]
		WHF	DDF	WF	RF	SSF	SF	MF	ESF	FR	LE	EX	I/F	SE	FL	OT	
Saguinus leucopus	Neotropics	+							+	+							VU
Saguinus oedipus	Neotropics	+	+						+	+		+	+	+			EN
Saimiri ustus	Neotropics	+							+	+			+				VU
Saimiri vanzolinii	Neotropics	+					+			+			+				
Saimiri boliviensis	Neotropics	+							+	+			+	+			
Saimiri oerstedii	Neotropics	+	+		+					+	+		+	+			EN
Saimiri sciureus	Neotropics	+							+	+			+				

[1] Species from Groves (2001), supplemented with information from Brandon-Jones *et al.* (2004) and Rylands *et al.* (2000).

[2] Habitat and diet data taken from Chapman *et al.* (1999), Gupta and Chivers (1999), Sussman (1999), Rowe (1996), Ganzhorn *et al.* (1997), Garbut (1999), and Eeley and Foley (1999). WHF = wet/humid forests, DDF = deciduous/dry forests, WF = woodland forests, RF = riparian forests, SSF = spiny forests/scrub, SF = swamp forests, MF = montane forests, and ESF = Edge/secondary forests.

[3] FR = fruits, LE = leaves, EX = exudate/gum, I/F = insects/fauna, SE = seeds, FL = flowers, OT = others (grasses, bark, fungi, buds).

[4] Threat follows the IUCN (2005) categories; CR = critically endangered, EN = endangered, VU = vulnerable.

REFERENCES

Albrecht, G. H., Jenkins, P. D., and Godfrey, L. R. 1990, Ecogeographic size variation among the living and subfossil prosimians of Madagascar. *Amer. J. Primatol.* 22:1–50.

Albrecht, G. H. and Miller, J. M. A. 1993, Geographic variation in primates: A review with implications for interpreting fossils, in: Kimbel, W. H. and Martin, L. B., eds., *Species, Species Concepts, and Primate Evolution*, Plenum Press, New York, pp. 123–161.

Arita, H. T., Robinson, J. G., and Redford, K. H. 1990, Rarity in Neotropical forest mammals and its ecological correlates. *Cons. Biol.* 4:181–192.

Avise, J. C. 2000 *Phylogeography: The History and Formation of Species*, Harvard University Press, Cambridge, Massachusetts.

Avise, J. C. and Walker, D. 1998, Pleistocene phylogeographic effects on avian populations and the speciation process. *Proc. R. Soc. Lond.*:457–463.

Ayres, J. M. 1986, Uakaries and Amazonian Flooded Forests, Unpublished Ph.D. dissertation, Cambridge University, Cambridge, England.

Ayres, J. M. and Clutton-Brock, T. H. 1992, River boundaries and species range size in Amazonian primates. *Amer. Nat.* 140:531–537.

Baranga, D. 2004, Forest fragmentation and primates' survival status in non-reserved forests of the 'Kampala area', Uganda. *Afr. J. Ecol.* 42:70–77.

Bates, H. W. 1863, *The Naturalist on the River Amazons*, J. Murray, London.

Bates, J. M., Hackett, S. J., and Cracraft, J. 1998, Area-relationships in the Neotropical lowlands: an hypothesis based on raw distributions of Passerine birds. *J. Biogeog.* 25:783–793.

Bauer, J. T. and Peterson, A. T. 2005, Visualizing environmental correlates of species geographical range limits. *Diversity and Distributions* 11:275–278.

Berggren, A., Birath, B., and Kindvall, O. 2002, Effect of corridors and habitat edges on dispersal behavior, movement rates, and movement angles in Roesel's Bush-Cricket (*Metrioptera roeseli*). *Cons. Biol.* 16:1562–1569.

Biedermann, R. 2003, Body size and area-incidence relationships: is there a general pattern? *Global Ecol Biogeography* 12:381–387.

Boecklen, W. J. 1997, Nestedness, biogeographic theory, and the design of nature reserves. *Oecologia* 112:123–142.

Böhm, M. and Mayhew, P. J. 2005, Historical biogeography and the evolution of the latitudinal gradient of species richness in the Papionini (Primata: Cercopithecidae). *Biological Journal of the Linnean Society* 85:235–246.

Booth, A. H. 1958, The Niger, the Volta and the Dahomy Gap as geographic barriers. *Evol.* 12:48–62.

Bourliere, F. 1985, Primate communities: Their structure and role in tropical ecosystems. *Int. J. Primatol.* 6:1–26.

Brandon-Jones, D. 1996, The Asian Colobinae (Mammalia: Cercopithecidae) as indicators of Quaternary climatic change. *Biol. J. Linn. Soc.* 59:327–350.

Brandon-Jones, D., Eudey, A. A., Geissmann, T., Groves, C. P., Melnick, D. J., Morales, J. C., Shekelle, M., and Stewart, C. B. 2004, Asian primate classification. *Int. J. Primatol.* 25:97–164.

Bridle, J. R., Pedro, P. M., and Butlin, R. K. 2004, Habitat fragmentation and biodiversity: testing for the evolutionary effects of refugia. *Evol.* 58:1394–1396.

Brooks, D. R. 1990, Parsimony analysis in historical biogeography and coevolution: Methodological and theoretical update. *Syst. Zool.* 39:14–30.

Brown, J. 1987, Conclusions, synthesis, and alternative hypotheses, in: Whitmore, T. C. and Prance, G. T. eds., *Biogeography and Quaternary History in Tropical America*, Oxford University Press, Oxford, pp. 175–195.

Brown, J. H. 1984, On the relationship between abundance and distribution of species. *Amer. Nat.* 124:253–279.

Brown, J. H. 1995, *Macroecology*, University of Chicago Press, Chicago.

Brummitt, N. and Lughadha, E. N. 2003, Biodiversity: Where's hot and where's not. *Cons. Biol.* 17:1442–1448.

Bush, M. B. 1994, Amazonian speciation: a necessarily complex model. *J. Biogeog.* 21:5–17.

Bush, M. B., Colinvaux, P. A., Wiemann, M. C., Piperno, D. R., and Liu, K. B. 1990, Late pleistocene temperature depression and vegetation change in Ecuadorian Amazonia. *Quaternary Research* 4:330–345.

Cadenasso, M. L. and Pickett, S. T. A. 2001, Effects of edge structure on the flux of species into forest interiors. *Cons. Biol.* 15:91–97.

Caparella, A. P. 1992, Neotropical avian diversity and riverine barriers. *Acta Congress on International Ornithology* 20:307–316.

Cartmill, M. 1972, Arboreal adaptations and the origin of the order Primates, in: Tuttle, R. ed., *The Functional and Evolutionary Biology of Primates*, Aldine-Atherton, Chicago, pp. 97–122.

Cartmill, M. 1992, New views on primate origins. *Evol. Anthro.* 1:105–111.

Channell, R. and Lomolino, M. V. 2000, Trajectories to extinction: spatial dynamics of the contraction of geographical ranges. *J. Biogeog.* 27:169–179.

Chapman, C. A., Gautier-Hion, A., Oates, J. F., and Onderdonk, D. A. 1999, African primate communities: Determinant of structure and threats to survival, in: Fleagle, J. G., Janson, C. H., and Reed, K., eds., *Primate Communities*, Cambridge University Press, Cambridge, pp. 1–37.

Chapman, C. A. and Peres, C. A. 2001, Primate conservation in the new millennium: The role of scientists. *Evol. Anthro.* 10:16–33.

Chapman, C. A., Struhsaker, T. T., and Lambert, J. E. 2005, Thirty years of research in Kibale National Park, Uganda, reveals a complex picture for conservation. *Int. J. Primatol.* 26:539–555.

Chen, J., Franklin, J. F., and Spies, T. A. 1992, Vegetation responses to edge environments in old-growth Douglas-fir forests. *Ecol. Appl.* 2:387–396.

Cheverud, J. M. and Moore, A. J. 1990, Subspecific morphological variation in the saddle-back tamarin (*Saguinus fuscicollis*). *Amer. J. Primatol.* 21:1–16.

Chiarello, A. G. 1999, Effects of fragmentation of the Atlantic forest on mammal communities in south-eastern Brazil. *Biol. Cons.* 89:71–82.

Chivers, D. J. 1980, *Malayan Forest Primates: Ten Years' Study in Tropical Rain Forest*, Plenum Press, New York.

Colinvaux, P. 1987, Amazon diversity in light of the paleoecological record. *Quat. Sci. Rev.* 6:93–114.

Colinvaux, P. A., de Oliveira, P. E., Moreno, J. E., Miller, M. C., and Bush, M. B. 1996, A long pollen record from lowland Amazonia: Forest and cooling in glacial times. *Science* 274:85–88.

Collins, A. C. and Dubach, J. M. 2000, Biogeographic and ecological forces responsible for speciation in *Ateles*. *Int. J. Primatol.* 21:421–444.

Colyn, M. and Deleporte, P. 2002, Biogeographic analysis of central African forest guenons, in: Glenn, M. E. and Cords, M., eds., *The Guenons: Diversity and Adaptation in African Monkeys*, Kluwer Acadademic/Plenum Publishing, New York, pp. 61 78.

Connell, J. 1961, The influence of interspecific competition and other factors on the distribution of the barnacle *Chthamalus stellatus*. *Ecol.* 42:710–723.

Conroy, C. J., Demboski, J. R., and Cook, J. A. 1999, Mammalian biogeography of the Alexander Archipelago of Alaska: a north temperate nested fauna. *J. Biogeog.* 26:343–352.

Coppeto, S. A. and Harcourt, A. H. 2005, Is a biology of rarity in primates yet possible? *Biod. and Cons.* 14:1017–1022.

Cortés-Ortiz, L., Bermingham, E., Rico, C., Rodriguez-Luna, E., Sampaio, I., and Ruiz-Garcia, M. 2003, Molecular systematics and biogeography of the Neotropical monkey genus, *Alouatta*. *Mol. Phylog. Evol.* 26:64–81.

Cowlishaw, G. 1999, Predicting the pattern of decline of African primate diversity: An extinction debt from historical deforestation. *Cons. Biol.* 13:1183–1193.

Cowlishaw, G. and Dunbar, R. 2000, *Primate Conservation Biology*, University of Chicago Press, Chicago.

Cowlishaw, G. and Dunbar, R. I. M. 1999, Community Ecology, in: Fleagle, J. G., Janson, C. H., and Reed, K. eds., *Primate Communities*, Cambridge University Press, Cambridge.

Cowlishaw, G. and Hacker, J. E. 1997, Distribution, diversity, and latitude in African primates. *Amer. Nat.* 150:505–512.

Cox, B. and Moore, P. D. 2005 *Biogeography: An Ecological and Evolutionary Approach*, Blackwell Publishing, Malden, MA.

Cracraft, J. 1988, Deep-history biogeography: retrieving the historical pattern of evolving continental biotas. *Syst. Zool.* 37:221–236.

Cracraft, J. and Prum, R. O. 1988a, Patterns and processes of diversification: speciation and historical congruence in some neotropical birds. *Evolution* 42:603–620.

Cracraft, J. and Prum, R. O. 1988b, Patterns and processes of diversification: speciation and historical congruence in some neotropical birds. *Evol.* 42:603–350.

Craw, R. C. and Weston, P. 1984, Panbiogeography: a progressive research program? *Syst. Zool.* 33:1–33.

Crisci, J. V., Katinas, L., and Posadas, P. 2003, *Historical Biogeography: An Introduction*, Harvard University Press, Boston.

Croizat, L. 1958, *Panbiogeography, Vol 1 The New World*, Published by the author, Caracas.

Croizat, L. 1976, *Biogeographia analtica y sintetica (Panbiogeografia) de las Americas*, Biblioteca de la Academia de Cinecias Fisicas, Matematicas y Naturales, Caracas.

Croizer, R. H. 1997, Preserving the information content of species: genetic diversity, phylogeny, and conservation worth. *Ann. Rev. Ecol. Syst.* 28:243–268.

Cropp, S. J., Larson, A., and Cheverud, J. M. 1999, Historical biogeography of tamarins, genus *Saguinus:* The molecular phylogenetic evidence. *Amer. J. Phys. Anth.* 108:65–89.

Da Silva, J. M. C. and Oren, D. C. 1996, Application of parsimony analysis of endemicity in Amazonian biogeography: an example with primates. *Biol. J. Linn. Soc.* 59:427–437.

Darlington, P. J. 1957, *Zoogeography: the geographical distribution of animals*, New American Library, New York.

Darwin, C. 1859, *The Origin of the Species by Means of Natural Selection, or the Preservation of Favoured Races in the Struggle of Life*, Harvard University Press, Cambridge, Mass.

Defler, T. R. 1994, *Callicebus torquatus* is not a white-sand specialist. *Amer. J. Primatol.* 33:149–154.

Dehgan, A. 2003, The behavior of extinction: Predicting the incidence and local extinction of lemurs in fragmented habitat in southeastern Madagascar, Unpublished Ph.D. dissertation, The University of Chicago, Chicago, IL.

Desdevises, Y., Legendre, P., Azouzi, L., and Morand, S. 2003, Quantifying phyloge-netically structured environmental variation. *Evol.* 57:2647–2652.

Diamond, A. W. and Hamilton, A. C. 1980, The distribution of forest passerine birds and Quaternary climatic change in tropical Africa. *J. Zool, Lond* 191:379–402.

Disotell, T. R. and Raaum, R. L. 2002, Molecular timescale and gene tree incongruence in the guenons, in: Glenn, M. E. and Cords, M., eds., *The Guenons: Diversity and Adaptation in African Monkeys*, Kluwer Acadademic/Plenum Publishing, New York, pp. 27–36.

Du Puy, D. and Moat, J. 1998, Vegetation mapping and classification in Madagascar (using GIS): implications and recommendations for the conservation of biodiversity, in: Huxley, C. R., Lock, J. M., and Cutler, D. F., eds., *Chorology, Taxonomy & Ecology of the Floras of Africa and Madagascar*, Royal Botanic Gardens, Kew, pp. 97–117.

Dunbar, R. I. M. 1998, Impact of global warming on the distribution and survival of the gelada baboon: a modelling approach. *Glob. Chang. Biol.* 4:293–304.

Eeley, H. A. C. and Foley, R. A. 1999, Species richness, species range size and ecolog-ical specialisation among African primates: Geographical patterns and conservation implications. *Biod. and Cons.* 8:1033–1056.

Eeley, H. A. C. and Laws, M. J. 1999, Large-scale patterns of species richness and species range size in anthropoid primates, in: Fleagle, J. G., Janson, C. H., and Reed, K., eds., *Primate Communities*, Cambridge University Press, Cambridge, pp. 191–219.

Eisenberg, J. F. 1979, Habitat, economy, and society: some correlations and hypotheses for the neotropical primates, in: Bernstein, I. S. and Smith, E. O., eds., *Primate Ecology and Human Origins: Ecological Influences on Social Organization*, Garland STPM Press, New York, pp. 215–262.

Eisenberg, J. F. 1981, *The Mammalian Radiations: An Analysis of Trends in Evolution, Adaptation, and Behavior*, University of Chicago Press, Chicago.

Eisenberg, J. F. 1989, *Mammals of the Neotropics: Panama, Colombia, Venezuela, Guyana, Suriname, French Guiana*, The University of Chicago Press, Chicago.

Emmons, L. H. 1999, Of mice and monkeys: Primates as predictors of mammal com-munity richness, in: Fleagle, J. G., Janson, C. H., and Reed, K., eds., *Primate Com-munities*, Cambridge University Press, Cambridge, pp. 75–89.

Endler, J. A. 1977, *Geographic Variation, Speciation, and Clines*, Princeton University Press, Princeton, NJ.

Endler, J. A. 1982, Pleistocene forest refuges: Fact or fancy?, in: Prance, G. T. ed., *Biolog-ical Diversification in the Tropics*, Columbia University Press, New York, pp. 641–657.

Estrada, A., Anzures D., A., and Coates-Estrada, R. 1999, Tropical rain forest frag-mentation, howler monkeys (*Alouatta palliata*), and dung beetles at Los Tuxtlas, Mexico. *Amer. J. Primatol.* 48:253–262.

Evans, B. J., Supriatna, J., Andayani, N., Setiadi, M. I., Cannatella, D. C., and Melnick, D. J. 2003, Monkeys and toads define areas of endemism on Sulawesi. *Evol.* 57:1436–1443.

Fagan, W. F., Cantrell, R. S., and Cosner, C. 1999, How habitat edges change species interactions. *The American Naturalist* 153:165–182.

Fagan, W. F., Fortin, M.-J., and Soykan, C. 2003, Integrating edge detection and dynamic modeling in quantitative analyses of ecological boundaries. *BioScience* 53:730–738.

Faith, D. P. 1992a, Conservation evaluation and phylogenetic diversity. *Biol. Cons.* 61:1–10.

Faith, D. P. 1992b, Systematics and conservation: on predicting the feature diversity of subsets of taxa. *Cladistics* 8:361–373.

Faith, D. P. 1993, Biodiversity and systematics, the use and misuse of divergence information in assessing taxonomic diversity. *Pac. Cons. Biol.* 1:53–57.

Faith, D. P. 1994a, Genetic diversity and taxonomic priorities for conservation. *Biol. Cons.* 68:69–74.

Faith, D. P. 1994b, Phylogenetic diversity: a general framework for the prediction of feature diversity, in: Forey, P. L., Humphries, C. J., and Vane-Wright, R. I., eds., *Systematics and Conservation Evaluation*, Clarendon Press, Oxford, United Kingdom, pp. 251–268.

Faith, D. P. 1994c, Phylogenetic pattern and the quantification of organismal biodiversity. *Phil. Trans. Roy. Soc., Series B* 345:45–58.

Faith, D. P. 1996, Conservation priorities and phylogenetic pattern. *Cons. Biol.* 10:1286–1289.

Faith, D. P. 2002, Quantifying biodiversity: A phylogenetic perspective. *Cons. Biol.* 16:248–252.

Fernandez, M. H. and Vrba, E. S. 2005a, Body size, biomic specialization and range size of African large mammals. *J. Biogeog.* 32:1243–1256.

Fernandez, M. H. and Vrba, E. S. 2005b, Rapoport effect and biomic specialization in African mammals: revisiting the climatic variability hypothesis. *J. Biogeog.* 32:903–918.

Findley, J. S. 1993, *Bats: a Community Perspective*, Cambridge University Press, Cambridge, MA.

Fleagle, J. G., Janson, C. H., and Reed, K. E., 1999, Spatial and temporal scales in primate community structure, in: Fleagle, J. G., Janson, C. H., and Reed, K. E., eds., *Primate Communities*, Cambridge University Press, Cambridge, pp. 284–288.

Fleagle, J. G. and Mittermeier, R. A. 1980, Locomotor behavior, body size, and comparative ecology of seven Surinam monkeys. *Amer. J. Phys. Anth.* 52:301–314.

Fleagle, J. G. and Reed, K. E. 1996, Comparing primate communities: A multivariate approach. *J. Hum. Evol.* 30:489–510.

Fleagle, J. G. and Reed, K. E. 1999, Phylogenetic and temporal perspectives on primate ecology, in: Fleagle, J. G., Janson, C. H., and Reed, K. E., eds., *Primate Communities*, Cambridge University Press, Cambridge, pp. 92–115.

Fleagle, J. G. and Reed, K. E. 2004, The evolution of primate ecology: patterns of geography and phylogeny, in: Anapol, F., German, R. Z., and Jablonski, N. G., eds., *Shaping Primate Evolution: Form, Function and Behavior*, Cambridge University Press, New York, pp. 353–367.

Fortin, M.-J., Drapeau, P., and Jacquez, G. M. 1996, Quantification of the spatial co-occurrences of ecological boundaries. *Oikos* 77:51–60.

Fortin, M.-J., Keitt, T. H., Maurer, B. A., Taper, M. L., Kaufman, D. M., and Blackburn, T. M. 2005, Species' geographic ranges and distributional limits: pattern analysis and statistical issues. *Oikos* 108:7–17.

Froehlich, J. W., Suprianata, J., and Froehlich, P. H. 1991, Morphometric analyses of *Ateles*: Systematics and biogegraphic implications. *Amer. J. Primatol.* 25:1–22.

Futuyma, D. J. 1998, *Evolutionary Biology*, Sinauer Associates, Sunderland, MA.

Ganzhorn, J., Goodman, S. M., Ramanamanjato, J. B., Rakotondravony, D., Rakotosamimanana, B., and Vallan, D. 2001, Effects of fragmentation and assessing minimum viable populations in Madagascar, in: Rheinwald, G., ed., *Isolated vertebrate communities in the tropics*, Museum Alexander Koenig, Bonn, pp. 265–272.

Ganzhorn, J. U. 1992, Leaf chemistry and the biomass of folivorous primates in tropical forests. *Oecologia* 91:540–547.

Ganzhorn, J. U. 1994, Lemurs as indicators for habitat change, in: Thierry, B., Anderson, J. R., Roeder, J. J., and Herrenschmidt, N., eds., *Current Primatology, Vol. 1: Ecology and Evolution*, University of Louis Pasteur, Strasbourg, pp. 51–56.

Ganzhorn, J. U. 1995, Low-level forest disturbances effects on primary production, leaf chemistry, and lemur populations. *Ecol.* 76:2084–2096.

Ganzhorn, J. U. 1997, Test of Fox's assembly rule for functional groups in lemur communities in Madagascar. *J. Zool.* 241:533–542.

Ganzhorn, J. U. 1998, Nested patterns of species composition and its implications for lemur biogeography in Madagascar. *Fol. Primatol.* 69:332–341.

Ganzhorn, J. U. 2002, Distribution of a folivorous lemur in relation to seasonally varying food resources: integrating quantitative and qualitative aspects of food characteristics. *Oecologia* 131:427–435.

Ganzhorn, J. U. and Eisenbeiss, B. 2001, The concept of nested species assemblages and its utility for understanding effects of habitat fragmentation. *Basic and Applied Ecology* 2.87 99

Ganzhorn, J. U., Goodman, S. M., and Dehgan, A. 2003, Effects of forest fragmentation on small mammals and lemurs, in: Goodman, S. M. and Benstead, J. P., eds., *The Natural History of Madagascar*, University of Chicago Press, Chicago, pp. 1228–1234.

Ganzhorn, J. U., Langrand, O., Wright, P. C., O'Connor, S. O., Rakotosamimanana, B., Feistner, A. T. C., and Rumpler, Y. 1996/1997, The state of lemur conservation in Madagascar. *Primate. Conserv.* 17:70–86.

Ganzhorn, J. U., Malcolmber, A., Adrianantoanina, O., and Goodman, S. M. 1997, Habitat characteristics and lemur species richness in Madagascar. *Biotropica* 29:331–343.

Ganzhorn, J. U., Wright, P. C., and Ratsimbazafy, H. J. 1999, Primate Communities: Madagascar, in: Fleagle, J. G., Janson, C. H., and Reed, K., eds., *Primate Communities*, Cambridge University Press, Cambridge, pp. 75–89.

Garbut, N. 1999, *Mammals of Madagascar*, Pica Press, Sussex.

Gascon, C. G., Lougheed, S. C., and Bogart, J. P. 1996, Genetic and morphological variation in *Vanzolinius discodactylus*: a test of the river barrier hypothesis of speciation. *Biotropica* 28:376–387.

Gascon, C. G., Lougheed, S. C., and Bogart, J. P. 1998, Patterns of genetic population differentiation in four species of Amazonian frogs: A test of the riverine barrier hypothesis. *Biotropica* 30:104–119.

Gaston, K. J. 1994, *Rarity*, Chapman Hall, New York.

Gaston, K. J., Blackburn, T. M., and Spicer, J. I. 1998, Rapoport's rule: time for an epitath? *TREE* 13:70–74.

Gause, G. F. 1934, *The Struggle for Existence*, Williams and Williams, Baltimore.

Gee, J. H. R. and Giller, P. S., eds., 1987, *Organization of Communities*, Blackwell Scientific, Oxford.

Gentry, A. H. 1989, Speciation in tropical forests, in: Holm-Nielson, I. C. and Balslev, H., eds., *Tropical Forests*, Academic Press, New York, pp. 113–134.

Godfrey, L., Junger, W., Reed, K. E., Simons, E. L., and Chatrath, P. S. 1997, Primate subfossils inferences about past and present primate community structure, in: Goodman, S. and Patterson, B., eds., *Natural Change and Human-Induced Change in Madagascar*, Smithsonian Institution Press, Washington, D.C., pp. 218–256.

Godfrey, L. R., Jungers, W. L., Simons, E. L., Chatrath, P. S., and Rakotosamimanana, B. 1999, Past and present distributions of lemurs in Madagascar, in: Rakotosamimanana, B., Rasamimanana, H., Ganzhorn, J. U., and Goodman, S. M., eds., *New Directions in Lemur Studies*, Kluwer Academic/Plenum, New York, pp. 19–53.

Godfrey, L. R., Samonds, K. E., Jungers, W. L., Sutherland, M. R., and Irwin, M. T. 2004, Ontogenetic correlates of diet in Malagasy lemurs. *Amer. J. Phys. Anth.* 123:250–276.

Godfrey, L. R., Sutherland, M. R., Petto, A. J., and Boy, D. S. 1990, Size, space, and adaptation in some subfossil lemurs from Madagascar. *Amer. J. Phys. Anth.* 81:45–66.

Goldberg, T. L. 1996, Genetics and Biogeography of East African Chimpanzees (*Pan troglodytes schweinfurthii*), Unpublished Ph.D. dissertation, Harvard University.

Goldberg, T. L. and Ruvolo, M. 1997, Molecular phylogenetics and historical biogeography of east African chimpanzees. *Biol. J. Linn. Soc.* 61:301–324.

Gonder, M. K., Oates, J. F., Disotell, T. R., Forstner, M. R. J., Morales, J. C., and Melnick, D. J. 1997, A new west African chimpanzee subspecies? *Nature* 388:337.

Goodman, S. M. and Benstead, J. 2005, Updated estimates of biotic diversity and endemism for Madagascar. *Oryx* 39:73–77.

Goodman, S. M. and Ganzhorn, J. 2003, Biogeography of lemurs in the humid forests of Madagascar: the role of elevational distribution and rivers. *J. Biogeog.* 31:47–56.

Goodman, S. M. and Raselimanana, A. 2003, Hunting of wild animals by Sakalava of the Menabe region: a field report from Kirindy-Mite. *Lemur News* 8:4–6.

Grandidier, A. 1875–1921, *Histoire physique, naturelle, et politique de Madagascar*, Hachette, Paris.

Green, G. M. and Sussman, R. W. 1990, Deforestation history of the eastern rain forests of Madagascar from satellite images. *Science* 248:212–215.

Groves, C. P. 2001, *Primate Taxonomy*, Smithsonian Institution Press, Washington, DC.

Grubb, P. 1990, Primate geography in the Afro-tropical forest biome, in: Peters, G. and Hutterer, R., eds., *Vetebrates in the Tropics*, Museum Alexander Koenig, Bonn, pp. 187–214.

Grubb, P. 1999, Evolutionary processes implicit in distribution patterns of modern African mammals, in: Bromage, T. G. and Schrenk, F., eds., *African Biogeography, Climate Change, and Human Evolution*, Oxford University Press, New York, pp. 150–164.

Grubb, P., Butynski, T. M., Oates, J. F., Bearder, S. K., Disotell, T. R., Groves, C. P., and Struhsaker, T. T. 2003, Assessment of the diversity of African primates. *Int. J. Primatol.* 24:1301–1357.

Gupta, A. K. and Chivers, D. J. 1999, Biomass and use of resources in south and southeast Asian primate communities, in: Fleagle, J. G., Janson, C. H., and Reed, K., eds., *Primate Communities*, Cambridge University Press, Cambridge, pp. 38–54.

Haffer, J. 1982, General aspects of the refuge theory, in: Prance, G. T., ed., *Biological Diversification in the Tropics*, Columbia University Press, New York, pp. 6–24.

Hamilton, A., Taylor, D., and Howard, P. 2001, Hotspots in African forests as quaternary refugia, in: Weber, W., White, L. J. T., Vedder, A., and Naughton-Treves, L., eds., *African Rain Forest Ecology and Conservation: An Interdisciplinary Perspective*, Yale University Press, New Haven, pp. 57–67.

Hanski, I. 1982, Dynamics of regional distribution: the core and satellite species hypothesis. *Oikos* 38:210–221.

Hanski, I. and Gyllenberg, M. 1997, Uniting two general patterns in the distribution of species. *Science* 275:397–400.

Hanski, I., Kouki, J., and Halkka, A. 1993, Three explanations of the positive relationship between distribution and abundance of species, in: Ricklefs, R. E. and Schluter, D., eds., *Species Diversity in Ecological Communities: Historical and Geographical Perspectives*, University of Chicago Press, Chicago, pp. 108–116.

Harcourt, A. H. 1999, Biogeographic relationships of primates on South-East Asian islands. *Global. Ecol. Biog.* 8:55–61.

Harcourt, A. H. 2000a, Coincidence and mismatch of biodiversity hotspots: a global survey for the order, primates. *Biol. Cons.* 93:163–175.

Harcourt, A. H. 2000b, Latitude and latitudinal extent: a global analysis of the Rapoport effect in a tropical mammalian taxon: primates. *J. Biogeog.* 27:1169–1182.

Harcourt, A. H. 2004, Are rare primate taxa specialists or simply less studied? *J. Biogeog.* 31:57–61.

Harcourt, A. H., Coppeto, S. A., and Parks, S. A. 2002, Rarity, specialization and extinction in primates. *J. Biogeog.* 29:445–456.

Harcourt, A. H., Coppeto, S. A., and Parks, S. A. 2005, The distribution-abundance (density) relationship: its form and causes in a tropical mammal order, Primates. *J. Biogeog.* 32:565–579.

Harcourt, A. H. and Doherty, D. A. 2005, Species-area relationships of primates in tropical forest fragments: a global analysis. *J. Appl. Ecol.* 42:630–637.

Harcourt, A. H. and Schwartz, M. W. 2001, Primate evolution: A biology of holocene extinction and survival on the southeast Asian Sunda Shelf Islands. *Amer. J. Phys. Anth.* 114:4–17.

Harper, K. A., Macdonald, S. E., Burton, P., Chen, J., Brosofsky, K. D., Saunders, S., Euskirchen, E. S., Roberts, D., Jaiteh, M., and Esseen, P.-A. 2005, Edge influence on forest structure and composition in fragmented landscapes. *Cons. Biol.* 19:1–15.

Haugaasen, T. and Peres, C. A. 2005, Mammal assemblage structure in Amazonian flooded and unflooded forests. *J. Trop. Ecol.* 21:133–145.

Heard, S. B. and Mooers, A. Ø. 2000, Phyllogenetically patterned speciation rates and extinction risks change the loss of evolutionary history during extinctions. *Proc. Roy. Soc. Lond, Series B* 267:613–620.

Henderson, A. 1995, *The Palms of the Amazon*, Oxford University Press, New York.

Hennig, W. 1966, *Phylogenetic Systematics*, University of Illinois, Urbana.

Hershkovitz, P. 1968, Metachromism or the principle of evolutionary change in mammalian tegumentary colors. *Evol.* 22: 556–575.

Hershkovitz, P. 1977, *Living New World Monkeys (Platyrrhini) With an Introduction to Primates*, University of Chicago Press, Chicago.

Hershkovitz, P. 1984, Taxonomy of the squirrel monkeys, genus *Saimiri* (Cebidae, Platyrrhini): A preliminary report with the description of a hitherto unnamed form. *Amer. J. Primatol.* 6:257–312.

Hershkovitz, P. 1988, Origin, speciation, and distribution of South American titi monkeys, genus *Callicebus* (Family Cebidae, Platyrhini). *Proc. Nat. Acad. Sci., Philad* 140:240–272.

Hesse, R., Allee, W. C., and Schmidt, K. P. 1937, *Ecological Animal Geography*, John Wiley & Sons, New York.

Heyer, W. R. and Maxson, L. R. 1982, Distribution, relationships, and zoogeography of lowland frogs, in: Prance, G. T., ed., *Biological Diversification in the Tropics*, Columbia University Press, New York, pp. 375–388.

Hill, W. C. O. 1969, The nomenclature, taxonomy and distribution of chimpanzees, in: Bourne, G. H., ed., *The Chimpanzee*, Karger, Basel, pp. 22–49.

Hovenkamp, P. 1997, Vicariance events, not areas, should be used in biogeographical analysis. *Cladistics* 13:67–79.

Huggett, R. J. 2004, *Fundamentals of Biogeography*, Routledge, New York.

Humphries, C. J. and Parenti, L. R. 1999, *Cladistic Biogeography*, Oxford University Press, Oxford.

Huston, M. A. 1996, *Biological Diversity: The Coexistence of Species on Changing Landscapes*, Cambridge University Press, Cambridge.

Hutchinson, G. E. 1957, Concluding remarks. *Cold Spring Harbor Symposium on Quantitative Biology* 22:415–427.

Irschick, D., Dyer, L., and Sherry, T. W. 2005, Phylogenetic methodologies for studying specialization. *Oikos* 110:404–408.

Isaac, N. J. B., Mallet, J., and Mace, G. M. 2004, Taxonomic inflation: its influence on macroecology and conservation. *TREE* 19:464–469.

IUCN, 2005, *IUCN Red List of Threatened Species*, Gland, Switzerland: IUCN Species Survival Commission.

Janson, C. H. and Chapman, C. 1999, Resources and primate community structure, in: Fleagle, J. G., Janson, C. H., and Reed, K., eds., *Primate Communities*, Cambridge University Press, Cambridge, pp. 237–267.

Jensen-Seaman, M. I. and Kidd, K. K. 2001, Mitochondrial DNA variation and biogeography of eastern gorillas. *Molecular Ecology* 10:2241–2247.

Jernvall, J. and Wright, P. C. 1998, Diversity components of impending primate extinctions. *Proc. Nat. Acad. Sci., USA* 95:11279–11283.

Johns, A. D. and Skorupa, J. P. 1987, Responses of rain-forest primates to habitat disturbance: a review. *Int. J. Primatol.* 8:157–191.

Johnson, A., Singh, S., Duangdala, M., and Hedemark, M. 2005, The western black crested gibbon *Nomascus concolor* in Laos: new records and conservation status. *Oryx.* 39:311–317.

Jones, C. B. 1997, Rarity in primates: implications for conservation. *Mastozoología Neotropical* 4:35–47.

Jones, T., Ehardt, C. L., Butynski, T. M., Davenport, T. R. B., Mpunga, N. E., Machaga, S. J., and De Luca, D. W. 2005, The highland mangabey *Lophocebus kipunji*: a new species of African monkey. *Science* 308:1161–1164.

Julliot, C. and Simmen, B. 1998, Food partitioning among a community of neotropical primates. *Fol. Primatol.* 69:43–44.

Jungers, W. L., Godfrey, L. R., Simons, E. L., and Chatrath, P. S. 1995, Subfossil Indri indri from the Ankarana Massif of northern Madagascar. *Amer. J. Phys. Anth.* 97:357–366.

Kay, R. F., Madden, R. H., van Schaik, C., and Higdon, D. 1997, Primate species richness is determined by plant productivity: Implications for conservation. *Proc. Nat. Acad. Sci., USA* 94:13023–13027.

Kingdon, J. 1971, *East African Mammals An Atlas of Evolution in Africa*, Academic Press, London.

Kinzey, W. G. 1982, Distribution of primates and forest refuges, in: Prance, G. T., ed., *Biological Diversification in the Tropics*, Columbia University Press, New York, pp. 455–482.

Kinzey, W. G. and Gentry, A. H., 1979, Habitat utilization in two species of Callicebus, in: Sussman, R. W., ed., *Primate Ecology: Problem-Oriented Field Studies*, John Wiley and Sons, New York, pp. 89–100.

Knapp, S. and Mallet, J. 2003, Refuting refugia? *Science* 300:71–72.

Kobayashi, S. and Langguth, A. 1999, A new species of titi monkey, *Callicebus* Thomas, from north-eastern Brazil (Primates, Cebidae). *Revista Brasileira de Zoologia* 16:531–551.

Laurance, M. F., Lovejoy, T. E., Vasconcelos, H. L., Bruna, E. M., Didham, R. K., Stouffer, P. C., Gascon, C., Bierregaard, O., Laurance, S., and Sampaio, E. 2002, Ecosystem decay of Amazonian forest fragments: a 22-year investigation. *Cons. Biol.* 16:605–618.

Laurance, W. F. 2000, Do edge effects occur over large spatial scales? *TREE* 15:134–135.

Laurance, W. F., Laurance, S. G., Ferreira, L. V., Rankin-de Merona, J. M., Gascon, C., and Lovejoy, T. E. 1997, Biomass collapse in Amazonian forest fragments. *Science* 278:1117–1118.

Laurance, W. F. and Yensen, E. 1991, Predicting the impacts of edge effects in fragmented habitats. *Biol. Cons.* 57:205–219.

Laws, M. J., and Eeley, H. A. C. 2000, Are local patterns of anthropoid primate diversity related to patterns of diversity at larger scale? *J. Biogeog.* 27:1421–1435.

Legendre, P., Dale, M. R. T., Fortin, M.-J., Gurevitch, J., Hohn, M., and Myers, D. 2002, The consequences of spatial structure for the design and analysis of ecological field surveys. *Ecography* 25:601–616.

Lehman, S. M. 1999, Biogeography of the Primates of Guyana, Unpublished Ph.D. dissertation, Washington University, St. Louis, MO.

Lehman, S. M. 2000, Primate community structure in Guyana: A biogeographic analysis. *Int. J. Primatol.* 21: 333–351.

Lehman, S. M. 2004, Distribution and diversity of primates in Guyana: Species-area relationships and riverine barriers. *Int. J. Primatol.* 25:73–95.

Lehman, S. M. in press, Conservation biology of Malagasy Strepsirhines: A phylogenetic approach. *Amer. J. Phys. Anth.*

Lehman, S. M., Mayor, M., and Wright, P. C. 2005, Ecogeographic size variations in sifakas: A test of the resource seasonality and resource quality hypotheses. *Amer. J. Phys. Anth.* 126:318–328.

Lehman, S. M., Rajaonson, A., and Day, S. 2006 a, Lemur responses to edge effects in the Vohibola III Classified Forest, Madagascar. *Amer. J. Primatol.* 68(3):293–299.

Lehman, S. M., Rajoanson, A., and Day, S. 2006-b, Edge effects and their influence on lemur distribution and density in southeast Madagascar. *Amer. J. Phys. Anth.* 129(2):232–241.

Lehman, S. M., Ratsimbazafy, H. J., Rajaonson, A., and Day, S. 2006-c, Decline in the distribution of *Propithecus diadema edwardsi* and *Varecia variegata variegata* in southeast Madagascar. *Oryx.* 40(1):108–111.

Lehman, S. M. and Wright, P. C. 2000, Preliminary description of the conservation status of lemur communities in the Betsakafandrika region of eastern Madagascar. *Lemur News* 5:23–25.

Lehtinen, R. M., Ramanamanjato, J.-B., and Raveloarison, J. G. 2003, Edge effects and extinction proneness in a herpetofauna from Madagascar. *Biod & Cons* 12:1357–1370.

Lidicker, W. Z. 1999, Responses of mammals to habitat edges: an overview. *Land Ecol* 14:333–343.

Lomolino, M. V. 2000, Ecology's most general, yet protean pattern: the species-area relationship. *J. Biogeog.* 27:17–26.

Lomolino, M. V., Riddle, B. R., and Brown, J. H. 2005, *Biogeography*, Sinauer Associates, Sunderland, MA.

Losos, J. B., Jackman, T. R., Larson, A., de Queiroz, K., and Rodríguez-Schettino, L. 1998, Contingency and determinism in replicated adaptive radiations of island lizards. *Science* 27:2115–2118.

Lotka, A. J. 1925, *Elements of Physical Biology*, Williams and Watkins, Baltimore.

Lovejoy, T. E., Bierregaard, R. O., Rylands, A. B., Malcolm, J. R., Quintela, C. E., Harper, L. H., Brown, K. S., Powell, A. H., Powell, V. N., Shubart, H. O. R., and Hays, M. B., 1986, Edge and other effects of isolation on Amazon forest fragments, in: Soule', M. E., ed., *Conservation Biology: the Science of Scarcity and Diversity.*, Sinauer Associates, Sunderland, MA, pp. 257–285.

MacArthur, R. 1972, *Geographical ecology*, Princeton University Press, Princeton, N.J.

MacArthur, R. H. and Wilson, E. O. 1967, *The Theory of Island Biogeography*, Princeton University Press, Princeton.

Malcolm, J. R. 1994, Edge effects in central Amazonian forest fragments. *Ecol.* 75:2438–2445.

Marsh, L. K., ed., 2003, *Primates in Fragments: Ecology and Conservation*, Kluwer Academic/Plenum Publishers, New York.

Martin, R. D. 1972, Adaptive radiation and behavior of Malagasy lemurs. *Philisophical Transactions of the Royal Society London (Series B)* 264:295–352.

Martin, R. D. 1993, Primate origins: Plugging the gaps. *Nature* 363:223–234.

Matthiae, P. E. and Stearns, F., 1981, Mammals in forest islands in southestern Wisconsin, in: Burgess, R. and Sharpe, D., eds., *Forest Island Dynamics in Man-Dominated Landscapes*, Springer-Verlag, New York, pp. 55–66.

May, R. M. 1990, Taxonomy as destiny. *Nature* 347:129–130.

Mayor, M., Sommer, J. A., Houck, M. L., Zaonarivelo, J. R., Wright, P. C., Ingram, C., Engel, S. R., and Louis, E. E. 2004, Specific status of *Propithecus* spp. *Int. J. Primatol.* 25:875–900.

Mayr, E. 1942, *Systematics and the Origin of Species From the Viewpoint of a Zoologist*, Columbia University Press, New York.

Mayr, E. and O'Hara, R. J. 1986, The biogeographic evidence supporting the pleistocene forest refuge hypothesis. *Evol.* 40:55–67.

Mbora, D. N. M. and Meikle, D. B. 2004, Forest fragmentation and the distribution, abundance and conservation of the Tana river red colobus (*Procolobus rufomitratus*). *Biol. Cons.* 118:67–77.

Medeiros, M. A., Barros, R. M. S., J.C., P., Nagamachi, C. Y., Ponsa, M., Garcia, F., and Egozcue, J. 1997, Radiation and speciation of spider monkeys, genus *Ateles*, from the cytogenetic viewpoint. *Amer. J. Primatol.* 42:167–178.

Meijaard, E. and Nijman, V. 2003, Primate hotspots on Borneo: Predictive value for general biodiversity and the effects of taxonomy. *Cons. Biol.* 17:725–732.

Meyers, D. M., Rabarivola, C., and Rumpler, Y. 1989, Distribution and conservation of Sclater's lemur: implications of a morphological cline. *Primate. Conserv.* 10:77–81.

Miller, L., Savage, A., and Giraldo, H. 2004, Quantifying remaining forested habitat within the historic distribution of the cotton-top tamarin (*Saguinus oedipus*) in Colombia: implications for long-term conservation. *Amer. J. Primatol.* 64:451–457.

Mittermeier, R. A., Gil, P. R., Hoffman, M., Pilgrim, J., Brooks, T., Goettsch Mittermeier, C., Lamoreux, J., and da Fonseca, G. A. B. 2005 *Hotspots Revisited: Earth's Biologically Richest and Most Endangered Terrestrial Ecoregions*, Conservation International, Washington, DC.

Mittermeier, R. A., Myers, N., Gil, P. R., and Mittermeier, C. G. 1999 *Hotspots: Earth's biologically richest and most endangered terrestrial ecoregions*, CEMEX, Conservation International and Agrupacion Sierra Madre, Mexico, Washington, DC.

Mittermeier, R. A., Myers, N., Thomsen, J. B., da Fonesca, G. A. B., and Olivieri, S. 1998, Biodiversity hotspots and major tropical wilderness areas: approaches to setting conservation priorities. *Cons. Biol.* 12:516–520.

Mittermeier, R. A., Tattersall, I., Konstant, W. R., Meyers, D. M., and Mast, R. B. 1994, *Lemurs of Madagascar*, Conservation International, Washington, DC.

Morin, P. A., Moore, J. J., Chakraborty, R., Jin, L., Goodall, J., and Woodruff, D. S. 1994, Kin selection, social structure, gene flow, and the evolution of chimpanzees. *Science* 265:1193–1201.

Murcia, C. 1995, Edge effects in fragmented forests: implications for conservation. *TREE* 10:58–62.

Myers, A. A. and Giller, P. S., eds., 1988, *Analytical Biogeography: An Integrated Approach to the Study of Animal and Plant Distribution*, Chapman and Hall, London.

Myers, N. 1988, Threatened biotas: 'hotspots' in tropical forests. *Environmentalist* 8:187–208.

Myers, N., Mittermeier, R., Mittermeier, C. G., Fonseca, G. A. B., and Kent, J. 2000, Biodiversity hotspots for conservation priorities. *Nature* 403:853–858.

Nelson, B. W., Ferreira, C. A. C., da Silva, M. F., and Kawaski, M. L. 1990, Endemism centres, refugia and botanical collection density in Brazilian Amazonia. *Nature* 345:714–716.

Norconk, M. A. 1997, Seasonal variations in the diets of white-faced and bearded sakis (*Pithecia pithecia* and *Chiropotes satanas*) in Guri Lake, Venezuela, in: Norconk, M. A., Rosenberger, A. L., and Garber, P. A., eds., *Adaptive Radiations of Neotropical Primates*, Plenum Press, 1996, pp. 403–426.

Norconk, M. A. and Grafton, B. W. 2003, Changes in forest composition and potential feeding tree availability on a small land-bridge island in Lago Guri, Venezuela, in: Marsh, L. K., ed., *Primates in Fragments: Ecology and Conservation*, Kluwer Academic/Plenum Publishers, New York, pp. 211–227.

Norconk, M. A., Sussman, R. W., and Phillips-Conroy, J. 1997, Primates of Guayana Shield Forests: Venezuela and the Guianas, in: Norconk, M. A., Rosenberger, A. L., and Garber, P. A., eds., *Adaptive Radiations of Neotropical Primates*, Plenum Press, New York, pp. 69–86.

Oka, T., Iskandar, E., and Ghozali, D. I. 2000, Effects of forest fragmentation on the behavior of Bornean gibbons. *Ecological Studies* 140:229–241.

Onderdonk, D. A. and Chapman, C. A. 2000, Coping with fragmentation: the primates of Kibale National Park, Uganda. *Int. J. Primatol.* 21:587–611.

Owens, I. P. F. and Bennett, P. M. 2000, Quantifying biodiversity: A phenotypic perspective. *Cons. Biol.* 14:1014–1022.

Paciulli, L. M. 2004, The effects of logging, hunting, and vegetation on the densities of the Pagai, Mentawai Island primates (Indonesia), Unpublished Ph.D. dissertation, SUNY-Stony Brook, Stony Brook.

Pagel, M. D., May, R. M., and Collie, A. R. 1991, Ecological aspects of the geographical distribution and diversity of mammalian species. *Amer. Nat.* 137:791–815.

Passamani, M. and Rylands, A. B. 2000, Home range of a Geoffroy's marmoset group, *Callithrix geoffroyi* (primates, Callitrichidae) in south-eastern Brazil. *Rev. Brasil. Biol.* 60:275–281.

Pastor-Nieto, R. and Williamson, D. K. 1998, The effect of rainfall seasonality on the geographic distribution of neotropical primates. *Neotrop. Prim.* 6:7–14.

Patterson, B. D. 1987, The principle of nested subsets and its implication for biological conservation. *Cons. Biol.* 1:323–334.

Patterson, B. D. and Atmar, W. 1986, Nested subsets and the structure of insular mammalian faunas and archipelagos. *Biol. J. Linn. Soc.* 28:65–82.

Patton, J. L., da Silva, M. N. F., Lara, M., and Mustrangi, M. A. 1997, Diversity, differentiation, and the historical biogeography of nonvolant mammals of the neotropical forests, in: Laurance, W. F. and Bierregaard, O., eds., *Tropical Forest Remnants: Ecology, Management, and Conservation of Fragmented Communities*, The University of Chicago Press, Chicago, pp. 455–465.

Patton, J. L., Da Silva, M. N. F., and Malcolm, J. R. 1994, Gene geneaology and differentiation among arboreal spiny rats (Rodentia: Echimyidae) of the Amazon Basin: a test of the riverine barrier hypothesis. *Evol.* 48:1314–1323.

Peres, C. 1988, Primate community structure in western Brazilian Amazonia. *Primate. Conserv.* 9:83–87.

Peres, C. 1993a, Notes on the primates of the Jaruá River, western Brazilian Amazonia. *Fol. Primatol.* 61:97–103.

Peres, C. 1993b, Structure and spatial organization of an Amazonian terra firme forest primate community. *J. Trop. Ecol.* 9:259–276.

Peres, C. A. 1997, Primate community structure at twenty western Amazonian flooded and unflooded forests. *J. Trop. Ecol.* 13:381–405.

Peres, C. A. 1999, Effects of hunting and forest types on Amazonian communities, in: Fleagle, J. G., Janson, C. H., and Reed, K., eds., *Primate Communities*, Cambridge University Press, Cambridge, pp. 268–283.

Peres, C. A. 2000, Evaluating the impact and sustainability of subsistence hunting at multiple Amazonian forest sites, in: Robinson, J. G., and Bennett, E. L., eds., *Hunting for Sustainability in Tropical Forests*, Columbia University Press, New York, pp. 31–56.

Peres, C. A. and Dolman, P. M. 2000, Density compensation in neotropical primate communities: evidence from 56 hunted and nonhunted Amazonian forests of varying productivity. *Oecologia* 122:175–189.

Peres, C. A. and Janson, C. H., 1999, Species coexistence, distribution, and environmental determinants of neotropical primate richness: A community-level zoogeographic analysis, in: Fleagle, J. G., Janson, C. H., and Reed, K., eds., *Primate Communities*, Cambridge University Press, Cambridge, pp. 55–74.

Peres, C. A., Patton, J. L., and da Silva, M. N. 1996, Riverine barriers and gene flow in Amazonian saddle-back tamarins. *Fol. Primatol.* 67:113–124.

Peterson, A. T. 2001, Predicting species' geographic distributions based on ecological niche modeling. *Condor.* 103:599–605.

Por, F. D. 1978, *Lessepian Migration The Influx of Red Sea Biota into the Mediterranean by way of the Suez Canal*, Springer Verlag, Berlin.

Prance, G. T. 1987, Biogeography of neotropical plants, in: Whitmore, T. C. and Prance, G. T., eds., *Biogeography and Quaternary History in Tropical America*, Oxford University Press, Oxford, pp. 46–65.

Prendergast, J. R., Quinn, R. M., Lawton, J. H., Eversham, B. C., and Gibbons, D. W. 1993, Rare species, the coincidence of diversity hotspots and conservation strategies. *Nature* 365:335–337.

Pressey, R. L., Possingham, H. P., and Margules, C. R. 1996, Optimality in reserve selection algorithms: when does it matter and how much? *Biol. Cons.* 76:259–267.

Preston, F. W. 1962, The canonical distribution of commonness and rarity. *Ecol.* 43:185–215.

Pugesek, B. H., Tomer, A., and von Eye, A., eds., 2002, *Structural Equation Modeling: Applications in Ecological and Evolutionary Biology*, Cambridge University Press, Cambridge.

Radispiel, U. and Raveloson, H. 2001, Preliminary study on the lemur communities at three sites in dry deciduous forest in the Reserve Naturelle d'Ankaranfantsika. *Lemur News* 6:22–24.

Räsänen, M. E., Salo, J., and Kalliola, R. J. 1987, Fluvial perturbance in the western Amazon basin: Regulation by long-term sub-Andean tectonics. *Science* 238:1398–1401.

Rasoloarison, R. M., Goodman, S. M., and Ganzhorn, J. U. 2000, Taxonomic revision of mouse lemurs (*Microcebus*) in the western portions of Madagascar. *Int. J. Primatol.* 21:963–1019.

Ratsimbazafy, J. 2002, On the brink of extinction and the process of recovery: Responses of black-and-white ruffed lemurs (*Varecia variegata variegata*) to disturbance in Manombo Forest, Madagascar, Unpublished Ph.D. dissertation, Stony Brook University, Stony Brook, NY.

Ravosa, M. J., Meyers, D. M., and Glander, K. E. 1993, Relative growth of the limbs and trunk in sifakas: Heterochronic, ecological, and functional considerations. *Amer. J. Phys. Anth.* 92:499–520.

Ravosa, M. J., Meyers, D. M., and Glander, K. E. 1995, Heterochrony and the evolution of ecogeographic size variation in Malagasy sifakas, in: McNamara, K. J., ed., *Evolutionary Change and Heterochrony*, John Wiley & Sons, New York, pp. 261–276.

Reed, K. E. 1999, Population density of primates in communities: Differences in community structure, in: Fleagle, J. G., Janson, C. H., and Reed, K. E., eds., *Primate Communities*, Cambridge University Press, Cambridge, pp. 116–140.

Reed, K. E. and Fleagle, J. G. 1995, Geographic and climate control of primate diversity. *Proc. Nat. Acad. Sci., USA* 92:7874–7876.

Remsen, J. V. and Parker, T. A. 1983, Contribution of river-created habitats to bird species richness in Amazonia. *Biotropica* 15:223–231.

Ricklefs, R. E. and Schluter, D., eds., 1993, *Species diversity in ecological communities: historical and geographical perspectives*, University of Chicago Press, Chicago.

Ries, L., Fletcher, R. J., Battin, J., and Sisk, T. D. 2004, Ecological responses to habitat edges: mechanisms, models, and variability explained. *Ann. Rev. Ecol. Syst.* 35:491–522.

Robinson, J. G. 1986, Seasonal variation in use of time and space by the wedge-capped capuchin monkey *Cebus olivaceus*: Implications for foraging theory. *Smithsonian Contributions to Zoology* 431:1–60.

Rodrigues, M. T. 1991, Herpetofauan das dunas interiores do Rio Sao Francisco, Bahia, Brasil I Introduçao a area e descriçao de um novo genero de microteiidos (*Calyptommatus*) com notas sobre sue ecologia, distribuiçao e especiçao (Sauria, Teiidae). *Papeis Avul Zooologica* 37:285–320.

Ron, S. R. 2000, Biogeographic area relationships of lowland neotropical rainforest based on raw distributions of vertebrate groups. *Biol. J. Linn. Soc.* 71:379–402.

Rosenzweig, M. L. 1995, *Species Diversity in Space and Time*, Cambridge University Press, Cambridge.

Rowe, N. 1996, *The Pictorial Guide to the Living Primates*, Pogonias Press, East Hampton.

Ruggiero, A. 1994, Latitidinal correlates of the sizes of mammalian geographical ranges in South America. *J. Biogeog.* 21:545–559.

Rylands, A. 1998, Old World primates: New species and subspecies. *Oryx.* 32:88–89.

Rylands, A. B. 1996, Habitat and the evolution of social and reproductive behavior in Callitrichidae. *Amer. J. Primatol.* 38:5–18.

Rylands, A. B., and de Faria, D. S., 1993, Habitats, feeding ecology, and home range size in the genus *Callithrix*, in: Rylands, A. B. ed., *Marmosets and Tamarins: Systematics, Behaviour, and Ecology*, Oxford University Press, Oxford, pp. 262–272.

Rylands, A. B., Mittermeier, R. A., and Konstant, W. R. 2002, Species and subspecies of primates described since 1990. *Lemur News* 7:5–6.

Rylands, A. B., Schneider, H., Langguth, A., Mittermeier, R. A., Groves, C. P., and Rodríguez-Luna, E. 2000, An assessment of the diversity of the New World Primates. *Neotrop. Prim.* 8:61–93.

Salo, J., Kalliola, R., Hakkinen, I., Makinen, Y., Niemela, P., Puhakka, M., and Coley, P. 1986, River dynamics of Amazon lowland forest. *Nature* 322:254–258.

Schall, J. J. and Pianka, E. R. 1978, Geographical trends in numbers of species. *Science* 201:679–686.

Schoener, T. W. 1988, Ecological Interactions, in: Myers, A. A. and Giller, P. S., eds., *Analytical Biogeography: An Integrated Approach to the Study of Animal and Plant Distribution*, Chapman and Hall, London, pp. 255–295.

Scott, J. M., Heglund, P. J., and Morrison, M. L. 2002, *Predicting species occurrences: issues of accuracy and scale*, Island Press, Washington, DC.

Sechrest, S., Brooks, T. M., da Fonseca, G. A. B., Konstant, W. R., Mittermeier, R. A., Purvis, A., Rylands, A. B., and Gittleman, J. L. 2002, Hotspots and the conservation of evolutionary history. *Proc. Nat. Acad. Sci., USA* 99:2067–2071.

Sick, H., 1967, Rios e enchentas na Amazonia como obstaculo para a avifauna, in: Lent, H., ed., *Atas do Simposio Sobre a Biota Amazonica*, Conselho Nacional de Pesquisas, Rio de Janeiro, pp. 495–520.

Silander, J. A. and Antonovics, J. 1982, Analysis of interspecific interactions in a coastal plant community – a perturbation approach. *Nature* 298:557–560.

Silva, J. S. and Noronha, M. A. 1998, On a new species of bare–eared marmoset, genus *Callithrix* Erxleben, 1777, from central Amazonia, Brazil (Primates: Callitrichidae). *Goeldiana Zoologia* 21:1–28.

Simpson, G. G. 1965, *The Geography of Evolution*, Chilton, Philadelphia.

Soini, P. 1986, A synecological study of a primate community in the Pacaya-Samiria National Reserve, Peru. *Primate Conserv.* 7:63–71.

Stevens, G. C. 1989, The latitudinal gradient in geographic range: how so many species coexist in the tropics. *Amer. Nat.* 133:240–246.

Sussman, R. W. 1991, Primate origins and the evolution of angiosperms. *Amer. J. Primatol.* 23:209–223.

Sussman, R. W. 1999, *Primate Ecology and Social Structure: Lorises, Lemurs, Tarsiers*, Pearson Custom Publishing, Needham Heights.

Sussman, R. W., Green, G. M., Porton, I., Andrianasolondraibe, O. L., and Ratsirarson, J. 2003, A survey of the habitat of *Lemur catta* in southwestern and southern Madagascar. *Primate. Conserv.* 19:23–31.

Tattersall, I. 1982, *The Primates of Madagascar*, Columbia University Press, New York.

Terborgh, J. and van Schaik, C. P., 1987, Convergence vs nonconvergence in primate communities, in: Gee, J. H. R. and Giller, P. S. eds., *Organization of Communities, Past and Present*, Blackwell Science Publications, Oxford, pp. 205–226.

Terborgh, J. W. 1971, Distribution on environmental gradients: Theory and a preliminary interpretation of distributional patterns in the avifauna of the Cordillera Vilcabamba, Peru. *Ecol* 52:26–36.

Thalmann, U. and Geissmann, T. 2000, Distribution and geographic variation in the western woolly lemur (*Avahi occidentalis*) with description of a new species (*A. unicolor*). *Int. J. Primatol.* 21:915–941.

Thalmann, U. and Rakotoarison, N. 1994, Distribution of lemurs in central western Madagascar, with a regional distribution hypothesis. *Fol. Primatol.* 63:156–161.

Thiollay, J. M. 1994, Structure, density, and rarity in an Amazonian rainforest bird community. *J. Trop. Ecol.* 10:449–481.

Tuomisto, H. and Ruokolainen, K. 1997, The role of ecological knowledge in explaining biogeography and biodiversity in Amazonia. *Biod. and Cons.* 6:347–357.

Tutin, C. E. G., Ham, R. M., White, L. J. T., and Harrison, M. J. S. 1997, The primate community of the Lope Reserve, Gabon: Diets, responses to fruit scarcity, and effects on biomass. *Amer. J. Primatol.* 42:1–24.

van Roosmalen, M. 1998, A new species of marmoset in the Brazilian Amazon. *Neotrop. Prim.*:90–91.

van Roosmalen, M. G. M., van Roosmalen, T., and Mittermeier, R. A. 2002, A taxonomic review of the titi monkeys, genus *Callicebus* Thomas, 1903, with the description of two new species, *Callicebus bernhardi* and *Callicebus stephennashi*, from Brazilian Amazonia. *Neotrop. Prim.* 10:1–52.

van Roosmalen, M. G. M., van Roosmalen, T., Mittermeier, R. A., and Rylands, A. B. 2000, Two new species of marmoset, genus *Callithrix* Erxleben, 1777 (Callitrichidae, Primates), from the Tapajos/Madeira Interfluvium, south central Amazonia, Brazil. *Neotrop. Prim.* 8:2–18.

Vane-Wright, R. I., Humphries, C. J., and Williams, P. H. 1991, What to protect? Systematics and the agony of choice. *Biol. Cons.* 55:235–254.

Vanzolini, P. E., and Williams, E. E. 1970, South American anoles: the geographic differentation and evolution of the *Anolis chrysolepis* species group (Sauria, Iguanidae). *Arq. Zool. Sao Paulo* 19:1–298.

Vazquez, D. P. and Simberloff, D. 2002, Ecological specialization and susceptibility to disturbance: conjectures and refutations. *The American Naturalist* 159:606–623.

Volterra, V. 1926, Variations and fluctuations in the number of individuals of animal species living together, in: Chapman, R. N. ed., *Animal Ecology*, McGraw-Hill, New York, pp. 409–448.

Wagner, M. 1868, *Die Darwin'sche Theorie und das Migrations-gesetz der Organismen*, Duncker and Humboldt, Leipzig.

Wallace, A. R. 1853, *A Narrative of Travels on the Amazon and Rio Negro*, Revee, London.

Wallace, R. B., Painter, R. L. E., Taber, A. B., and Ayres, J. M. 1996, Notes on a distributional river boundry and southern range extension for two species of Amazonian primate. *Neotrop. Prim.* 4:10–13.

Waser, P. M. 1986, Interactions among primate species, in: Smuts, B. B., Cheney, D. L., Seyfarth, R. M., Wrangham, R. W., and Struhsaker, T. T., eds., *Primate Societies*, University of Chicago Press, Chicago, pp. 210–226.

Watson, J. E. M., Whittaker, R. J., and Dawson, T. P. 2004, Habitat structure and proximity to forest edge affect the abundance and distribution of forest-dependent birds in tropical coastal forests of southeastern Madagascar. *Biol. Cons.* 120:311–327.

Webb, C. O., Ackerly, D. D., McPeek, M. A., and Donoghue, M. J. 2002, Phylogenies and community ecology. *Ann. Rev. Ecol. Syst.* 33:475–505.

White, F. J., Overdorff, D. J., Balko, E. A., and Wright, P. C. 1995, Distribution of ruffed lemurs (*Varecia variegata*) in Ranomafana National Park, Madagascar. *Fol. Primatol.* 64:124–131.

Whittaker, R. J., Araujo, M. B., Paul, J., Ladle, R. J., Watson, J. E. M., and Willis, K. J. 2005, Conservation Biogeography: assessment and prospect. *Diversity and Distributions* 11:3–23.

Wiens, J. A. 1989, *The Ecology of Bird Communities*, Cambridge University Press, Cambridge.

Wiens, J. J. and Graham, C. R. 2005, Niche conservatism: integrating evolution, ecology, and conservation biology. *Annual Review of Ecology, Evolution, and Systematics* 36.

Wiley, E. O. 1988, Vicariance biogeography. *Ann. Rev. Ecol. Syst.* 19:513–542.

Williams, C. B. 1964, *Patterns in the Balance of Nature and Related Problems in Quantitative Ecology*, Academic Press, New York.

Williams, P. H., Humphries, C. J., and Vane-Wright, R. I. 1991, Measuring biodiversity: Taxonomic relatedness for conservation priorities. *Aust. J. Syst. Bot.* 4:665–679.

Willig, M. R., Kaufman, D. M., and Stevens, R. D. 2003, Latitudinal gradients of biodiversity: pattern, process, scale, and synthesis. *Annual Review of Ecology, Evolution, and Systematics* 34:273–309.

Woodroffe, R. and Ginsberg, J. R. 1998, Edge effects and the extinction of populations inside protected areas. *Science* 280:2126–2128.

Wright, P. C. 1999, Lemur traits and Madagascar ecology: Coping with an island environment. *Yrbk. Phys. Anth.* 42:31–42.

Wright, P. C. and Jernvall, J. 1999, The future of primate communities: A reflection of the present?, in: Fleagle, J. G., Janson, C. H., and Reed, K., eds., *Primate Communities*, Cambridge University Press, Cambridge, pp. 295–309.

Yiming, L., Niemela, J., and Dianmo, L. 1998, Nested distribution of amphibians in the Zhoushan archipelago, China: can selective extinction cause nested subsets of species? *Oecologia* 113:557–564.

Zagt, R. J., Marinus, J. A., and Werger, J. A. 1997, Spatial components of dispersal and survival for seeds and seedlings of two codominant tree species in the tropical rain forest of Guyana. *J. Trop. Ecol.* 38:343–355.

Zanette, L. 2000, Fragment size and the demography of an area-sensitive songbird. *J. Anim. Ecol.* 69:458–470.

Zheng, D. and Chen, J. 2000, Edge effects in fragmented landscapes: a generic model for delineating area of edge influences (D-AEI). *Ecological Modeling* 132:175–190.

Neotropics

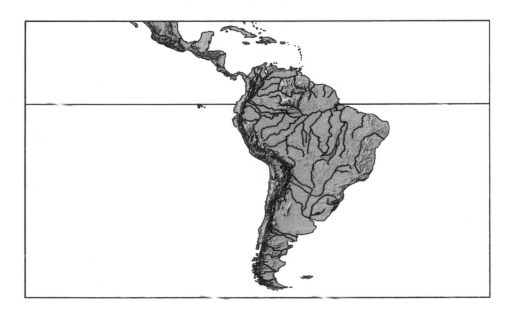

N eotropical primate habitats extend through almost 60 degrees of latitude and 19 countries, from Mexico to the subtropical regions of northern Argentina and southern Brazil (Terborgh and Andresen, 1998). Mountains and rivers dominate primate habitats throughout the Neotropics. The northern mountains of Central America are an extension of the western mountain system of North America while the ranges in southern Central America are outliers of the Andes Mountains of South America. The central region of Central America is an active zone of volcanoes, and contains the Nicaraguan Depression, which includes lakes Managua and Nicaragua. Forest habitats vary from low-altitude coastal dry forests to the central high-altitude cloud forests. Rivers are important biogeographic barriers in South America (Ayres and Clutton-Brock, 1992). The major rivers of South America flow from headwaters in the West to a mouth in the East, the largest being the Amazon with its many tributaries, but also the Orinoco

Primate Biogeography, edited by Shawn M. Lehman and John G. Fleagle.
Springer, New York, 2006.

of Venezuela and the Parana and Plata in the South. However, the main rivers
of the Guianan Shield run South to North. The Guianan Shield represents a
unique biogeographic region on the Atlantic seaboard of South America. Its
landmass of 1,800,000 km^2 is defined by the Orinoco River, the Amazon River,
and the Atlantic Ocean. Most of the Shield is covered by the Venezuelan Highlands, Guyana, Suriname, and French Guiana; but it also includes small part of
Columbia and northern Brazil.

Neotropical primates are classified as the infraorder Platyrrhini. Platyrrhines
are typically divided into two families: Cebids and Atelids. The number of
species and phylogenetic relationships among taxa are sources of ongoing revi-
sions (e.g. Rylands *et al.*, 2000). However, primate diversity tends to be highest
in western Amazonia and then decreases moving eastwards into the Guianas and
Central America (Peres and Janson, 1999).

The papers in this section address questions concerning the historical biogeography and distribution of primates in Central America and Guyana. In "Genetic Evidence on the Historical Biogeography of Central American Howler
Monkeys," Julie Ellsworth and Guy Hoelzer test the hypothesis that a small
number of howlers colonized Central America unidirectionally from northern
South America. Their analyses support the colonization of Central America by
mantled howler monkeys from northern South America via a series of founder
events. Contrary to previous research, Ellsworth and Hoelzer suggest that mantled and black howler monkeys probably represent independent invasions into
Central America. These data are important for understanding the historical
biogeography of many Neotropical mammals.

In "Nested Distribution Patterns and the Historical Biogeography of the
Primates of Guyana," Shawn Lehman sought to determine if primate diversity reflects a hierarchical pattern of species composition in Guyana. Lehman
documented a strong pattern of nestedness as well as a significant correlation
between species composition and intersite distances. Thus, there may be interspecific differences in the ability of primates to cross rivers and then colonize
habitats. However, the observed pattern may also represent species extinctions
due to climatic variation in western Guyana.

In "Ecological Biogeography of Primates in Guyana," Shawn Lehman,
Robert Sussman, Jane Phillips-Conroy, and Waldyke Prince examine the relationship between primate diversity and abundance as a factor of habitat selection and interspecific associations. Their analyses indicate the biogeographic
importance of riparian forests, which are prone to human disturbance, for six of

the eight primate species in Guyana. They suggest that contrary to other South American forest sites, *terra firme* forests may not contain enough fruiting trees to support all eight species during periods of fruit scarcity in Guyana. They also documented that brown and wedge-capped capuchins have a negative pattern of interspecific association, which may indicate high levels of interspecific competition. The complexity of biogeographic patterns for such a small country has important implications for researchers considering biogeographic studies at broader levels in the Neotropics.

REFERENCES

Ayres, J. M. and Clutton-Brock, T. H. 1992, River boundaries and species range size in Amazonian primates. *Amer. Nat.* 140:531–537.

Peres, C. A. and Janson, C. H., 1999, Species coexistence, distribution, and environmental determinants of neotropical primate richness: A community-level zoogeographic analysis. In: Fleagle, J. G. Janson, C. H. and Reed K. eds., *Primate Communities.* Cambridge University Press, Cambridge, pp. 55–74.

Rylands, A. B., Schneider, H., Lamgguth, A., Groves, C. P., and Rodriguez-Luna, E. 2000, An assessment of the diver sity of New World Primates. *Neotrop. Primates* 8:61–93.

Terborgh, J. and Andresen, E. 1998. The composition of Amazonian forests: patterns at local and regional scales. *J. Trop. Ecol.* 14:645–664.

CHAPTER TWO

Nested Distribution Patterns and the Historical Biogeography of the Primates of Guyana

S. M. Lehman

ABSTRACT

I investigated if primate species assemblages exhibit nestedness in Guyana. In a nested pattern, individual species have a strong tendency to be present in all assemblages of equal or greater size than the smallest one in which they occur. I conducted 1,725 km of surveys to determine primate species composition and distribution patterns at sixteen survey sites in Guyana. The resulting dataset showed a strong pattern of nestedness in the distribution of Guyanese primates, and differed significantly from random species assemblages generated using Monte Carlo simulations. Species similarities between sites was significantly but weakly negatively correlated with distance between sites. These assemblage patterns may be due to interspecific variations in the ability of some primate species to cross rivers as well as to species extirpations in western Guyana. The absence of wedge-capped capuchins at four sites, which the model predicted should be occupied by this species, may be due to interspecific competition with brown capuchins.

S. M. Lehman • Department of Anthropology, University of Toronto, Toronto, Ontario, Canada M5S 3G3

Primate Biogeography, edited by Shawn M. Lehman and John G. Fleagle.
Springer, New York, 2006.

Key Words: Biogeography, community structure, Guyana, primates, surveys

INTRODUCTION

Numerous studies have revealed that variations in species assemblages can reflect nested distribution patterns at the landscape level (Cook and Quin, 1995, 1998; Boecklen, 1997; Ganzhorn, 1998; Hansson, 1998; Wright *et al.*, 1998; Yiming *et al.*, 1998; Bruun and Moen, 2003; Heino and Soininen, 2005; Sada *et al.*, 2005). In a nested pattern, individual species have a strong tendency to be present in all assemblages of equal or greater size than the smallest one in which they occur (Atmar and Patterson, 1993). Nestedness results from selective extirpations such that species will disappear from different habitats in roughly the same order (Patterson, 1991). Conversely, Cook and Quin (1995) suggested that nested patterns represent differential colonization abilities of species. For example, Ganzhorn (1998) documented that species-poor lemur communities represent nested subsets of species-rich communities in both eastern humid and western dry forests in Madagascar. However, there was a distance effect of species similarity only for lemur communities in western Madagascar. Ganzhorn (1998) suggested that this pattern of differential colonization reflected selective species extinctions from a common species pool in eastern Madagascar. In western dry forests, lemurs dispersed north and south from the SW part of the island. Subsequent genetic and biogeographic analyses of mouse lemurs supports a north-south pattern of speciation and dispersal in western dry forests (Yoder *et al.*, 2000). Thus, nestedness models can provide important information on both ecological and historical biogeographic processes. However, no studies have investigated nestedness as a model for primate assemblages in South America.

Although nestedness has been detected in numerous species assemblages (e.g., Fernandez-Juricic, 2000; Puyravaud *et al.*, 2003; Greve *et al.*, 2005), there tends to be some species that are either present at sites not predicted by the model or absent at sites where they are predicted to exist (Cook and Quin, 1998; Wright *et al.*, 1998; Puyravaud *et al.*, 2003). Four ecological mechanisms are responsible for these species-specific departures from the model predictions (Atmar and Patterson, 1993). First, postisolation immigration of new species into the site may generate idiosyncratic distributions. Second, these distributions may also be the result of competitive exclusion. For example, generalist primates may be excluded from larger sites dominated by competitively superior

Table 1. Primate species found in Guyana

Species	Common name	Local name(s)
Alouatta seniculus	Red howler monkey	Baboon
Ateles paniscus	Guianan red-faced spider monkey	Kwatta
Cebus albifrons[a]	White-fronted capuchins	Unknown
Cebus apella	Brown capuchin	Blackjack, corn monkey
Cebus olivaceus	Wedge-capped capuchin	Ring tail
Chiropotes satanas	Brown bearded saki	Besa
Pithecia pithecia	White faced saki	Moon monkey, hurawea
Saguinus midas	Golden handed tamarin	Marmoset
Saimiri sciureus	Common squirrel monkey	Monkey-monkey, squirrel

[a] Not used in further analyses due to lack of data on distribution or density.

specialists (Thiollay, 1994; Ganzhorn, 1997). These generalist species may then be relegated to small peripheral sites. Third, the distributions may result from the presence of a fundamental disjunction in the historical evolution of community structure. Last, the presence of unique ecogeographic features, such as rivers, in the region of some sites may influence species closely associated with such features.

The primates of Guyana represent a unique opportunity to test the nestedness model. Of the nine primate species in Guyana (Table 1), only three—red howler monkeys, wedge-capped capuchins, and white faced sakis—are found throughout the country (Muckenhirn *et al.*, 1975; Sussman and Phillips-Conroy, 1995; Norconk *et al.*, 1997; Lehman, 2004b). The other six species are found in only some parts of Guyana. This biogeographic pattern is remarkable given that some primate species, such as brown capuchins (*Cebus apella*) and squirrel monkeys (*Saimiri sciureus*), with limited distributions in Guyana are amongst the widest ranging of all platyrrhines (Thorington, 1985; Eisenberg, 1989; Brown and Zunino, 1990; Wallane *et al.*, 1996).

In this paper I investigate if primate species assemblages reflect nestedness in Guyana. Specifically, I address the following questions: (1) if patterns of nestedness do occur, are they the result of primate extirpations or colonization and (2) how do the observed distribution patterns of Guyanese primates relate to historical biogeographical processes?

METHODS

Guyana is a small country of 215,500 km^2 situated on the northeastern coast of South America, between 56° 20' and 61° 23' west and 1° 10' and 8° 35'

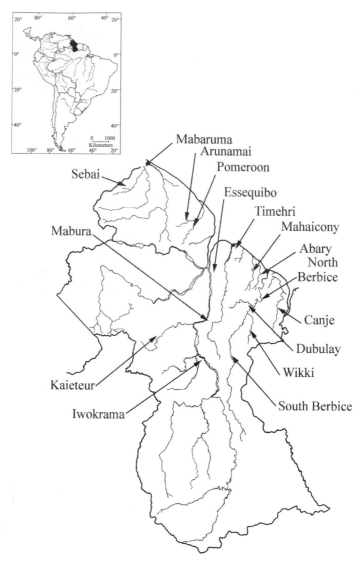

Figure 1. Location of study sites used in the analyses of nested subsets for primates in Guyana.

north (Figure 1). The climate is tropical with a high mean daily temperature of 25.7 °C (ter Steege, 1993). Temperature is highest in September and October and is lowest in December and January. Mean annual precipitation is between 2,000 and 3,400 mm, and is neither evenly distributed throughout the year or throughout the country (ter steege, 1993). There are generally two wet seasons and two dry seasons. Much of the annual rainfall comes during the summer rainy season, from May to mid-August. There is a shorter rainy season from November to January. The long dry season begins in mid-August and runs through to November or December. This season is characterized by monthly rainfall of less than 200 mm. The short dry season is usually from February to April.

I surveyed the distribution and diversity of primates in forests and along rivers at sixteen sites in Guyana (Figure 1). Complete description of each site can be found in Lehman (1999, 2000, 2004b). Survey data were collected during three periods: (1) November 1994 to June 1995, (2) September 1995 to June 1996, and (3) June to August 1997. These periods cover all four seasons in Guyana. Surveys were conducted throughout the day from 0500 to 1900 hours.

While surveying forests, I used randomly selected and predetermined tran sect lines. Although most studies on the distribution of animals use only random selection of transects (e.g., Anderson *et al.*, 1979; Krebs, 1989; Buckland *et al.*, 1993), I also used predetermined transect lines to ensure that biogeographic features, such as rivers that may be barriers to dispersal, were included in the data set (Peres, 1999). Predetermined transect lines often ran along paths in the forest to maximize survey time in remote areas. Transects were measured and marked every 10 m with numbered blocks or flagging tape before surveys were conducted. Two types of surveys were conducted: (1) unique and (2) repeat. Unique surveys were made along transects, such as trails or riverbanks, where one to two transits were made during a census. During repeat surveys, I conducted more than two transits of a transect line. Repeat surveys were conducted along paths at five locations: (1) Timehri; (2) Dubulay Ranch; (3) Kaieteur Falls National Park; (4) Mabura Hill Ecological Reserve; and (5) Sebai River. I walked slowly along transects at a rate of 1.0 km/h, stopping every 10 min to listen for the sounds of movement in the forest. I also surveyed riparian forests by paddling slowly (1.5–2.0 km/h) along riverbanks, either alone or with the assistance of local guides. Randomly selected areas were chosen on river banks for land surveys. However, it is illegal to cut trails in protected areas (e.g., Kaieteur Falls National Park, Mabura Hill Forest Reserve, and Iwokrama Forest Reserve). Thus, established trails were used in these protected areas.

During surveys, data were recorded on: (1) primate species; (2) time of day; (3) weather; (4) vegetation height; (5) general height of group; (6) number of animals in group; (7) cue by which animals detected; (8) activity; (9) perpendicular distance from the transect [meters]; (10) sighting angle; and (11) habitat type. When a primate group was seen, a standardized time of 10 min was spent observing the behavior of individuals in the group. *Ad libitum* notes on behavior, obvious individual physical characteristics, and vocalizations were also collected. The location of primate groups seen during surveys was determined using LANDSAT-5 satellite photographs, 1:50,000 topographic maps of the region, and a Magellan NAV 5000D GPS. Habitat descriptions were made using soil features, a vegetation map (Huber *et al.*, 1995), various monographs on Guyanese flora (van Roosmalen, 1985; de Granville, 1988; Mennega *et al.*, 1988; Lindeman and Mori, 1989; ter Steege, 1990; Comiskey *et al.*, 1993; ter Steege, 1993), and LANDSAT-5 satellite imagery of survey areas.

I created a presence–absence dataset of primate species composition in Guyana. Because there are few data on the biogeography of white-fronted capuchins in Guyana (Barnett *et al.*, 2000), they were not used in further analyses. Following Atmar and Patterson (1993), if two sites contained similar species composition, the one site was removed from the dataset to avoid unnecessary duplication of biogeographic data (i.e., South Berbice, Wikki). Thus, presence–absence data were taken from 14 sites in eastern ($N = 8$) and western ($N = 6$) Guyana. NESTCALC software was used to sort the dataset from high to low for site diversity (top to bottom) and species diversity (left to right). NESTCALC also calculates a statistical test value T of the order (nestedness) or disorder (lack of nestedness) in the dataset (Atmar and Patterson, 1993). The test value T ranges from 0 (complete order) to 100 (complete disorder). In an ordered dataset, every site contains a proper subset of the species at all of the sites above it. As T increases, complete disorder approaches and the biogeography of the sites or species in question become unpredictable. The observed T value was then compared to a distribution of values generated by Monte Carlo simulations (Atmar and Patterson, 1993). Every program was run 1000 times to generate 1000 random primate faunas.

A geometric extinction line, which represents the line of smoothest transition (Figure 2), was calculated for the dataset. This line separates the occupied area of the dataset from the unoccupied area. Species absence above the line is defined as unexpected, as is a species presence below the line (Atmar and Patterson, 1993).

NESTCALC was then used to calculate idiosyncratic T values by site and by species (Atmar and Patterson, 1993). Unexplained species presence or absence lead to specifically higher T values than the complete dataset. Such elevated T values may indicate that the species in question was influenced by a

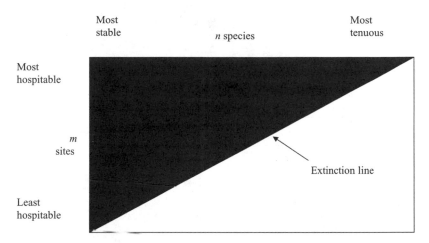

Figure 2. Example of a perfectly nested (ordered) dataset.

biogeographic event different from that affecting the other species. Following Ganzhorn (1998), pairwise similarities of species composition between sites were described using Jaccard's index (J). This index is given by

$$J = \frac{t}{a + b - t}$$

where t is the number of species occurring in both sites, a is the number of species at site A and b is the number of species at site B.

Spearman rank correlations (r_s) were used to determine the relationship between species similarities and intersite distance. Statistical analyses were conducted using SPSS 11.5 and the alpha level was set at 0.05 for all analyses.

RESULTS

Figure 3 shows the dataset and idiosyncratic temperatures for eight primate species at 14 sites in Guyana. The dataset has a T value of 14.04, indicating a pattern of nestedness in the distribution of Guyanese primates. The observed dataset differs significantly from random species assemblages generated using Monte Carlo simulations (mean T of 1000 simulations = 44.61 ± 8.51, p = 0.0001). One primate species (*C. olivaceus*) and four sites (Canje, Dubulay, Timehri, and Abary) exhibited T values that departed from the total metric for the dataset.

Communities of primates at all sites in Guyana exhibited similarities between 0 and 100% (mean and SD of Jaccard's index: 0.48 ± 0.22; $N = 98$). Figure 4

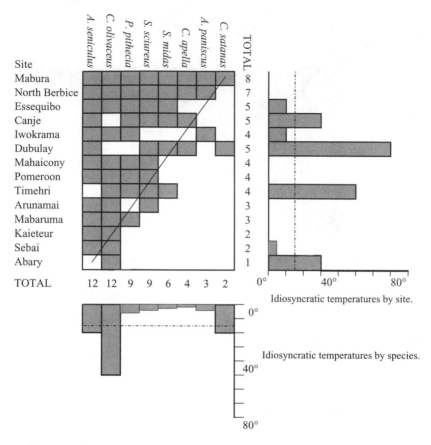

Figure 3. Nested patterns and idiosyncratic temperatures for eight primate species at fourteen sites in Guyana.

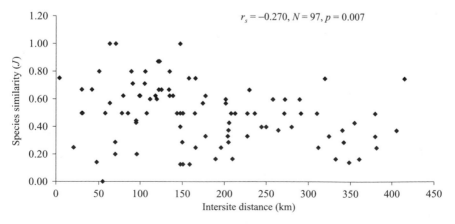

Figure 4. Correlation between Jaccard's index for similarity of species composition and intersite distance for sixteen sites in Guyana.

shows the relationship between community structure and distance between sites in Guyana. Species similarities was negatively correlated with distance between sites in Guyana ($r_s = -0.270$, $N = 97$, $p = 0.007$).

DISCUSSION

The primates of Guyana exhibit a strong pattern of nestedness, which may be the result of habitat characteristics. Specifically, eastern and western Guyana contained all species of the common species pool (i.e., eight species) but sites in western Guyana began losing species. Climatic fluctuations may also relate to species extirpations in western Guyana. Reduced rainfall and lower world temperatures occurred during the last glacial period (Colinvaux, 1987; Colinvaux et al., 1996). Palynological studies by van der Hammen and colleagues (van der Hammen, 1963; Wijmstra and van der Hammen, 1966; van der Hammon and Absy, 1994) found that coastal areas of northern Guyana and Suriname were covered with dry grass savanna during the Pleniglacial period (ca. 21,000–14,000 year B.P.). Models of rainfall and forest area during this period indicate that large tracts of rain forest existed only in extreme NW Guyana and the middle section of eastern Guyana (Figure 5). Despite the presence of a forest refuge in NW Guyana, it is unlikely that most primates could have existed in the area. This refuge may have been flooded swamp forest and swamp woodlands, as it is today. Few primates in NE South America exist in these habitats (Mittermeier and van Roosmalen, 1981; Eisenberg, 1989; Lindeman and Mori, 1989). Thus, the forest refuge in eastern Guyana may be the site in which primates survived climatic fluctuations during the Pleniglacial period.

The statistically significant but weak distance effect on species similarities indicates that there may have been colonization of some sites in Guyana from a species pool (i.e., forest refuge) in the eastern portion of the country. Primates may have dispersed into Guyana from areas outside the country, such as northern Brazil (Lehman, 1999). Despite the possibility for recolonization of sites in western Guyana from the refuge in eastern Guyana and from northern Brazil, rivers may have limited the colonization abilities of many primate species. Rivers have an important role in delimiting the distribution of primates in Guyana (Lehman, 2004b). Primates dispersing out of eastern Guyana would have been faced with a series of large rivers (e.g., Essequibo, Rupununi,

Figure 5. Location of heaviest rainfall and approximate rain forest in Guyana during Pleniglacial period (based on van der Hammen and Absy, 1994).

and Mazaruni) that must be crossed to recolonize western Guyana. The body weight and foraging behavior of a primate are important factor sinfluencing its ability to cross a river. Ayres (1986) found a positive correlation between the size of a river and the maximum body weight of the largest primate whose distribution was limited by a river. Thus, large rivers can limit the distribution of all primates, but small rivers may not limit the distribution of large-bodied primates. Once a river has been crossed successfully, a primate must also be able to cope with a variety of vegetation types within the new region. Although forest habitats in eastern Guyana tend to be similar across major rivers, there is considerable habitat variation between river banks in western Guyana (ter Steege, 1993; Funk, 1995; Huber *et al.*, 1995; Ek, 1997; Barnett *et al.*, 2000). These habitat variations are due to elevational changes that occur from SW Guyana up through the Pakaraima Mountains and then down into the alluvial floodplains of NW Guyana. Generalized foragers should be most likely to survive river crossings into western Guyana because they are not limited in their dietary requirements (Ayres and Clutton-Brock, 1992; Goodman and Ganzhorn, 2003; Lehman, 2004b). Therefore, primate species with most generalized diets are found throughout much of the country (e.g., *A. seniculus* and *S. sciureus*) whereas primates with more specialized dietary and habitat requirements have a smaller geographic distribution limited to eastern Guyana (e.g., *S. midas* and *C. satanas*).

Other biogeographic factors relate to differences in primate assemblages between western and eastern Guyana. Reduced rainfall during the Pleniglacial period may have enlarged savannas in present-day western Guyana (Rupununi and Pakaraima savannas). Eastern Guyana contains fewer and smaller savanna regions. Furthermore, western Guyana is considerably more mountainous than eastern Guyana. High montane habitats (500–800 m elevation) and shrubland/scrub habitats (1000–2400 m elevation) in this region support few primates (Lehman, 1999). These habitats may have expanded downslope as the climate dried during the Late Pleistocene, further reducing forest areas in western Guyana. The stochastic fluctuation of rain forest and monkey populations in western Guyana may have resulted in local extirpations and brought about the present pattern of discontinuous primate distribution in this country. This scenario is supported by the fact that in western Guyana, species diversity decreases northward, with only three species (red howler monkeys, wedge-capped capuchins, and white faced sakis) surviving in the extreme northwest region of the country (Lehman, 1999). Therefore, climatic variation during the Pleistocene may have reduced forest habitats in Guyana and ultimately reduced the number of primate species found in the western half of the country. Climatic change during the Pleniglacial period described herein has been cited as a significant force in the biogeography of primate taxa in South America,

Africa, Asia, and Madagascar (Froehlich *et al.*, 1991; Brandon-Jones, 1996; Ganzhorn, 1998; Jablonski, 1998). For example, Brandon-Jones (1996) analyzed the biogeography of Asian colobines and concluded that Quaternary climatic change played an essential role in delimitating primate populations in this region.

Despite the strong nestedness for primate assemblages in Guyana, there was an unexplained absence for wedge-capped capuchins at four sites (Dubulay Ranch, Canje River, Timehri, and Abary River). It is doubtful that the absence of wedge-capped capuchins from these sites was the result of postisolation immigration or disjunct historical evolution. Forests in some parts of eastern Guyana were not adversely effected by Quaternary climatic changes. In fact, the four survey sites where wedge-capped capuchins were absent are located near to the proposed eastern forest refuge (Figure 5). Thus, it is doubtful that wedge-capped capuchins would be absent from regions that experienced the least forest disturbance since the Quaternary. Unique ecogeographic features, such as rivers, are also unlikely to have caused the observed idiosyncratic temperatures. Wedge-capped capuchins are found in forests bordering rivers, such as the Essequibo River, that are much larger than the Berbice River, where it is absent. Moreover, survey results are unlikely to be related to sampling error (i.e., animals present but not seen). I conducted repeat surveys at two of the sites (Timehri and Dubulay Ranch). Furthermore, other researchers have noted the absence of wedge-capped capuchins at these sites (Sussman and Phillips-Conroy, 1995). Instead, the unexpected absence of wedge-capped capuchins from sites in eastern Guyana may be due to competitive exclusion. My data on the community ecology of the primates of Guyana indicate that a negative pattern of interspecific association exists between wedge-capped capuchins and brown capuchins (Lehman, 2000). Furthermore, sighting rates for both species were reduced in areas of sympatry compared to allopatric areas Lehman *et al.* (this volume). Therefore, the absence of wedge-capped capuchins may be the result of interspecific competition with brown capuchins.

If wedge-capped capuchins and brown capuchins are competing at sympatric sites, then the question arises as to which competitive process, contest or scramble, is involved (Terborgh, 1986; Janson, 1987; van Schaik and van Noordwijk, 1988). Contest competition typically involves agonistic interactions over access to scarce food resources (Koenig *et al.*, 1998; Ganzhorn, 1999; Iwanaga and Ferrari, 2002). This form of competition has been observed rarely between wedge-capped capuchins and brown capuchins in French Guyana, Surinam, and Guyana (Muckenhirn *et al.*, 1975; Mittermeier, 1977; Sussman and Phillips-Conroy, 1995; Simmon and Sabatier, 1996; Lehman, 1999). Between group scramble competition results in resource depression or depletion (Janson and van Schaik, 1988). Scramble competition may have a negative impact on primate densities because of low food availability. This impact may be particularly pronounced in the Guianas because the forests are located on nutrient poor soils (ter Steege, 1993). As a result, the forests are characterized by low plant species diversity as well as low fruit and leaf production compared to other sites in South America (ter Steege, 1993; Chale, 1996; Terborgh and Andresen, 1998; Toriola *et al.*, 1998). A recent study of primate species richness in South America by Kay *et al.* (1997) found that plant productivity was the ecological variable most strongly correlated with primate species richness. Thus, low plant productivity in Guyanese forests may reduce the diversity of feeding niches and result in scramble competition between wedge-capped and brown capuchins. Further studies of the diet and habitat use of these capuchins are needed to test this hypothesis.

SUMMARY

The primates of Guyana exhibit a strong pattern of nestedness. Specifically, individual species are present in all assemblages of equal or greater size than the smallest one in which they occur. This nestedness may be the result of species extirpations in western Guyana during the last Pleniglacial period (ca. 21,000–14,000 year B.P.). Colonization may have occurred for primates dispersing from eastern to western Guyana. However, large rivers and montane habitats would have limited primate dispersal to generalized foragers (e.g., *A. seniculus* and *S. sciureus*) in western Guyana. The unexpected absence of wedge-capped capuchins from sites in eastern Guyana may be due to competitive exclusion by brown capuchins.

ACKNOWLEDGMENTS

For permission to conduct my study I thank Dean Indirjit Ramdas, Dean Catherine Cox, Mr. Phillip daSilva, Mr. John Caesar, Dr. Karen Pilgrim, Office of the President, University of Guyana, Department of Biology at the University of Guyana, Ministry of Amerindian Affairs, Ministry of Health, National Parks Commission, Tropenbos Guyana, Demarara Timbers Ltd., Iwokrama

Rain Forest Reserve, and Wildlife Division of the Department of Health. I greatly appreciate the support of Drs. Vicki Funk and Carol Kelloff of the Biological Diversity of the Guianas Program at the Smithsonian Institution. I thank the many local people for assistance with data collection. I am grateful to Alexander Mendes, the Mendes family, and the staff at Dubulay Ranch for their friendship, hospitality, and support. I thank Robert Sussman, John Fleagle, Jane Phillips-Conroy, Charles Hildeboldt, Charles Janson, Richard Smith, and the reviewers for comments on an earlier draft of this manuscript. My research was supported in part by the Lincoln Park Zoo Scott Neotropic Fund, the Biological Diversity of the Guianas Program of the Smithsonian Institution, USAID, the Ministry of Finance of the Government of Guyana, the Global Environmental Fund of the World Bank, Connaught Foundation, and NSERC.

REFERENCES

Anderson, D. R., Laake, J. L., Crain, B. R., and Burnham, K. P. 1979, Guidelines for line transect sampling of biological populations. *J. Wildl. Manag.* 43:70–78.

Atmar, W. and Patterson, B. D. 1993, The measure of order and disorder in the distribution of species in fragmented habitats. *Oecologia* 96:373–382.

Ayres, J. M. 1986, Uakaries and Amazonian Flooded Forests. Unpublished Ph.D. dissertation, Cambridge University, Cambridge, England.

Ayres, J. M. and Clutton-Brock, T. H. 1992, River boundaries and species range size in Amazonian primates. *Am. Nat.* 140:531–537.

Barnett, A., Shapely, B., Lehman, S. M., Mayor, M., Henry, E., Benjamin, P., McGarrill, M., and Nagala, R. 2000, Primate records from the Potaro Plateau, western Guyana. *Neotrop. Primates* 8:35–40.

Boecklen, W. J. 1997, Nestedness, biogeographic theory, and the design of nature reserves. *Oecologia* 112:123–142.

Brandon-Jones, D. 1996, The Asian Colobinae (Mammalia: Cercopithecidae) as indicators of Quaternary climatic change. *Biol. J. Linnaen Soc.* 59:327–350.

Brown, A. D. and Zunino, G. E. 1990, Dietary variability in *Cebus apella* in extreme habitats: Evidence for adaptability. *Folia Primatol.* 54:187–195.

Bruun, H. H. and Moen, J. 2003, Nested communities of alpine plants on isolated mountains: Relative importance of colonization and extinction. *J. Biogeogr.* 30:297–303.

Buckland, S. T., Burnham, K. P., Anderson, D. R., and Laake, J. L. 1993, *Density Estimation using Distance Sampling*, Chapman Hall, London, England.

Colinvaux, P. 1987, Amazon diversity in light of the paleoecological record. *Quaternary Sci. Rev.* 6:93–114.

Colinvaux, P. A., de Oliveira, P. E., Moreno, J. E., Miller, M. C., and Bush, M. B. 1996, A long pollen record from lowland Amazonia: Forest and cooling in glacial times. *Science* 274:85–88.

Comiskey, J., Dallmeier, F., Aymard, G., and Hanson, A. (1993). *Biodiversity Survey of Kwakwani, Guyana*, The Smithsonian Institution/Man and the Biosphere Biological Diversity Program, Washington, DC.

Cook, R. R. and Quin, J. F. 1995, The influence of colonization in nested species subsets. *Oecologia* 102:413–424.

Cook, R. R. and Quin, J. F. 1998, An evaluation of randomization models for nested species subset analysis. *Oecologia* 113:584–592.

de Granville, J.-J. 1988, Phytogeographical characteristics of the Guianan forests. *Taxon* 37:578–594.

Eisenberg, J. F. 1989, *Mammals of the Neotropics: Panama, Colombia, Venezuela, Guyana, Suriname, French Guiana*, The University of Chicago Press, Chicago.

Ek, R. C. 1997, *Botanical Diversity in the Tropical Rain Forest of Guyana*, Tropenbos-Guyana Programme, Georgetown, Guyana.

Fernandez-Juricic, E. 2000, Bird community composition patterns in urban parks of Madrid: The role of age, size and isolation. *Ecol. Res.* 15:373–383.

Froehlich, J. W., Suprianata, J., and Froehlich, P. H. 1991, Morphometric analyses of *Ateles*: Systematics and biogegraphic implications. *Am. J. Primatol.* 25:1–22.

Funk, V. 1995, A Preliminary Analysis of the Biological Diversity of Guyana. Center for the Study of Biological Diversity and the Smithsonian Institution, Georgetown, Guyana.

Ganzhorn, J. U. 1997, Test of Fox's assembly rule for functional groups in lemur communities in Madagascar. *J. Zool.* 241:533–542.

Ganzhorn, J. U. 1998, Nested patterns of species composition and its implications for lemur biogeography in Madagascar. *Folia Primatol.* 69:332–341.

Goodman, S. M. and Ganzhorn, J. 2003, Biogeography of lemurs in the humid forests of Madagascar: The role of elevational distribution and rivers. *J. Biogeog.* 31:47–56.

Greve, M., Gremmen, N. J. M., Gaston, K. J., and Chown, S. L. 2005, Nestedness of Southern Ocean island biotas: Ecological perspectives on a biogeographical conundrum. *J. Biogeog.* 32:155–168.

Hansson, 1998, Nestedness as a conservation tool: Plants and birds of oak-hazel woodland in Sweden. *Ecol. Lett.* 1:142–145.

Heino, J. and Soininen, J. 2005, Assembly rules and community models for unicellular organisms: Patterns in diatoms of boreal streams. *Freshwater Biol.* 50:567–577.

Huber, O., Funk, V., and Gharbarran, G. 1995, Vegetation map of Guyana. Centre for the Study of Biological Diversity, Georgetown.

Iwanaga, S. and Ferrari, S. F. 2002, Geographic distribution and abundance of woolly (*Lagothrix cana*) and spider (*Ateles chamek*) monkeys in southwestern Brazilian Amazonia. *Am. J. Primat.* 56:57–64.

Jablonski, N. 1998, The response of Catarrhine primates to Pleistocene environmental fluctuations in East Asia. *Primates* 39:29–37.

Janson, C. H. 1987, Food competition in brown capuchin monkeys (*Cebus apella*): Quantitative effects of group size and tree productivity. *Behaviour* 14:53–76.

Koenig, A., Beise, J., Chalise, M. K., and Ganzhorn, J. 1998, When female should contest for food—testing hypotheses about resource density, distribution, size, and quality with Hanuman langurs (*Presbytis entellus*). *Behav. Ecol. Sociobiol.* 42:225–237.

Krebs, C. J. 1989, *Ecological Methodology*, Harper Collins, New York.

Lehman, S. M. 1999, Biogeography of the Primates of Guyana. Unpublished Ph.D. dissertation, Washington University, St. Louis, MO.

Lehman, S. M. 2000, Primate community structure in Guyana: A biogeographic analysis. *Int. J. Primatol.* 21: 333–351.

Lehman, S. M. 2004a, Biogeography of the primates of Guyana: Effects of habitat use and diet on geographic distribution. *Int. J. Primatol.* 25:1225–1242.

Lehman, S. M. 2004b, Distribution and diversity of primates in Guyana: Species-area relationships and riverine barriers. *Int. J. Primatol.* 25:73–95.

Lindeman, J. C. and Mori, S. A. 1989, The Guianas. in: Campbell, D. G. and Hammond, H. D., eds., *Floristic Inventory of Tropical Countries: The Status of Plant Systematics, Collections, and Vegetation, plus Recommendations for the Future*, New York Botanical Gardens, New York, pp. 375–390.

Mennega, E. A., Tammen-de Rooij, W. C. M., and Jansen-Jacobs, M. J. 1988, *A Check-List of Woody Plants of Guyana*, Stichting Tropenbos, Wageningen.

Mittermeier, R. A. and van Roosmalen, M. G. M. 1981, Preliminary observations on habitat utilization and diet in eight Suriname monkeys. *Folia Primatol.* 36:1–39.

Muckenhirn, N. A., Mortenson, B. K., Vessey, S., Fraser, C. E. O., and Singh, B. 1975, Report of a primate survey in Guyana: Pan American Health Organization.

Norconk, M. A., Sussman, R. W., and Phillips-Conroy, J. 1997, Primates of Guayana shield forests: Venezuela and the Guianas, in: Norconk, M. A., Rosenberger, A. L., and Garber, P. A., eds., *Adaptive Radiations of Neotropical Primates*, Plenum Press, New York, pp. 69–86.

Patterson, B. D. 1991, The integral role of biogeographic theory in the conservation of tropical forest diversity, in: Mares, M. A. and Schmidly, D. J., eds., *Latin American Mammalogy: History, Biodiversity, and Conservation*, University of Oklahoma Press, Norman, OK, pp. 124–149.

Peres, C. A. 1999, General guidelines for standardizing line-transect surveys of tropical forest primates. *Neotrop. Primates* 7:11–16.

Puyravaud, J.-P., Dufour, C., and Aravajy, S. 2003, Rain forest expansion mediated by successional processes in vegetation thickets in the Western Ghats of India. *J. Biogeog.* 30:1067–1080.

Sada, D. W., Fleishman, E., and Murphy, D. D. 2005, Associations among spring-dependent aquatic assemblages and environmental and land use gradients in a Mojave Desert mountain range. *Divers. Distrib.* 11:91–99.

Sussman, R. W. and Phillips-Conroy, J. 1995, A survey of the distribution and diversity of the primates of Guyana. *Int. J. Primat.* 16:761–792.

ter Steege, H. 1990, *A Monograph of Wallaba, Mora, and Greenheart*, Stichting, Tropenbos, Wageningen, The Netherlands.

ter Steege, H. 1993, *Patterns in Tropical Rain Forest in Guyana*, Stichting Tropenbos, Wageningen, The Netherlands.

Terborgh, J. 1986, Keystone plant resources in the tropical forests, in: Soule, M. E., ed., *Conservation Biology: The Science of Scarcity and Diversity*, Sinauer, Sunderland, MA, pp. 330–344.

Thiollay, J. M. 1994, Structure, density, and rarity in an Amazonian rainforest bird community. *J. Trop. Ecol.* 10:449–481.

Thorington, R. W., Jr. 1985, The taxonomy and distribution of squirrel monkeys (*Saimiri*), in: Rosenblum, L. A. and Coe, C. L., eds., *Handbook of Squirrel Monkey Research*, Plenum Press, New York, pp. 1–33.

van der Hammen, T. 1963, A palynological study on the Quaternary of British Guiana. *Leidse Geol. Mededel.* 29:125–180.

van der Hammon, T. and Absy, M. L. 1994, Amazonia during the last glacial. *Palaeogeog. Palaeoclimat. Palaeoecol.* 109:247–261.

van Roosmalen, M. G. M. 1985, *Fruits of the Guianan Flora*, Utrecht University, Utrecht.

van Schaik, C. P. and van Noordwijk, M. A. 1988, Scramble and contest in feeding competition among female long-tailed macaques (*Macaca fascicularis*). *Behaviour* 105:77–98.

Wallace, R. B., Painter, R. L. E., Taber, A. B., and Ayres, J. M. 1996, Notes on a distributional river boundry and southern range extension for two species of Amazonian primate. *Neotrop. Primates* 4:10–13.

Wijmstra, T. A. and van der Hammen, T. 1966, Palynological data on the history of tropical savannas in northern South America. *Leidse Geolog. Mededel.* 38:71–83.

Wright, D. H., Patterson, B. D., Mikkelson, G. M., Cutler, A., and Atmar, W. 1998, A comparative analysis of nested subsets patterns of species composition. *Oecologia* 113:1–20.

Yiming, L., Niemela, J., and Dianmo, L. 1998, Nested distribution of amphibians in the Zhoushan archipelago, China: Can selective extinction cause nested subsets of species? *Oecologia* 113:557–564.

Yoder, A. D., Rasoloarison, R. M., Goodman, S. M., Irwin, J. A., Atsalis, S., Ravosa, M. J., and Ganzhorn, J. U. 2000, Remarkable species diversity in Malagasy mouse lemurs (Primates, *Microcebus*), in: *Proceedings of the National Academy of Sciences, USA* 97:11325–11330.

CHAPTER THREE

Genetic Evidence on the Historical Biogeography of Central American Howler Monkeys

Julie A. Ellsworth and Guy A. Hoelzer

ABSTRACT

The study described in this chapter aimed to elucidate the historical biogeography of howler monkeys in Central America. We expected to find evidence supporting an invasion from a common ancestor of the three species proceding northward from South America into Central America, with mantled howlers and black howlers being sister species that diverged after the northward invasion of a common source population. We examined patterns of variation at eight microsatellite loci across three populations of mantled howler monkeys ranging from Southern Mexico to Panama, and one population each of black howlers (Belize) and red howlers (Venezuela). The data reveal a broad pattern of declining genetic variation from south to north in mantled howlers, and a closest relationship between the two most northerly sampling sites, consistent with the hypothesis of an historical invasion from the south. These populations are also genetically distinctive, indicating limited gene flow among them. Another result that matched our a priori expectations was that the red howler population exhibited the

Julie A. Ellsworth and Guy A. Hoelzer • University of Nevada at Reno, USA

Primate Biogeography, edited by Shawn M. Lehman and John G. Fleagle.
Springer, New York, 2006.

81

greatest genetic diversity in our comparisons. We were surprised, however, to find that black howlers were the most genetically distinctive population in our data set, suggesting that they are not the sister species of mantled howlers. We suggest to hypotheses that could explain this result. First, black howlers may have decended from a different invasion of Central America that predated the one leading to mantled howlers. Second, black howlers may have arrived in Central America via a different route than that taken by the ancestors of matled howlers. Specifically, they may have used the islands of the Caribbean archipelago as stepping stones to reach their current location without moving up the Isthmus of Panama.

Key Words: Historical biogeography, microsatellites, howler monkeys, Alouatta

New World monkeys diverged from the Old World primate lineage over 35 million years ago. It is most probable that their ancestors rafted across the South Atlantic from Africa to the New World (Fleagle, 1988; Flynn *et al.*, 1995; Trtkova *et al.*, 1995). The modern howler monkey clade is estimated to have arisen in the last five million years (Schneider *et al.*, 1993). There are eight frequently recognized species of howler monkeys (genus *Alouatta*): the red-handed howler (*A. belzebul*), the black-and-gold howler (*A. caraya*), the brown howler (*A. fusca = guariba*, see Rylands and Brandon-Jones, 1998), the Bolivian red howler (*A. sara*), the red howler (*A. seniculus*), the Coiba Island howler (*A. coibensis*), the black howler (*A. pigra*), and the mantled howler (*A. palliata*; Wolfheim 1983, Groves 1993, Rowe 1996; but see Groves, 2001, and Cortes-Ortiz *et al.*, 2003). Five of these species have relatively large ranges in South America, two are restricted to small areas of Central America (*A. coibensis* and *A. pigra*), and only the mantled howler is found in both Central and South America (Wolfheim, 1983; Groves, 1993; Rowe, 1996; Figure 1).

Mantled howler monkeys range throughout most of Central America, from southern Mexico south through Panama, into northern South America west of the Andes mountains along coastal Ecuador and Colombia, and possibly into northwestern Peru (Groves, 1993; Rylands *et al.*, 1995; Rowe, 1996). The mantled howler monkey species is divided into three subspecies corresponding to the northern (*A. p. mexicana* in Mexico and Guatemala), central (*A. p. palliata* in Nigaragua, Honduras, El Salvador, Costa Rica, and western Panama), and southern (*A. p. aequatorialis* in eastern Panama, Colombia, Ecuador, and Peru) extents of its range (Rylands *et al.*, 1995). The other Central American howlers (i.e., the Coiba Island howler and the black howler) were at one time

Figure 1. Map showing the estimated ranges of the eight howler monkey species (genus *Alouatta*).

considered subspecies of the much wider ranging mantled howler. The Coiba Island howler was elevated from a mantled howler subspecies based on differences in the dermal ridges of the hands and feet (Froehlich and Froehlich, 1987), however, the species designation is controversial (Rowe, 1996; Cortes-Ortiz *et al.*, 2003). Its range includes Coiba Island (518 km^2) and Jicaron Island (13 km^2) off the coast of Panama (*A. coibensis coibensis*), and the Azuero peninsula on the adjacent mainland (*A. coibensis trabeata*, Rylands *et al.*, 1995). Both islands have been subjected to extreme hunting and logging and complete clear-cutting was predicted by 2000 (Froehlich and Froehlich, 1987). Although recent surveys have recorded howlers on both Coiba Island and the Azuero peninsula (E. Bermingham and L. Cortes-Ortiz, personal communication), the species is listed as critically endangered by IUCN (Rowe, 1996).

The black howler monkey is restricted to a relatively small area in the Yucatan peninsula of Mexico, central and northern Guatemala, Belize, and perhaps into Honduras (Wolfheim, 1983; Rylands *et al.*, 1995; Figure 1). This range overlaps somewhat with that of the mantled howler monkey (Smith, 1970; Wolfheim, 1983; Rylands *et al.*, 1995). Black howlers were elevated from a subspecies of mantled howlers based on differences in cranial morphology, dentition, and pelage; they have bigger heads, different tooth cusp patterns, and darker, softer hair than mantled howlers (Smith, 1970; Groves, 1993). Although the taxonomic distinctiveness of the black howler monkey has been disputed (Rowe, 1996), recent studies of mtDNA have supported its distinctiveness (Cortes-Ortiz et al., 2003). Black howlers have undergone a series of recent population crashes since the 1930s due to hurricanes and a yellow fever epidemic (James et al., 1997; Behie and Pavelka, 2005) and are listed as vulnerable by IUCN (IUCN, 2005).

Central American howlers are believed to have originated via range expansion of northern South American populations into the region following the formation of the Isthmus of Panama (Fleagle, 1988). The final closure of the Isthmus of Panama, which unites the American continents, is estimated to have occurred about 3 million years ago (Coates et al., 1992). However, recent research suggests that the rise of the isthmus "was not as much an event as a process" (Knowlton and Weigt, 1998), which has probably resulted in intermittent periods of divided and connected lands possibly over the past 18 million years. Of the howler species endemic to South America, the red howler is believed to be the most closely related to the Central American howler clade because it is the most widespread and inhabits northern South America (Figure 1; Rowe, 1996) including east of the Andes mountains in Colombia, Venezuela, Guyana, Suriname, French Guiana, Ecuador, Peru, and Brazil (Rylands et al., 1995; Rowe, 1996).

In this study, microsatellite markers (Ellsworth and Hoelzer, 1998) were used to investigate the degree of genetic variability and relatedness among mantled howler monkey populations across Central America. These intraspecific patterns can be influenced by both ancient and recent processes (Avise, 2000). For example, current genetic relationships among populations can be explained by analysing how the populations originally formed (e.g., historical vicariance or colonization pattern) and/or by using contemporary processes, such as studying the extent of recent genetic exchange (e.g., gene flow or dispersal patterns) or fluctuations in effective population size (Hartl and Clark, 1997). Based on their suspected colonization history, we expected mantled howlers to exhibit a northward clinal loss of genetic variation, and that the northernmost and the southernmost populations would be the least similar. These expectations were based on three suppositions: (1) mantled howlers colonized Central America unidirectionally from northern South America (i.e., the source gene pool is in the south, if it remains at all); (2) small numbers of migrants were involved in the northward colonization (i.e., a series of founder events would have reduced genetic variability in a south to north cline); and (3) the Central American populations have not attained large enough population sizes or have not had sufficient time, to regenerate high levels of genetic variability. Furthermore, unless gene flow into and throughout Central America has been high and/or the colonization history is different from the above scenario, variation in the mantled howler monkey gene pool should be geographically structured.

The same microsatellite loci (Ellsworth and Hoelzer, 1998) were used to survey mantled howlers, black howlers, and red howlers. The controversy concerning the taxonomic distinctiveness of black howlers revolves, in part, around the assumption of a single howler monkey invasion into Central America. Although black and mantled howlers have distinguishing morphological features, this assumption has led to the hypothesis that black howlers are a very recent offshoot of northern mantled howler populations (Fleagle, 1988; Rowe, 1996). If black howlers were recently isolated from mantled howlers, these species should be genetically similar. Black howlers should be particularly linked to northern mantled howler populations in phylogenetic analyses. Furthermore, if mantled howlers and black howlers were part of a single Central American invasion, and if these lineages evolved at the same rate, then both species would differ genetically from red howler monkeys by the same degree.

METHODS

Samples

Samples were obtained from 8 Mexican mantled howlers (mex), 89 Costa Rican mantled howlers (cr), 20 Panamanian mantled howlers (pan), 28 Belizean black howlers (blk), and 6 Venezuelan red howlers (red) for a total of 151 adult individuals. The Mexican samples came from two locations, Cascajal ($n = 4$) and Villa Isla ($n = 4$). The Costa Rican mantled howlers, the Panama mantled howlers, and the black howler samples each came from single locations (see Figure 2). The red howler samples came from two locations, El Frio ($n = 3$) and Pinero ($n = 3$). Samples were collected from those captured between 1994 and 1998.

We obtained several different tissue samples of wild howler monkeys from numerous sources. Hair and/or blood samples were collected from Vera Cruz,

Figure 2.   Map of the sampling sites (see Methods).

Mexico (*A. palliata mexicana*; Site 1; E. Rodriguez Luna and L. Cortes-Ortiz, personal communication), La Pacifica, Costa Rica (*A. palliata palliata*; Site 2; K. Glander, personal communication), Barro Colorado Island, Panama (*A. palliata palliata*; Site 3; D. DeGusta and K. Milton, personal communication), Bermuda Landing, Belize (*A. pigra*; Site 4; R. James and K. Glander, personal communication), and Hato Masaguaral, Venezuela (*A. seniculus*; Site 5; T. Pope, personal communication; Figure 2).

Genotyping

DNA was obtained from tissue samples of wild howler monkeys using standard extraction methods (Sambrook *et. al.*, 1989 for blood, and Morin *et al.*, 1994 for hair). Each were genotyped at eight microsatellite loci (Ellsworth and Hoelzer, 1998). PCR reactions contained the following: 50 ng DNA template, 4 pmol of each primer, one of which was either radioactively labeled ($\gamma - ^{33}$P dATP) or fluorescently labeled, 4 nmol of each dNTP, 0.625 units of AmpliTaq Gold (PE Applied Biosystems), 60 mM Tris-HCl, 15 mM $(NH_4)SO_4$, and

1.5 mM $MgCl_2$. PCR reactions were heated to 95°C for 10 min, and then subjected to 30 cycles of 1 min at 95°C, 1 min at the optimal annealing temperature (determined by the OLIGO® program), and 1 min at 72°C. PCR products were either run manually on 6% denaturing polyacrylamide gels or on an automated sequencer (ABI 310) until single base differences in length could be resolved.

Analyses of Genetic Variability

To assess genetic variability, we calculated the number of alleles, the allele frequencies, the numbers and frequencies of unique alleles, and the heterozygosities for all populations at each locus. Our null expectation for heterozygosity assumed a random, frequency-weighted association of the observed alleles (Nei, 1987). Deviations from Hardy-Weinberg equilibrium were investigated for each locus using an exact test based on a Markov chain iteration (Guo and Thompson, 1992), as conducted by GENEPOP (web version) with the null hypothesis being equilibrium (Raymond and Rousset, 1995).

Analyses of Genetic Differentiation, Phylogeny, and Gene Flow

Estimations of genetic differentiation among populations were conducted in three ways, using F_{ST} (Wright, 1965), R_{ST} (Slatkin, 1995), and $(\Delta\mu)^2$ (Goldstein et al., 1995). F-statistics, their confidence intervals, and their significance values were calculated using FSTAT (Versions 1.2 and 2.8; Goudet, 1995). F_{ST}, F_{IS}, and F_{IT} were estimated with modifications recommended by Weir and Cockerham (1984), which corrects for incomplete sampling of individuals within populations. F-statistic values estimated with this method can range from -1 to 1. Bootstrap 95% confidence intervals were calculated for each by resampling the data with replacement 15,000 times. While bootstrapping cannot yield confidence intervals in the traditional sense, these ranges inform us about the stability of the estimation when the data are perturbed. F-statistic significance values were calculated via 1000 random permutations of the data. This method estimates the probability of obtaining by chance a value as large or larger than the observed value, given the genetic variation in the data set (Goudet, 1995).

Rho, an unbiased estimator of Slatkin's (1995) R_{ST} that corrects for differences in variance between loci and differences in sample sizes among populations, was calculated using RSTCALC (Version 2.2, Goodman, 1997; Rho values also range from −1 to 1). R_{ST} was developed specifically to analyze microsatellite data and incorporates relative allele sizes into the analysis by assuming a step-wise model of microsatellite length evolution (Slatkin, 1995). The F_{ST} analysis assumes the infinite alleles model (IAM) of evolution. Random permutations ($n = 1000$) were used to determine if Rho values across loci were significantly different from zero and bootstrapping was used to calculate 95% bootstrap confidence intervals (Goodman, 1997).

The RSTCALC software (Goodman, 1997) also was used to estimate ($\Delta\mu^2$), the squared difference of the mean allele sizes between populations (Goldstein et al., 1995). This genetic distance measure assumes a stepwise model of microsatellite length evolution. Mutations of microsatellite loci usually make small changes to the number of repeat copies in the sequence; thus the similarity in microsatellite length between alleles can provide information about the amount of time that has passed since two populations shared a common ancestor (Goldstein et al., 1995). PAUP* version 4.0b2 for the Macintosh was used to construct neighbor-joining trees representing the evolutionary relationships among the populations based on the ($\Delta\mu^2$) genetic distance matrix.

Rates of gene flow were estimated by calculating the effective number of migrants per generation between population pairs (Nm) in two ways, using F_{ST} (Wright, 1951) and private alleles data (Slatkin, 1985). According to Wright's Island Model of migration, the number of migrants per generation is related to F_{ST} via the formula $F_{ST} = 1/4\,Nm + 1$, where N is the effective population size and m is the effective migration rate; the product Nm is the effective number of migrants per generation. However, this model assumes large subpopulations of equal size and that the population is at equilibrium between genetic drift and gene flow. Slatkin's (1985) method uses private alleles at multiple loci to estimate the effective number of migrants. This method is based on the idea that private alleles are likely to attain high frequency within a population only when Nm is low. Nm was estimated from the frequencies of private alleles with modifications recommended by Barton and Slatkin (1986) that correct for differences in sample sizes of populations. The Island Model suggests that values of Nm greater than one indicate the homogenizing influence of gene flow that has overridden the diversifying effects of mutation and genetic drift, whereas Nm less than one suggests the converse (Avise, 2000). Recent simulation-based studies suggest that values of Nm as great as six can permit substantial divergence under a model of isolation-by-distance (Gavrilets et al., 2000).

RESULTS

Genetic Variability

Between 6 and 12 alleles were found per locus across five populations (mex, cr, pan, blk, red), for a mean of 8.9 alleles per locus and a total of 71 alleles across eight loci. There were four instances (out of 63 possible) in which sequentially ordered allele sizes were separated by more than one repeat unit. For individual populations, between 1 and 8 alleles were found per locus, for a mean of 3.8 alleles per population per locus ($n = 40$). Mexican mantled howlers exhibited 1–5 alleles per locus (mean 1.6) and mean heterozygosity of 0.14. The Costa Rican population had 3–6 alleles per locus (mean 4.4) and a mean heterozygosity of 0.33. The Panama population had 3–7 alleles per locus (mean 4.9) and a mean heterozygosity of 0.51 (Tables 1 and 2).

Comparing groups at the species level, mantled howler monkeys exhibited between four and nine alleles per locus (mean 6.3), and a mean heterozygosity of 0.35. Black howler monkeys had between one and seven alleles per locus (mean 3.8) and a mean heterozygosity of 0.45. Red howler monkeys had between one and eight alleles per locus (mean 4.3) and a mean heterozygosity of 0.56 (Tables 1 and 2).

Costa Rican mantled howlers deviated significantly ($p < 0.05$) from Hardy-Weinberg equilibrium expectations at two loci (D6S260 and D14S51), Panama deviated at one locus (D14S51), black howlers at two loci (Ap74 and D5S111), and red howlers at one locus (D6S260).

The mean number of alleles found per locus per individual was calculated for each species to assess the relative allelic diversities among species. Red howler monkeys exhibited significantly higher allelic diversity than mantled or black howlers, with a mean of 0.71 alleles found per locus per individual, compared to 0.165 and 0.134 for mantled and black howlers, respectively (Figure 3).

Relationship among Mantled Howler Populations

F_{ST} and Rho values were relatively consistent with each other and when evaluated across all eight loci indicated significant partitioning of genetic variation among mantled howler monkey populations ($F_{ST} = 0.280$, $p < 0.001$, and Rho = 0.118, $p < 0.001$; Table 3).

Estimates of the effective number of migrants (Nm) between population pairs suggest higher gene flow between the Mexican and Costa Rican mantled howler

Table 1. Allele frequencies for three mantled howler populations, one black howler population, and one red howler population across eight microsatellite loci

Alleles

Ap6

	175	177	179	181	183	185
Mex		1.00				
CR		0.96	0.01		0.03	0.01
Pan		0.25	0.65		0.05	0.05
Blk		0.05		0.86	0.09	
Red	0.17	0.75			0.08	

Ap68

	175	185	187	191	193	195	197	199	201	203
Mex					1.00					
CR		0.19	0.01	0.08	0.72					0.01
Pan			0.29	0.41	0.28			0.03		
Blk		0.40				0.04	0.30		0.25	
Red	0.02						0.25	0.17	0.25	0.33

Ap74

	141	145	147	149	151	155	157	159
Mex				1.00				
CR			0.21	0.76	0.03			
Pan	0.15	0.10	0.55	0.18	0.03			
Blk				0.61	0.39			
Red		0.40	0.17			0.17	0.23	

D5S111

	157	159	161	163	165	167	169	171	177	179	181
Mex				1.00							
CR			0.01	0.95	0.04						
Pan		0.02		0.90		0.03	0.08	0.02			
Blk	0.02		0.02				0.34	0.50			0.46
Red				0.13		0.33	0.17		0.25	0.17	0.08

	169	171	175	177	179	181	183	185	187	189	191	193
D6S260												
Mex				0.69	0.31							
CR				0.38		0.28	0.34					
Pan				0.15		0.24		0.20	0.39			
Blk	0.14	0.20	0.03	0.02	0.15	0.11	0.15	0.18	0.01	0.03		
Red		0.17	0.03		0.17		0.08		0.08	0.17	0.08	0.17
D8S165	118	129	135	137	139	141	143	145				
Mex						1.00						
CR				0.01		0.88		0.03				
Pan			0.08	0.15		0.68	0.11					
Blk	1.00											
Red		0.33			0.03	0.58	0.05					
D14S51	136	138	140	142	144	146	148	152				
Mex	0.13		0.06	0.19		0.56	0.08					
CR	0.06	0.04		0.01	0.04	0.86						
Pan		0.05	0.05	0.35	0.43	0.13	0.06	0.01				
Blk				0.55		0.45						
Red		1.00										
D17S804	155	157	159	161	163	165	169	171				
Mex			1.00									
CR		0.03	0.64	0.20	0.11	0.05						
Pan	0.02	0.35	0.50	0.03	0.08							
Blk	0.39		0.43	0.05	0.13							
Red				0.17	0.17	0.17	0.25	0.25				

Table 1. The total number of alleles, the mean number of alleles per locus, the number of unique alleles, and the mean observed and expected heterozygosity across eight microsatellite loci for three populations of mantled howlers, one population of black howlers, and one population of red howlers

	Mexican mantled ($n = 8$)	Costa Rican mantled ($n = 89$)	Panama mantled ($n = 20$)	All mantled ($n = 117$)	Black ($n = 28$)	Red ($n = 6$)
Total number of alleles	13	35	39	50	30	34
Mean number of alleles per locus	1.6	4.4	4.9	6.3	3.8	4.3
Number of unique alleles	1	3	4	8	8	13
Mean observed heterozygosity	0.14	0.33	0.51	0.35	0.45	0.56
Mean expected heterozygosity	0.14	0.34	0.59	0.37	0.50	0.65

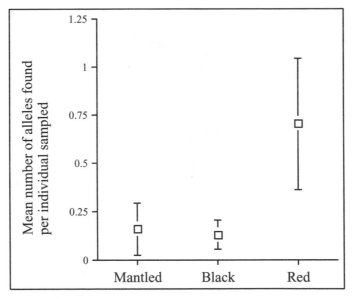

Figure 3. Mean and standard deviation of the number of alleles found per individual sampled averaged across eight microsatellite loci for mantled howlers (three populations sampled at eight loci, $n = 24$), black howlers ($n = 8$ loci), and red howlers ($n = 8$ loci), there is a significant difference among groups, $p < 0.001$, one-way ANOVA; red howlers are significantly different from mantled and black howlers, Tukey multiple comparisons).

Table 2. F_{ST} and Rho values across three mantled howler populations (mex, cr, and pan) for eight microsatellite loci. All values are significant ($p < 0.003$)

	Ap6	Ap68	Ap74	D5S111	D6S260	D8S165	D14S51	D17S804	All loci
F_{ST}	0.696	0.243	0.336	0.021	0.093	0.096	0.450	0.124	0.280
Rho	0.406	0.017	0.437	0.062	0.045	0.162	0.134	0.077	0.118

monkeys than between the Costa Rican and Panamanian populations, or between the Mexican and Panamanian populations (Table 4).

Relationships among Howler Species

F_{ST} and Rho values were relatively consistent with each other and when evaluated across all eight loci indicated very high and significant genetic structure among mantled, black, and red howler monkeys ($F_{ST} = 0.443$, $p < 0.001$, Rho $= 0.709$, $p < 0.001$; Table 5).

Genetic distance estimates $(\Delta\mu)^2$ were low between pairs of mantled howler monkey populations (ranging between 1.28 and 2.42), higher between mantled and red howler population pairs (40.80 and 44.75), and highest between mantled and black howler population pairs (76.10 and 84.77; Table 6).

There was substantial overlap in the specific alleles found in different populations and species. The frequency with which populations had the same most common allele across eight loci ranged between 0.13 and 1.0, depending on the locus; however, the most common allele was different at every locus when comparing red and black howler populations (Table 6).

The phylogenetic analyses showed that the three mantled howler monkey populations (mex, cr, and pan) are tightly clustered. Surprisingly, red howlers appeared to be more closely related to this cluster than black howlers based on relative branch lengths (Figures 4 and 5).

Table 3. Estimates of the effective number of migrants per generation among mantled howler monkey populations based on F_{ST} (above the diagonal) and private alleles (below the diagonal)

	Mexico	Costa Rica	Panama
Mexico		2.14	0.48
Costa Rica	0.89		0.55
Panama	0.35	0.77	

Table 4. F_{ST} and Rho values across three species of howler monkey (mantled, black, and red) for eight microsatellite loci. All values are significant ($p < 0.001$)

	Ap6	Ap68	Ap74	D5S111	D6S260	D8S165	D14S51	D17S804	All loci
F_{ST}	0.668	0.382	0.236	0.675	0.205	0.764	0.362	0.117	0.443
Rho	0.667	0.393	0.286	0.608	0.044	0.944	0.538	0.408	0.709

DISCUSSION

Geographical Structure in the Gene Pool of Mantled Howler Monkeys

Mantled howler monkeys were expected to exhibit a northward decline of genetic diversity if their current genetic structure is primarily influenced by their suspected colonization history. Based on this study, the Costa Rican population of mantled howler monkeys is clearly less genetically diverse than the Panama population; in fact, it appears to be one of the least genetically variable, well sampled, sexually-reproducing populations known. The Costa Rican population exhibited fewer total alleles, lower mean number of alleles per locus, fewer unique alleles, and lower mean heterozygosity than the Panama population, even though more than four times as many individuals were sampled in Costa Rica than in Panama (Table 2). These data are consistent with the predicted clinal loss of genetic diversity in mantled howler monkeys.

Despite the relatively small sample of Mexican howlers analyzed in this study, the data strongly support the trend of decreasing genetic variation with northern

Table 5. Genetic distance estimates [$(\Delta^2\mu)$; below the diagonal], the number of shared alleles (above the diagonal), and the frequency of having the same most common allele across eight loci (above the diagonal in parentheses) between pairs of populations for eight microsatellite loci

	Mexico ($n = 8$)	Costa Rica ($n = 89$)	Panama ($n = 20$)	Red ($n = 6$)	Black ($n = 28$)
Mexico		11 (1.0)	9 (0.38)	2 (0.25)	7 (0.25)
Costa Rica	1.28		25 (0.38)	11 (0.25)	16 (0.25)
Panama	2.42	1.83		18 (0.13)	16 (0.13)
Red	40.80	43.06	44.75		9 (0)
Black	84.77	83.45	76.10	75.31	

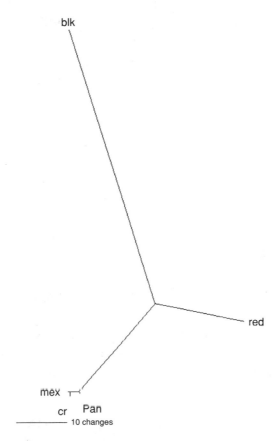

Figure 4. Unrooted neighbor-joining tree representing the phylogenetic relationships among three populations of mantled howler monkeys, one red howler population, and one black howler population.

latitude. The Mexican howlers exhibited the fewest number of alleles, the lowest mean number of alleles per locus, the fewest unique alleles, and the lowest mean heterozygosity among the mantled howler populations (Table 2). Furthermore, the Mexican samples, which were taken from two locations, were monomorphic at six out of eight microsatellite loci (allele frequency = 1.0, thus heterozygosity = 0.0; Table 1), which means that the eight sampled individuals had the same homozygous genotype at six microsatellite loci. Given our allele frequency estimates for the Costa Rican population, the probability of a Costa Rican individual being homozygous at the six loci with the highest individual allele frequencies is 0.14. The probability of eight sequentially sampled individuals sharing this six locus homozygous genotype in the Costa Rican population is very small indeed (7.5×10^{-6}). The Mexican population

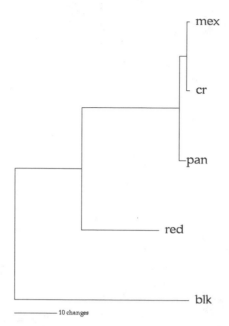

Figure 5. Midpoint rooted neighbor-joining tree with corresponding branch lengths estimating the phylogenetic relationship among three populations of mantled howler monkeys, one population of red howlers, and one population of black howlers.

of mantled howlers appears to be even less variable than the Costa Rican population. Thus, the three sampled mantled howler monkey populations fulfill the expectation of a northward gradient loss of variability.

The genetic data were also consistent with the expectation that the northernmost (i.e., Mexico) and the southernmost (i.e., Panama) populations of mantled howler monkeys are the least genetically similar. This prediction was evaluated in two ways, based on the number of shared alleles and the degree of genetic distance between pairs of the sampled populations. The Mexican and Panamanian populations shared the fewest alleles (nine, compared to 11 between Mexico and Costa Rica, and 25 between Costa Rica and Panama) and had the highest genetic distance between them (2.42, compared to 1.28 and 1.83 for the CostaRica/Mexico and Costa Rica/Panama comparisons, respectively; Table 5). Furthermore, the Mexican and Costa Rican populations are most closely related according to the phylogenetic reconstruction, although all three mantled howler monkey populations are tightly clustered (Figures 4 and 5).

We also expected that a colonization history dominated by a series of founder events would result in genetic structure among populations of mantled howler monkeys. The estimated F_{ST} and Rho values ($F_{ST} = 0.280$, Rho $= 0.118$), which indicate high and significant genetic structure among mantled howler monkey populations, were consistent with this prediction (Table 3). These measures were approximately two to ten times higher than estimates of genetic structure among mantled howler social groups at the local scale ($F_{ST} = 0.025$, Rho $= 0.046$; Ellsworth, 2000). Thus, the genetic variability in mantled howler monkeys is partitioned among populations across the species range.

Gene flow among mantled howler monkey populations appears to be generally low. Five out of six estimates of the effective number of migrants per generation (Nm) between populations were less than one. Estimates from two methods (i.e., F_{ST} and private alleles) showed Nm to be lowest between Mexico and Panama and highest between Mexico and Costa Rica (Table 4). These migration estimates are positively correlated with distances between the populations, but are perhaps influenced by contemporary patterns of habitat fragmentation, which have caused population size reductions and genetic drift. Although gene flow may have been higher across Central America in the past, it was apparently not sufficient to erode the patterns produced by historical colonization. It is also important to consider that the historical influence on genetic estimates of effective migration rates compromises the validity of such estimates regarding current patterns of migration when the landscape has been recently altered.

Northern populations of the mantled howler monkey exhibited exceedingly little genetic variability (Tables 1 and 2). Despite the expectation of a northward gradient in loss of genetic variability in this species, the degree of homogeneity was surprising. Recent and severe reductions in effective population sizes have probably exacerbated historically low levels of genetic variation. For example, the yellow fever epizootic that reduced black howler numbers (James et al., 1997) also had an extreme impact on mantled howler populations in both Costa Rica and Panama (Baldwin, 1976; Milton, 1982). However, the deficiency of genetic variability, and high genetic similarity between populations as geographically distant as Mexico and Costa Rica are so great as to suggest the very recent spread of mantled howlers throughout this region. If mantled howler monkeys existed throughout their current range for millions of years, even a dramatic population crash, or a series of them, is unlikely to have resulted in the similarly reduced gene pools apparent in these sites.

Evaluating the relative importance of historical vs. recent processes in determining current genetic patterns can be difficult (Avise, 2000). In this case, while recent reductions in effective population size may help to explain the lack of genetic variability in this species, it does not explain the regional patterns predicted to emerge as a result of colonization history. Thus, the observed intraspecific patterns are best explained as the consequence of mantled howler monkeys colonizing Central America from northern South America via a series of founder events.

Relationships among Species and the Distinctiveness of Black Howlers

In contrast with northern populations of mantled howlers, estimates of genetic variability indicate that red howler monkeys, the alleged most recent common ancestor to the Central American howlers, are extremely diverse (Pope, 1996). Despite a relatively small sample size, red howlers exhibited the second highest total number of alleles and mean number of alleles per locus and the highest mean heterozygosity (Table 2). When sample size is taken into account, red howlers have a higher allelic diversity than black or mantled howlers (Figure 3). This high level of genetic variability is consistent with red howlers having a large species range with a correspondingly large effective population.

Estimates of F_{ST} and Rho ($F_{ST} = 0.443$, Rho $= 0.709$) indicate that most of the sampled genetic variability exists among the three sampled howler species (mantled, black, and red howlers; Table 5), and suggest that these species are quite different from one another. Phylogenetic analysis, the number of shared alleles, and the frequency of sharing the same most common allele across loci, all indicate that black howler monkeys are very distinct from the closely clustered mantled howler monkey populations (Table 6; Figures 4 and 5). In fact, the data show that black howlers share the fewest alleles, have the lowest frequency of sharing the same most common allele, and the greatest genetic distance with the Mexican, or northernmost, population of mantled howlers (Table 6). These genetic data support species designation for black howler monkeys.

Mantled and black howlers were predicted to be equally related to red howlers if they both descended from a single lineage that invaded Central America. However, the data show that mantled and black howlers are much more different than expected based on that scenario. The black howler population is approximately twice as genetically distant from red howlers as the mantled howler populations (Table 6). In our results, mantled and black howlers do not even form a clade relative to red howlers (Figure 5). The mantled howler populations appear more closely related to the sampled red howlers than they are to black howlers (Figure 5). In this case, the rooted topology appears quite robust, because there are no nodes in the proximity of the midpoint. However, our results contrast with those of Cortes-Ortiz *et al.* (2003) based on mtDNA who found that the Central American howlers formed a distinct clade. These contrasting results indicate that our understanding of the biogeographic history of black howler monkeys remains unresolved.

Central American Colonization History

Our genetic data from microsatellites support the colonization of Central America by mantled howler monkeys from northern South America via a series of founder events. However, the degree of genetic distinctiveness between mantled howlers and black howlers, and the phylogenetic results, argue strongly against the hypothesis that black howlers were isolated recently from mantled howlers. Although their current geographic distributions suggest a common colonization history (see Cortes-Ortiz, 2003), our results suggest that mantled and black howler monkeys probably represent independent invasions into Central America.

If both species arrived in Central America via the Isthmus of Panama, black howler monkeys probably arrived first. Black howlers, or their ancestors, could have occupied a larger range throughout the region with subsequent range reduction following the invasion of mantled howlers. If monkeys were unable to cross into Central America until the final closure of the Isthmus of Panama, mantled and black howlers must have originated from different source populations. However, if monkeys were able to colonize the region before the final formation of the Isthmus, black howlers may have become isolated after crossing into Central America during an earlier period of land connection or proximity (Knowlton and Weigt, 1998). A study of the divergence times between Panamanian and Costa Rican fresh water fishes suggested that reduced sea levels in the Miocene, five to seven million years ago, may have created the land connection required by both freshwater fishes and howler monkeys to invade from South America (Bermingham and Martin, 1998).

Alternatively, black howlers or their ancestors could have reached Central America via island hopping from northeastern South America through the Caribbean island archipelago. Fossil evidence shows that monkeys existed on at least some of the islands (i.e., Cuba and Jamaica) in the past. However, how they got there and how the fossil specimens are related to modern Platyrrhines is debated (Fleagle, 1999). The Caribbean fossils are quite different from extant Platyrrhines, although they appear most similar to modern *Ateles* (spider monkeys), *Cebus* (capuchin monkeys), and *Alouatta* (howler monkeys, MacPhee and Iturralde-Vincent, 1995). A much broader genetic analysis of extant red howler populations might provide evidence that black howlers originated from eastern red howler ancestors via the Lesser Antilles; but, even if this was the case, subsequent gene flow among red howler populations could have obfuscated the historical paraphyly of the species (Avise, 2000). Regardless of how black howlers came to occupy their current range in Central America, the phylogeographic history of this region is more complex than previously thought.

Conservation Implications

We found very little genetic variability in populations of a relatively abundant and wide-ranging species, the mantled howler monkey (listed as lower risk by IUCN; Rowe, 1996), north of the Isthmus of Panama. The genetic variability that does exist in mantled howler monkeys is partitioned among populations (Table 3), and all populations appear to have unique alleles (Table 2). Thus, in order to preserve mantled howler genetic diversity, populations from across the species range will require protection. The lack of genetic diversity in these mantled howler populations is more severe than in the black howler, a more range-restricted species with IUCN endangered status.

There are 13 other species of primate sympatric with the mantled howler monkey in portions of its geographic range, with four species endemic to Central America and 2 species endemic to the northern Andes region along coastal Colombia and Ecuador (Rylands *et al.*, 1995). If the genetic patterns found in mantled howlers are representative of patterns in other Central American primates, northern species and/or populations north of the Isthmus of Panama should receive special attention of conservationists. In addition to the lack of genetic variation, the complex colonization history of this region suggested by the distinctiveness of the black howler monkeys points to the possibility of other cases in which northern populations thought to be recent offshoots of more widely distributed Central American species may be genetically unique and deserving of species designation and protection.

REFERENCES

Avise, J. C. 2000, *Phylogeography: The History and Formation of Species*, Harvard University Press, Cambridge, MA.

Baldwin, L. A. 1976, Vocalizations of howler monkeys (*Alouatta palliata*) in southwestern Panama. *Folia Primatol.* 26:81–108.

Barton, N. H. and Slatkin, M. 1986, A quasi-equilibrium theory of the distribution of rare alleles in a subdivided population. *Heredity* 56:409–415.

Behie A. M. and Pavelka, M. S. M. 2005, The short-term effects of a hurricane on the diet and activity of black howlers (*Alouatta pigra*) in Monkey River, Belize. *Folia Primatol.* 76:1–9.

Bermingham, E. and Martinl, A. 1998, Comparative mtDNA phylogeography of neotropical freshwater fishes: Testing shared history to infer the evolutionary landscape of lower Central America. *Mol. Ecol.* 7:499–517.

Coates, A. G., Jackson, J. B. C., Collins, L. S., Cronin, T. M., Dowsett, H. J., Bybell, L. M., Jung, P., and Obando, J. A. 1992, Closure of the Isthmus of Panama: the near-shore marine record of Costa Rica and western Panama. *Geol. Soc. Amer. Bull.* 104:814–828.

Cortes-Ortiz, L., Bermingham, E., Rico, C., Rodriguez-Luna, E., Sampaio, I., and Ruiz- Garcia, M. 2003, Molecular systematics and biogeography of the Neotropical monkey genus, *Alouatta. Mol. Phylogenet. Evol.* 26:64–81.

Ellsworth, J. A. and Hoelzer, G. A. 1998, Characterization of microsatellite loci in a New World primate, the mantled howler monkey (*Alouatta palliata*). *Mol. Ecol.* 7:657–666.

Ellsworth, J. A. 2000, Molecular evolution, social structure, and phylogeography of the mantled howler monkey (*Alouatta palliata*). Dissertation, University of Nevada at Reno, p. 179.

Fleagle, J. G. 1988, *Primate Adaptation and Evolution*, Academic Press, New York.

Fleagle, J. G. 1999, *Primate Adaptation and Evolution, 2nd Edn.*, State University of New York, Stony Brook, New York.

Flynn, J. J., Wyss, A. R., Charrier, R., and Swisher, C. C. 1995, An early Miocene anthropoid skull from the Chilean Andes. *Nature* 373:603–607.

Froehlich, J. W. and Froehlich, P. H. 1987, The status of Panama's endemic howling monkey. *Primate Conserv.* 8:58–62.

Gavrilets, S., Li, H. and Vose, M. D. 2000, Patterns of parapatric speciation. *Evolution* 54:1126–1134.

Goldstein, D. B., Linares, A. R., Cavalli–Sforza, L. L., and Feldman, M. W. 1995, An evaluation of genetic distances for use with microsatellite loci. *Genetics* 139:463–471.

Goodman, S. J. 1997, R_{ST} Calc: A collection of computer programs for calculating estimates of genetic differentiation from microsatellite data and determining their significance. *Mol. Ecol.* 6:881–885.

Goudet, J. 1995, FSTAT, Version 1.2: A computer program to calculate F-statistics. *J. Hered.* 86:485–486.

Groves, C. P. 1993, Order primates, in: Wilson, D. E. and Reader, D. M., eds., *Mammalian Species of the World: A Taxonomic and Geographic Reference, 2nd edn.* Smithsonian Institution Press, Washington, DC, pp. 243–277.

Groves, C. P. 2001, *Primate Taxonomy.* Smithsonian Institution Press, Washington, DC.

Guo, S. W. and Thompson, E. A. 1992, Performing the exact test of Hardy-Weinberg proportions for multiple alleles. *Biometrics* 48:361–372.

Hartl, D. L. and Clark, A. G. 1997, *Principles of Population Genetics, 3rd Edn.* Sinauer Associates, Inc., Sunderland, MA.

IUCN 2005, *IUCN Red List of Threatened Species,* IUCN Species Survival Commission, Gland, Switzerland.

James, R. A., Leberg, P. L., Quattro, J. M., and Vrijenhoek, R. C. 1997, Genetic diversity in black howler monkeys (*Alouatta pigra*) from Belize. *Amer. J. Phys. Anthropol.* 102:329–336.

Knowlton, N. and Weigt, L. A. 1998, New dates and new rates for divergence across the Isthmus of Panama. *Proc. Roy. Soc. Lond., Ser. B* 265:2257–2263.

MacPhee, R. D. E. and Iturralde-Vincent, M. A. 1995, Earliest monkey from Greater Antilles. *J. Hum. Evol.* 28:197–200.

Milton, K. 1982, Dietary quality and demographic regulation in a howler monkey population, in: Rand, A. S. and Windsor, D. M., eds., *The Ecology of the Tropical Forest, Seasonal Rhythms and Long-Term Change,* Smithsonian Institution Press, Washington, DC, pp. 273–289.

Morin, P. A., Wallis, J., Moore, J. J., and Woodruff, D. S. 1994, Paternity exclusion in a community of wild chimpanzees using hypervariable simple sequence repeats. *Mol. Ecol.* 3:469–478.

Nei, M. 1987. *Molecular Evolutionary Genetics,* Columbia University Press, New York.

Pope, T. R. 1996, Socioecology, population fragmentation, and patterns of genetic loss in endangered primates, in: Avise, J. C. and Hamrick, J. L., eds., *Conservation Genetics: Case Histories from Nature,* Chapman and Hall, New York, pp. 119–159.

Raymond, M. and Rousset, F. 1995, GENEPOP (Version 1.2): Population genetics software for exact tests and ecumenicism. *J. Hered.* 86:248–249.

Rowe, N. 1996, *The Pictorial Guide to the Living Primates,* Pogonias Press, East Hampton, New York.

Rylands, A. B. and Brandon-Jones, D. 1998, Scientific nomenclature of the red howlers from the northeastern Amazon in Brazil, Venezuela, and the Guianas. *Int. J. Primatol.* 19:879–905.

Rylands, A. B., Mittermeier, R. A., and Rodriguez Luna, E. 1995, A species list for the New World primates (Platyrrhini): Distribution by country, endemism, and conservation status according to the Mace-Lande System. Neotrop. *Primates* 3 (Suppl.):113–160.

Sambrook, J. E., Fritsch, R., and Maniatis, T. 1989, *Molecular Cloning: A Laboratory Manual, 2nd Edn.* Cold Spring Harbor Press, New York.

Schneider, H., Schneider, M. P. C., Sampaio, I., Harada, M. L., Stanhope, M., Czelusniak, J., and Goodman, M. 1993, Molecular phylogeny of the New World monkeys (Platyrrhini, Primates). *Mol. Phyl. Evol.* 2:225–242.

Slatkin, M. 1985, Rare alleles as indicators of gene flow. *Evolution* 39:53–65.

Slatkin, M. 1995, A measure of population subdivision based on microsatellite allele frequencies. *Genetics* 139:457–462.

Smith, J. D. 1970, The systematic status of the black howler monkey, *Alouatta pigra* (Lawrence). *J. Mammal.* 51:358–369.

Trtkova, K., Mayer, W. E., O'hUigin, C., and Klein, J. 1995, Mhc-DRB Genes and the origin of New World Monkeys. *Mol. Phyl. Evol.* 4:408–419.

Weir, B. S. and Cockerham, C. C. 1984, Estimating F-statistics for the analysis of population structure. *Evolution* 38:1358–1370.

Wolfheim, J. H. 1983, *Primates of the World: Distribution, Abundance, and Conservation*, University of Washington Press, Seattle, Washington, DC.

Wright, S. 1951, The genetical structure of populations. *Ann. Eugen.* 15:323–354.

Wright, S. 1965, The interpretation of population structure by F statistics with special regard to systems of mating. *Evolution* 19:395–420.

CHAPTER FOUR

Ecological Biogeography of Primates in Guyana

S.M. Lehman, R.W. Sussman,
J. Phillips-Conroy, and W. Prince

ABSTRACT

One of the goals of ecological biogeography is to determine correlates to species diversity and abundance in biological communities. Although large-scale disturbances, such as deforestation, have been linked to declining mammal population sizes in tropical forests, the effects of less severe forms of natural disturbances (flooding, black water swamps) and anthropogenic disturbances (logging, hunting, and agriculture) are poorly understood. Moreover, interspecific associations may influence the presence or absence of primate species. We used data from 2108 km of primate surveys we conducted from 1994–1997 to determine the ecological correlates of biogeography in eight primate species in Guyana. Our data indicate the importance of riparian forests for understanding the biogeography of six of the eight primate species in Guyana. Edge-related variations in leaf quality may explain higher sighting rates for howler monkeys in agricultural areas. Insect abundance may relate to the higher sighting rates for golden-handed tamarins in swamp forests. Sighting rates for brown capuchins were positively correlated with

S.M. Lehman • Department of Anthropology, University of Toronto, 100 St. George Street, Toronto, Ontario M5S 3G3, Canada.   **R.W. Sussman** • Department of Anthropology, Washington University, St. Louis, MO 63130.   **J. Phillips-Conroy** • Department of Anthropology, Washington University St. Louis, MO 63130; Department of Anatomy and Neurobiology, Washington University Medical School, St. Louis, MO 63110.   **W. Prince** • Iwokrama International Centre for Rain Forest Conservation and Development, Georgetown, Guyana, South America.

Primate Biogeography, edited by Shawn M. Lehman and John G. Fleagle.
Springer, New York, 2006.

flooding intensities, which may be the result of the abundance of palm species used as a keystone resource in seasonally inundated habitats. Although wedge-capped and brown capuchins are sympatric at some sites, analyses of species composition across all survey sites indicate that these monkeys have a negative pattern of interspecific association. Sighting rates of both species were lower at sites where they were found to be sympatric, but reduced sighting rates were particularly noticeable in wedge-capped capuchins. The combined effects of natural disturbances, anthropogenic disturbances, and interspecific associations strongly influence primate biogeography in Guyana.

Key Words: Ecological biogeography, Guyana, community structure, surveys

INTRODUCTION

The ecological biogeography of many primate species is influenced by natural (e.g., flooding, tree falls) and anthropogenic disturbances (e.g., logging, agricultural development, and hunting). Natural disturbances occur at many sites in South America, in that, highly seasonal rainfall increases water level by up to 15 m in many river systems (Ayres, 1986; Ayres and Clutton-Brock, 1992). The resulting flooding of forests affects ranging patterns and diet in many primate species (e.g., Peres, 1997; Peres and Janson, 1999; Lehman, 2000; Bennett *et al.*, 2001). Furthermore, there may be species-specific responses to natural habitat disturbance. For example, it has been suggested that spider monkeys (*Ateles* sp.) and bearded sakis (*Chiropotes* sp.) are sensitive to flooding (Johns and Skorupa, 1987; Peres, 1990; Robinson and Bennett, 2000). Although large-scale disturbances, such as deforestation, have been linked to declining mammal population sizes in tropical forests, the effects of low intensity anthropogenic disturbances are poorly understood (Vazquez and Simberloff, 2002). Understanding the manner in which primates respond to changes in their habitats is of obvious importance for biogeographers and conservation biologists.

There tends to be spatial variations in the abundance of a primate species. For example, species-specific sighting rates can vary between sites (Peres and Janson, 1999; Peres and Dolman, 2000). If habitat characteristics do not vary between sites, then these differences can provide indirect information on community dynamics (Thiollay, 1994; Ganzhorn, 1997; Davies *et al.*, 2001; Cushman and McGarigal, 2004). Some Neotropical primates also form interspecific associations. For example, squirrel monkeys (*Saimiri* sp.) often form polyspecific associations with capuchins (*Cebus* sp.) and/or howler monkeys (*Alouatta* sp.) at many sites in South America (e.g., Terborgh, 1983; Podolsky, 1990; Pontes, 1997; Lehman, 1999). The presumed benefits of positive associations are greater foraging efficiency and predator avoidance (Terborgh, 1983; Norconk, 1990a, 1990b; Terborgh, 1990). In negative associations, it is assumed that ecologically similar species cannot coexist because of competition for food resources (e.g., Lotka, 1925; Gause, 1934; Connell, 1961). Interspecific competition may be particularly intense between species that mutually exploit similar keystone resources (Tutin *et al.*, 1997). Although sympatric species employ different dietary strategies and modes of habitat use to avoid interspecific competition (Fleagle and Mittermeier, 1980; Mittermeier and van Roosmalen, 1981), evidence still exists for negative association patterns (Rodman, 1973; Terborgh, 1983; Peres, 1993). For example, it has been suggested that *C. apella* may displace congeners at some sites in Guyana (Sussman and Phillips-Conroy, 1995; Lehman, 1999; Lehman, 2000).

The aim of our paper was to determine ecological correlates to primate biogeography in Guyana. Specifically, we sought to address the following questions: (1) are there species-specific patterns of habitat selection, (2) how do anthropogenic and natural disturbances affect primate biogeography, and (3) what are the patterns of community structure and polyspecific associations at the beta level?

METHODS

Location and Climate

Guyana is a small country of 215,500 km^2 situated on the northeastern coast of South America, between 56°20′ and 61°23′ west and 1°10′ and 8°35′ north (Figure 1). The climate is tropical with a high mean daily temperature of 25.7°C (ter Steege, 1993b). Mean annual precipitation is between 2,000 and 3,400 mm (ter Steege, 1993b). There are generally two wet seasons (May to mid-August and November to January) and two dry seasons (mid-August to November or December and February to April).

Forest Habitats

We used four habitat types in our analyses of primate habitat use: basimontane, *terra firme*, riparian, and swamp forests. These four habitat types were classified

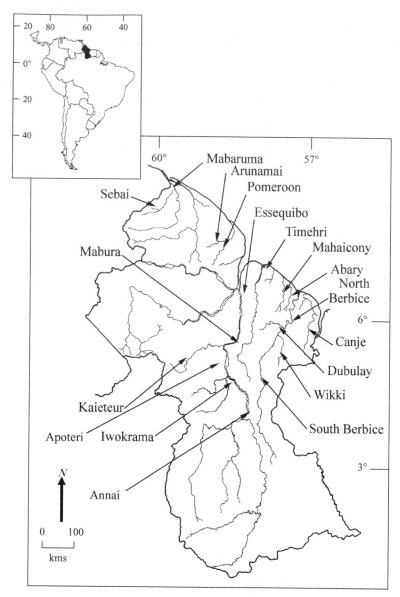

Figure 1. Location of Guyana and 18 study sites surveyed for primate diversity and habitat characteristics.

based on specific vegetation types outlined in Huber *et al.*, (1995). Basimontane habitats are located predominantly in the highland regions of western Guyana, near the border with Venezuela and Brazil (Maguire, 1972; Huber *et al.*, 1995). Other basimontane forests, similar in general floristic composition, are found in southern Guyana. The flora of these regions is also characterized as having extremely high species diversity and abundance (Maguire *et al.*, 1953; Maguire,

1972; Hoffman, 1992). *Terra firme* forest covers approximately 80% of Guyana (Henkel, 1994). This forest is characterized as tall (20–40 m), evergreen lowland forest dominated by kakaralli (*Eschweilera* sp.), kabukalli (*Goupia glabra*), kautabali (*Licania* sp.), baromalli (*Castostemma* sp.), and greenheart (*Chlorocardium rodiei*). Riparian forests are found along the banks of the main rivers in Guyana (e.g., Essequibo, Berbice, Mazaruni, Cuyuni, and Demarara). These habitats contain *Mora excelsa*, *Pterocarpus* sp., and *Carapa* sp. Swamp forests are located primarily in the NW and NE sections of Guyana. These forests are dominated by corkwood (*Pterocarpus* sp.) and white cedar (*Tabebuia insignis*).

Survey Methods

We used data from 2,108 km of primate surveys that we conducted from 1994–1997 at 18 sites in Guyana (ter Steege, 1993b; Sussman and Phillips-Conroy, 1995; Lehman, 2000). Table 1 shows the specific survey sites, associated census distances (km), and estimated disturbance and hunting levels. Disturbance levels

Table 1. Natural and anthropogenic disturbances and survey effort for eighteen survey sites in Guyana

No.	Site	Key sources of disturbance[a]	Hunting/Trapping pressures[b]	Total census distance (km)
1	Essequibo	L (l)	–	80
2	Timehri	F (h), BW (h), A (h), L (m)	M	203
3	Mahaicony	F (h), BW (l), A (l)	L	108
4	Abary	F (h), BW (h), A (h), L (h)	M	37
5	North Berbice	F (m), L (m), A (m)	M	120
6	Canje	F (m), BW (m), L (l), A (l)	H	127
7	Dubulay	F (m), BW (l)	–	270
8	Wikki	F (m), L (h)	H	51
9	South Berbice	F (l)	L	136
10	Mabura	–	–	200
11	Apoteri	F (m), L (l)	L	171
12	Iwokrama		L	58
13	Annai	A (l)	L	13
14	Kaieteur	–	L	129
15	Sebai	F (h), BW (h), A (l)	L	256
16	Mabaruma	F (m), BW (h), L (m), A (m)	L	32
17	Arunamai	F (h), BW (m), A (m)	M	12
18	Pomeroon	F (h), BW (l), A (h)	H	105
TOTAL				2108

[a] Indicates most important natural or anthropogenic sources of forest disturbance: (F) prolonged seasonal flooding, (BW) blackwater palm swamps, (L) selective logging, (A) agriculture. Classes of intensity are: none (–), light (l), moderate (m), and heavy (h).

[b] Classes of hunting/trapping pressures are: (–) none, (L) light, (M) moderate, and (H) heavy.

(flooding, blackwater swamps, selective logging, and agriculture) were assessed during surveys and repeat visits to the survey areas. Hunting and trapping pressures were estimated with data collected during interviews with local people and wildlife trappers, direct observations of hunting/trapping, and accounts from social scientists working in the survey areas (Lehman, 1999). Following Peres (1997, 1999), intensities of natural and anthropogenic disturbances were estimated using a subjective four-point scale: none, light, moderate, and heavy.

Geographic Information System

Detailed maps of the geographic range of each species in Guyana were produced and then measured with a digitizer tablet (Figure 2). Using published descriptions of habitat and elevation preferences for each species (Peres, 1997, 1999b; Lehman, 2004), habitats deemed unsuitable for Guyanese primates were excluded from estimates of range size (savannas and meadows, montane shrublands, high-tepui forests, upper montane forests, lakes, mining areas, and urban centers).

Analyses and Statistics

We computed Shannon-Weiner diversity indices (H') for each primate species in the four main forest habitats (basimontane, *terra firme*, riparian, and seasonally flooded blackwater swamps). We used a sighting rate of the number of groups censused per 100 km surveyed. Sighting distances were used only for sites at which each species was surveyed rather than using the total sighting distance. Spearman rank correlations (r_s) were used to determine how variations in species-specific sighting rates covaried with patterns of disturbance at the beta level. Linear regression analyses were used to determine if survey effort (i.e., survey distance per site) influenced species diversity and the total number of group surveyed per site. Thus, the regression models would determine if our results were an artifact of differential survey intensities rather than disturbance intensities.

We used two association types in our analysis of ecological correlates to primate community structure: polyspecific associations and site associations. Polyspecific groups were defined as two or more groups of different species feeding or traveling within 20 m of each other (Mittermeier, 1977). The association had to occur for at least 10 min to be scored as a polyspecific group.

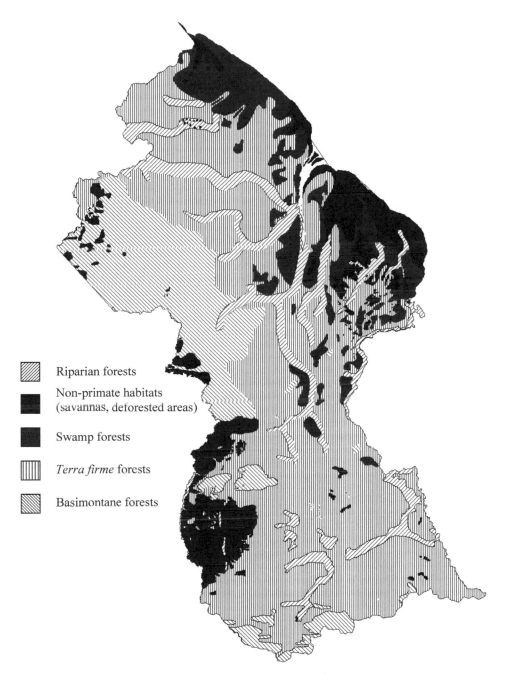

Figure 2. Location of the four main habitat types used in our study of the ecological biogeography of eight primate species in Guyana.

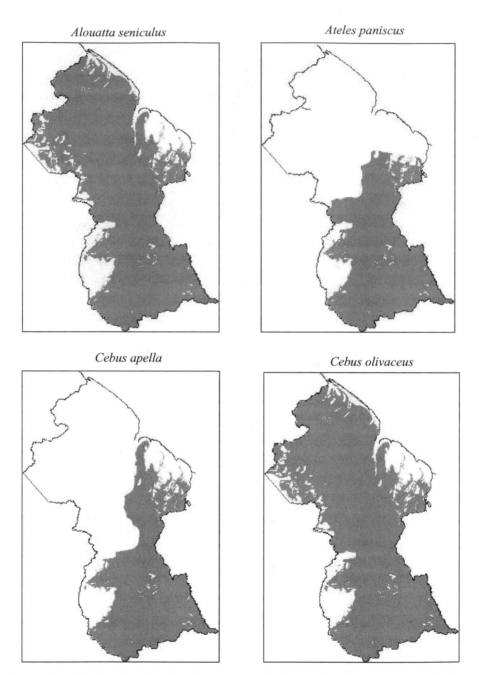

Figure 3. Distribution of eight primate species in Guyana based on surveys and habitat use. Lehman 2004.

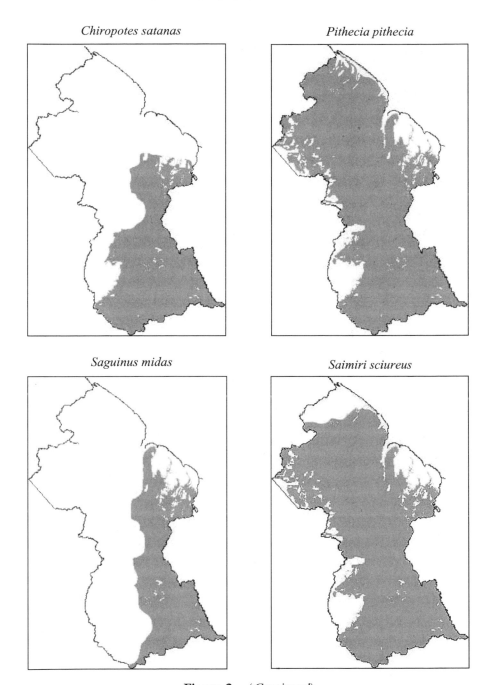

Figure 3. (*Continued*)

Following Schluter (1984), a site association for two or more species was analyzed if the species were observed during the same census of a transect line at a
site. Site associations were used to produce association indices. A null association model which employs a variance ratio (VR) was used to test for significant
associations between species pairs (Schluter, 1984). The variance ratio is given
by:

$$VR = S_T^2 / \sigma_T^2$$

Where S_T^2 is the total sample variance for the occurrences of the S species in
the sample and σ_T^2 is the variance in total species number. If $VR > 1$, then the
species exhibit a positive association. If $VR < 1$, then a negative net association
is suggested. A statistic (W) was then computed to determine if deviations from
1 were significant. This statistic is given by:

$$W = (N)(VR)$$

If, for example, the species are not associated, then there is a 90% probability
that W lies between limits given by the following chi-square distribution:

$$\chi^2_{.05, N} < W < \chi^2_{0.95, N}$$

The Ochiai index (O_i) was used to measure the degree of association between
pairs of species (Ochiai, 1957). In this analysis, a presence-absence matrix was
computed for each species pair at each survey site. A chi-squared test statistic
was then used to test the null hypothesis of independence in the 2×2 table. A
Yate's correction was calculated to avoid biased values resulting from low cell
expectations.

Statistical analyses were conducted using SPSS 11.5, and the alpha level was
set at 0.05 for all analyses.

RESULTS

A total of 64.1% ($N = 141$) of primate groups were seen in riparian forests
(Table 2). Of the 220 primate groups for which we have habitat data, 15.4%
($N = 34$) were seen in swamp forests and 18.6% ($N = 41$) were censused
in *terra firme* forests. Few primate groups were seen in basimontane forest ($N = 4$ groups). Although most species preferred riparian forests, spider
monkeys and bearded sakis were each sighted most often in *terra firme* forest. Golden handed tamarins were surveyed most frequently in swamp forests.

Table 2. Habitat use by eight primate species in Guyana

| Species | Group sighting frequency by forest habitat type[a] | | | | | Habitat diversity index (H') |
	Basimontane	Terra firme	Riparian	Swamp	Total	
A. paniscus	0	4	2	0	6	0.92
A. seniculus	1	16	33	7	57	1.44
C. apella	0	3	22	5	30	1.09
C. olivaceus	3	6	11	6	26	1.86
C. satanas	0	4	1	0	5	0.72
P. pithecia	0	3	16	2	21	1.02
S. midas	0	4	3	7	14	1.49
S. sciureus	0	1	53	7	61	0.63
Total	4	41	141	34	220	1.35

[a] Sighting frequencies listed only for surveys in which habitat type could be accurately determined. Thus, habitat sighting frequency is sometimes less than total sighting frequency for each species.

Wedge-capped capuchins ($H' = 1.86$), golden handed tamarins ($H' = 1.49$), and howler monkeys ($H' = 1.44$) had the highest habitat diversity indices. Squirrel monkeys ($H' = 0.92$) and bearded sakis ($H' = 0.72$) were the most conservative in their use of the four main habitat types.

Squirrel monkeys (5.41 groups/100 km, $N = 72$) had the highest sighting rates (Table 3). The total sighting rate for brown capuchins (4.09 groups/100 km, $N = 37$) was almost double the rate for wedge-capped capuchins (2.29 groups/100 km, $N = 31$). Red howler monkeys were sighted frequently during surveys (3.18 groups/100 km, $N = 59$). Golden-handed tamarins had a sighting rate of 2.67 groups/100 km ($N = 25$). White-faced sakis had a total sighting rate of 2.05 groups/100 km ($N = 24$). Sighting rates were lowest for spider monkeys (1.21 groups/100 km, $N = 11$) and bearded sakis (sighting rate = 1.56 groups/10 km, $N = 13$). The total primate sighting rate was 12.90 groups/100 km ($N = 272$). There was no correlation between survey distance and either species diversity ($R = 0.407$, ANOVA $F_{0.09[1,16]} = 3.17$) or total number of groups sighted per site ($R = 0.422$, ANOVA $F_{0.08[1,16]} = 3.46$), indicating that biogeographic patterns in our data are not an artifact of survey effort.

Table 4 shows the ecological correlates to sighting rates in the eight primate species we surveyed in Guyana. *A. seniculus* sighting rates were positively correlated with varying intensities of agriculture. For *C. apella*, there was a significant positive correlation between sighting rates and flooding intensities. Sighting rates for *S. midas* covaried with the intensity of swamps.

Table 3. Sighting rates (number of groups/100 km surveyed) for eight primate species at eighteen sites in Guyana

	Site	As	Ap	Ca	Co	Cs	Pp	Sm	Ss	Total
						Sighting rate[a]				
1	Essequibo	6.25	–	–	5.00	–	1.25	2.50	1.25	16.25
2	Timehri	–	–	–	2.96	–	1.97	3.94	1.48	10.34
3	Mahaicony	4.63	–	–	5.56	–	10.19	–	15.74	36.11
4	Abary	–	–	–	5.41	–	–	–	–	5.41
5	North Berbice	11.67	1.67	8.33	0.83	–	1.67	1.67	14.17	40.00
6	Canje	3.15	–	6.30	–	–	0.79	3.15	9.45	22.83
7	Dubulay	2.59	0.37	3.70	–	2.22	–	2.96	2.96	14.81
8	Wikki	1.96	–	7.84	–	–	–	–	–	9.80
9	South Berbice	2.21	0.74	2.21	–	0.74	0.74	0.74	1.47	8.82
10	Mabura	1.00	1.00	1.00	0.50	1.00	0.50	–	–	5.00
11	Apoteri	1.17	2.34	–	–	1.75	–	–	2.34	7.60
12	Iwokrama	1.72	–	–	1.72	1.72	1.72	–	–	6.90
13	Annai	–	7.69	–	7.69	–	–	–	–	15.38
14	Kaieteur	0.78	–	–	2.33	–	–	–	–	3.10
15	Sebai	1.56	–	–	0.78	–	–	–	–	2.34
16	Mabaruma	6.25	–	–	3.13	–	3.13	–	–	12.50
17	Arunamai	25.00	–	–	8.33	–	–	–	8.33	41.67
18	Pomeroon	4.76	–	–	1.90	–	0.95	–	6.67	14.29
Total		3.18	1.21	4.09	2.29	1.56	2.05	2.67	5.41	12.90

[a] As = *Alouatta seniculus*, Ap = *Ateles paniscus*, Ca = *Cebus apella*, Co = *Cebus olivaceus*, Cs = *Chiropotes satanas*, Pp = *Pithecia pithecia*, Sm = *Saguinus midas*, Ss = *Saimiri sciureus*. - indicates species not seen at that site.

Table 4. Spearman rank correlations between natural (flooding and swamps) and anthropogenic (logging, agriculture, and hunting/trapping) disturbance intensities and primate sighting rates for eighteen survey sites in Guyana

Species[a]	Flooding	Swamp	Logging	Agriculture	Hunting/ Trapping
A. paniscus	−0.247 (0.63)	−0.655 (0.15)	0.372 (0.46)	0.541 (0.26)	−0.525 (0.28)
A. seniculus	0.423 (0.11)	0.377 (0.16)	0.368 (0.17)	**0.683 (0.005)**	−0.304 (0.27)
C. apella	**0.845 (0.03)**	0.034 (0.94)	0.626 (0.18)	0.676 (0.14)	−0.706 (0.11)
C. olivaceus	0.206 (0.49)	0.194 (0.52)	0.016 (0.95)	0.244 (0.42)	−0.153 (0.61)
C. satanas[b]	0.632 (0.25)	0.707 (0.18)	0.354 (0.55)	NA	−0.289 (0.63)
P. pithecia	0.389 (0.23)	0.538 (0.08)	0.272 (0.41)	0.509 (0.11)	−0.110 (0.74)
S. midas	0.698 (0.12)	**0.941 (0.005)**	0.359 (0.48)	0.516 (0.29)	−0.353 (0.49)
S. sciureus	0.276 (0.38)	0.101 (0.75)	−0.204 (0.52)	0.220 (0.49)	−0.067 (0.83)

[a] Statistically significant correlations in bold.
[b] NA indicates no variation in intensity for *C. satanas* (all sightings at sites with no agriculture).

Eighteen polyspecific groups were observed during surveys (Table 5). Although most groups involved only two species, one polyspecific group was observed that comprised three species (squirrel monkeys, brown capuchins, and howler monkeys). Squirrel monkeys were observed to form polyspecific groups more often than the other species, accounting for 94.1% ($N = 16$) of total associations. Squirrel monkeys formed polyspecific groups with four primate species: brown capuchins ($N = 10$), red howlers ($N = 3$), wedge-capped capuchins ($N = 2$), and white-faced sakis ($N = 1$). The only polyspecific group not involving squirrel monkeys was one observed between red howlers and wedge-capped capuchins. Table 6 shows interspecific association indices for 28 pairwise combinations of primate species in Guyana. There was a significant trend towards positive species associations for primates censused at the eighteen survey sites (Schluter's variance test for overall association, $V = 4.12$, $W = 57.34$, $p = 0.001$). Four primate pairs showed significant positive patterns of interspecific associations. The highest degree of positive association was between spider monkeys and brown bearded sakis ($O_i = 1.00$), golden handed tamarins and squirrel monkeys ($O_i = 0.79$), and brown capuchins with bearded sakis ($O_i = 0.77$) and spider monkeys ($O_i = 0.77$). Brown and wedge-capped capuchins had a significant negative pattern of interspecific associations. Brown capuchins had a total sighting rate of 4.28 groups/100 km at the four sites (Wikki River, South Berbice, Canje River, and Dubulay Ranch) where they were observed but wedge-capped capuchins were absent (Table 7). Wedge-capped capuchins had a sighting rate of 2.81 groups/100 km at eleven allopatric sites (Timehri, Mahaicony, Abary, Essequibo, Iwokrama Rain Forest Reserve, Kaieteur Falls National Park, Annai, Pomeroon, Arunamai, Mabaruma, and Sebai). The average sighting rate of brown capuchins (3.75 groups/100 km) was almost six times greater than that of wedge-capped capuchins (0.63 groups/ 100 km) at the two sympatric sites (North Berbice and Mabura).

DISCUSSION

Our data indicate the importance of riparian forests for understanding the biogeography of six of the eight primate species in Guyana. Peres (1999), in his study of the effects of forest type on primate community structure in Amazonia, reported the pervasive influence of floodplain forests on primate distribution and ecology. He found that total primate density and biomass in floodplain forests were twice that of *terra firme* forests. Ultimately, primate community

Table 5. Patterns of polyspecific groups observed during surveys at eighteen survey sites in Guyana

Species[a]	A. seniculus	A. paniscus	C. apella	C. olivaceus	C. satanas	P. pithecia	S. midas	S. sciureus
A. seniculus	–	0	0	1	0	0	0	3
A. paniscus	0	–	0	0	0	0	0	0
C. apella	0	0	–	0	0	0	0	10
C. olivaceus	5.8	0	0	–	0	0	0	2
C. satanas	0	0	0	0	–	0	0	0
P. pithecia	0	0	0	0	0	–	0	1
S. midas	0	0	0	0	0	0	–	0
S. sciureus	17.7	0	58.8	11.8	0	5.8	0	–

[a] Numbers above the diagonal refer to frequency of association. Numbers below the diagonal refer to frequency as percentage of total associations.

Table 6.   Interspecific association indices and test statistics between eight primate species at eighteen sites in Guyana

Species pair	Association type[a]	χ^{2b}		Yates χ^{2b}		O_i^c
A. seniculus & A. paniscus	+	0.97		0.04		0.50
A. seniculus & C. apella	+	2.73		1.07		0.67
A. seniculus & C. olivaceus	+	0.14		0.14		0.80
A. seniculus & C. satanas	+	0.88		0.42		0.51
A. seniculus & P. pithecia	+	3.69		1.64		0.78
A. seniculus & S. midas	−	0.01		0.60		0.50
A. seniculus & S. sciureus	+	0.51		0.10		0.69
A. paniscus & C. apella	+	8.12	*	4.66	*	0.77
A. paniscus & C. olivaceus	+	0.58		0.04		0.41
A. paniscus & C. satanas	+	16.88	*	12.31	*	1.00
A. paniscus & P. pithecia	+	0.41		0.00		0.41
A. paniscus & S. midas	+	2.16		0.60		0.52
A. paniscus & S. sciureus	+	0.53		0.06		0.47
C. apella & C. olivaceus	−	4.75	*	7.85	*	0.26
C. apella & C. satanas	+	8.12	*	4.66	*	0.77
C. apella & P. pithecia	+	2.29		1.02		0.71
C. apella & S. midas	+	2.80		1.19		0.60
C. apella & S. sciureus	+	2.29		0.00		0.47
C. olivaceus & C. satanas	−	0.14		1.23		0.33
C. olivaceus & P. pithecia	+	1.33		0.33		0.71
C. olivaceus & S. midas	−	0.87		2.42		0.39
C. olivaceus & S. sciureus	−	0.00		0.33		0.61
C. satanas & P. pithecia	+	0.41		0.00		0.41
C. satanas & S. midas	+	2.16		0.60		0.52
C. satanas & S. sciureus	+	0.26		0.16		0.39
P. pithecia & S. midas	+	1.33		0.33		0.53
P. pithecia & S. sciureus	+	1.00		0.25		0.63
S. midas & S. sciureus	+	7.27	*	4.65	*	0.79

[a] Sign indicates direction of the species association (+ = pair of species occurred together more often than expected if independent, − = pair of species occurred together less often than expected if independent).
[b]*indicates significance at $p < 0.05$ and d.f. = 1.
[c] Ochiai's index of association.

Table 7.   A comparison of sighting rates (# groups/100 km survey distance) for *Cebus apella* and *Cebus olivaceus* at sympatric and allopatric sites in Guyana

	Sighting rate		
Species	Allopatric sites	Sympatric sites	Total
C. apella	4.28	3.75	4.09
C. olivaceus	2.81	0.63	2.29
Total	3.34	2.19	3.01

structure may be the result of variation in the amounts of macronutrients available to plants in flooded and non-flooded forests. Peres (1999) noted that *terra firme* forests tend to be under extremely tight nutrient budgets because they rest upon nutrient-poor Ultisol soils. The Ultisol soils of Amazonia are very similar to the soils of the Guiana Shield, which underlie forests in Guyana. The Guiana Shield is a Precambrian eroded base that is the remaining exposed section of Gondwanaland (Gibbs and Barron, 1993). In the coastal zone, the Precambrian base is overlain by Quaternary marine silts forming the alluvial coastal plain or lowlands. The uplands of the Pakaraima Mountains in western Guyana consist of sandstone table mountains, known as *tepuis* in Venezuela. Some upland areas, principally along the Venezuelan border in western Guyana, consist of schist, quartzite, and conglomerate of the Orapu-Bonidoro series (Gibbs and Barron, 1993). The interior region of the country has geosynclines of the Paracama series. The soils in this region are red, clayey, and ferrallitic; some areas are covered by white sand. The southern area of Guyana is predominantly peneplain on a crystalline base. The soils in each of these non-flooded areas are severely impoverished because they lack the seasonal influxes of nutrient-rich alluvial sediments. Concomitantly, *terra firme* forests in Guyana have low levels of floral diversity and abundance of plant families that are valuable food resources for primates (Comiskey *et al.*, 1993; ter Steege, 1993b; Ek, 1997; Terborgh and Andresen, 1998; Lehman, 2000). For example, Terborgh and Andresen (1998) analyzed floristic patterns in tree plots at 29 sites in South America. They found that *terra firme* plots in Guyana had the lowest abundance of trees in the plant family Moraceae. This plant family contains tree species that are critical food resources during periods of low resource abundance for many platyrrhines (Terborgh, 1983; Pontes, 1997). Each of the eight primate species in Guyana eat at least some fruit, and three—spider monkeys, bearded sakis, and white-faced sakis—are among the most frugivorous of all South American primates (van Roosmalen, 1987; Norconk, 1997). Thus, *terra firme* forests in Guyana may not contain enough fruiting trees to support all eight species, particularly during periods of fruit scarcity.

We found that spider monkeys prefer *terra firme* forest habitats and that they tend to respond negatively but not significantly, in terms of covariates to their sighting rates, to natural disturbances in Guyana. Muckenhirn *et al.*, (1975) observed a similar pattern of habitat use in that only 36% ($N = 8$) of the 22 groups of spider monkeys they saw were censused in riparian forests in Guyana. Spider monkeys should be the species most likely to use flooded forests because they have a diet composed largely of fruits (van Roosmalen, 1987), and *terra firme* forests are characterized by low fruit abundance and low species diversity for fruiting trees in Guyana (ter Steege, 1993b). The question arises then as to why spider monkeys rarely range into flooded forests in Guyana. Although flooded forests tend to have low fruit production on an annual basis (e.g., Ayres, 1986; Peres, 1997; Terborgh and Andresen, 1998), they do contain some tree species, such as *Ficus* (Moraceae), that have mast fruitings. Ahumada *et al.* (1998) found that long-haired spider monkeys (*Ateles belzebuth*) in Colombia exploit these mast fruiting trees in flooded forests when fruit resources are at critically low levels in *terra firme* forests. One *Ficus* tree in flooded forests was the fifth most important fruit species in the annual diet of the spider monkeys. However, few mast fruiting trees exist in flooded forests in Guyana, particularly in the important Moraceae plant family (Comiskey *et al.*, 1993; ter Steege, 1993b; Terborgh and Andresen, 1998). Terborgh and Andresen (1998) found that 1-hectare plots outside Guyana contained on average 24.1 trees in the family Moraceae whereas only one Moraceae tree was found in five *terra firme* and riparian plots in Guyana. Although we have few data on habitat use among spider monkeys, we suggest that a lack of large fruiting trees in flooded forests, particularly those in the plant family Moraceae, may limit this primate species to only infrequent use of this habitat. Testing this hypothesis will require detailed information on density estimates and diet in spider monkeys as well as more extensive botanical sampling in flooded and *terra firme* forests in Guyana.

Sighting rates for *C. apella* were positively correlated with flooding intensities. This correlation is likely due to differences in the diversity and density of palms between flooded and non-flooded forests in Guyana. Palm species, such as *Astrocaryum* sp. and *Attalea* sp., represent a critical food resource for brown capuchins during the dry season when fruit is scarce (Kiltie, 1980; Terborgh, 1983). *Terra firme* forests in Guyana are characterized by having some of the lowest densities of palm species (4.6 trees/hectare) in South America (Ahumada *et al.*, 1998; Terborgh and Andresen, 1998). Palms exist at higher densities in flooded than non-flooded forests in Guyana (Davis and Richards, 1934; Terborgh and Andresen, 1998). Therefore, *C. apella* may be selecting habitats that contain the highest densities of palms.

We suggest that edge-related variations in the quality of leaves may explain the increased sighting rates of howler monkeys in agricultural areas. Forest edges are dynamic zones characterized by the penetration, to varying depths and intensities, of conditions from the surrounding environment (matrix) into the forest interior (Malcolm, 1994). Edge effects have been shown to have positive influences on the distribution and density of many species of plants, animals, and insects (e.g., Malcolm, 1997; Ries *et al.*, 2004; Lehman *et al.*, 2006). Young leaves are the dominant food category for red howler monkeys (e.g., Gaulin and Gaulin, 1982; Julliot and Sabatier, 1993). Because even young leaves tend to contain plant secondary compounds (Milton, 1980; Glander, 1982), howler monkeys are very selective in their foraging strategies. Although there are few data on how light levels influence plant chemistry in South American forests, Ganzhorn (1995) documented that low-intensity forest disturbance increased light levels in forests in Madagascar, which resulted in higher protein concentration in leaves. Elevated light levels have been documented near forest edges and agricultural areas in many Neotropical forests (Malcolm and Ray, 2000; Foggo *et al.*, 2001; Hill and Curran, 2001; Cochrane and Laurance, 2002; Laurance *et al.*, 2002). Thus, we hypothesize that the quality of leaves may be highest in forest edges near agricultural development, which would attract howler monkeys to these habitats. We also hypothesize that high sighting rates of golden-handed tamarins in swamp forests are a result of animals selecting habitats with a high abundance and diversity of insect prey. Swamp forests tend to contain a high density and diversity of insects (Howard, 2001). Golden-handed tamarins are one of the most insectivorous of the eight primate species in Guyana (Kessler, 1995; Simmon and Sabatier, 1996; Packa *et al.*, 1999). A clearer understanding of how howler monkeys and golden-handed tamarins respond to habitat variations will require increased survey sample sizes and detailed data on the feeding ecology of conspecific groups that range into edge and interior habitats.

We were surprised that hunting pressures had no statistical effect on primate diversity and abundance. This result may be an artifact of statistical analyses rather than a biogeographic pattern. Correlations organize sampling entities along a gradient or continuum, although some ecological variables, such as hunting, are neither linear nor unidirectional (McGarigal *et al.*, 2000). Species-specific hunting is particularly relevant to studies, such as ours, in which group size was not recorded during surveys. Thus, we make no statistical distinction between a large group of 10–12 spider monkeys and one under intense hunting pressures that contains only 2–3 individuals. Moreover, Sussman and Phillips-Conroy (1995) reported a decrease in primate densities, particularly for large-bodied species such as spider monkeys and brown capuchins, since work done by Muckenhirn *et al.*, in the early 1970's (Muckenhirn *et al.*,

1975; Muckenhirn and Eisenberg, 1978). Sussman and Phillips-Conroy (1995) attributed the density decrease to hunting pressures. This prediction seems logical given numerous reports and our observations on the excellent tracking and hunting abilities of local Amerindians (Beebe, 1925; Guppy, 1958; Muckenhirn *et al.*, 1975; Muckenhirn and Eisenberg, 1978; A.R.U., 1992; Lehman, 1999). Amerindians in Guyana prefer meat from spider monkeys and brown capuchins, as has been reported at many other sites in South America (Mittermeier, 1977; Johns and Skorupa, 1987; Peres, 1999a; Peres and Dolman, 2000; Laurance *et al.*, 2002). It is interesting to note that spider monkeys and bearded sakis were seen most often at or near sites that had some measure of protection for local animals (Mabura Hill, Dubulay Ranch, Iwokrama). Although spider monkeys were seen near the large Amerindian settlement of Apoteri, the animals were very shy and existed at low densities. Hunting of spider monkeys and brown capuchins may have been so successful in the past that populations of these species are susceptible to localized extirpation. Therefore, detailed information on hunting practices by local people must be compared to longitudinal data on the demographics and ecology of primate populations and those in habitats with lower hunting intensities in Guyana.

Our data on patterns of polyspecific associations are similar to those found by researchers in other regions of South America (Thorington, 1968; Klein and Klein, 1973, 1975; Mittermeier, 1977; Terborgh, 1983; Pontes, 1997). In these studies, squirrel monkeys were most likely to be in the company of other species, and that the most common association was with brown capuchins. Our data corroborate those of Mittermeier (1977) in that spider monkeys, white-faced sakis, and golden-handed tamarins rarely form polyspecific groups. However, brown-bearded sakis associated with brown capuchins and squirrel monkeys in Suriname whereas no polyspecific associations were observed for brown-bearded sakis in Guyana. Differences in association patterns for brown-bearded sakis between the studies conducted in Suriname and Guyana may be due to low sighting rates in Guyana.

Despite reports that brown and wedge-capped capuchins compete with each other (Eisenberg, 1979) and that they are morphologically, ontogenetically, and behaviorally similar (Moynihan, 1976; Terborgh, 1983; Eisenberg, 1989; Ford and Hobbs, 1994), they are sympatric at some sites in Guyana, Suriname (Mittermeier, 1977), French Guiana (Simmon and Sabatier, 1996; Youlatos, 1998) and northern Amazonia (Pontes, 1997). Sussman and Phillips-Conroy (1995) hypothesized that where the two species cooccur, wedge-capped capuchins may be found at lower densities than brown capuchins. Although we found that brown and wedge-capped capuchins cooccurred at two sites (Berbice River and Mabura Hill), analyses of species composition across all survey sites indicate that these monkeys have a negative pattern of interspecific association. Sighting rates of both species were lower at sites where they were found to be sympatric, but reduced sighting rates were particularly noticeable in wedge-capped capuchins. Thus, as predicted by Sussman and Phillips-Conroy (1995), wedge-capped capuchins may be sensitive to the presence of a congener.

Our findings are clear for those areas and ecological/biogeographical factors that we have been able to analyze. However, further data are needed, particularly on the influence of abiotic factors, such as rainfall, and hunting pressures in influencing primate distribution and diversity. We also require data on the community structure of yet unsurveyed/unsampled areas of SE Guyana and the western highlands. Hunting pressures may have only a minor effect on the diversity and abundance of the primate community but a strong effect on certain preferred game species (e.g., spider monkeys and brown capuchins).

SUMMARY

We investigated how natural and anthropogenic disturbances as well as community structure influenced habitat use and sighting rates for eight primate species in Guyana. Our data indicate the importance of riparian forests for understanding the biogeography of six of the eight primate species in Guyana (*A. seniculus*, *C. apella*, *C. olivaceus*, *P. pithecia*, *S. midas*, and *S. sciureus*). These primates may exploit food resources in riparian forests because *terra firme* forests do not contain enough fruit trees to support all taxa throughout the year. Sighting rates for brown capuchins were positively correlated with flooding intensities, which may be the result of the abundance of palm species used as a keystone resource in seasonally inundated habitats. Edge-related variations in leaf quality may explain higher sighting rates for howler monkeys in agricultural areas. Insect abundance may relate to the higher sighting rates for golden-handed tamarins in swamp forests. Although wedge-capped and brown capuchins are sympatric at some sites, analyses of species composition across all survey sites indicate that these monkeys have a negative pattern of interspecific association. Sighting rates of both species were lower at sites where they were found to be sympatric, but reduced sighting rates were particularly noticeable in wedge-capped capuchins.

ACKNOWLEDGMENTS

For permission to conduct our study we thank Dean Indirjit Ramdas, Dean Catherine Cox, Mr. Phillip daSilva, Mr. John Caesar, Dr. Karen Pilgrim, Office of the President, University of Guyana, Department of Biology at the University of Guyana, Ministry of Amerindian Affairs, Ministry of Health, National Parks Commission, Tropenbos Guyana, Demarara Timbers Ltd., Iwokrama Rain Forest Reserve, and Wildlife Division of the Department of Health. We greatly appreciate the support of Drs. Vicki Funk and Carol Kelloff of the Biological Diversity of the Guianas Program at the Smithsonian Institution. We thank the local people for assistance in data collection. We are grateful to Alexander Mendes, the Mendes family, and the staff at Dubulay Ranch for their friendship, hospitality, and support. This project was supported in part by the Lincoln Park Zoo Scott Neotropic Fund, National Science Foundation, the Biological Diversity of the Guianas Program of the Smithsonian Institution, the National Geographic Society, USAID, the Ministry of Finance of the Government of Guyana, the Global Environmental Fund of the World Bank, and the Connaught Foundation.

REFERENCES

A.R.U. 1992, *The Material Culture of the Wapishana of the South Rupununi Savannas in 1989*, Amerindian Research Unit, University of Guyana, Georgetown.

Ahumada, J. A., Stevenson, P. R., and Quinones, M. J. 1998, Ecological response of spider monkeys to temporal variation in fruit abundance: The importance of flooded forest as a keystone habitat. *Primate Conservation* 18:10–14.

Ayres, J. M. 1986, Uakaries and Amazonian Flooded Forests. Unpublished Ph.D. dissertation, Cambridge University, Cambridge, England.

Ayres, J. M. and Clutton-Brock, T. H. 1992, River boundaries and species range size in Amazonian primates. *Am. Nat.* 140:531–537.

Beebe, W. 1925, Studies of a tropical jungle One-quarter of a square mile of jungle at Kartabo, British Guiana. *Zoologica* 6:1–193.

Bennett, C. L., Leonard, S., and Carter, S. 2001, Abundance, diversity, and patterns of distribution of primates on the Tapiche River in Amazonian Peru. *Am. J. Primat.* 54:119–126.

Chapman, C. A. and Chapman, L. J. 1996, Mixed-species primate groups in the Kibale forest: Ecological constraints on association. *Int. J. Primat.* 17:31–50.

Cochrane, M. A. and Laurance, M. F. 2002, Fire as a large-scale edge effect in Amazonian forests. *J. Trop. Ecol.* 18:311–325.

Comiskey, J., Dallmeier, F., Aymard, G., and Hanson, A. 1993, *Biodiversity Survey of Kwakwani, Guyana*, The Smithsonian Institution/Man and the Biosphere Biological Diversity Program, Washington, DC.

Connell, J. 1961, The influence of interspecific competition and other factors on the distribution of the barnacle Chthamalus stellatus. *Ecology* 42:710–723.

Cushman, S. A. and McGarigal, K. 2004, Patterns in the species-environment relationship depend on both scale and choice of response variables. *Oikos* 105:117–124.

Davies, K. F., Melbourne, B. A., and Margules, C. R. 2001, Effects of within-and between-patch processes on community dynamics in a fragmented experiment. *Ecology* 82:1830–1846.

Davis, T. A. and Richards, P. W. 1934, The vegetation of Moraballi Creek, British Guiana: An ecological study of a limited tropical rain forest. *J. Ecol.* 22:106–155.

Ek, R. C. 1997, *Botanical Diversity in the Tropical Rain Forest of Guyana*, Tropenbos-Guyana Programme, Georgetown, Guyana.

Fleagle, J. G. and Mittermeier, R. A. 1980, Locomotor behavior, body size, and comparative ecology of seven Surinam monkeys. *Am. J. Phys. Anthropol.* 52:301–314.

Foggo, A., Ozanne, C. M. P., Spreight, M. R., and Hambler, C. 2001, Edge effects and tropical forest canopy invertebrates. *Plant Ecol.* 153:347–359.

Ganzhorn, J. U. 1995, Low-level forest disturbance effects on primary production, leaf chemistry, and lemur populations. *Ecology* 76:2084–2096.

Ganzhorn, J. U. 1997, Test of Fox's assembly rule for functional groups in lemur communities in Madagascar. *J. Zool.* 241:533–542.

Gaulin, S. J. C. and Gaulin, C. K. 1982, Behavioral ecology of *Alouatta seniculus* in Andean cloud forest. *Int. J. Primat.* 3:1–52.

Gause, G. F. 1934, *The Struggle for Existence*, Williams and Williams, Baltimore, MD.

Gibbs, A. K. and Barron, C. N. 1993, *The Geology of the Guiana Shield*, Oxford University Press, New York.

Glander, K. E. 1982, The impact of plant secondary compounds on primate feeding behavior. *Yearbook of Phy. Anthropol.* 25:1–18.

Guppy, N. L. 1958, *Wai Wai—Through the Forests North of the Amazon*, John Murray, London.

Henkel, T. 1994, Expedition Reports: Smithsonian Institution Department of Botany.

Hill, J. L. and Curran, P. J. 2001, Species composition in fragmented forests: Conservation implications of changing forest area. *Appl. Geog.* 21:157–174.

Hoffman, B. 1992, Expedition Reports: Smithsonian Institution Department of Botany.

Howard, F. W. 2001, The animal class Insecta and the plant family Palmae, in: Howard, F. W., Moore, D., Giblin-Davis, R. M., and Abad, R. G., eds., *Insects on Palms*, Oxford University Press, Oxford, UK, pp. 1–32.

Huber, O., Funk, V., and Gharbarran, G. 1995, *Vegetation Map of Guyana*. Centre for the Study of Biological Diversity, Georgetown.

Johns, A. D. and Skorupa, J. P. 1987, Responses of rain-forest primates to habitat disturbance: A review. *Int. J. Primatol.* 8:157–191.

Julliot, C. and Sabatier, D. 1993, Diet of the red howler monkey (*Alouatta seniculus*) in French Guiana. *Int. J. Primatol.* 14:527–550.

Kessler, P. 1995, Preliminary field study of the red-handed tamarin, *Saguinus midas*, in French Guiana. *Neotrop. Primates* 3:20–22.

Kiltie, R. 1980, Seed Predation and Group Size in Rain Forest Peccaries. Unpublished Ph.D. dissertation, Princeton University.

Klein, L. L. and Klein, D. J. 1973, Observations on two types of Neotropical primate intertaxa associations. *Am. J. Phys. Anthropol.* 38:649–653.

Klein, L. L. and Klein, D. J. 1975, Social and ecological contrasts between four taxa of neotropical primates, in: Tuttle, R. H., ed., *Socioecology and Psychology of Primates*. The Hague, Mouton, pp. 59–85.

Laurance, M. F., Lovejoy, T. E., Vasconcelos, H. L., Bruna, E. M., Didham, R. K., Stouffer, P. C., Gascon, C., Bierregaard, O., Laurance, S., and Sampaio, E. 2002, Ecosystem decay of Amazonian forest fragments: A 22-year investigation. *Conservation Biol.* 16:605–618.

Lehman, S. M. 1999, Biogeography of the Primates of Guyana. Unpublished Ph.D. dissertation, Washington University, St. Louis, MO.

Lehman, S. M. 2000, Primate community structure in Guyana: A biogeographic analysis. *Int. J. Primat.* 21:333–351.

Lehman, S. M. 2004, Biogeography of the primates of Guyana: Effects of habitat use and diet on geographic distribution. *Int. J. Primat.* 25:1225–1242.

Lehman, S. M., Rajoanson, A., and Day, S. (2006). Edge effects and their influence on lemur distribution and density in southeast Madagascar. *Am. J. Phys.Anthropol.* 126:318–328.

Lotka, A. J. 1925, *Elements of Physical Biology*. Williams and Watkins, Baltimore, MD.

Maguire, B. 1972, The botany of the Guayana higland, IX (Plant Taxonomy). *Mem. NY Botan. Gardens* 23:1–832.

Maguire, B., Cowan, R. S., and Wurdack, J. J. 1953, The botany of the Guayana highland. *Mem. NY Botan. Gardens* 8:87–96.

Malcolm, J. R. 1994, Edge effects in central Amazonian forest fragments. *Ecology* 75:2438–2445.

Malcolm, J. R. 1997, Biomass and diversity of small mammals in Amazonian forest fragments, in: Laurance, W. F. and Bierregaard, O., eds., *Tropical Forest Remnants: Ecology, Management, and Conservation of Fragmented Communities*, The University of Chicago Press, Chicago, pp. 207–221.

Malcolm, J. R. and Ray, J. C. 2000, Influence of timber extraction routes on central African small-mammal communities, forest structure, and tree diversity. *Conservation Biol.* 14:1623–1638.

McGarigal, K., Cushman, S., and Stafford, S. 2000, *Multivariate Statistics for Wildlife and Ecology Research*, Springer, New York.

Milton, K. 1980, *The Foraging Strategy of Howler Monkeys*, Columbia University Press, New York.

Mittermeier, R. A. 1977, Distribution, Synecology, and Conservation of Surinam Monkeys. Unpublished Ph.D. dissertation, Harvard University, Boston, MA.

Mittermeier, R. A. and van Roosmalen, M. G. M. 1981, Preliminary observations on habitat utilization and diet in eight Suriname monkeys. *Folia Primatol.* 36:1–39.

Muckenhirn, N. A. and Eisenberg, J. F. 1978, The status of primates in Guyana and ecological correlations for neotropical primates, in: Chivers, D. J. and Lane-Petter, W., eds., *Recent Advances in Primatology*, Vol. 2: *Conservation*, Academic Press, London.

Muckenhirn, N. A., Mortenson, B. K., Vessey, S., Fraser, C. E. O., and Singh, B. 1975, Report of a primate survey in Guyana: Pan American Health Organization.

Norconk, M. A. 1990a, Introductory remarks: Ecological and behavioral correlates of polyspecific primate troops. *Am. J. Primat.* 21:81–85.

Norconk, M. A. 1990b, Mechanisms promoting stability in mixed *Saguinus mystax* and *S fuscicollis* troops. *Am. J. Primatol.* 21:159–170.

Norconk, M. A. 1997, Seasonal variations in the diets of white-faced and bearded sakis (*Pithecia pithecia* and *Chiropotes satanas*) in Guri Lake, Venezuela, in: Norconk, M. A., Rosenberger, A. L. and Garber, P. A., eds., *Adaptive Radiations of Neotropical Primates*, 1996, Plenum Press, New York, pp. 403–426.

Ochiai, A. 1957, Zoogeographic studies on the soleoid fishes found in Japan and its neighbouring regions. *Bull. Japan. Soc. Sci. Fisheries* 22:526–530.

Packa, K. S., Henry, O., and Sabatier, D. 1999, The insectivorous-frugivorous diet of the golden-handed tamarin (*Saguinus midas midas*) in French Guiana. *Folia Primatol.* 70:1–7.

Peres, C. 1990, Effects of hunting on western Amazonian primate communities. *Biol. Conserv.* 54:47–59.

Peres, C. 1993, Structure and spatial organization of an Amazonian terra firme forest primate community. *J. Trop. Ecol.* 9:259–276.

Peres, C. A. 1997, Primate community structure at twenty western Amazonian flooded and unflooded forests. *J. Trop. Ecol.* 13:381–405.

Peres, C. A. 1999a, Effects of hunting and forest types on Amazonian communities, in: Fleagle, J. G., Janson, C. H. and Reed, K., eds., *Primate Communities*, Cambridge University Press, Cambridge, pp. 268–283.

Peres, C. A. 1999b, General guidelines for standardizing line-transect surveys of tropical forest primates. *Neotrop. Primates* 7:11–16.

Peres, C. A. and Dolman, P. M. 2000, Density compensation in neotropical primate communities: evidence from 56 hunted and nonhunted Amazonian forests of varying productivity. *Oecologia* 122:175–189.

Peres, C. A. and Janson, C. H. 1999, Species coexistence, distribution, and environmental determinants of neotropical primate richness: A community-level zoogeographic analysis, in: Fleagle, J. G., Janson, C. H. and Reed, K., eds., *Primate Communities*, Cambridge University Press, Cambridge, pp. 55–74.

Podolsky, R. D. 1990, Effects of mixed-species assocation on resource use by Saimiri sciureus and Cebus apella. *Am. J. Primatol.* 21:147–158.

Pontes, A. R. M. 1997, Habitat partitioning among primates in Maracá Island, Roraima, northern Brazilian Amazonia. *Int. J. Primatol.* 18:131–157.

Ries, L., Fletcher, R. J., Battin, J., and Sisk, T. D. 2004, Ecological responses to habitat edges: mechanisms, models, and variability explained. *Ann Rev. Ecol. Syst.* 35:491–522.

Robinson, J. G. and Bennett, E. L. eds., 2000, *Hunting for Sustainability in Tropical Forests*, Columbia University Press, New York.

Rodman, P. S. 1973, Synecology of Bornean primates I A test for interspecific interactions in spatial distribution of five species. *Am. J. Phys. Anthropol.* 38:655–660.

Rowe, N. 1996, *The Pictorial Guide to the Living Primates*, Pogonias Press, East Hampton, New York.

Schluter, D. 1984, A variance test for detecting species assocations, with some example applications. *Ecology* 65:998–1005.

Simmon, B. and Sabatier, D. 1996, Diets of some French Guianan Primates: Food comsumption and food choices. *Int. J. Primatol.* 17:661–694.

Sussman, R. W. and Phillips-Conroy, J. 1995, A survey of the distribution and diversity of the primates of Guyana. *Int. J. Primatol.* 16:761–792.

ter Steege, H. 1993a, *A Monograph of Wallaba, Mora, and Greenheart*, Wageningen, Stichting Tropenbos, Wageningen, The Netherlands.

ter Steege, H. 1993b, *Patterns in Tropical Rain Forest in Guyana*, Wageningen, Stichting Tropenbos, Wageningen, The Netherlands.

Terborgh, J. 1983, *Five New World Primates: A Study in Comparative Ecology*, Princeton, Princeton University Press, Princeton, NJ.

Terborgh, J. 1990, Mixed flocks and polyspecific associations: costs and benefits of mixed groups to birds and monkeys. *Am. J. Primatol.* 21:87–100.

Terborgh, J. and Andresen, E. 1998, The composition of Amazonian forests: patterns at local and regional scales. *J. Trop. Ecol.* 14:645–664.

Thiollay, J. M. 1994, Structure, density, and rarity in an Amazonian rainforest bird community. *J. Trop. Ecol.* 10:449–481.

Thorington, R. W., Jr. 1968, Observations of squirrel monkeys in a Colombian forest, in: Rosenblum, L. A. and Cooper, R. W., eds., *The Squirrel Monkey*, Academic Press, New York, pp. 69–85.

Tutin, C. E. G., Ham, R. M., White, L. J. T., and Harrison, M. J. S. 1997, The primate community of the Lope Reserve, Gabon: Diets, responses to fruit scarcity, and effects on biomass. *Am. J. Primatol.* 42:1–24.

van Andel, T. R. 2003, Floristic composition and diversity of three swamp forests in northwest Guyana. *Plant Ecol.* 167:293–317.

van Roosmalen, M. G. M. 1987, Diet, feeding behavior and social organization of the Guianan black spider monkey (*Ateles paniscus paniscus*). *Int. J. Primatol.* 8:421.

Vazquez, D. P. and Simberloff, D. 2002, Ecological specialization and susceptibility to disturbance: Conjectures and refutations. *Am. Nat.* 159:606–623.

Africa

The area covered by tropical forests in Africa inhabited by primates is greater that of any other major biogeographical region, extending from the lower border of the Sahara desert to the southern tip of the continent (e.g. Chapman et al., 1999). However, African forests are generally drier than those of Madagascar or Asia and include a wider range of forest types. In addition they show considerable fragmentation due to both historical factors and human activities (see Chapman et al., 1999; Kingdon, 1990). The major rivers show no consistent pattern of drainage with the Nile draining north into the Mediterranean Sea, the Niger and the Congo flowing west into the Atlantic Ocean and the Zambezi and the Limpopo flowing East into the Indian Ocean. Much of the biogeographic patterning of the continent is determined by these rivers as well as regional topographic features including the Atlas Mountains in the North, the Ethiopian Plateau in the Northeast and the rift valley extending from North to South through the eastern side of the continent.

Primate Biogeography, edited by Shawn M. Lehman and John G. Fleagle.
Springer, New York, 2006.

Primates have a long history in Africa and many clades seem to have originated there (see Fleagle and Gilbert, this volume; Rossie and Seiffert, this
volume). The most recent assessment of African Primate diversity recognizes
79 species in 21 genera (Grubb et al., 2003).There are two groups of strepsirrhines, the galagos which are endemic to Africa and the lorises that are found
in Africa and Asia. Both subfamilies of Old World monkeys, colobines and cercopithecines, are widespread in Africa and three genera of apes originated in
Africa—chimpanzees, gorillas and humans. As in other biogeographical regions,
some taxa (e.g. *Mandrillus*) are very localized while others (e.g. *Papio*) have
ranges that extend over much of the continent (Eely and Lawes 1999).

The papers in this section investigate three of the most widespread taxa
of African primates. In Chapter 5 "Contrasting Phylogeographic Histories of
Chimpanzees in Nigeria and Cameroon: A Multi-Locus Genetic Analysis,"
Katy Gonder and Todd Disotell examine the genetic support for the widely
accepted subdivision of chimpanzees (*Pan troglodytes*) into three subspecies
and the location of the most significant biogeographical barriers to gene flow
in chimpanzees. They find that studies of the HRV1 region of mitochondrial
DNA support a division of chimpanzees into a western group and an eastern
group separated by the Sanga River. They suggest that the eastern subspecies
of chimpanzees, *Pan troglodytes schweinfurthi* is a very recent radiation. However studies of STR genotypes indicate a more complex picture of chimpanzee
differentiation, especially west of the Sanga River.

In Chapter 6, Jason Kamilar examines the relationship between ecological diversity and morphological diversity among baboons, one of the most widespread
primates of Africa. He finds that ecological differences among baboon populations do not correlate with traditional subspecific groupings based on morphology. Rather ecological differences seem to follow a latitudinal cline similar to
the pattern Frost et al. (2003) found in analysis of cranial morphology. These
results support the view that baboon populations are best regarded as subspecies
rather than distinct species.

In Chapter 7, McGraw and Fleagle use cranial morphology to investigate
the relationship between mandrills and drills (genus *Mandrillus*) and their sister
taxon, mangabeys of the genus *Cercocebus*. They show that there is considerable
diversity of the facial morphology within the genus *Cercocebus*. Some species are
more similar to *Lophocebus* while others are more similar to mandrills and drills.
Within the genus *Cercocebus*, the Collared Mangabey, *Cercocebus torquatus* is
most similar to drills and mandrills in many aspects of craniofacial morphology.

These morphological results support Grubb's (1982) biogeographic hypothesis that *Cercocebus torquatus* is the most derived species of *Cercocebus* and suggests that the origin of drills and mandrills lies in this species of mangabey.

REFERENCES

Chapman, C. A., Gautier-Hion, A., Oates, J. F., and Onderdonk, D. A. 1999, African primate communities: Determinants of structure and threats to survival. In: Fleagle, J. G., Janson, C. H., and Reed, K. E., eds, *Primate Communities,* Cambridge University Press, Cambridge, pp. 1–37 .

Eely, H. A. C. and Lawes, M. J. 1999, Large–scale patterns of species richness and species range size in anthropoid primates In: Fleagle, J. G., Janson, C. H., and Reed, K. E eds, *Primate Communities*, Cambridge University Press, Cambridge, pp. 191–219.

Frost, S., Marcus, L. F., Bookstein F. L., Reddy D. P., and Delson, E. (2003). Cranial allometry, Phylogeography, and systematics of large-bodied papionins (Primates: Cercopithecinae) inferred from geometric morphometric analysis of landmark data. *Anatom. Rec.* 275A:1048–1072.

Grubb, P. 1982, Refuges and dispersal in the speciation of African mammals. In: Prance, G. T., ed., *Biological Diversification in the Tropics*, Columbia University Press, New York, pp. 537–553.

Grubb, P., Butynski, T. M., Oates, J. F., Bearder, S. K., Disotell, T. R., Groves, C. O., and Struhsaker, T. M. 2003, Assessment of the Diversity of African Primates. *Int. J. Primatol.* 24:1301–1357.

Kingdon, J. 1990, *Island Africa*, William Collins and Sons, London.

CHAPTER FIVE

Contrasting Phylogeographic Histories of Chimpanzees in Nigeria and Cameroon: A Multi-Locus Genetic Analysis

M. Katherine Gonder and Todd R. Disotell

ABSTRACT

As many as four geographically distinct subspecies of chimpanzees (*Pan troglodytes*) may persist across sub-Saharan Africa, but little is known about the geographic boundaries that delimit these populations. Genetic studies of the first hypervariable region (HVRI) of mitochondrial (mt)DNA of wild chimpanzees suggest that the Sanaga River may important in delineating chimpanzee populations in western Africa from those western

M. Katherine Gonder • Department of Anthropology, Graduate School and Hunter College of the City, University of New York, New York, NY 10021; Department of Biology, University of Maryland, College Park, MD 20742. **Todd R. Disotell** • Department of Anthropology, New York University, New York, NY 10003.

Primate Biogeography, edited by Shawn M. Lehman and John G. Fleagle.
Springer, New York, 2006.

equatorial Africa. However, the HVRI represents only a single realization of the evolutionary process. Here we present microsatellite, or Single Tandem Repeat (STR), genotypes of wild chimpanzees from Nigeria and Cameroon to complement and expand upon previous studies of chimpanzee mtDNA. We observed a different but compatible pattering of genetic diversity between the STR loci and the HVRI sequences. Generally, our analyses of these data suggest that the Sanaga River has played an important, but not exclusive, role in delimiting chimpanzees from western Africa from those in western equatorial Africa. The significance of the Niger River and of the Dahomey Gap in limiting chimpanzee distribution patterns remain equivocal. Additional multi-locus sampling of chimpanzees near these putative biogeographic boundaries may more clearly resolve the roles they have played in recent chimpanzee evolution.

Key Words: Chimpanzees, chimpanzee subspecies, phylogeography, population genetics, microsatellites, STR

INTRODUCTION

Despite the inevitably different histories and dispersal abilities of separate lineages, African rainforest primates cluster into discrete communities which reflect a shared historical relationship with the forests they inhabit (Haffer, 1969, 1982; Mayr and O'Hara, 1986). Nigeria and Cameroon are evolutionary hot spots, with high species diversity and high species endemism (Kingdon, 1989; Sayer *et al.*, 1992; Hacker *et al.*, 1998). Several disjunct Pleistocene refuges persisted in the vicinity of Nigeria and Cameroon (Figure 1a). In particular, the Niger delta harbored a small isolated refuge. In addition, the complex distribution patterns of forest taxa in eastern Nigeria and Gabon imply that two refuges may have existed in the area. The northern refuge may have covered the area east of the Cross River to Mount Cameroon to the Sanaga River. The southern refuge persisted south of the Sanaga River into Gabon (Grubb, 1982, 1990; Hamilton, 1988; Oates, 1988; Kingdon, 1989; Maley, 1991, 1996).

Geographic barriers (Figure 1b) have also influenced the dispersal patterns of the rainforest fauna in western Africa and western equatorial Africa (Grubb, 1982, 1990; Chapman, 1983; Oates, 1988). To the west, the Cavally, Sassandra, Bandama, and Comoé Rivers limit the distributions of several taxa. Further to the east the Volta River and the Dahomey Gap (a dry-forest zone covering present-day eastern Ghana, Togo, and Benin) may have been important in the history of many lineages. In the vicinity of Nigeria and Cameroon, the Niger River, Cross River, Sanaga River, and the Cameroon Highlands have

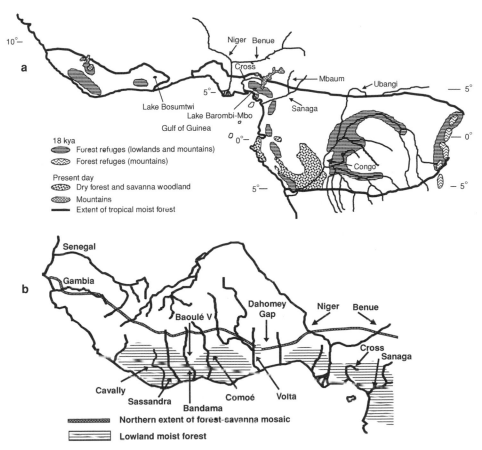

Figure 1. Biogeography of western Africa. (a) a consensus of probable locations and configuration of forest refuges after the last period of glacial maximum (Maley, 1996); (b) proposed dispersal barriers in western and western equatorial Africa (Oates, 1988).

been proposed to have influenced the distribution of many taxa (Booth, 1958a, 1958b; Schiøtz, 1967; Moreau, 1969; Robbins, 1978; Oates, 1988; Maley, 1996).

The complex biogeographic history of western Africa and western equatorial Africa is likely to have influenced the recent evolution of chimpanzees. This influence may be reflected by the nomenclature schemes that were proposed for chimpanzees during early explorations. Of the 38 named chimpanzee taxa, 18 were thought to inhabit the vicinity of Nigeria and Cameroon (Hill, 1967, 1969). These taxa have been recognized based largely on geographical criteria, without reference to the historical relationships that exist among different chimpanzee populations and without direct evidence supporting the geographical

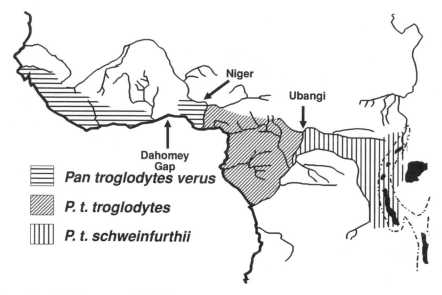

Figure 2.   Widely accepted distribution pattern of chimpanzees (*Pan troglodytes*) after Hill (1967, 1969) and Schwarz (1934).

distributions of these taxa. Currently, chimpanzees are widely recognized as belonging to a single species, *Pan troglodytes*, that has been further divided into three subspecies: west African *P. t. verus*, central African *P.t. troglodytes*, and east African *P. t. schweinfurthii* (Figure 2). The distribution patterns of chimpanzees have *P. t. verus* ranging from Senegal to western Nigeria, with this subspecies divided into two populations by the Dahomey Gap. The lower Niger River, in Nigeria, has been proposed to separate *P. t. verus* and *P. t. troglodytes*. *P. t. troglodytes* range from eastern Nigeria to the Ubangi River in the Democratic Republic of Congo (the former Zaire), and as far south as the Congo River. The Ubangi River is believed to separate *P. t. troglodytes* and *P.t. schweinfurthii*. *P. t. schweinfurthii* ranges from the Ubangi River and as far east as the western Rift Valley (Schwarz, 1934; Hill, 1967, 1969; Teleki, 1989; Groves, 1993).

One reason that it has been difficult to place chimpanzees into a larger historical context may be due to the high morphological variation and large overlap of phenotypic traits observed in this species, particularly in western equatorial Africa (Hill, 1967, 1969; Shea and Coolidge, 1988; Uchida, 1992; Groves *et al.*, 1993). Several genetic analyses of captive chimpanzees also have documented high levels of genetic diversity among chimpanzees across their range (Deinard and Kidd, 1996, 1999, 2000; Kaessmann *et al.*, 1999, 2001; Ebersberger *et al.*, 2002; Stone *et al.*, 2002). These studies complement several genetic surveys of wild chimpanzee populations that have greatly enhanced our understanding of the recent evolutionary history of this species (Morin, 1992; Goldberg, 1996; Gagneux, 1998; Gonder, 2000). These studies have relied on analyses of small selectively neutral loci such as, the first hypervariable region (HVRI) of (mt) mitochondrial DNA which does not undergo recombination and is maternally inherited (Gray *et al.*, 1999). The cumulative findings of these genetic surveys suggest that, in contrast to their widely accepted distribution pattern, the geographical boundaries proposed to delimit different chimpanzee lineages may need to be revised. In addition, these studies suggest that one key to understanding the phylogeographic history of chimpanzees may lie in Nigeria and Cameroon.

Figures 3a and 3b illustrate two geographical distributions of chimpanzees that correspond to the cumulative findings of these genetic surveys. Analyses of the HVRI locus most strongly support the geographical distribution shown in Figure 3a. Chimpanzees appear to be divided into two deeply divergent lineages (Gonder *et al.*, 1997; Gonder, 2000), who shared a common ancestor between 900–200 thousand years ago (kya) (Morin *et al.*, 1994; Goldberg and Ruvolo, 1997; Goldberg, 1998; Gagneux *et al.*, 1999; Stone *et al.*, 2002). The western African lineage is composed mostly of chimpanzees from Upper Guinea, Nigeria, and western Cameroon. The central African lineage is composed mostly of chimpanzees from southern Cameroon, western equatorial Africa, and eastern Africa. The ranges of these lineages converge in central Cameroon near the Sanaga River. This genetic differentiation has probably existed between chimpanzees for at least the past few hundred thousand years. The Sanaga River probably has played an important, if not exclusive, role in limiting gene flow between these lineages (Gonder *et al.*, 1997; Gonder, 2000; Gonder *et al.*, 2006).

The relationships between chimpanzees within these two lineages are more complex. Chimpanzees in western equatorial Africa and in eastern Africa have been widely recognized as belonging to separate subspecies (Figure 3b). Some studies have supported the proposition that they belong to two separate groups (Morin *et al.*, 1994; Stone *et al.*, 2002). Most studies, however, have not found genetic evidence that clearly separates chimpanzees in western equatorial Africa from those in eastern Africa (Goldberg and Ruvolo, 1997; Goldberg, 1998; Gagneux *et al.*, 1999, 2001; Gonder, 2000; Gonder *et al.*, 2006). In fact, chimpanzees in eastern Africa may share a very recent relationship with those in western equatorial Africa (Goldberg, 1998; Gonder, 2000; Gagneux *et al.*,

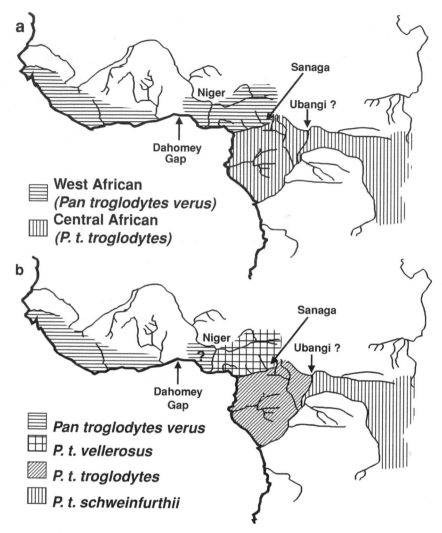

Figure 3. Two subspecies arrangements inferred from the pan-African HVRI genetic database (Gonder 2000; Gonder *et al.* 2006). The evidence most strongly supports the division shown in panel A in which chimpanzee populations are separated into two subspecies that reflects the ancient western and central Africa division of chimpanzees: *P. t. vellerosus* and *P. t. troglodytes*. Panel B incorporates evidence suggesting further subdivisions of chimpanzees may be warranted. In this distribution four chimpanzee subspecies would be recognized: Upper Guinea *P. t. verus*, western African *P. t. vellerosus*, western equatorial African *P. t. troglodytes*, and eastern African *P. t. schweinfurthii*.

2001; Gonder *et al.*, 2006). For example, chimpanzees in western equatorial Africa and those in eastern Africa do not form two monophyletic groups in HVRI gene trees that correspond to their geographical distributions. Moreover, two HVRI sequences of chimpanzees from Cameroon are nearly identical to HVRI sequences of chimpanzees residing in eastern Africa (Gagneux *et al.*, 2001; Gonder *et al.*, 2006).

Chimpanzees in western equatorial Africa possess the highest genetic diversity and shared a last common mitochondrial ancestor approximately 311 kya (Morin *et al.*, 1994; Kaessmann *et al.*, 1999; Stone *et al.*, 2002; Gonder *et al.*, 2006). Chimpanzees from western equatorial Africa may, therefore, be ancestral to all other chimpanzee populations (Morin *et al.*, 1994; Gagneux *et al.*, 1999, 2001; Gonder, 2000; Gonder *et al.*, 2006). In addition, populations from western equatorial Africa also may have persisted there throughout the Pleistocene (Gonder, 2000; Stone *et al.*, 2002). In contrast, chimpanzees in eastern Africa shared a last common mitochondrial ancestor approximately 117 kya. These chimpanzees may have colonized the forests of eastern Africa most recently. This colonization does not appear to have been influenced by any geographic barriers or by geographic distance. The expansion of chimpanzees into eastern African forests was probably rapid, following a population bottleneck that may have occurred during the last period of maximum glaciation (Stone *et al.*, 2002). Analyses of the HVRI sequence data also suggest that there has been extensive gene flow among chimpanzee populations in eastern Africa. Identical HVRI haplotypes of chimpanzees in eastern Africa are shared between populations separated by up to 700 km (Goldberg and Ruvolo, 1997; Goldberg, 1998).

There is some evidence to suggest that further subdivisions within the western African lineage may be warranted. Phylogenetic analyses of the HVRI data separates this lineage into an Upper Guinea chimpanzee group and a group composed of chimpanzees from eastern Nigeria and western Cameroon (Figure 3b). These lineages in western Africa appear to be more isolated from each other and are more isolated from chimpanzees in western equatorial African forests than those in eastern Africa. In addition, chimpanzees in Upper Guinea and those in eastern Nigeria and western Cameroon may have shared a last common mitochondrial ancestor 207 kya and 233 kya, respectively (Gonder *et al.*, 1997; Gonder, 2000; Gonder *et al.*, 2006).

Chimpanzees from Upper Guinea possess deeply divided HVRI haplotypes, but the clustering patterns of these haplotypes do not appear to be correlated with geographic isolation or with distance between populations (Gagneux, 1998; Gagneux et al., 1999). Gene flow appears to have been extensive among chimpanzee populations located in Upper Guinea. Identical mtDNA sequences are present in populations there separated by upto 1000 km (Gagneux et al., 1997b). Similarly, there also appears to have been considerable gene flow between chimpanzees in eastern Nigeria and western Cameroon, and some limited gene flow between the western African and central African chimpanzee lineages in the vicinity of the Sanaga River. The relationship between chimpanzees in western Nigeria and other western African chimpanzees is unresolved. Samples obtained from chimpanzees located in western Nigeria suggest that the Niger River or that the Dahomey Gap may further delimit chimpanzee populations in western Africa (Gonder, 2000; Gonder et al., 2006). However, chimpanzees are rare in western Nigeria (Agbelusi, 1994; Oates, 1996) and have not been sampled in western Ghana which prohibits further analysis on a finer scale.

This scenario of recent chimpanzee evolution is largely based on analyses of mtDNA. The mitochondrial genome is small (approximately 16,500 base pairs (bp) in length) and is 0.00006% the size of the total human genome. This genome is maternally inherited and does not undergo recombination. Therefore, its effective population size is one-fourth that of nuclear autosomes (Kocher and Wilson, 1991; Tamura and Nei, 1993; Stoneking, 1994; Gray et al., 1999). In some cases, this fact results in mtDNA having a higher probability of accurately reflecting the true species tree (Moore, 1995). However, several equally likely gene trees can often be inferred from small regions of mtDNA like the HVRI, particularly when the numbers of taxa analyzed is large (Maddison et al., 1992). Moreover, the maternal mode of inheritance of the mitochondrial genome combined with the female-biased dispersal pattern observed in chimpanzees (Goodall, 1986) makes it difficult to detect differences in male- and female-mediated gene flow using mtDNA alone.

Several properties of the HVRI locus can confound phylogenetic analyses. The locus is small, occupying only 2.5% of the mitochondrial genome. The HVRI locus mutates rapidly and is subject to homoplasy (Maddison et al., 1992). Mutations are not randomly distributed across the length of the locus making rate heterogeneity an important issue in calculating divergence estimates from this locus (Meyer et al., 1999; Stoneking, 2000). Nuclear mitochondrial pseudogenes occur frequently among hominoids (Collura and Stewart, 1995; Zischler et al., 1998). Some of these pseudogenes contain regions that are homologous to the HVRI (Mourier et al., 2001; Tourmen et al., 2002). However, it is unlikely that the chimpanzee HVRI data contains pseudogenes (Gonder *et al.*, 2006).

In addition, different phylogenetic histories of chimpanzees have been proposed in studies that have relied on samples obtained from captive chimpanzees and in those that included different loci. However, the contrasting patterns of genetic variation found in these studies have been attributed to predating the diversification of chimpanzee lineages (Kaessmann *et al.*, 1999), being evidence of selection (Wise *et al.*, 1997; de Groot *et al.*, 2002) or resulting from sex-specific differences in dispersal patterns (Wise *et al.*, 1997). These caveats render a species' history of chimpanzees based nearly exclusively on the HVRI locus tentative at best.

A complete analysis of a species' phylogeographic history should include multiple genetic systems, preferably including loci that are independent and that have different mutation rates (Hedrick, 1999; Avise, 2000; Hare, 2001). Microsatellites, or short tandem repeat (STR) loci, have been widely used to complement phylogeographic studies using mtDNA because they occur ubiquitously throughout animal genomes, mutate rapidly, are nearly selectively neutral, and are inherited in a Mendelian fashion. Moreover, analyses of only a few STR loci sometimes can resolve the relationships between populations (Jarne and Lagoda, 1996; Goldstein and Pollock, 1997). Therefore, we have genotyped DNA samples of wild chimpanzees from Nigeria and Cameroon at 10 STR loci to complement the HVRI sequence database. The purpose of our study was to examine the apportionment of genetic diversity found at 10 STR loci between chimpanzees in Nigeria and Cameroon. Specifically, our goal was to determine whether the Sanaga River has influenced the distribution of chimpanzees in Nigeria and Cameroon.

METHODS

Study Locations and Sampling

Study locations (Figure 4) were chosen to maximize sampling coverage across potential biogeographic barriers including: west and east of the Niger River, north and south of the Cross River, either side of the Cameroon Highlands following the border of Nigeria and Cameroon, north and south of the Sanaga River, east of the Mbaum River (Figure 4, map location 13), and into the Congo basin forest expanse near the Dja River (map location 16). All genetic

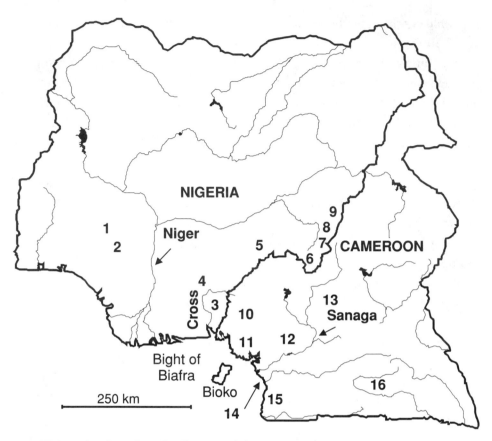

Figure 4. Sampling distribution of chimpanzees in Nigeria and Cameroon.

material was obtained noninvasively, from hairs shed in chimpanzee sleeping nests, from hairs plucked from captive individuals, and from samples excised from preserved skins following widely used sample collection protocols (Morin, 1992; Goldberg, 1996; Gagneux, 1998; Gonder, 2000).

Genetic Analysis

DNA extraction. Between 2 and 10 DNA extractions were performed on each nest hair collection. DNA was isolated from shed and plucked hair following well-established protocols used in other studies (Walsh *et al.*, 1991; Goldberg, 1996; Gonder, 2000). DNA from preserved skin samples was obtained using a PUREGENE™ DNA Extraction Kit (Gentra Systems, Inc.).

STR Genotyping. Several fluorescently labeled human STR primers were examined for variation among the chimpanzee samples, including di-, tri-, and

Table 1. STR loci used to genotype chimpanzees for this study

Marker	Locus	Size (bp)	Repeat unit
Mfd 3	APOA2 (D1)	120–162	AC
Mfd23	D16S265	78–104	AC
HumPla2a	HumPla2a	65–98	AAT
D4S1652	G00-228-893	114–162	ATCT
D7S1809	G00-364-834	202–242	AGGA
D9S303	G00-218-317	142–194	GATA × CAGA
D11S1984	G00-364-803	104–196	CAAA
D13S317	CHLC.GATA7G10.415	155–255	TATC × ATCT
D16S539	G00-228-700	152–168	ACAG × GATA
D20S470	G00-364-824	167–303	TTCC × CCTT × CCTT × TC

tetra-nucleotide repeats, in addition to loci with more complex repeat motifs (Table 1). PCR amplifications for all samples at these loci were performed using standard touchdown PCR protocols used in other studies (Don *et al.*, 1991; Rithidech *et al.*, 1997; Morin *et al.*, 1998; Gonder, 2000). Alleles for each sample were assessed using either the ABI™ 310 Genetic Analyzer or the ABI Prism® 377 DNA Sequencer.

Allelic drop out (the amplification of only one of two alleles at a heterozygous locus) and scoring of false alleles (the amplification of two alleles at a homozygous locus from small amounts of contaminants in DNA extracts) occur frequently in genotypes obtained from degraded DNA isolated from non-invasive sample sources (Gagneux *et al.*, 1997a; Goosens *et al.*, 1998; Constable *et al.*, 2001). Because these problems can greatly affect population diversity estimates (Balloux *et al.*, 2000; Gagneux, 2001), we attempted to verify allele sizes presented in this study by several independent PCR amplifications. Wherever possible, we closely followed the "multiple-tubes" protocol described elsewhere (Navidi *et al.*, 1992; Taberlet *et al.*, 1996). Unlike studies using feces and cheek swabs as sources of DNA (Constable *et al.*, 2001; Vigilant *et al.*, 2001) we could not follow the "multiple-tubes" procedure precisely due to the small amounts of DNA in each extract; and DNA from each hair sample could only be extracted once.

We amplified each STR locus via PCR from at least two DNA extractions from each nest hair collection. Allele sizes for all samples were confirmed by at least two, and up to ten, independent amplifications for each extract. Homozygous samples were confirmed by additional independent amplifications. Of all the genotypes included in this study, roughly 20%, 50%, and 30% were confirmed by 7–10, 3–6, and 2 separate PCR amplifications, respectively. Conflicting allele sizes for all samples were either re-processed until the ambiguities were resolved or were removed from the study. Allele sizes for each locus were determined by combining ABI^{TM} software analysis (GeneScan$^{®}$ and Genotyper$^{®}$ and discrimination by visual inspection using criteria described elsewhere (Gonder, 2000).

Statistical Analysis

Diversity statistics. All statistical analyses were performed using the computer program Arelquin (Schneider *et al.*, 2000). Several descriptive statistics were calculated for the STR database. Genetic diversity, observed heterozygosity, expected heterozygosity, and deviations from Hardy-Weinberg null expectations within each population were calculated based on the mean number of different alleles per locus. The probability of deviation from Hardy-Weinberg Equilibrium (HWE) was tested using a Markov chain method for each locus and each population (Guo and Thompson, 1992).

AMOVA. An Analysis of Molecular Variance (AMOVA) framework was applied to the STR dataset to examine the apportionment of genetic diversity and to evaluate the influence of putative biogeographic boundaries on chimpanzee distribution patterns. AMOVA partitions multilocus STR genetic variation into differences: between individuals at the allelic level [ϕ_{IS}], within populations [ϕ_{IT}], among populations within regions [ϕ_{SC}], and among regions [ϕ_{CT}]. In the last partition, regions are exclusive groups of populations that are defined by *a priori* criteria. This property of the AMOVA framework allows a specific geographic structure of the data to be superimposed on a sample population, thereby enabling direct tests of phylogeographic hypotheses about the population in study (Excoffier *et al.*, 1992; Michalakis and Excoffier, 1996). The significance of all observed ϕ-statistics were determined by comparing the observed value to random permutations of the data set (Schneider *et al.*, 2000).

Two statistical approaches, F-statistics (F_{ST}) and R-statistics (R_{ST}), were applied to the STR data using the AMOVA framework. F_{ST} were corrected for unequal sample sizes with modifications proposed by Weir and Cockerham (1984). In this model, F_{ST} values can range from -1 to 1. Departures from Hardy-Weinberg expectations are indicated by positive fixation indices which suggest that microevolutionary processes may be acting to produce population substructure. The true value of the test statistic is zero, but negative variance components can occur in the absence of genetic structure, when samples sizes are limited or when more alleles are shared between rather than within populations (Weir, 1996a, 1996b).

Size differences between alleles at STR loci are presumably due to repeat unit size differences that accumulate in a stepwise fashion where repeat units are lost or added during the mutation process. R_{ST} incorporates information about divergence time between relative allele sizes and assume a stepwise mutation model. R_{ST} tends to perform better than F_{ST} for STR data as long as these loci accumulate mutations following the stepwise mutation model (Slatkin, 1995; Balloux et al., 2000). However, homoplasy probably occurs frequently because these loci mutate rapidly from 10^{-2} to 10^{-5} mutations per generation depending on repeat motif and evolutionary constraints at a given locus (Valdes et al., 1993; Goldstein et al., 1995; Chakraborty et al., 1997; Schug et al., 1998).

Different potential regional geographic structures were tested using all 10 STR loci. In addition, each locus was analyzed separately by AMOVA for each potential phylogeographic division of the STR data using both the F_{ST} and R_{ST} models. Loci were analyzed separately to distinguish the relative effects of locus-specific evolutionary forces, to determine the relationship of each locus to overall allelic diversity, to evaluate violations of model assumptions, and to compare observed patterns of genetic structure (Bossart and Powell, 1998; Hare, 2001).

Individual genotypes were prepared for analysis by AMOVA by dividing populations on either side of the Sanaga River, the Cross River, and the Cameroon Highlands. For the Sanaga River division, samples were divided into two groups: eastern Nigeria and western Cameroon north of the Sanaga River (Figure 4, map locations 3–12), and Cameroon south and east of the Sanaga River (map locations 13–16). For the Cross River division also, samples were divided into two groups: west and north of the Cross River (map locations 4–9) and south and east of the Cross River (map locations 3, 10–16). For the Cameroon Highlands division, samples were divided along the Nigeria-Cameroon border: west of the Cameroon highlands (map locations 3–9) and east of the Cameroon Highlands (map locations 10–16). All samples from western Nigeria were excluded from these analyses due to the possible influence of the Niger River on chimpanzee distribution patterns. Sample sizes for populations in western Nigeria were too small to permit the test of the influence of the Niger River on chimpanzee population structure.

RESULTS

Genetic Diversity

One hundred and fifteen chimpanzees from Nigeria and Cameroon were geno-typed for up to 10 different STR loci listed in Table 1. STR diversity statistics are summarized in Appendices 1 and 2. The multilocus STR database revealed a high level of genetic diversity among chimpanzees in Nigeria and Cameroon. The 10 STR loci had an average of 8 alleles, with a range of 5 to 15 alleles per locus. With few exceptions, alleles for a given locus were not limited to any single population or group of populations. Heterozygosity estimates ranged widely from 17–96%, with an average heterozygosity estimate of 74%. In most cases, observed heterozygosity estimates did not differ significantly from expected levels of heterozygosity or from Hardy-Weinberg null expectations. There was also no detectable geographical pattern or trend for any locus for those populations that did deviate from expected Hardy-Weinberg proportions.

Apportionment of Genetic Diversity among Chimpanzees in Nigeria and Cameroon

Table 2 summarizes AMOVA results for different possible subdivisions of the multilocus data set. In each regional partition, the bulk of the genetic diversity was distributed among chimpanzees populations and among individuals. Negative ϕ_{CT} values were obtained for both the Cross River (F_{ST}, $\phi_{CT} = -0.012$; R_{ST}, $\phi_{CT} = -0.054$) and Cameroon Highlands (F_{ST}, $\phi_{CT} = -0.031$; R_{ST}, $\phi_{CT} = -0.034$) regional divisions. These regional divisions also produced very high variance components. In contrast, the Sanaga River division produced positive ϕ_{CT} values in both the F_{ST} and R_{ST} AMOVAs, and in the case of the R_{ST} AMOVA, a much lower variance component (F_{ST}, $\phi_{CT} = 0.026$; R_{ST}, $\phi_{CT} = 0.022$). In addition, the Sanaga River regional partition accounted for 2.24% (R_{ST}) to 2.53% (F_{ST}) of the observed regional variation. High variance indicates a loss of statistical power, while negative variance components can indicate an absence of population genetic structure or that more alleles are shared between rather than within populations (Cockerham, 1969; Weir, 1996a, 1996b). Consequently, the higher variances and negative ϕ_{CT} values

Table 2. AMOVA for hypothetical population structures of chimpanzees in Nigeria and Cameroon

		Conventional F_{ST} genotype frequencies				Standardized R_{ST} sum of squared differences			
		Variance	%Total	p^a	ϕ-statistics	Variance	%Total	p^a	ϕ-statistics
Sanaga River	$\sigma_a^{2\S}, \phi_{CT}$	0.040	2.53	>0.120	0.026	18.469	2.24	>0.316	0.022
	$\sigma_b^{2\S}, \phi_{SC}$	−0.12	0	>1.000	−0.076	−41.135	0	>0.797	−0.051
	$\sigma_c^{2\S}, \phi_{IT}$	−0.188	0	<1.00	−0.162	366.292	44.34	=0.000	0.416
	$\sigma_d^{2\S}, \phi_{IS}$	1.819	97.47	>0.100	−0.109	482.467	53.42	=0.000	0.432
Cross River	σ_a^2, ϕ_{CT}	−0.018	0	>0.771	−0.012	−43.327	0	>0.947	−0.054
	σ_b^2, ϕ_{SC}	−0.076	0	>0.100	−0.049	−12.478	0	>0.586	−0.015
	σ_c^2, ϕ_{IT}	−0.178	0	<1.000	−0.178	395.951	49.63	=0.000	0.426
	σ_d^2, ϕ_{IS}	1.802	100	>0.100	−0.110	457.629	50.37	=0.000	0.464
Cameroon Highlands	σ_a^2, ϕ_{CT}	−0.047	0	>0.918	−0.031	−27.275	0	>0.878	−0.034
	σ_b^2, ϕ_{SC}	−0.066	0	>1.000	−0.042	−21.187	0	>0.675	−0.025
	σ_c^2, ϕ_{IT}	−0.173	0	<1.000	−0.188	399.184	49.38	=0.000	0.437
	σ_d^2, ϕ_{IS}	1.802	100	>0.100	−0.106	457.629	50.62	=0.000	0.466

§ Denotes variance among regions (σ_a^2), among populations within regions (σ_b^2), within populations (σ_c^2), and within individuals in populations (σ_d^2).

^{a} Denotes the probability of obtaining an empirical value =, <, or > than the null distribution of random values.

associated with the Cross River and Cameroon Highlands divisions make them less likely candidates as phylogeographic boundaries for chimpanzees in Nigeria and Cameroon compared to the Sanaga River.

Variance among regions was not significantly different from chance by 16,000 random permutations of these data for any division of the STR data. However, in the F_{ST} AMOVA a regional division of the data at the Sanaga River approached statistical significance revealing only a 12% probability that sampling error could account for the genetic differentiation observed at the regional level. There was a significant lack of genetic structure among populations within regions, implying that a high level of gene flow has occurred between populations within regions. Among individuals within populations there was a significant excess of genetic diversity. At the gametic phase of individual genotypes (i.e., between alleles), there was a statistically significant lack of genetic differentiation. The lack of genetic differentiation at the allelic level may be due to problems with allelic dropout that persisted despite multiple attempts to obtain amplicons via PCR (Gagneux *et al.*, 1997a; Goosens *et al.*, 1998). In addition, negative variance components may indicate that more alleles are shared between rather than within populations (Weir, 1996a, 1996b). Consequently, the high proportion of shared alleles between individuals may largely explain the high number of negative variance components detected by AMOVA among these chimpanzees.

Each locus also was analyzed separately for each potential phylogeographic division of the STR data, using both the F_{ST} and R_{ST} models. Representative results of these single locus AMOVAs from the division of the data by the Sanaga River are shown Table 3. Seven of these loci reveal that most of the genetic differences among chimpanzees in Nigeria and Cameroon were attributable to differences between individual genotypes, and did not reveal significant population differentiation at the regional level. These loci included di-, tri-, tetra-, and more complex nucleotide repeat motifs. The D7, D9, and D11 loci all have tetra-nucleotide repeat motifs. These loci revealed significant regional subdivision at the Sanaga River when the data were permuted by AMOVA using the R_{ST} model. (D7S1809 $\phi_{CT} = 0.384$; D9S303, $\phi_{CT} = 0.153$; and D11S1984, $\phi_{CT} = 0.317$). The regional division of the data at the Sanaga River accounted for 15.32%–38.41% of the variation detected at these loci. Migration rates calculated from the these AMOVA values were high but were difficult to interpret due to the hypervariability of the STR data.

Table 3. AMOVA for 10 STR loci of chimpanzees in Nigeria and Cameroon*

Locus	Component	Conventional F_{ST} genotype frequencies				Standardized R_{ST} sum of squared differences			
		Variance	%Total	p^a	φ-statistics	Variance	%Total	p^a	φ-statistics
APOA2	$\sigma^2_{a^s}, \phi_{CT}$	0.000	0.07	>0.317	0.001	-0.630	0	>0.824	-0.045
	$\sigma^2_{b^s}, \phi_{SC}$	0.015	3.78	>0.008	0.037	2.148	15.29	¿0.029	0.146
	$\sigma^2_{c^s}, \phi_{TT}$	-0.008	0	<0.430	0.018	5.275	37.56	=0.000	0.484
	$\sigma^2_{d^s}, \phi_{IS}$	0.392	96.22	>0.670	-0.021	7.250	47.15	=0.000	0.421
D4S1652	σ^2_a, ϕ_{CT}	0.004	0.82	>0.151	0.008	-4.467	0	>0.927	-0.015
	σ^2_b, ϕ_{SC}	0.006	1.36	>0.250	0.013	-13.172	0	>0.688	-0.043
	σ^2_c, ϕ_{TT}	0.129	28.84	=0.000	0.310	95.790	31.77	<0.003	0.259
	σ^2_d, ϕ_{IS}	0.309	68.98	=0.000	0.295	223.345	68.23	>0.001	0.300
D7S1809†	σ^2_a, ϕ_{CT}	0.010	2.32	>0.192	0.023	65.265	38.41	>0.003	0.384
	σ^2_b, ϕ_{SC}	0.041	9.68	>0.012	0.010	6.357	3.74	>0.253	0.061
	σ^2_c, ϕ_{TT}	0.130	30.93	=0.000	0.429	64.599	37.99	=0.000	0.801
	σ^2_d, ϕ_{IS}	0.239	57.07	=0.000	0.351	33.739	19.86	=0.000	0.657
D9S303†	σ^2_a, ϕ_{CT}	0.005	1.14	>0.243	0.011	26.100	15.32	>0.027	0.153
	σ^2_b, ϕ_{SC}	0.005	1.17	>0.245	0.012	-16.614	0	>0.910	-0.115
	σ^2_c, ϕ_{TT}	0.167	36.55	=0.000	0.389	118.263	59.64	=0.000	0.749
	σ^2_d, ϕ_{IS}	0.277	61.15	=0.000	0.374	42.667	25.04	=0.000	0.735
D11S1984†	σ^2_a, ϕ_{CT}	0.020	4.33	>0.110	0.043	130.679	31.76	>0.040	0.317
	σ^2_b, ϕ_{SC}	0.037	7.98	>0.087	0.083	-44.879	0	>0.711	-0.160
	σ^2_c, ϕ_{TT}	0.289	63.08	=0.000	0.754	224.548	43.65	<0.000	0.754
	σ^2_d, ϕ_{IS}	0.113	24.61	=0.000	0.719	101.162	24.59	>0.015	0.689
HUMPLA2A	σ^2_a, ϕ_{CT}	-0.002	0	>0.580	0.007	-0.641	0	>0.692	-0.025
	σ^2_b, ϕ_{SC}	0.007	2.17	>0.217	0.023	-0.128	0.76	>0.397	-0.007
	σ^2_c, ϕ_{TT}	0.177	52.99	=0.000	0.552	21.84	70.60	=0.000	0.734
	σ^2_d, ϕ_{IS}	0.149	44.84	=0.000	0.545	9.093	28.64	=0.000	0.718

(Continued)

Table 3. (*Continued*)

	Conventional F_{ST} genotype frequencies				Standardized R_{ST} sum of squared differences			
	Variance	%Total	p^a	ϕ-statistics	Variance	%Total	p^a	ϕ-statistics
D13S317								
σ_a^2, ϕ_{CT}	−0.003	0	>0.757	−0.006	−0.003	0	>0.754	−0.005
σ_b^2, ϕ_{SC}	−0.003	0	>0.570	−0.006	−0.003	0	>0.570	−0.006
σ_c^2, ϕ_{TT}	0.192	42.57	=0.000	0.414	0.198	42.57	=0.000	0.414
σ_d^2, ϕ_{TS}	0.264	57.84	=0.000	0.421	0.264	57.84	=0.000	0.421
D16S265								
σ_a^2, ϕ_{CT}	−0.009	0	>0.959	−0.023	−5.497	0	>0.901	−0.038
σ_b^2, ϕ_{SC}	0.027	6.66	=0.000	0.065	15.711	10.85	>0.008	0.105
σ_c^2, ϕ_{TT}	−0.067	0	<0.976	−0.123	−36.021	0	<0.908	−0.178
σ_d^2, ϕ_{TS}	0.452	93.34	>0.992	−0.174	170.581	89.15	>0.978	−0.267
D16S539								
σ_a^2, ϕ_{CT}	0.000	0.13	>0.338	0.001	−0.17	0	>0.822	−0.006
σ_b^2, ϕ_{SC}	−0.003	0	>0.700	−0.010	−0.366	0	>0.676	−0.013
σ_c^2, ϕ_{TT}	0.009	2.79	<0.327	0.019	−3.554	0	<0.920	−0.145
σ_d^2, ϕ_{TS}	0.299	97.08	>0.288	0.028	32.235	100	>0.891	−0.124
D20S470								
σ_a^2, ϕ_{CT}	−0.002	0	>0.564	−0.004	−23.370	0	>0.831	−0.026
σ_b^2, ϕ_{SC}	0.009	1.98	>0.080	0.020	11.773	1.32	>0.353	0.013
σ_c^2, ϕ_{TT}	0.113	24.54	=0.000	0.262	671.960	72.57	=0.000	0.739
σ_d^2, ϕ_{TS}	0.340	73.48	=0.000	0.249	233.340	26.11	=0.000	0.742

* Populations were divided into two regions, (1) eastern Nigeria and Cameroon north and (2) Cameroon south of the Sanaga River.
§ Denotes variance among regions (σ_a^2), among populations within regions (σ_b^2), within populations (σ_c^2), and within individuals in populations (σ_d^2).
† Denotes significant differentiation at the regional level.

DISCUSSION

Comparison of the HVRI and STR Databases

The apportionment of genetic diversity among chimpanzees and their inferred phylogeographic histories based on the HVRI sequence data and the STR loci are compatible. However, there are a number of striking similarities and contrasts in the patterns of variation found between the HVRI data and the STR data among chimpanzees in Nigeria and Cameroon. Both data sets document a high level of genetic diversity among chimpanzees in the area. The HVRI data suggest that chimpanzee populations are divided at the Sanaga River into two deeply divergent lineages: a western African lineage and a central African lineage (Gonder, 2000; Gonder et al., 2006).

In contrast, the multilocus STR data only weakly supports a division of chimpanzees at the Sanaga River. In addition, the bulk of the genetic differences were attributable to differences between sampling populations and between individual chimpanzees. However, the multilocus STR analyses suggest that a division of chimpanzees on either side of the Sanaga River is more likely than for other putative biogeographic boundaries in Nigeria and Cameroon. Separate AMOVA analyses of each STR locus provide more conclusive evidence that the Sanaga River may have been important in recent chimpanzee evolution. Three loci revealed significant differentiation between chimpanzees limited to different sides of the Sanaga River. Analyses of the STR database imply, therefore, that chimpanzees form a relatively panmictic population in Nigeria and Cameroon with only a weak differentiation at the Sanaga River.

Limitations of the STR Database

There are several reasons why these discrepancies exist between the HVRI and STR data. Several of the DNA extracts failed to produce unambiguous allele sizes at six of the ten loci, despite repeated attempts to obtain amplicons via PCR. In addition, the genotypes of some chimpanzees may be inaccurate because problems with allelic dropout and false alleles may have persisted despite the fact that we extensively repeated PCR amplifications to verify allele sizes. Both of these problems can greatly affect genetic diversity estimates (Balloux et al., 2000). For example, the negative ϕ-statistics calculated by AMOVA from the STR may reflect problems with sample size and with analytical strategy, and not high levels of gene flow or recent historical association (Weir, 1996a, 1996b).

More generally, STR loci may be of limited value in studies addressing phylogeographic questions over a large time scale. There appear to be constraints on maximum allele size, on the longevity of mutational processes, and on the purity of repeat series between lineages (Garza *et al.*, 1995; Goldstein *et al.*, 1995; Crouau-Roy *et al.*, 1996; Garza and Freimer, 1996; Goldstein and Pollock, 1997). These problems increase the likelihood that there are STR alleles in this dataset that are identical in size but not in descent. Homoplasy in STR allele sizes can affect the accuracy of distance measures that assume time linearity, such as R_{ST} (Balloux *et al.*, 2000). Further, F_{ST} and R_{ST} values can be greatly reduced if homoplasy occurs frequently in STR data (Goldstein and Pollock, 1997; Hedrick, 1999).

We chose a wide range of STR loci for study which presumably are subject to different mutational constraints that may characterize the relationships between populations on different time scales. The three tetra-nucleotide repeat STR loci used in this study that revealed a significant division of chimpanzees at the Sanaga River mutate 1.5 to 2.5 times more slowly than di- or tri- nucleotide repeats STR loci and are presumably less subject to homoplasy (Chakraborty *et al.*, 1997). Since the genetic differentiation between chimpanzees on either side of the Sanaga River may have existed for at least the last 200 ky, homoplasy at the rapidly evolving di- and tri- nucleotide STR loci may make these populations appear to have a more recent historical relationship than they actually share (Hedrick, 1999; Balloux *et al.*, 2000). Similar findings have been reported in a variety of organisms (Blanquer-Maumont and Crouau-Roy, 1995; Garza *et al.*, 1995; Garza and Freimer, 1996; Doyle *et al.*, 1998; Schug *et al.*, 1998; Taylor *et al.*, 1999). The tetra-nucleotide repeat STRs show the most promise for more clearly delineating the relationships between these populations. However, up to 19 tetra-nucleotide repeat STR loci may be necessary to fully resolve the recent evolutionary history of chimpanzees in Nigeria and Cameroon (Goldstein *et al.*, 1999).

Chimpanzee Phylogeography in Nigeria and Cameroon

Recent chimpanzee evolution in Nigeria and Cameroon has been very complex. The overall picture that emerges from the HVRI data and the STR data is that the Sanaga River appears to have been very important in separating chimpanzee populations in western Africa from those in western equatorial Africa for at least the last 200 ky. Surprisingly, little is known about the history of the Sanaga River. However, the area surrounding the Sanaga is geographically complex and has had a rich forest history. The area contains a number of important biogeographic barriers (Oates, 1988) and three Pleistocene refuges that likely persisted in Nigeria and Cameroon (Maley, 1996).

In particular, the area between the Cross River in southeastern Nigeria and the Sanaga River appears to have exerted considerable influence in limiting the distributions of other primates in Nigeria and Cameroon. *Cercopithecus sclateri*, for example, occurs from the east bank of the Niger River and the eastern portions of the Niger Delta to the west bank of the Cross River in southeastern Nigeria (Oates *et al.*, 1992). *C. erythrotis camerunensis* and *Mandrillus leucophaeus* extend only to the east bank of the Cross. Other primates limited to the east of the Cross River, the Cameroon highlands and the island of Bioko include: *Cercopithecus preussi*, *C. pogonias pogonias,* two *Procolobus badius* subspecies, and *Cercocebus albigena* (Grubb, 1982, 1990; Oates, 1988, 1996; Kingdon, 1989). Distributional data for several primates suggest that the Sanaga River is also an important boundary between *Arctocebus calabarensis/A. aureus, Euoticus pallidus/E. elegantulus, Cercopithecus erythrotis/C. cephus, C. nictitans martini/C. n. nictitans, C. pogonias pogonias/C. p. grayi,* and *Mandrillus leucophaeus/M. sphinx.* The Sanaga also limits the distribution of *Colobus satanas* (Grubb, 1982, 1990; Oates, 1988, 1996; Kingdon, 1989). Thus the Sanaga or a combination of geographic boundaries as well as variation in forest size and composition during the Pleistocene in the vicinity of the Sanaga have influenced chimpanzee distribution patterns in the area.

The HVRI data also show that the Niger River or the Dahomey Gap played an important role in separating chimpanzee populations in Upper Guinea from those in eastern Nigeria. The Niger and the Gap was important in limiting the distribution of other taxa in the region. The Niger River, along with its large delta, appears to have been important in the evolution of other primates in the region. The Niger limits *Arctocebus calabarensis, Euoticus pallidus,* and *Galago alleni* (Booth, 1958a, 1958b; Grubb, 1982, 1990; Oates, 1988; Groves *et al.*, 1993).

The Dahomey Gap is presently about 400 km wide but during periods of maximum glaciation it may have been as much as 1400 km wide, extending from western Ivory Coast to the Cameroon Highlands in the east (Maley, 1996) which may have isolated chimpanzees in Upper Guinea from those further east. The Gap also limits other animals in Upper Guinea from those in western equatorial Africa (Booth, 1958a, 1958b; Schiøtz, 1967; Moreau, 1969). Areas furthest from the Gap (Liberia and Gabon) contain far more endemic groups than regions closer to the Gap. In addition, Booth (1958a, 1958b) observed that several species of Primates, Sciuromorpha, Artiodactyla, and Hyracoidea occurred frequently in forested areas west and east of the Dahomey Gap but were greatly reduced inside the Gap. In some cases, animals especially dependent upon lowland rainforest, such as some *Cercopithecus* species, *Procolobus*, and *Cercocebus*, are limited in their distribution by the Gap (Oates, 1988).

The Niger River and the Dahomey Gap probably have played a less important role in recent chimpanzee evolution than the Sanaga River. Unfortunately, we cannot directly address the significance of the Niger or the Gap in delimiting chimpanzee populations in western Africa with the genetic data that are currently available. Sample sizes of chimpanzees from western Nigeria were too small to be included in this study. In addition, chimpanzee populations from western Ghana have not been sampled. Additional samples of chimpanzees from western Nigeria and from western Ghana are urgently needed to directly address the roles of the Dahomey Gap and of the Niger River in recent chimpanzee evolution in western Africa.

SUMMARY

Several genetic surveys of wild chimpanzees (*Pan troglodytes*) have revealed that in contrast to the widely accepted distribution pattern proposed for this species, the geographical boundaries proposed to delimit chimpanzee subspecies may need to be revised. However, these studies have relied on sequences of the first hypervariable region (HVRI) of mitochondrial (mt) DNA, which is a small locus that is subject to homoplasy. This study presents microsatellite, or short tandem repeat (STR), genotypes of wild chimpanzees to complement previous studies. This database includes a geographically comprehensive genetic survey of chimpanzees ($n = 115$) from Nigeria and Cameroon. The patterning of genetic diversity at these STR loci was compatible with but different from the patterns found in the HVRI sequence database. Analyses of the HVRI database show that Sanaga River in central Cameroon is an important phylogeographic boundary separating chimpanzees in western Africa from those in western equatorial Africa. In contrast, the STR database only weakly supported a division of chimpanzees at the Sanaga. Differences in the patterning of genetic diversity in the HVRI sequences and the STR genotypes may be attributed to high homoplasy, locus-specific evolutionary forces or poor performance of test statistics when model assumptions are violated. These findings highlight the importance of including genetic data from multiple loci in reconstructing the phylogeographic history of chimpanzees. The HVRI database suggests that the Niger River or the Dahomey Gap may further delimit chimpanzee populations in western Africa. Additional samples of chimpanzees in western Nigeria and western Ghana are needed to fully address the significance of the Niger River and of the Dahomey Gap in limiting chimpanzee distributions in western Africa.

AKNOWLEDGMENTS

John Oates made valuable comments on early drafts of this manuscript and provided significant assistance during data collection and analysis. We would like to thank the governments of Nigeria and Cameroon, the Nigerian Conservation Foundation, the *Pandrillus* organization in Nigeria and Cameroon, *Pronatura International*, the Wildlife Conservation Society and the World Wide Fund for Nature for their support during sample collection in Nigeria and Cameroon by MKG. This research was supported by the L.S.B. Leakey Foundation, Primate Conservation, Inc., the National Science Foundation (Dissertation Improvement Award, Graduate Research Fellowship and NYCEP Research Training Grant) and the Wenner-Gren Foundation. All samples were exported from Africa under CITES exportation permits (Nigeria: FEPA/LSN/68/T/39; Cameroon: 0172/PE/MINEF/DFAP/SL/SLP), and were imported under U.S. CITES (US810330) and USDA (97-418-2) importation permits.

APPENDIX 1. Allele frequencies for 10 STR Loci of Chimpanzees in Nigeria and Cameroon

APOA2

Locus	120	124	126	128	130	132	134	136	138	140	142	148	150	154	162
WN	0.07				0.21	0.29	0.14	0.09		0.07			0.07	0.07	0.07
EN + WC		0.02	0.03	0.06	0.25	0.35	0.12		0.03	0.03	0.01	0.01			
SC	0.03	0.01	0.07	0.07	0.14	0.30	0.16	0.08	0.03	0.11	0.01	0.01			

D4S1652

Locus	114	118	122	126	130	134	138	142	146	148	150	154	158	162
WN		0.30	0.20			0.10	0.10	0.20	0.20					
EN + WC	0.04		0.21	0.15	0.07	0.16	0.16	0.06	0.03	0.01	0.03		0.01	
SC	0.04	0.17	0.31	0.02	0.10	0.05	0.05	0.07	0.07		0.02			

D7S1809

Locus	202	206	210	214	218	222	226	230	234	238	242
WN	0.17	0.33		0.25	0.25						
EN + WC	0.14	0.05	0.45	0.14	0.09	0.03	0.03	0.03	0.02	0.02	
SC		0.03	0.26	0.06	0.12	0.03	0.12	0.06		0.12	0.21

D9S303

Locus	142	146	150	154	158	162	166	168	170	174	176	178	180	182	184	186	190	194
WN									0.50	0.50								
EN + WC		0.05	0.08	0.08		0.18	0.08		0.13	0.10		0.08		0.05		0.08	0.13	0.08
SC	0.02	0.10	0.06	0.04		0.12	0.08	0.12	0.32	0.06	0.10		0.27		0.06			

D11S1984

Locus	104	124	128	144	164	168	176	180	184	188	192	196
WN					0.33					0.33	0.33	
EN + WC	0.06	0.03	0.03	0.03	0.16	0.28	0.06	0.06	0.06	0.16	0.06	
SC						0.07	0.10	0.27	0.33	0.13	0.03	0.07

(Continued)

APPENDIX 1. (*Continued*)

Alleles

HUMPLA2A

Locus	65	77	80	83	86	89	92	95	98
WN			0.17		0.33	0.50			
EN + WC	0.02	0.01	0.09	0.06	0.55	0.07	0.07	0.11	0.01
SC		0.10		0.06	0.58	0.10		0.06	0.12

D13S317

Locus	155	167	171	175	179	183	187	191	195	199	203	215	223	227	247	251	255
WN					0.50	0.50											
EN + WC	0.06	0.06	0.05	0.10	0.05	0.06	0.15	0.21	0.08	0.05		0.05	0.02	0.03	0.03		
SC		0.05		0.12	0.19	0.14	0.12	0.21	0.02	0.02	0.02				0.02	0.02	0.05

D16S265

Locus	78	80	82	84	90	92	94	96	100	102	104	106	108	110	112	114
WN	0.10	0.10		0.20				0.10		0.10		0.20	0.10	0.10		
EN + WC		0.18	0.04	0.04	0.01	0.03					0.03	0.22	0.29	0.06	0.03	0.06
SC	0.02	0.17	0.04	0.02			0.02		0.01	0.02		0.27	0.35	0.04		0.04

D16S529

Locus	152	156	160	164	168
WN	0.61	0.10	0.05	0.24	
EN + WC	0.60	0.11	0.09	0.21	
SC	0.63	0.09	0.01	0.27	

D20S470

Locus	167	203	207	211	215	227	243	247	251	255	259	263	267	271	275	279	283	287	291	295	303
WN	0.13	0.01																			
EN + WC	0.10	0.01	0.01		0.05	0.05	0.03	0.05	0.01	0.01	0.10	0.14	0.29	0.14	0.14	0.14	0.14	0.10	0.08	0.04	0.03
SC	0.10	0.01	0.01	0.06		0.07		0.04	0.09		0.11	0.07	0.21	0.08		0.01		0.01			

* Samples were pooled into the following partitions: (1) western Nigeria (WN, map locations 1–2) and samples from preserved skins that were collected in western Nigeria; (2) eastern Nigeria and western Cameroon north of the Sanaga River (EN + WC, Figure 4, map locations 3–12), and (3) Cameroon south and east of the Sanaga River (SC, map locations 13–16).

APPENDIX 2. Diversity statistics for 10 STR Loci of Chimpanzees in Nigeria and Cameroon

Map Loc. (Figure 4)		APOA2	D4S1652	D7S1809	D9S303	D11S1984	HUMPLA2A	D13S317	D16S265	D16S539	D20S470
1*	Alleles	6	5	4	2	3	3	No data	7	3	5
	Genes	12	10	10	8	6	6		8	10	6
	$\hat{h}$	0.85	0.87	0.78	0.57	0.80	0.73		0.96	0.51	0.93
	HWE p	0.39	0.06	0.01	0.08	0.07	0.19		1.00	0.33	1.00
3–4	Alleles	5	6	2	5	No data	2	5	No data	4	5
	Genes	20	20	12	10		6	8		20	14
	$\hat{h}$	0.65	0.84	0.17	0.87		0.53	0.89		0.54	0.73
	HWE p	0.49	0.19	1.00	0.00		0.20	0.01		0.14	0.01
5	Alleles	4	9	5	7	6	3	4	8	4	12
	Genes	20	20	12	14	14	16	12	20	22	16
	$\hat{h}$	0.75	0.89	0.80	0.91	0.87	0.43	0.77	0.79	0.71	0.96
	HWE p	0.42	0.01	0.02	0.00	0.00	0.01	0.20	0.01	0.50	0.36
6	Alleles	6	7	No data	4	No data	2	No data	3	4	4
	Genes	10	8		4		4		4	8	4
	$\hat{h}$	0.88	0.96		1.00		0.50		0.83	0.64	1.00
	HWE p	0.63	0.17		1.00		1.00		1.00	0.44	1.00
7–9	Alleles	6	9	5	5	5	4	7	4	3	6
	Genes	18	14	10	8	10	12	16	10	18	10
	$\hat{h}$	0.80	0.93	0.82	0.86	0.67	0.74	0.90	0.78	0.52	0.89
	HWE p	0.05	0.52	0.35	0.65	0.63	0.27	0.00	1.00	1.00	0.13
10	Alleles	2	3	3	No data	5	3	No data	4	5	
	Genes	10	6	4		10	4		8	8	
	$\hat{h}$	0.36	0.80	0.83		0.84	0.83		0.64	0.86	
	HWE p	1.00	0.46	0.34		0.35	0.33		0.43	0.08	

(*Continued*)

APPENDIX 2. (*Continued*)

Map Loc. (Figure 4)		APOA2	D4S1652	D7S1809	D9S303	D11S1984	HUMPLA2A	D13S317	D16S265	D16S539	D20S470
11	Alleles	6	No data	5	No data	No data	6	6	5	5	8
	Genes	7		8			18	14	18	22	14
	ĥ	0.93		0.86			0.75	0.85	0.78	0.70	0.87
	HWE p	0.05		0.84			0.83	0.09	0.63	0.83	0.87
12	Alleles	7	No data	5	No data		5	6	7	5	7
	Genes	12		10			16	8	16	18	12
	ĥ	0.91		0.82			0.53	0.93	0.83	0.72	0.93
	HWE p	1.00		0.07			0.01	0.02	0.25	0.03	0.14
13	Alleles	6	3	5	5	No data	3	9	6	5	11
	Genes	18	10	8	10		14	14	18	28	22
	ĥ	0.83	0.60	0.86	0.84		0.67	0.93	0.81	0.72	0.88
	HWE p	0.12	1.00	0.09	0.36		0.01	0.00	0.30	0.03	0.64
14	Alleles	6	5	4	8	6	4	7	3	3	8
	Genes	18	8	12	14	10	20	10	14	18	16
	ĥ	0.77	0.86	0.68	0.90	0.89	0.50	0.91	0.69	0.50	0.88
	HWE p	0.97	0.09	0.01	0.21	0.01	0.03	0.04	0.48	0.63	0.10
15	Alleles	6	10	4	5	4	3	4	5	4	8
	Genes	22	18	8	12	8	10	12	6	20	14
	ĥ	0.79	0.92	0.77	0.72	0.82	0.67	0.77	0.93	0.60	0.92
	HWE p	0.00	0.00	0.78	1.00	0.03	0.05	0.09	0.21	0.05	0.00
16	Alleles	5	3	3	8	3	3	4	5	4	9
	Genes	18	6	6	14	10	8	6	14	22	20
	ĥ	0.78	0.73	0.73	0.82	0.71	0.71	0.87	0.73	0.46	0.92
	HWE p	0.27	0.20	1	0.00	0.01	0.03	0.07	0.33	0.40	0.01

*Includes samples from map location 1 and preserved skins obtained in western Nigeria.
ĥ, Denotes heterozygosity measured as the mean number of alleles per locus.

REFERENCES

Agbelusi, E. A. 1994, Wildlife conservation in Ondo State. *The Nigerian Field* 59:73–83.

Avise, J. C. 2000, *Phylogeography: The History and Formation of Species*, Harvard University Press, Cambridge, MA.

Balloux, F., Brunner, H., Lugon-Moulin, N., Hausser, J., and Goudet, J. 2000, Microsatellites can be misleading: An empirical and simulation study. *Evolution* 54:1414–1422.

Blanquer-Maumont, A. and Crouau-Roy, B. 1995, Polymorphism, monomorphism, and sequences in conserved microsatellites in primate species. *J. Mol. Evol.* 41:492–497.

Booth, A. H. 1958a, The Niger, the Volta, and the Dahomey Gap as geographic barriers. *Evolution* 12:48–62.

Booth, A. H. 1958b, The zoogeography of West African primates: A review. *Bulletin de l'I.F.A.N.* 20A:587–622.

Bossart, J. L. and Powell, D. P. 1998, Genetic estimates of population structure and gene flow: Limitations, lessons, and new directions. *Trends Ecol. Evol.* 13:202–206.

Chakraborty, R., Kimmel, M., Stivers, D. N., Davison, L. J., and Deka, R. 1997, Relative mutation rates at di-, tri-, and tetranucleotide microstallite loci. *Proc. Nat. Acad. Sci. USA* 94:1041–1046.

Chapman, C. A. 1983, Speciation of tropical rainforest primates of Africa: Insular biogeography. *Afr. J. Ecol.* 21:297–308.

Cockerham, C. C. 1969, Variance of gene frequencies. *Evolution* 23:72–84.

Collura, R. V. and Stewart, C. B. 1995, Insertions and duplications of mtDNA in the nuclear genome of old world monkeys and hominoids. *Nature* 378:485–489.

Constable, J. L., Ashley, M. V., Goodall, J., and Pusey, A. 2001, Noninvasive paternity assignment in Gombe chimpanzees. *Mol. Ecol.* 10:1279–1300.

Crouau-Roy, B., Service, S., Slatkin, M., and and Freimer, N. 1996, A fine-scale comparison of the human and chimpanzee genomes–linkage, linkage disequilibrium and sequence-analysis. *Hum. Mol. Genet.* 5:1131–1137.

de Groot, N. G., Otting, N., Doxiadis, G. G., Balla-Jhagjhoorsingh, S. S., Heeney, J. L., van Rood, J. J., Gagneux, P., and Bontrop, R. E. 2002, Evidence for an ancient selective sweep in the MHC class I gene repertoire of chimpanzees. *Proc. Natl. Acad. Sci. USA* 99:11748–11753.

Deinard, A. and Kidd, K. 1996, Indentifying "wild" levels of genetic diversity within Pan paniscus and Pan troglodytes. *16th Congress of the International Primatological Society* pp. 784.

Deinard, A., and Kidd, K. 1999, Evolution of a HOXB6 intergenic region within the great apes and humans. *J. Hum. Evol.* 36:687–703.

Deinard, A. S. and Kidd, K. 2000, Identifying conservation units within captive chimpanzee populations. *Am. J. Phys. Anthropol.* 111:25–44.

Don, R. H., Cox, P. T., Wainwright, B. J., Baker, K., and Mattick, J. S. 1991, 'Touchdown' PCR to circumvent spurious priming during gene amplification. *Nucleic Acids Res.* 19:4008.

Doyle, J. J., Morgante, M., Tingey, S. V., and Powell, W. 1998, Size homoplasy in chloroplast microsatellites of wild perennial relatives of soybean (Glycine subgenus Glycine). *Mol. Biol. Evol.* 15:215–218.

Ebersberger, I., Metzler, D., Schwarz, C., and Paabo, S. 2002, Genomewide comparison of DNA sequences between humans and chimpanzees. *Am. J. Hum. Genet.* 70:1490–1497.

Excoffier, L., Smouse, P. E., and Quattro, J. M. 1992, Analysis of molecular variance inferred from metric distances among DNA haplotypes: Applications to human mitochondrial DNA restriction data. *Genetics* 131:479–491.

Gagneux, P. 1998, *Population Genetics of West African Chimpanzees* (*Pan troglodytes verus*), University of Basel, Basel, SW.

Gagneux, P., Boesch, C., and Woodruff, D. S. 1997a, Microsatellite scoring errors associated with noninvasive genotyping based on nuclear DNA amplified from shed hair. *Mol. Ecol.* 6:861–868.

Gagneux, P., Woodruff, D. S., and Boesch, C. 2001, Retraction: Furtive maiting in female chimpanzees. *Nature* 414:508.

Gagneux, P., Gonder, M. K., Goldberg, T. L., and Morin, P. A. 2001, Gene flow in wild chimpanzee populations: What genetic data tell us about chimpanzee movement over space and time. *Philos. Trans. R. Soc. Lond. B. Biol. Sci.* 356:889–897.

Gagneux, P., Wills, C., Gerloff, U., Tautz, D., Morin, P. A., Boesch, C., Fruth, B., Hohmann, G., Ryder, O. A., and Woodruff, D. S. 1999, Mitochondrial sequences show diverse evolutionary histories of African hominoids. *Proc. Nat. Acad. Sci. USA* 96:5077–5082.

Gagneux, P., Woodruff, D. S., and Boesch, C. 1997b, Furtive mating in female chimpanzees. *Nature* 387:358–359.

Garza, J. C. and Freimer, N. B. 1996, Homoplasy for size at microsatellite loci in humans and chimpanzees. *Genome Res.* 6:211–217.

Garza, J. C., Slatkin, M., and Freimer, N. B. 1995, Microsatellite allele frequencies in humans and chimpanzees, with implications for constraints on allele size. *Mol. Biol. Evol.* 12:594–603.

Goldberg, T. L. 1996, *Genetics and Biogeography of East African Chimpanzees* (*Pan troglodytes schweinfurthii*). Ph.D., Harvard University, Cambridge, MA.

Goldberg, T. L. 1998, Biogeographic predictors of genetic diversity in populations of East African chimpanzees (*Pan troglodytes schweinfurthii*). *Int. J. Primatol.* 19:237–254.

Goldberg, T. L. and Ruvolo, M. 1997, The geographic apportionment of mitochondrial genetic diversity in east African chimpanzees, Pan troglodytes schweinfurthii. *Mol. Biol. Evol.* 14:976–984.

Goldstein, D. B., Linares, A. R., Feldman, M. W., and Cavalli-Sforza, L. L. 1995, An evaluation of genetic distances for use with microsatellite loci. *Genetics* 139:463–471.

Goldstein, D. B. and Pollock, D. D. 1997, Launching microsatellites: A review of mutation processes and methods of phylogenetic inference. *J. Hered.* 88: 335–342.

Goldstein, D. B., Roemer, G. W., Smith, D. A., Reich, D. E., Bergmen, A., and Wayne, R. K. 1999, The use of microsatellite variation to infer population structure and demographic history in a natural model system. *Genetics* 151:797–801.

Gonder, M. K. 2000, *Evolutionary Genetics of Chimpanzees (Pan troglodytes) in Nigeria and Cameroon*. Ph.D. Dissertation, City University of New York, New York.

Gonder, M. K., Disotell, T. R., and Oates, J. F. 2006, New genetic evidence on the evolution of chimpanzee populations, and implications for taxonomy. *Int. J. Primatol.* 27(4).

Gonder, M. K., Oates, J. F., Disotell, T. R., Forstner, M. R., Morales, J. C., and Melnick, D. J. 1997 A new west African chimpanzee subspecies? *Nature* 388:337.

Goodall, J. 1986, *The Chimpanzees of Gombe: Patterns of Behavior*, Belknap Press of Harvard University Press, Cambridge, MA.

Goosens, B., Waits, L. P., and Taberlet, P. 1998, Plucked hair samples as a source of DNA: Reliability of dinucleotide microsatellite genotyping. *Mol. Ecol.* 7:1237–1241.

Gray, M. W., Burger, G., and Lang, B. F. 1999, Mitochondrial evolution. *Science* 283:1476–1481.

Groves, C. P. 1993, Order primates, in: Wilson, D. E., and Reader, D. M., eds., *Mammalian Species of the World: A Taxonomic and Geographic Reference*, Smithsonian Institution Press, Washington, DC, pp. 243–277.

Groves, C. P., Westwood, C., and Shea, B. T. 1993, Unfinished business: Mahalanobis and a clockwork orang. *J. Hum. Evol.* 22:22–37.

Grubb, P. 1982, Refuges and dispersal in the speciation of African forest mammals, in: Prance, G. T., ed., *Biological Diversification in the Tropics*, Academic Press, New York, pp. 537–553.

Grubb, P. 1990, Primate geography in the Afro-tropical forest biome, in: Peters, G., and Hutterer, R., eds., *Vertebrates in the Tropics*, Museum Alexander Koenig, Bonn, pp. 187–214.

Guo, S. W. and Thompson, E. A. 1992, Performing the exact test of Hardy-Weinberg proportion for multiple alleles. *Biometrics* 48:361–372.

Hacker, J. E., Cowlinshaw, G., and Williams, P. H. 1998, Patterns of African primate diveristy and their evaluation for the selection of conservation areas. *Biol. Conserv.* 84:251–262.

Haffer, J. 1969, Speciation in Amazon forest birds. *Science* 165:131–137.

Haffer, J. 1982, General aspects of the refuge theory, in: Prance, G. T., ed., *Biological Diversification in the Tropics*, Columbia University Press, New York, pp. 6–24.

Hamilton, A. C. 1988, Guenon evolution and forest history, in: Gautier-Hion, A., Bourlière, F., Gautier, J. P. and Kingdon, J., eds., *A Primate Radiation: Evolutionary Biology of the African Guenons*, Cambridge University Press, Cambridge, MA, pp. 13–34.

Hare, M. P. 2001, Prospects for nuclear gene phylogeography. *Trends in Ecology and Evolution* 16:700–706.

Hedrick, P. W. 1999, Perspective: Highly variable loci and their interpretation in evolution and conservation. *Evolution* 53:313–318.

Hill, W. C. O. 1967, The taxonomy of the genus Pan, in: Stark, D., Schneider, D., and Kuhn, H., eds., *Progress in Primatology*, Fischer, Stutgart, New York, pp. 47–54.

Hill, W. C. O. 1969, The nomenclature, taxonomy, and distribution of chimpanzees, in: Bourne, G. H., ed., *The Chimpanzee*, Karger, Basel, pp. 22–49.

Jarne, P. and Lagoda, J. L. 1996, Microsatellites, from molecules to populations and back. *Trends in Ecology and Evolution* 11:424–429.

Kaessmann, H., Wiebe, V., and Paabo, S. 1999, Extensive nuclear DNA sequence diversity among chimpanzees. *Science* 286:1159–1162.

Kaessmann, H., Wiebe, V., Weiss, G., and Paabo, S. 2001, Great ape DNA sequences reveal a reduced diversity and an expansion in humans. *Nat. Genet.* 27:155–156.

Kingdon, J. 1989, *Island Africa*, Princeton University Press Princeton, NJ.

Kocher, T. D. and Wilson, A. C. 1991, Sequence evolution of mitochondrial DNA in human and chimpanzees, in: *Evolution of Life: Fossils, Nucleotides and Culture*, Springer, Tokoyo, pp. 391–413.

Maddison, D. R., Ruvolo, M., and Swofford, D. L. 1992, Geographic origins of human mitochondrial DNA: Phylogenetic evidence from control region sequences. *Syst. Biol.* 41:111–124.

Maley, J. 1991, The African rainforest vegetation and paleoenvironments during the late Quaternary. *Climatic Change* 19:79–98.

Maley, J. 1996, The African rain forest–main characteristics of changes in vegetation and climate from the Upper Cretaceous to the Quaternary. *Proc. Roy. Soc. Edin.* 104B:31–73.

Mayr, E. and O'Hara, R. J. 1986, The biogeographic evidence supporting the Pleistocene forest refuge hypothesis. *Evolution* 40:55–67.

Meyer, S., Weiss, G., and von Haesler, A. 1999, Pattern of nucleotide substitution and rate heterogeneity in the hypervariable regions I and II of human mtDNA. *Genetics* 152:1103–1110.

Michalakis, Y. and Excoffier, L. 1996, A generic estimation of population subdivision using distances between alleles with special reference for microsatellites. *Genetics* 142:1061–1064.

Moore, W. S. 1995, Inferring phylogenies from mtDNA variation: Mitochondrial-gene trees versus nuclear gene trees. *Evolution* 49:718–726.

Moreau, R. E. 1969, Climatic changes and the distribution of forest vegetation in Africa. *J. Zool. Soc. Lond.* 158:39–61.

Morin, P. A. 1992, *Population Genetics of Chimpanzees.* Ph.D., University of California, San Diego.

Morin, P. A., Mahboubi, P., Wedel, S., and Rogers, J. 1998, Rapid screening and comparison of human microsatellite markers in baboons: Allele size is conserved, but allele number is not. *Genomics* 53:12–20.

Morin, P. A., Moore, J. J., Chakraborty, R., Jin, L., Goodall, J., and Woodruff, D. S. 1994, Kin selection, social structure, gene flow, and the evolution of chimpanzees. *Science* 265:1193–1201.

Mourier, T., Hansen, A. J., Willerslev, E., and Arctander, P. 2001, The Human Genome Project reveals a continuous transfer of large mitochondrial fragments to the nucleus. *Mol. Biol. Evol.* 18:1833–1837.

Navidi, W., Arnheim, N., and Waterman, M. S. 1992, A multiple-tubes approach for accurate genotyping of very small DNA samples by using PCR: Statistical considerations. *Am. J. Hum. Genet.* 50:347–359.

Oates, J. F. 1988, The distribution of Cercopithecus monkeys in West African forests, in: Gautier-Hion, A., Boulière, F., Gautier, J. P., and Kingdon, J., eds., *A Primate Radiation: Evolutionary Biology of the African Guenons,* Cambridge University Press, Cambridge, MA, pp. 79–103.

Oates, J. F. 1996, *African Primates: Status Survey and Conservation Action Plan.* Gland, Switzerland, IUCN/SSC Primate Specialist Group., International Union for Conservation of Nature and Natural Resources. Species Survival Commission.

Oates, J. F., Anadu, P. A., Gadsby, E. L., and Werre, J. L. 1992, Sclater's Guenon. *National Geographic Research and Exploration* 8:476–491.

Rithidech, K. N., Dunn, J. J., and Gordon, C. R. 1997, Combining multiplex and touchdown PCR to screen murine microsatellite polymorphisms. *Biotechniques* 23:36, 40, 42, 44.

Robbins, C. B. 1978, The Dahomey Gap–A reevaluation of its significance as a faunal barrier to West African high forest mammals. *Bulletin of the Carnegie Museum of Natural History* 6:168–174.

Sayer, J., Harcourt, C. S., and Collins, N. M., eds. 1992, *Conservation Atlas of Tropical Forests, Africa,* McMillian, New York.

Schiøtz, A. 1967, The treefrogs (Rhacophoridae) of West Africa. *Zoologica Musei Hauniensis* 25:1–346.

Schneider, S., Kueffer, J. M., Roessli, D., and Excoffier, L. 2000, *Arelequin: A Software for Population Genetic Data Analysis*. Genetics and Biometry Laboratory, University of Geneva, Geneva.

Schug, M. D., Hutter, C. M., Wetterstrand, K. A., Gaudette, M. S., Mackay, T. F. C., and Aquardro, C. F. 1998, The mutation rates of di-, tri and tetranucleotide repeats in Drosophila melanogater. *Mol. Biol. Evol.* 15:1751–1760.

Schwarz, E. 1934, On the local races of chimpanzee. *Annals & Magazine of Natural History, London* 13:576–583.

Shea, B. T. and Coolidge, H. J. 1988, Craniometric differentiation and systematics in the genus Pan. *J. Hum. Evol.* 17:671–685.

Slatkin, M. 1995, A measure of population subdivision based on microsatellite allele frequencies. *Genetics* 139(1):457–462.

Stone, A. C., Griffiths, R. C., Zegura, S. L., and Hammer, M. F. 2002, High levels of Y-chromosome nucleotide diversity in the genus Pan. *Proc. Natl. Acad. Sci. USA* 99:43–48.

Stoneking, M. 1994, Mitochondrial DNA and human evolution. *J. Bioenerg. Biomembr.* 26:251–259.

Stoneking, M. 2000, Hypervariable sites in the mtDNA control region are mutational hotspots. *Am. J. Hum. Genet.* 67:1029–1032.

Taberlet, P., Griffin, S., Goossens, B., Questiau, S., Manceau, V., Escaravage, N., Waits, L. P., and Bouvet, J. 1996, Reliable genotyping of samples with very low DNA quantities using PCR. *Nucleic Acids Res.* 24:3189–3194.

Tamura, K., and Nei, M. 1993, Estimation of the number of nucleotide substitutions in the control region of mitochondrial DNA in humans and chimpanzees. *Mol. Biol. Evol.* 10:512–526.

Taylor, J. S., Sanny, J. S., and Breden, F. 1999, Microsatellite allele size homoplasy in the guppy (*Poecilia reticulata*). *J. Mol. Evol.* 48:245–247.

Teleki, G. 1989, Population status of wild chimpanzees (*Pan troglodytes*) and threats to survival. in: Heltne, P. G., and Marquardt, L. A., eds., *Understanding Chimpanzees*, Harvard University Press, Cambridge, MA, pp. 312–353.

Tourmen, Y., Baris, O., Dessen, P., Jacques, C., Malthiery, Y., and Reynier, P. 2002, Structure and chromosomal distribution of human mitochondrial pseudogenes. *Genomics* 80:71–77.

Uchida, A. 1992, *Intra-Species Variation among the Great Apes: Implications for Taxonomy of Fossil Hominoids*. Ph.D., Harvard University, US.

Valdes, A. M., Slatkin, M., and Freimer, N. B. 1993, Allele frequencies at microsatellite loci: the stepwise mutation model revisited. *Genetics* 133:737–749.

Vigilant, L., Hofreiter, M., Siedel, H., and Boesch, C. 2001, Paternity and relatedness in wild chimpanzee communities. *Proc. Natl. Acad. Sci. USA* 98:12890–12895.

Walsh, P. S., Metzger, D. A., and Higuchi, R. 1991, Chelex 100 as a medium for simple extraction of DNA for PCR-based typing from forensic material. *Biotechniques* 10:506–513.

Weir, B. 1996a, *Genetic Data Analysis*, Sinaeur Associates, Inc., Sunderland, MA.

Weir, B. 1996b, Intraspecific differentiation, in: Hillis, D. M., Moritz, C., and Mable, B. K., eds., *Molecular Systematics*, Sinaeur Associates, Inc., Sunderland, MA, pp. 385–406.

Weir, B. S. and Cockerham, C. C. 1984, Estimating F-Statisics for the analysis of population structure. *Evolution* 38:1358–1370.

Wise, C. A., Sraml, M., Rubinsztein, D. C., and Easteal, S. 1997, Comparative nuclear and mitochondrial genome diversity in humans and chimpanzees. *Mol. Biol. Evol.* 14:707–716.

Zischler, H., Geisert, H., and Castersana, J. 1998, A hominoid-specific nuclear insertion of the mitochondrial d-loop: Implications for reconstructing ancestral mitochondrial sequences. *Mol. Biol. Evol.* 15:463–469.

CHAPTER SIX

Geographic Variation in Savanna Baboon (*Papio*) Ecology and its Taxonomic and Evolutionary Implications

Jason M. Kamilar

ABSTRACT

Jolly (1993) stated that the degree of ecological niche separation among closely related taxa may help to distinguish their evolutionary relationships since ecological divergence is often thought of as a characteristic of true biological species. Based on qualitative data, Jolly (1993) hypothesized that there is little niche separation among savanna baboon forms and therefore suggested that they are a single species. In addition, a recent study by Frost and colleagues (2003) found that baboon cranial morphology covaried with latitude that also suggests a single species designation. This present study quantitatively examined the ecological niche space of savanna baboons to test Jolly's hypothesis and to examine how their ecological variation varied with geography. To investigate this idea, previously published long-term data were accumulated from over twenty savanna

Jason M. Kamilar • Interdepartmental Doctoral Program in Anthropological Sciences, Stony Brook University. Stony Brook, New York, 11794-4364

Primate Biogeography, edited by Shawn M. Lehman and John G. Fleagle.
Springer, New York, 2006.

baboon populations. Variables from four categories were used to quantify their niche space: 1) Environment, 2) Diet, 3) Activity budget, and 4) Social organization. A discriminant function and principal components analysis was conducted for each dataset, and confirmed that savanna baboon subspecies inhabit significantly distinct environments, yet display a statistically non-significant difference in their diet, activity budget, and social organization. In addition, a hierarchical cluster analysis revealed that savanna baboon ecology followed a latitudinal cline. Therefore, the results of these analyses cannot falsify Jolly's hypothesis that there is little ecological niche separation among baboon taxa.

Key Words: ecogeography, biogeography, species concepts, speciation, niche, intraspecific

INTRODUCTION

Identifying and defining species has been a problem in biology for many years. Discussion of the species problem has made a resurgence recently due to the frequent disparity between the phylogeny and taxonomy of taxa (Hey, 2001; Sites and Marshall, 2003) and the importance of identifying species for conservation purposes (Isaac *et al.*, 2004). Baboon (*Papio*) taxonomy is one of the most contentious issues in primatology. Ecologically, savanna baboons are parapatrically distributed in a variety of habitat types, while consuming a broad array of dietary items, and demonstrating a wide range of behavioral activity patterns (Altmann, 1974; Jolly, 1993; Barton *et al.*, 1996; Henzi and Barrett, 2003). In addition to their ecological diversity, savanna baboons are quite varied in their body size, pelage color, craniodental anatomy, and other morphological traits (Hill, 1967; Jolly and Brett, 1973; Hayes *et al.*, 1990; Frost *et al.*, 2003).

The ecological and morphological geographic variation in savanna baboons is quite high compared to other primate taxa and is one reason that contributes to the uncertainty surrounding their taxonomy and evolutionary history. The two major taxonomic hypotheses, a single or multispecies classification, depend on the type of data and species concept utilized. The distinct morphological traits present in each baboon taxon lends support to a multiple species arrangement as defined by the phylogenetic species concept (Fleagle, 1999; Groves, 2001; Grubb *et al.*, 2003). This species concept relies on the idea that species display a unique combination of traits distinct from other such organisms within the context of ancestry and decent (Cracraft, 1987; Kimbel and Rak, 1993). The phylogenetic species concept is commonly implemented by paleontologists since the available data of extinct animals are limited to anatomical structures.

Alternatively, the biological species concept defines a species as a group of individuals that interbreed or can potentially interbreed, and are reproductively isolated from other such groups (Mayr, 1942). The biological species concept is probably the most objective species concept since its definition relies on measuring gene flow among populations, yet it is often difficult to implement since genetic data are difficult to obtain in many circumstances. More recently, with the advent of molecular techniques, genetic data has been used to help solve this taxonomic puzzle. Molecular data from several baboon populations confirm the gene flow among baboon taxa, which would support the idea of a single baboon species if the biological species concept is employed (Rogers, 2000; Newman *et al.*, 2004). The seminal paper by Jolly (1993) combines these two species concepts by labeling baboons "phylogenctic subspecies", acknowledging the phenotypic distinctiveness of each taxon, yet also accounting for the lack of reproductive isolation among them. I will adopt Jolly's (1993) definition of savanna baboon forms as subspecies for the purposes of this investigation.

Traditionally, morphological traits have been used to examine animal taxonomy because these data were readily available from museum specimens and can be quantified relatively easily. In addition, a predominant school of thought is that morphological characters are less labile than behavioral or ecological traits, and therefore more useful in reconstructing a phylogeny or taxonomy (Atz, 1970; Wilson, 1975; Baroni Urbani, 1989). Alternatively, several more recent studies have shown that behavioral and ecological traits often exhibit similar levels of homoplasy as morphological traits (de Queiroz and Wimberger, 1993; Proctor, 1996; Wimberger and de Queiroz, 1996; Doran *et al.*, 2002). This is not surprising, because much of an animal's behavior and ecology depends in part on morphological traits such as body mass, and feeding and locomotor adaptations (Fleagle, 1999; Alcock, 2001). Therefore, a species' ecological niche is also influenced by its evolutionary history (Fleagle and Reed, 1999) and may be an interesting line of evidence in investigating taxonomic questions. In fact, Mayr (1982) altered his definition of the biological species concept to clarify that a species is, "... a reproductive community of populations (reproductively isolated from others) that occupies a specific niche in nature."

The purpose of this chapter is to quantify the ecological variability in savanna baboons and place it in a geographic and taxonomic context. First, I examined the ecological variation within and between baboon subspecies in order to investigate whether baboons should be considered a single species. The logic for this analysis is based on Jolly's (1993) statement that the degree of ecological niche separation among taxa may help to distinguish their evolutionary relationships since ecological divergence is often thought of as a characteristic of true biological species. In addition, with respect to savanna baboons, Jolly stated that there is no niche separation or adaptive differences among the subspecies. This first analysis will quantitatively test Jolly's idea, where populations of a single species are expected to display similar niches, whereas populations from separate species should display distinctive ecological roles. Therefore, if the population's niche is defined by the environment in which they live, their diet, activity budget, and social organization, then there should be a significant difference among subspecies in these traits if they are truly separate species. Alternatively, a lack of significant differences in these traits should indicate a cohesive yet ecologically variable species. Second, the adaptive response of savanna baboon subspecies will be examined. If savanna baboons are a single species, then the effects of environmental factors on their diet, activity budget, and social organization should be similar. Alternatively, different species would be expected to display different responses to environmental characteristics. Lastly, the overall ecological similarity among savanna baboon subspecies was assessed in relation to their geographic distribution. This may also provide information regarding their taxonomic status. A recent paper by Frost *et al.* (2003) showed that the cranial morphometric variation of baboons follows a latitudinal cline which supports the genetic data of a single geographically varied species. An ecological distribution following a similar cline would corroborate this idea (Coyne and Orr, 2004; Fooden and Albrecht, 1993).

METHODS

Data Collection

Data were collected from published material from a total of 27 wild savanna baboon populations (Appendix 1). Data were gathered for 11 olive baboon populations, four yellow baboon populations, ten chacma baboon populations, and two guinea baboon populations. Hamadryas baboons were not included since comprehensive long-term data are not available. The variables included in the analyses were chosen because of two criteria: (1) their biological relevance to a baboon's niche, and (2) their availability in the published literature. Based on these criteria, the analyses included the following variables that were grouped into four datasets: (1) *Environment*: (a) mean annual rainfall, (b) number of dry months, (c) altitude, (d) number of sympatric cercopithecoids, (e) predation risk (as defined by Hill and Lee, 1998), and (f) latitude. The broad-scale variables in the *Environment* dataset are important in shaping the abundance and distribution of vegetation in a habitat, as well as other factors that have significant effects on primate diet, activity budget, and social organization (Murphy and Lugo, 1986; Janson, 1992; Bronikowski and Altmann, 1996; Chapman *et al.*, 1999). (2) *Diet*: annual percentage of (a) fruit/seeds, (b) leaves, (c) flowers, (d) fauna, and (e) underground items. (3) *Activity budget*: percentage of time spent (a) resting, (b) social, (c) feeding, and (d) moving. (4) *Social organization*: (a) group size, (b) number of adult males, (c) number of adult females, and (d) adult sex ratio. The definition of social organization used in this study follows Kappeler and van Schaik (2002), as, ". . . the size, sexual composition and spatiotemporal cohesion of a society." The spatiotemporal characteristics of the savanna baboon populations will not be included in the analyses since these data are rarely quantified by researchers, yet all populations are gregarious.

Social organization data were included in the analyses if the authors stated that group composition could be accurately determined. The diet and activity budget data used in this study were accumulated from sources with a research period of at least 10 months. Some populations were studied by more than one researcher and/or had data available for more than one social group, resulting in varied data produced for a single baboon taxon at a single study site (e.g., Amboseli). The mean value for these data was used in these cases.

Data Analyses

Two multivariate approaches were used to examine the amount of niche overlap among savanna baboon subspecies. Ideally, all datasets would be combined and entered into a single multivariate analysis, yet this would result in a reduced sample size since many populations do not have data for all variables. Consequently, to increase the sample size, each dataset was subjected separately to the multivariate analyses. It is important to note that another consequence of having incomplete datasets is that each analysis was not comprised of identical populations to represent the variation in each subspecies.

First, a discriminant function analysis (DFA) was conducted to examine explicitly the within versus among subspecies ecological variation. Investigations comparing within versus among taxa morphological variation have been conducted many times using this analytical technique (see Albrecht, 1976; Shea and Coolidge, 1988 for examples; Hayes et al., 1990; Froehlich et al., 1991; Albrecht and Miller, 1993; Ford, 1994). There is no a priori reason to suggest that ecological data should perform any differently. DFA is used to test for differences among groups by maximizing the differences among them. In addition, it examines whether the independent variables suitably predict the a priori group assignments while controlling for covariation among predictor variables (Tabachnick and Fidell, 1989; McGarigal et al., 2000). These group assignments are based on a priori knowledge of the partitioning of the samples, in this case, assigning a subspecies designation to each savanna baboon population. Guinea baboon populations were unclassified in the DFA since only one population had the available data for each dataset.

Two major assumptions of DFA are the multivariate normality of the data and that the variance-covariance matrices are homogenous among groups. The second assumption is the most critical and may lead to increased Type I or II error rates (depending on how sample size is related to variance) if not met (Tabachnick and Fidell, 1989; McGarigal et al., 2000). All variables were tested for normality using Shapiro-Wilk normality tests. Those variables failing normality tests were log transformed. In addition, Levene's test of homogeneity of variance was conducted for each variable. Testing the univariate homogeneity of variance is usually a good indicator of the homogeneity of variance-covariance matrices (McGarigal et al., 2000). The results of the DFA were examined more closely if it contained variables failing the Levene's test at the alpha level of 0.01. The results of the DFA were especially focused on the degree of subspecies overlap based on an examination of the discriminant function biplots (Gower and Hand, 1995). When DFA is used in this exploratory manner, the assumptions of the test can be relaxed (Tabachnick and Fidell, 1989). For each dataset, Pearson's correlations were conducted between the original variables and the discriminant functions to assess the importance of the original variables in distinguishing among the baboon groups.

Since discriminant function analyses have several statistical assumptions that may be difficult to check using a relatively small sample size, a principal components analysis (PCA) was conducted as a complementary technique. A PCA is a strictly exploratory technique and as such, has fewer statistical assumptions (Tabachnick and Fidell, 1989). A correlation matrix was used as the basis of each PCA. The savanna baboon populations were plotted in multidimensional space to examine the degree of ecological overlap among subspecies. For each dataset, Pearson's correlations were conducted between the original variables and the principal component axes to assess the correlation between the original variables and the principal components.

A series of analysis of covariance (ANCOVA) tests were used to examine whether savanna baboon subspecies responded to environmental forces in the same manner. An ANCOVA was conducted with each variable in the *Diet*, *Activity Budget*, and *Social organization* datasets as the dependent variable. For all ANCOVAs the savanna baboon subspecies acted as the categorical predictor variable and the variables in the *Environment* dataset as the covariates. The alpha level for these analyses was corrected with a Bonferroni adjustment (Sokal and Rohlf, 1995).

Finally, a hierarchical cluster analysis was implemented to examine the overall ecological similarity among the savanna baboon subspecies in a geographic context. The population mean for each baboon subspecies was calculated for each variable. All data were standardized using z scores. The average Euclidian distances among taxa were calculated and taxa were joined using the unweighted pair group method with arithmetic mean (UPGMA) (Tabachnick and Fidell, 1989).The cluster analysis included all variables from the *Social organization, Diet, and Activity budget* datasets. The *Environment* dataset was not included in the cluster analysis because the dendrogram produced from the cluster analysis was mapped onto a distribution map of savanna baboons to examine the biogeographical pattern of savanna baboon ecology. Including the variables from the *Environment* dataset would be logically circular since many of the variables are geographic in nature.

All analyses were conducted with SPSS 11.0 and Statistica 6.0 for Windows. A p-value of <0.05 was considered significant for the DFAs and Pearson's correlations.

RESULTS

The discriminant function analysis of the *Environment* dataset yielded a significant difference among taxa, with all of the populations being correctly grouped

Table 1.   Results of the discriminant function analyses

Dataset	Wilks' Lambda	Chi-square	df	p-value
Environment (with Latitude)	0.048	50.033	12	<0.001
Diet	0.212	13.941	10	0.176
Activity Budget	0.498	7.316	8	0.503
Social organization	0.383	11.993	8	0.152

into their *a priori* classifications (Wilks' Lambda = 0.048, $p < 0.001$, $df = 12$) (Table 1). Latitude is the most important variable in this analysis, clearly separating the chacma baboons from the remaining groups on the first axis (Figure 1a). The number of sympatric cercopithecoids at a site and mean annual rainfall were additional variables that contributed to distinguishing chacma populations from the yellow and olive baboons. Function two of the analysis best discriminated olive from yellow baboons. The most important variables that correlated with function two were predation risk and annual rain (Table 2). The guinea baboon population is most similar to the olive and yellow baboon populations with respect to their environmental characteristics.

The *Diet, Activity budget,* and *Social organization* DFAs did not yield statistically significant results ($p = 0.176$, $p = 0.503$, $p = 0.152$, respectively) (Table 1), yet the majority of the populations were correctly classified (Table 3). Examining the DFA biplots from these datasets showed that the amount of overlap among subspecies for the *Diet* dataset was not high, yet was moderate for the *Activity budget,* and *Social organization* datasets (Figure 1b–d). In the *Diet* biplot, the chacma baboon populations were found in the right half of the biplot, which indicated a high proportion of underground items in their diet (Figure 1b). The olive baboons generally occupied the left half of the graph, indicating lower levels of underground food items. Yellow baboon populations were intermediate on this dietary axis. The second function of this DFA correlated most strongly with the consumption of fruit/seed and underground food items. Olive and chacma baboons overlapped greatly in the Y-axis, but yellow baboons showed higher levels of underground item intake and lower levels of frugivory. The diet of the guinea baboon population was most similar to that of the chacma baboons.

The *Activity budget* DFA produced a first function which accounted for nearly 90% of the variation in the dataset and is negatively correlated with time spent social (Figure 1c). The yellow baboon populations, along with one

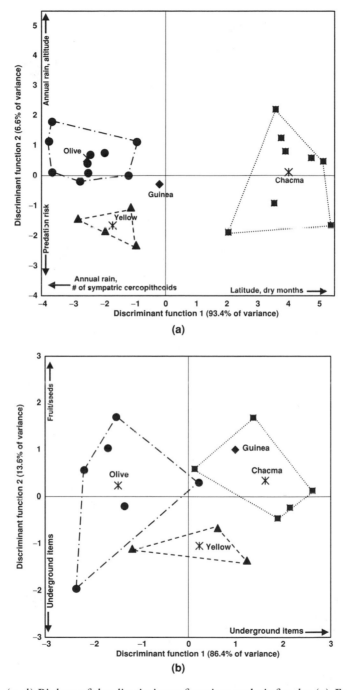

Figure 1. (a–d) Biplots of the discriminant function analysis for the (a) *Environment*, (b) *Diet*, (c) *Activity budget*, and (d) *Social organization* datasets. Symbols represent olive (•), yellow (▲), chacma (■), and guinea (♦) savanna baboon populations. The subspecies centroid (✶) as calculated from the DFA is also displayed.

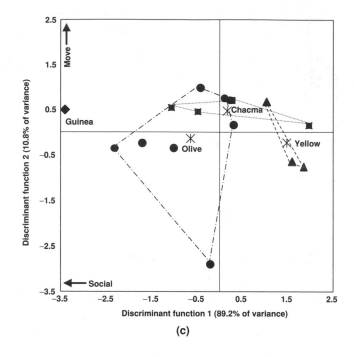

(c)

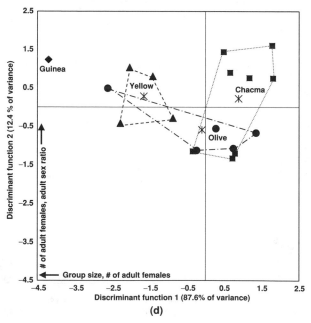

(d)

Figure 1. (*Continued*)

Table 2. Correlation coefficients between original variables and discriminant functions

Variable	Function 1	Function 2
Fruit/seed	0.012	0.554[a]
Leaves	−0.483	0.246
Underground items	0.551[a]	−0.562[a]
Flowers	0.163	−0.090
Fauna	−0.258	−0.356
Feed	0.488	0.275
Move	−0.189	0.534[a]
Rest	−0.172	−0.358
Social	−0.913[b]	0.085
Group size	−0.804[b]	0.425
# of males	−0.300	0.285
# of females	−0.797[b]	0.670[b]
Adult sex ratio	−0.136	0.558[a]
Annual rain	−0.418[a]	0.640[b]
Dry months	0.509[a]	−0.187
Altitude	−0.402	0.417[a]
Latitude	0.982[b]	0.122
Log (Sympatric cercopithecoids)	−0.705[b]	0.096
Predation risk	−0.383	−0.770[b]

[a]Correlation is significant at the 0.05 level (2-tailed).
[b]Correlation is significant at the 0.01 level (2-tailed).

chacma population (Drakensberg) devoted the least amount of time to social behavior. The olive baboons and the remaining chacma baboon populations displayed intermediate values for function one, with the lone guinea population (Assirik) located in the far left of the biplot indicating the highest level of time being social. The percentage of time spent moving significantly correlated with function two. The olive baboons varied considerably in relation to this axis,

Table 3. Percentage of correctly classified savanna baboon populations

Taxon	Environment	N[a]	Diet	N	Activity	N	Social Org	N	Taxon Mean	N
Olive	100.0	10/10	83.3	5/6	87.5	7/8	20.0	1/5	79.3	23/29
Yellow	100.0	4/4	66.7	2/3	66.7	2/3	75.0	3/4	78.6	11/14
Chacma	100.0	8/8	100.0	5/5	25.0	1/4	75.0	6/8	80.0	20/25
Dataset Mean	100.0	22/22	85.7	12/14	66.7	10/15	58.8	10/17	79.4	54/68

[a]The proportion of correctly classified populations

with yellow baboons displaying intermediate values. The chacma populations were less variable and displayed relatively high values. The single guinea baboon population at Assirik displayed an intermediate level of time allocated to moving.

The biplot from the *Social organization* DFA exhibited the most overlap among savanna baboon subspecies, especially between olive and chacma populations (Figure 1d). Mean group size and the number of adult females in a group negatively correlated with function one. Yellow baboons tended to have the highest values, followed by olive populations, with chacma baboons having the lowest scores. The guinea baboon population from Badi exhibited the highest function one score of all the savanna baboon populations.

Two variables from the *Environment* dataset, predation risk and the number of sympatric cercopithecoids, failed the Shapiro-Wilk normality test. Both variables were subsequently log transformed, yet transforming predation risk still yielded non-normal results. Consequently, the untransformed data were used in the analyses. The results of the Levene's tests showed that one variable (predation risk) exhibited significantly different variances among subspecies at the $p < 0.01$ level (Table 4). The violation of this DFA assumption most likely

Table 4. Results from Levene's test of homogeneity of variance

Variable	Levene's statistic	$df1$	$df2$	p-value
Annual rain	1.451	2	19	0.259
Dry months	0.050	2	19	0.951
Altitude	3.936	2	19	0.037
Latitude	0.738	2	19	0.491
Log (Sympatric cercopithecoids)	1.211	2	19	0.320
Predation risk	8.023	2	19	0.003
Fruit/seed	0.334	2	12	0.722
Leaves	6.476	2	12	0.012
Underground items	2.507	2	12	0.123
Flowers	1.272	2	12	0.315
Fauna	4.403	2	12	0.037
Feed	1.704	2	12	0.223
Move	1.542	2	12	0.253
Rest	4.708	2	12	0.031
Social	1.778	2	12	0.211
Group size	0.085	2	14	0.919
# of males	2.282	2	14	0.139
# of females	0.598	2	14	0.563
Adult sex ratio	1.321	2	14	0.298

did not have a substantial effect since predation risk was the least important variable separating the taxa on the first axis.

Although a PCA does not statistically test for differences among groups, the PCA biplots showed that there is a lack of differentiation among savanna baboon subspecies (Figure 2a–d). The PCA biplot of the *Environment* dataset (Figure 2a) displayed the least overlap among subspecies, similar to the DFA results. The PCA results of the remaining datasets (Figure b–d) displayed more overlap among subspecies compared to the DFA results. The eigenvalues for all principal components analyses are presented in Table 5, with the correlation coefficients between the original variables and the principal components listed in Table 6. The PCA analyses support the non-significant differences among savanna baboon subspecies for these traits. Overall, maximizing the differences among subspecies using the DFA analyses did not yield statistically significant results, and this was supported by the PCA biplots.

The ANCOVAs resulted in no significant difference among the *Diet*, *Activity budget*, and *Social organization* variables among subspecies (Table 7). These tests suggest that the baboon subspecies respond to environmental factors in a similar fashion and therefore show similar adaptive responses.

The results of the UPGMA cluster analysis demonstrated that chacma and yellow baboons were the most ecologically similar, followed by olive baboons, with guinea baboons being the most distinct taxon. When these results were plotted on a map displaying the geographic distribution of savanna baboons, ecological similarity followed a latitudinal cline (Figure 3).

DISCUSSION

The result of this study lends support to Jolly's (1993) conclusion, that savanna baboon subspecies are ecologically similar. There was a statistically significant difference among the savanna baboon subspecies in only one of the datasets (*Environment*). The results of the remaining DFA analyses showed that each subspecies displayed trends in certain niche characteristics, yet these differences were not sufficient to yield statistically significant results. In addition, the PCA analyses corresponded to the DFA tests showing lack of separation among the subspecies in the *Diet*, *Activity budget*, and *Social organization* datasets. These results show that the ecological variability among savanna baboon subspecies does not exceed the variation within subspecies. Finally, the ANCOVAs

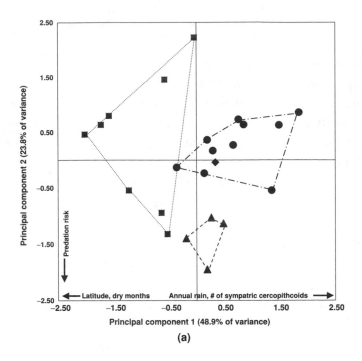

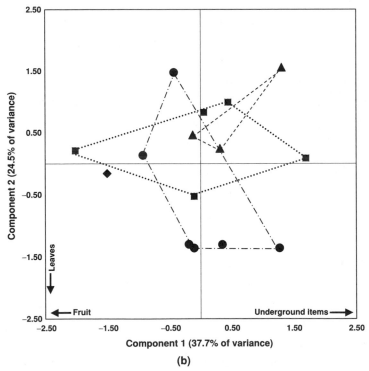

Figure 2. (a–d) Biplots of the principal components analysis of the (a) *Environment*, (b) *Diet*, (c) *Activity budget*, and (d) *Social organization* datasets. Symbols represent olive (●), yellow (▲), chacma (■), and guinea (♦) savanna baboon populations.

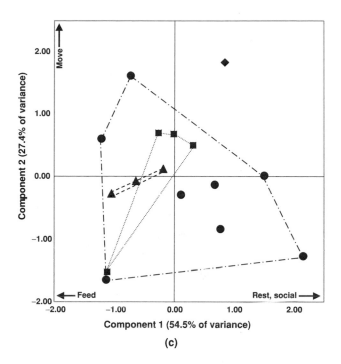

(c)

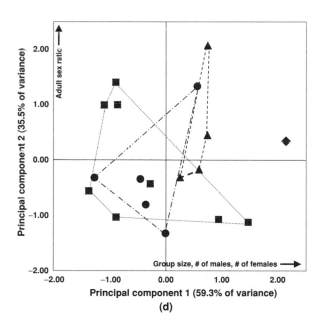

(d)

Figure 2. (*Continued*)

Table 5. Eigenvalues for the principal components analyses

Dataset	Comp1	Comp2	Comp3	Comp4	Comp5	Comp6
Environment (with Latitude)	2.977	1.274	0.929	0.399	0.253	0.168
Diet	1.885	1.225	1.046	0.805	0.039	–
Activity Budget	2.182	1.096	0.689	0.033	–	–
Social organization	2.373	1.421	0.154	0.051	–	–

suggest that the dietary, activity budget, and social organization characteristics of savanna baboon subspecies are shaped by environmental factors in a similar way.

A closer examination of the DFA results demonstrated that the Social Organization dataset had the least success at correctly predicting the subspecies of populations compared to the other datasets. This lower degree of variability in these variables may support Dunbar's proposal (1992) that baboons maintain their group size in varying environments by adjusting their activity budget.

Table 6. Correlation coefficients between the original variables and the principal components

Variable	Comp1	Comp2	Comp3	Comp4	Comp5	Comp6
Fruit/seed	−0.985[b]	0.089	−0.053	0.023	0.136	–
Leaves	0.492	−0.739[b]	−0.151	0.426	0.083	–
Underground items	0.779[b]	0.433	−0.215	−0.385	0.108	–
Flowers	0.101	−0.313	0.890[b]	−0.313	0.037	–
Fauna	0.235	0.621[a]	0.426	0.614[a]	0.023	–
Feed	−0.937[b]	−0.286	0.155	0.127	–	–
Move	−0.075	0.993[b]	0.079	0.054	–	–
Rest	0.911[b]	−0.127	−0.374	0.115	–	–
Social	0.684[b]	−0.113	0.720[b]	0.026	–	–
Group size	0.911[b]	0.286	−0.292	0.045	–	–
# of males	0.814[b]	−0.540[a]	0.169	0.130	–	–
# of females	0.933[b]	0.290	0.151	−0.148	–	–
Adult sex ratio	−0.094	0.981[b]	0.133	0.102	–	–
Annual rain	0.722[b]	0.459[a]	−0.388	0.031	0.310	−0.139
Dry months	−0.821[b]	−0.172	0.326	−0.285	0.330	−0.019
Altitude	0.444[a]	0.425[a]	0.745[b]	0.239	0.074	0.071
Latitude	−0.806[b]	0.320	−0.337	0.241	0.107	0.256
Log (Sympatric cercopithecoids)	0.898[b]	−0.108	−0.051	−0.315	0.044	0.278
Predation risk	0.357	−0.860[b]	−0.024	0.321	0.170	0.025

[a]Correlation is significant at the 0.05 level (2-tailed).
[b]Correlation is significant at the 0.01 level (2-tailed).

Table 7. ANCOVAs examining the relationship between *Environment* and *Diet*, *Activity budget*, and *Social organization* among savanna baboon subspecies

Variable	$F\,(df)$	p-value
Fruit/seed	0.874 (2, 4)	0.484
Leaves	1.045 (2, 4)	0.432
Underground items	1.776 (2, 4)	0.281
Flowers	0.311 (2, 4)	0.749
Fauna	0.779 (2, 4)	0.518
Feed	3.269 (2, 6)	0.109
Move	0.969 (2, 6)	0.432
Rest	2.352 (2, 6)	0.176
Social	1.187 (2, 6)	0.367
Group size	0.751 (2, 6)	0.512
# of males	2.506 (2, 6)	0.162
# of females	2.367 (2, 6)	0.175
Adult sex ratio	0.334 (2, 6)	0.729

Bonferroni corrected p-values are significant at the 0.01 level for the *Diet* dataset and 0.0125 for the *Activity budget* and *Social organization* datasets.

Although the results of this study showed that savanna baboon subspecies inhabit significantly different environments, their diet, activity budget, and social organization do not exhibit a corresponding distinctiveness. These results do not necessarily contradict the well established idea that environmental factors are an important influence in shaping primate behavior and ecology (Crook and Gartlan, 1966; Clutton-Brock and Harvey, 1977; van Schaik and van Hooff, 1983; Janson, 1992). The genetic cohesiveness of savanna baboon subspecies may be the cause of this disparity. Baboons are generally regarded as ecologically flexible (Post, 1981; Barton *et al.*, 1992; Barton and Whiten, 1993), yet there may be a limit to this flexibility due to gene flow. It has been shown that even low levels of genetic introgression among populations are enough to produce homogenizing effects (Ridley, 1997; Futuyma, 1998; Coyne and Orr, 2004).

Geographic Variation and Species Concepts

The idea that species exhibit a unique set of traits separate from other species is central to the phylogenetic species concept (Cracraft, 1987). The distinguishing morphological features displayed by each baboon subspecies include pelage, body size, and dental attributes (Hill, 1967; Jolly and Brett, 1973; Hayes *et al.*,

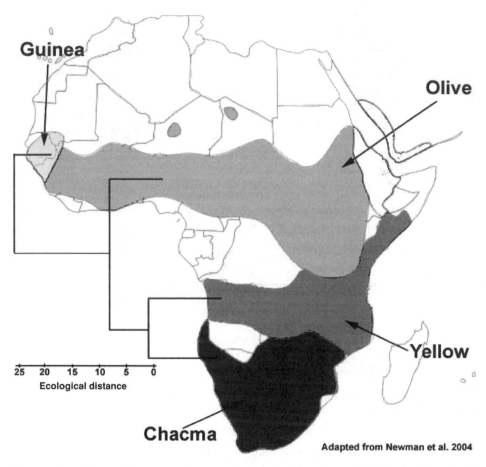

Figure 3. Geographic range of savanna baboon subspecies and their overall ecological niche similarity based on a UPGMA cluster analysis using a multivariate Euclidian distance matrix. Map adapted from Newman *et al.*, 2004.

1990). Yet, as Jolly (1993) aptly pointed out, suggesting that these taxa are full species would be ignoring the knowledge that there is gene flow among them. The hybridization of olive and hamadryas, and olive and yellow baboons has been well documented (Nagel, 1971; Nagel, 1973; Samuels and Altmann, 1986). The genetic cohesiveness of these taxa is evidenced by the production of non-sterile hybrids, yet the relative fitness of these hybrids is not known (Phillips-Conroy *et al.*, 1991; Woolley-Barker, 1999; Beehner and Bergman, 2003). On the basis of this genetic information, a single baboon species should be adopted by proponents of the biological species concept (Mayr, 1942).

Jolly (1993) solves the discordance between the two taxonomic arrangements by stating that baboons are a single polytypic species ("phylogenetic subspecies"). This idea acknowledges the distinctiveness of each taxon while accounting for the gene migration among them. The results of this study support Jolly's hypothesis. In addition, the allopatric nature of baboons suggests a recent divergence among populations. The amount of overlap among populations should increase with divergence time as populations achieve full species status and shift their geographic range (Losos and Glor, 2003). Therefore, if savanna baboon subspecies were indeed full species, we should expect more range overlap among them.

The idea that savanna baboons are a single species is also supported by ecological and biogeographical data within the ecological species concept. This species concept defines a species as a collection of allopatric populations occupying a more similar niche to each other than any other population in their geographic range (van Valen, 1976). The savanna baboon taxa displayed a relatively low degree of ecological niche separation. Other sympatric species that approximate the savanna baboon niche are the vervet monkey (*Chlorocebus aethiops*) and patas monkey (*Erythrocebus patas*) (Fleagle, 1999). A future study concurrently examining the quantitative niche space of savanna baboon, vervet monkey, and patas monkey populations may support this idea.

An interesting contrast to the idea that savanna baboons are a cohesive ecological species may be found with guenon monkeys (*Cercopithecus* spp.). Guenon species occupy a similar ecological niche to each other compared to other sympatric primate taxa, with lineages diverging relatively recently (Ruvolo, 1988; Struhsaker *et al.*, 1988; Disotell and Raaum, 2002). Yet, compared to baboons, an important distinction in guenons is that they are often found in sympatric associations with other *Cercopithecus* species while maintaining high degrees of reproductive isolation (but see Struhsaker *et al.*, 1988 for cases of hybridization; Detwiler, 2002), and thus maintain species' identities. This biogeographic contrast to baboons, and the resulting differences in the degree of reproductive isolation among closely related taxa, may be a vital clue to the idea that baboons are a single, ecological generalist species that occupy a broad niche, whereas guenons comprise many species that are ecological specialist, each occupying a narrow niche space (Kamilar, 2004).

Does Savanna Baboon Ecology Reflect their Evolutionary History?

A recent study examining the genetic relatedness of baboon populations has found that chacma baboons are the most basal lineage, followed by guinea, and hamadryas baboons, with the olive/yellow baboon clade diverging most recently (Newman *et al.*, 2004). The purpose of the study was not to investigate the possible taxonomic arrangement of baboons; rather it focused on the molecular relationships among baboon taxa. The branching pattern of this phylogenetic analysis does not correlate with the current geographical distribution of savanna baboon subspecies. The fact that the phylogeny of baboon taxa does not correlate with their geography suggests that the dispersal of individuals is not limited by geographic barriers (Epperson, 2004). In addition, the typology of the molecular phylogeny is not congruent with the phenogram representing the overall ecological similarity among these subspecies, which follows a latitudinal cline. The disparity between the evolutionary and ecological relationships may represent the effects of environmental traits in shaping the ecology of savanna baboons.

When comparing within versus among subspecies variation it is interesting that the lack of clear ecological differentiation among subspecies is in contrast to their morphological differences. Previous research examining morphological variation among baboon subspecies has noted that there is little overlap in the variation in odontometric traits (Hayes *et al.*, 1990) and pelage color (Hill, 1967). This discordance between morphological and ecological labiality may be unexpected to some. Traditionally, morphological data are often thought to be more highly conserved among closely related taxa, with behavioral and ecological traits more easily affected by environmental characteristics (Wilson, 1975). The ecological uniformity among subspecies may lend support to some more recent studies suggesting that behavioral and ecological traits do not display more homoplasy than morphological characters (de Queiroz and Wimberger, 1993; Proctor, 1996; Doran *et al.*, 2002).

Clinal Variation in Savanna Baboon Biology

The dendrogram produced by the cluster analysis and its projection on a map of Africa suggests that the ecological variation found in savanna baboons corresponds to a latitudinal cline. Biological clines have been observed in many non-primate taxa. Perhaps the best known is Bergmann's rule, where species in colder climates tend to exhibit larger body sizes. This has been illustrated in several mammalian taxa including kangaroos (*Macropus gigan-teus*) (Yomtov and Nix, 1986), kangaroo rats (*Dipodomys* spp.) (Baumgardner and Kennedy, 1993), and small carnivores (*Lupus* spp. and *Vulpes* spp.) (Rosenzweig, 1968).

In primates, investigations of the relationship between latitudinal and phenotypic variation are relatively rare. Yet, the existing studies that examined this relationship have produced interesting results. Fooden and Albrecht (1993) found that *Macaca fascicularis* skull size covaried with latitude throughout their range in southeast Asia. Additionally, an earlier study by Albrecht (1976) is especially relevant to this current study. Albrecht examined cranial variation in several macaque taxa distributed throughout Sulawesi to examine their taxonomy and evolution. Albrecht found that these macaques displayed discrete morphological breaks and did not display variation correlated with their latitudinal distribution. These results led the author to conclude that these taxa should indeed be recognized as full species. In a contrasting scenario to Albrecht's study, is the recent paper by Frost and colleagues (2003). These researchers showed that baboon cranial shape and size displayed variation along a north-south geographic cline. They argue that these results support a single species hypothesis, since multiple species should display some degree of character displacement in adjacent taxa and not a clinal pattern in morphology. The results of this current study support the findings of Frost and colleagues.

The concordance between the relationships of eco-behavioral and cranial morphometric similarity with latitude suggests that variation in savanna baboon biology may be tied to broad-scale climatic factors that correlate with latitude. In addition, the clinal relationship between latitude and ecology and cranial morphology may suggest that baboons are currently in an intermediate stage of parapatric speciation. The "clinal model" of parapatric speciation proposes that a single species has a continuous distribution through a variable environment and that the populations are locally adapted to their environmental conditions (Fisher, 1930). Eventually, enough local adaptations will evolve to produce reproductively isolated taxa that become full species (Endler, 1977). Unfortunately, we can not be sure if the ecological and morphological variation exhibited by baboons are adaptations to local environmental factors or the result of developmental plasticity with no substantial change in allele frequencies among populations (Foster and Endler, 1999). Perhaps further studies can address this question.

CONCLUSIONS

The results of this chapter showed that there is no clear differentiation among the ecologies of savanna baboon subspecies. Therefore, these results cannot reject Jolly's (1993) hypothesis that non-hamadryas baboon taxa lack ecological separation. In addition, the ecological variation that does exist corresponds to a latitudinal cline. This supports Jolly's (1993) concept that savanna baboons are currently in some intermediate stage of the speciation process where they can be best described as "phylogenetic subspecies". The ecological data support the hypothesis that savanna baboons are a single species. Although it is useful to have animals classified as species or subspecies, our current ideas about species concepts may not be sufficient to apply them to all organisms, with baboons possibly being a good example of this (Hey, 2001). Recent and future studies of baboon biology at the subspecific level may shed more light on the nature of extant baboon taxonomy and evolutionary history (e.g., Frost *et al.*, 2003; Kamilar, 2004; Newman *et al.*, 2004).

ACKNOWLEDGMENTS

I would like to thank John Fleagle and Shawn Lehman for inviting me to contribute to this volume. Charlie Janson, John Fleagle, Patricia Wright, Tim Newman, Jim Moore, Dennis O'Rourke, Anthony Olejniczak, Ryne Palombit, and Linda Wolfe provided stimulating discussions on various aspects of this topic. Charlie Janson, Charlie Nunn and three anonymous reviewers provided thoughtful comments that improved an earlier version of this manuscript. This research was supported by a National Science Foundation Dissertation Improvement Grant.

APPENDIX 1. Ecological data from savanna baboon populations

Site	Taxon	Mean group size	Mean # of males	Mean # of females	Mean adult sex ratio	Mean annual rainfall
Chololo	olive	43.30	5.30	10.70	2.02	476
Gilgil	olive	66.40	5.10	21.42	3.83	691
Gombe	olive	35.33	7.13	11.88	1.65	1450
Q. E. Nat Park	olive	45.00	9.50	10.50	1.11	746
Shai Hills	olive	23.70	3.20	7.10	2.22	1065
Budongo	olive	43.00	–	–	–	1495
Mulu	olive	22.00	–	–	–	1105
Awash	olive	55.00	–	–	–	550
Bole	olive	20.00	–	–	–	1100
Kibale	olive	–	–	–	–	1662
Masai Mara	olive	–	–	–	–	–
Amboseli	yellow	48.00	9.00	21.00	2.35	423
Tana	yellow	78.80	4.60	21.80	4.74	494
Mikumi	yellow	51.70	7.70	16.00	2.08	851
Ruaha	yellow	72.00	7.00	19.00	2.71	304
Cape Good Hope	chacma	31.80	2.80	12.80	4.57	631
De Hoop	chacma	30.50	2.50	10.00	4.00	428
Honnet/Transvaal	chacma	47.20	13.00	20.50	1.58	307
Namib(Kuiseb)	chacma	28.00	5.70	7.70	1.35	18
Moremi	chacma	69.50	14.30	19.70	1.38	457
Cape Point	chacma	85.00	–	–	–	631
Suikerbosrand	chacma	78.00	–	–	–	700
T. Leopard Park	chacma	34.30	3.00	11.80	3.93	85
Mkuzi	chacma	47.30	6.00	11.30	1.88	630
Drakensberg	chacma	19.20	3.40	6.60	1.94	1149
Badi	guinea	83.00	12.00	31.00	2.58	–
Assirik	guinea	184.00	–	–	–	954

(Continued)

APPENDIX 1. (*Continued*)

Site	# of dry months (<50 mm)	Altitude (meters)	Latitude	# of sympatric cercopithecoids	Predation risk
Chololo	8	1661	0	2	2
Gilgil	6	1765	0	1	2
Gombe	4	680	4	5	3
Q. E. Nat Park	4	990	0	6	1
Shai Hills	4	1000	5	2	2
Budongo	2	1050	1	4	2
Mulu	6	1050	9	1	2
Awash	9	1400	8	1	2
Bole	6	2000	9	3	2
Kibale	3	1500	0	7	2
Masai Mara	–	–	–	–	–
Amboseli	9	1140	1	1	3
Tana	8	55	1	4	3
Mikumi	6	550	7	3	3
Ruaha	3	1230	7	1	3
Cape Good Hope	10	10	31	0	1
De Hoop	11	10	34	0	1
Honnet/Transvaal	–	–	–	–	2
Namib(Kuiseb)	–	–	24	–	1
Moremi	7	300	19	1	3
Cape Point	8	10	34	0	1
Suikerbosrand	7	1600	27	1	1
T. Leopard Park	11	1060	22	1	2
Mkuzi	6	125	27	1	3
Drakensberg	5	2045	29	1	1
Badi	–	–	–	–	2
Assirik	3	150	13	2	2

Chololo	.40	.36	.15	.08
Gilgil	.51	.30	.08	.09
Gombe	.26	.19	.30	.11
Q. E. Nat Park	–	–	–	–
Shai Hills	.20	.18	.39	.23
Budongo	.59	.18	.06	.17
Mulu	.41	.25	.22	.15
Awash	.31	.25	.31	.12
Bole	.21	.25	.35	.16
Kibale	–	–	–	–
Masai Mara	–	–	–	–
Amboseli	.46	.26	.19	.09
Tana	–	–	–	–
Mikumi	.37	.26	.25	.06
Ruaha	.47	.24	.17	.05
Cape Good Hope	–	–	–	–
De Hoop	.40	.31	.16	.13
Honnet/Transvaal	–	–	–	–
Namib(Kuiseb)	–	–	–	–
Moremi	–	–	–	–
Cape Point	.34	.29	.26	.11
Suikerbosrand	–	–	–	–
T. Leopard Park	–	–	–	–
Mkuzi	.36	.30	.21	.12
Drakensberg	.57	.18	.17	.08
Badi	–	–	–	–
Assirik	.24	.37	.21	.19

(Continued)

APPENDIX 1. (*Continued*)

Site	% fruit/seed in diet	% leaves in diet	% underground items in diet	% flowers in diet	% fauna in diet	References
Chololo	.23	.27	.15	.21	.01	1, 2, 3, 4, 5, 6
Gilgil	.10	.53	.27	.03	.02	1, 2, 5, 6
Gombe	.49	.14	.07	.02	.13	1, 2, 7, 8
Q. E. Nat Park	–	–	–	–	–	9, 10
Shai Hills	.59	.08	.17	.05	.00	1, 2, 12, 18
Budongo	–	–	–	–	–	1, 13, 19
Mulu	–	–	–	–	–	1
Awash	–	–	–	–	–	1
Bole	.41	.41	.01	.12	.04	1, 11, 14, 17
Kibale	.72	.08	.00	.00	.20	14
Masai Mara	.46	.44	.08	.01	.01	15
Amboseli	.23	.15	.33	.04	.01	1, 14, 16, 17, 18, 19, 20
Tana	.53	–	–	.06	–	2, 14, 21
Mikumi	.43	.14	.12	.20	.10	1, 2, 22, 23
Ruaha	.16	.19	.52	.01	.09	1, 2, 23
Cape Good Hope	–	–	–	–	–	14
De Hoop	.35	.13	.42	.05	.05	24
Honnet/Transvaal	–	–	–	–	–	14, 25
Namib(Kuiseb)	–	–	–	–	–	14, 26
Moremi	–	–	–	–	–	2, 26, 27, 28
Cape Point	.42	.25	.16	.12	.03	1, 29
Suikerbosrand	.43	.08	.39	.07	.03	30
T. Leopard Park	–	–	–	–	–	14, 31
Mkuzi	.90	.06	.01	.03	.01	2, 32, 33
Drakensberg	.03	.26	.53	.14	.04	1, 2, 34
Badi	–	–	–	–	–	14
Assirik	.74	.09	.03	.09	.01	1, 35

(1) Lee, 1999, (2) Hill *et al.*, 2000, (3) Kenyatta, 1995, (4) Barton, 1989, (5) Harding, 1976, (6) Eley *et al.*, 1989, (7) Oliver and Lee, 1978, (8) Ransom, 1981, (9) Hill and Lee, 1998, (10) Spinage, 1970, (11) Dunbar and Dunbar, 1974, (12) Depew, 1983, (13) Patterson, 1976, (14) Fleagle *et al.*, 1999, (15) Popp, 1978, (16) Post *et al.*, 1980, (17) Murruthi, 1997, (18) Bronikowski and Altmann, 1996, (19) Alberts *et al.*, 1996, (20) Post, 1981, (21) Bentley-Condit and Smith, 1997, (22) Norton *et al.*, 1987, (23) Rasmussen, 1978, (24) Hill, 1999, (25) Stoltz and Saayman, 1970, (26) Hamilton *et al.*, 1975, (28) Cheney *et al.*, 2004, (29) Davidge, 1978, (30) Anderson, 1981, (31) Cowlishaw, 1999 (32) Ron *et al.*, 1996, (33) Gaynor, 1994, (34) Whiten *et al.*, 1987, (35) Sharman and Dunbar, 1982

REFERENCES

Alberts, S. C., Altmann, J., and Wilson, M. L. 1996, Mate guarding constrains foraging activity of male baboons. *Animal Behaviour* 51:1269–1277.

Albrecht, G. H. 1976, Methodological approaches to morphological variation in primate populations: the Celebesian macaques. *Yrbk. Phys. Anthrop.* 20:290–308.

Albrecht, G. H. and Miller, J. M. A. 1993, Geographic variation in primates: A review with implications for interpreting fossils, in: Kimbel, W. H. and Martin, L. B., eds., *Species, species concepts, and primate evolution*, Plenum Press, New York, pp. 123–161.

Alcock, J. 2001, *Animal behavior: An evolutionary approach*, Sinauer Associates, Sunderland.

Altmann, S. A. 1974, Baboons, space, time, and energy. *American Zoologist* 14:221–248.

Atz, J. W. 1970, The application of the idea of homology to behavior, in: Aronson, L. R., Tolbach, E., Lehrman, D. S. and Rosenblatt, J. S., eds., *Development and Evolution of Behavior*, San Fransisco, Freeman, pp. 53–74.

Baroni Urbani, C. 1989, Phylogeny and behavioural evolution in ants, with a discussion of the role of behaviour in evolutionary processes. *Ethol. Ecol. Evol.* 1:137–168.

Barton, R. A. 1989, Foraging strategies, diet and competition in olive baboons, University of St. Andrews.

Barton, R. A., Byrne, R. W., and Whiten, A. 1996, Ecology, feeding competition and social structure in baboons. *Behav. Ecol. Sociobiol.* 38:321–329.

Barton, R. A. and Whiten, A. 1993, Feeding competition among female olive baboons, *Papio anubis. Primate Behav.* 46:777–789.

Barton, R. A., Whiten, A., Strum, S. C., and Byrne, R. W. 1992, Habitat use and resource availability in baboons. *Animal Behav.* 43:831–844.

Baumgardner, G. D. and Kennedy, M. L. 1993, Morphometric variation in kangaroo rats (Genus *Dipodomys*) and its relationship to selected abiotic variables. *J. Mammal.* 74:69–85.

Beehner, J. and Bergman, T. 2003, Female reproductive strategies in a baboon hybrid zone, Awash National Park, Ethiopia. *Am. J. Phys. Anthropol.* Suppl 36:63.

Bentley-Condit, V. and Smith, E. 1997, Female reproductive parameters of Tana River yellow baboons. *Int. J. Primatol.* 18:581–596.

Bronikowski, A. M. and Altmann, J. 1996, Foraging in a variable environment: Weather patterns and the behavioral ecology of baboons. *Behav. Ecol. Sociobiol.* 39:11–25.

Chapman, C. A., Wrangham, R. W., Chapman, L. J., Kennard, D. K., and Zanne, A. E. 1999, Fruit and flower phenology at two sites in Kibale National Park, Uganda. *J. Trop. Ecol.* 15:189–211.

Clutton-Brock, T. H. and Harvey, P. H. 1977, Primate ecology and social organization. *J. Zool, London* 183:1–39.

Coyne, J. A. and Orr, H. A. 2004, *Speciation*, Sinauer, Sunderland.

Cracraft, J. 1987, Species concepts and speciation analysis, in: Ereshefsky, M. ed., *The Units of Evolution: Essays on the Nature of Species*, MIT Press, Cambridge, MA. pp. 93–120.

Crook, J. H. and Gartlan, J. S. 1966, Evolution of primate societies. *Nature* 210:1200–1203.

de Queiroz, A. and Wimberger, P. H. 1993, The usefulness of behavior for phylogeny estimation: Levels of homoplasy in behavioral and morphological characters. *Evolution* 47:46–60.

Depew, L. A. 1983, Ecology and behaviour of baboons (*Papio anubis*) in the Shai Hills Game Production Reserve, Cape Coast University, Ghana.

Detwiler, K. 2002, Hybridization between red-tailed monkeys (*Cercopithecus ascanius*) and blue monkeys (*C. mitis*) in east African forests, in: Glenn, M. and Cords, M., eds., *The Guenons: Diversity and Adaptation in African Monkeys*, Kluwer Academic/Plenum Publication, New York, pp. 79–97.

Disotell, T. and Raaum, R. 2002, Molecular timescale and gene tree incongruence in the guenons, in: Glenn, M. and Cords, M. eds., *The Guenons: Diversity and Adaptation in African Monkeys*, Kluwer Academic/Plenum Publication, New York, pp. 27–36.

Doran, D., Jungers, W., Sugiyama, Y., Fleagle, J., and Heesy, C. 2002, Multivariate and phylogenetic approaches to understanding chimpanzee and bonobo behavioral diversity, in: Boesch, C., Hohmann, G. and Marchant, L., eds., *Behavioural Diversity in Chimpanzees and Bonobos*, Cambridge University Press, New York, pp. 14–34.

Dunbar, R. I. M. 1992, Time: A hidden constraint on the behavioral ecology of baboons. *Behav. Ecol. Sociobiol.* 31:35–49.

Dunbar, R. I. M. and Dunbar, E. P. 1974, Ecology and population dynamics of *Colobus guereza* in Ethiopia. *Folia Primatol.* 21:188–208.

Endler, J. A. 1977, *Geographic Variation, Speciation, and Clines*, Princeton University Press, Princeton.

Epperson, R. 2004, *Geographical Genetics*. Princeton University Press, Princeton.

Fisher, R. A. 1930, *The Genetical Theory of Natural Selection, First Edition*. Oxford University Press, Oxford.

Fleagle, J. G. 1999, *Primate Adaptation and Evolution*. Academic Press, San Diego.

Fleagle, J. G., Janson, C. H., and Reed, K. E. eds., 1999, *Primate Communities*, Cambridge University Press, New York.

Fleagle, J. G. and Reed, K. E. 1999, Primate communities and phylogeny, in: Fleagle, J. G., Janson, C. H. and Reed, K. E. eds., *Primate Communities*, Cambridge University Press, New York.

Fooden, J. and Albrecht, G. H. 1993, Latitudinal and insular variation of skull size in crab-eating macaques (Primates, Cercopithecidae: *Macaca fascicularis*). *Amer. J. Phys. Anthropol.* 92:521–538.

Ford, S. M. 1994, Taxonomy and distribution of the owl monkey, in: Baer, J. F., Weller, R. E. and Kakoma, I., eds., Aotus: *The Owl Monkey*, Academic Press, San Diego, CA. pp. 1–57.

Foster, S. A. and Endler, J. A., 1999, *Geographic Variation in Behavior: Perspectives on Evolutionary Mechanisms*, Oxford University Press, Oxford.

Froehlich, J., Supriatna, J., and Froehlich, P. 1991, Morphometric analyses of *Ateles*: Systematic and biogeographic implications. *Am. J. Primatol.* 25:1–22.

Frost, S. R., Marcus, L. F., Bookstein, F. L., Reddy, D. P., and Delson, E. 2003, Cranial allometry, phylogeography, and systematics of large-bodied papionins (Primates: Cercopithecinae) inferred from geometric morphometric analysis of landmark data. *Anatom. Rec. Part a-Discoveries Mol. Cell. Evol. Biol.* 275A:1048–1072.

Futuyma, D. J. 1998, *Evolutionary Biology, 3rd ed.* Sinauer, Sunderland, MA.

Gower, J. C. and Hand, D. J. 1995, Biplots, Chapman and Hall/CRC, London.

Groves, C. 2001, *Primate taxonomy*, Smithsonian Institution Press, Washington, DC.

Grubb, P., Butynski, T., Oates, J., Bearder, S., Disotell, T., Groves, C., and Struhsaker, T. 2003, Assessment of the diversity of African primates. *Int. J. Primatol.* 24:1301–1357.

Hayes, V. J., Freedman, L., and Oxnard, C. E. 1990, The taxonomy of savannah baboons—An odontomorphometric analysis. *Am. J. Primatol.* 22:171–190.

Henzi, P. and Barrett, L. 2003, Evolutionary ecology, sexual conflict, and behavioral differentiation among baboon populations. *Evol. Anthropol.* 12:217–230.

Hey, J. 2001, The mind of the species problem. *Trends Ecol. Evol.* 16:326–329.

Hill, R. A. 1999, Ecological and demographic determinants of time budgets in baboons: Implications for cross-populational models of baboon socioecology, University of Liverpool, UK.

Hill, R. A. and Lee, P. C. 1998, Predation risk as an influence on group size in cercopithecoid primates: Implications for social structure. *J. Zool.* 245:447–456.

Hill, W. C. O. 1967, Taxonomy of the baboon, in: Vagtborg, H., ed., *The Baboon in Medical Research*, University of Texas Press, Austin, pp. 3–11.

Isaac, N. J. B., Mallet, J., and Mace, G. M. 2004, Taxonomic inflation: Its influence on macroecology and conservation. *Trends Ecol. Evol.* 19:464–469.

Janson, C. H. 1992, Evolutionary ecology of primate social structure, in: Smith, E. A. and Winterhalder, B. eds., *Evolutionary Ecology and Human Behavior*, Aldine de Gruyter, New York, pp. 95–130.

Jolly, C. J. and Brett, F. 1973, Genetic markers and baboon biology. *J. Med. Primatol.* 2:85–99.

Jolly, C. J. 1993, Species, subspecies, and baboon systematics, in: Kimbel, W. H. and
    Martin, L. B., eds., *Species, Species Concepts, and Primate Evolution*, Plenum, New
    York, pp. 67–107.

Kamilar, J. 2004, An ecological view of baboon (*Papio*) taxonomy with insights from
    forest guenons (*Cercopithecus* spp.). *Am. J. Primatol.* 62:57.

Kappeler, P. M. and van Schaik, C. P. 2002, Evolution of primate social systems. 23:707–
    740.

Kenyatta, C. G. 1995, Ecological and social constraints on maternal investment strate-
    gies, Dissertation, University College, London.

Kimbel, W. H. and Rak, Y. 1993, The importance of species taxa in paleoanthropology
    and an argument for the phylogenetic concept of the species category, in: Kimbel,
    W. H. and Martin, L., eds., *Species, Species Concepts, and Primate Evolution*, Plenum
    Press, New York, pp. 461–484.

Lee, P. C. 1999, *Comparative Primate Socioecology*, Cambridge University Press, New
    York.

Losos, J. B. and Glor, R. E. 2003, Phylogenetic comparative methods and the geography
    of speciation. *Trends Ecol. Evol.* 18:220–227.

Mayr, E. 1942, *Systematics and the Origin of Species*, Columbia University Press, New
    York.

Mayr, E. 1982, *The Growth of Biological Thought: Diversity, Evolution and Inheritance*,
    Belknap Press, Cambridge.

McGarigal, K., Cushman, S., and Stafford, S. 2000, *Multivariate Statistics for Wildlife
    and Ecology Research*, Springer, New York.

Murphy, P. G. and Lugo, A. E. 1986, Ecology of tropical dry forest. *Ann. Rev. Ecol.
    Syst.* 17:67–88.

Muruthi, P. 1997, Socioecological correlates of parental care and demography in savanna
    baboons. University Microfilms, Inc, Ann Arbor, MI.

Nagel, U. 1971, Social organization in a baboon hybrid zone, *Proc, 3rd Int. Cong.
    Primatol.* 3:48–57.

Nagel, U. 1973, A comparison of anubis baboons, hamadryas baboons and their hybrids
    at a species border in Ethiopia. *Folia Primatol.* 19:104–165.

Newman, T. K., Jolly, C. J., and Rogers, J. 2004, Mitochondrial phylogeny and sys-
    tematics of baboons (*Papio*). *Am. J. Phys. Anthropol.* 124:17–27.

Norton, G. W., Rhine, R. J., Wynn, G. W., and R.D. Wynn 1987, Baboon diet: A five
    year study of stability and variability in the plant feeding and habitat of the yellow
    baboons (*Papio cynocephalus*) of Mikumi National Park, Tanzania. *Folia Primatol.*
    48:78–120.

Oliver, J. and Lee, P. 1978, Comparative aspects of the behaviour of juveniles in two
    species of baboon in Tanzania, in: Chivers, D. J. and Herbert, J. eds., *Recent Advances
    in Primatology*, Academic Press, New York, pp. 151–153.

Patterson, J. D. 1976, Variations in the ecology and adaptation of Ugandan baboons *Papio cynocephalus anubis*, University of Calgary, Canada.

Phillips-Conroy, J. E., Jolly, C. J., and Brett, F. L. 1991, Characteristics of hamadryaslike male baboons living in anubis troops in the Awash hybrid zone, Ethiopia. *Am. J. Phys. Anthropol.* 86:353–386.

Popp, J. L. 1978, Male baboons and evolutionary principles, University of Harvard.

Post, D., Hausfater, G., and McCuskey, S. 1980, Feeding behavior of yellow baboons (*Papio cynocephalus*): Relationship to age, gender and dominance rank. *Folia Primatol.* 34:170–195.

Post, D. G. 1981, Activity patterns of yellow baboons (*Papio cynocephalus*) in the Amboseli National Park, Kenya. *Anim. Behav.* 29:357–374.

Proctor, H. 1996, Behavioral characters and homoplasy: Perception versus practice, *Homoplasy: The Recurrence of Similarity in Evolution*, Academic Press, New York, pp. 131–149.

Ransom, T. W. 1981, *Beach troop of the Gombe*, Lewisberg, Bucknell University Press, Lewisberg.

Rasmussen, D. R. 1978, Environmental and behavioural correlates of changes in range use in a troop of yellow (*Papio cynocephalus*) and a troop of olive (*Papio anubis*) baboons, University of California, Riverside.

Ridley, M., ed. 1997, *Evolution*, Oxford University Press, New York.

Rogers, J. 2000, Molecular genetic variation and population structure in *Papio* baboons, in: Whitehead, P. F. and Jolly, C. J., eds., *Old World Monkeys*, Cambridge University Press, Cambridge, pp. 57–76.

Rosenzweig, M. 1968, Strategy of body size in mammalian carnivores. *Am. Midland Nat.* 80:299–315.

Ruvolo, M. 1988, Genetic evolution in the African guenons, in: Gautier-Hion, A., Bourliere, F. and Gautier, J.-P., eds., *A Primate Radiation: Evolutionary Biology of the African Guenons*, Cambridge University Press, Cambridge, pp. 127–139.

Samuels, A. and Altmann, J. 1986, Immigration of a *Papio anubis* male into a group of *Papio cynocephalus* baboons and evidence for an *Anubis-Cynocephalus* hybrid zone in Amboseli, Kenya. *Int. J. Primatol.* 7:131–138.

Shea, B. and Coolidge, H., Jr 1988, Craniometric differentiation and systematics in the genus Pan. *J. Hum. Evol.* 17:671–685.

Sites, J. W. and Marshall, J. C. 2003, Delimiting species: A Renaissance issue in systematic biology. *Trends Ecol. Evol.* 18:462–470.

Sokal, R. R. and Rohlf, F. J. 1995, *Biometry: The Principles and Practice of Statistics in Biological Research*, W. H. Freeman and Company, New York.

Spinage, C. A. 1970, Population dynamics of uganda defassa waterbuck (*Kobus defassa* -Ugandae Neumann) in Queen-Elizabeth-Park, Uganda. *J. Anim. Ecol.* 39:51–&.

Struhsaker, T., Butynski, T., and Lwanga, J. 1988, Hybridization between redtail (*Cercopithecus ascanius schmidti*) and blue (*C. mitis stuhlmanni*) monkeys in the Kibale Forest, Uganda, in: Gautier-Hion, A., Bourliere, F., Gautier, J. P. and Kingdon, J., eds., *A Primate Radiation: Evolutionaty Biology of the African Guenons*, Cambridge University Press, Cambridge, pp. 477–497.

Tabachnick, B. G. and Fidell, L. S. 1989, *Using Multivariate Statistics*, Harper and Row, New York.

van Schaik, C. P., and van Hooff, J. A. R. A. M. 1983, On the ultimate causes of primate social systems. *Behaviour* 85:91–117.

van Valen, L. 1976, Ecological species, multispecies, and oaks. *Taxon* 25:233–239.

Wilson, E. O. 1975, *Sociobiology: The New Synthesis*, Harvard University Press, Cambridge, MA.

Wimberger, P. H. and de Queiroz, A. 1996, Comparing behavioral and morphological characters as indicators of phylogeny, in: Martins, E. P., ed., *Phylogenies and the Comparative Method in Animal Behavior*, Oxford University Press, Oxford, pp. 206–233.

Woolley-Barker, T. 1999, Social organization and genetic structure in baboon hybrid zone. Dissertation, New York University, New York City.

Yomtov, Y. and Nix, H. 1986, Climatological correlates for body size of 5 species of Australian mammals. *Biol. J. Linnean Soc.* 29:245–262.

CHAPTER SEVEN

Biogeography and Evolution of the *Cercocebus-Mandrillus* Clade: Evidence from the Face

W. Scott McGraw and John G. Fleagle

ABSTRACT

Numerous lines of evidence indicate that mangabeys are not a natural group and that terrestrial mangabeys (genus *Cercocebus*) are more closely related to mandrills and drills (genus *Mandrillus*) than they are to arboreal mangabeys (genus *Lophocebus*). Available field data indicate that *Cercocebus* mangabeys and *Mandrillus* share a foraging regime characterized by a reliance on hard object foods and habitual aggressive use of the forelimbs during foraging. These behaviors are reflected in the dentition and limb anatomy of terrestrial mangabeys and *Mandrillus* to the exclusion of *Lophocebus*, *Papio* and *Theropithecus*. In this study, we examine variation in several facial characters in mangabey skulls to test biogeographic hypotheses about interrelationships of the

W. Scott McGraw • Department of Anthropology, 114 Lord Hall, The Ohio State University, 124 West 17th Avenue, Columbus, OH 43210-1364   John G. Fleagle • Department of Anatomical Sciences, Health Sciences Center, Stony Brook University, Stony Brook, NY 11794-8081

Primate Biogeography, edited by Shawn M. Lehman and John G. Fleagle.
Springer, New York, 2006.

Cercocebus-Mandrillus clade. All mangabeys possess depressed cheekbones, however the extent of maxillary excavation is much less pronounced in all *Cercocebus spp.*, particularly in *C. torquatus*. Mandrills exhibit little suborbital excavation. In *Mandrillus* and *Cercocebus*, the paranasal ridges run medially towards the incisors while in *Lophocebus albigena*, they run towards the canines. The extent of nasal ridge development—a striking feature in male mandrills—varies considerably in *Cercocebus* but is most pronounced in *C. torquatus*. Additionally, *Mandrillus spp.* and *C. torquatus* exhibit virtually identical orientations and development of the temporal lines. Based on these cranial features, we suggest that *C. agilis* exhibits the primitive cranial morphology while the derived condition is shared by *C. torquatus* and *Mandrillus*. The hypothesis that *C. torquatus* is the sister taxon of *Mandrillus* is concordant with Grubb's (1978, 1982) hypothesis for the evolution and radiation of terrestrial mangabeys.

Key Words: Papionins, Mangabey, Mandrill, Sub-orbital fossa, Maxillary fossa, Nasal bones, Facial Morphology

INTRODUCTION

Mangabeys are large, long-limbed monkeys generally restricted to forested regions throughout sub-Saharan Africa. Most taxonomists today recognize nine mangabey species: *albigena, aterrimus, kipunji, atys, torquatus, galeritus, agilis, chrysogaster* and *sanjei* (Kingdon, 1997; Groves, 2001; Jones *et al.*, 2005; but see Grubb *et al.*, 2003). Field studies have shown that the behavioral ecology of these species is diverse. The most striking difference among them is substrate preference: two and perhaps three species (*albigena, aterrimus,* and *kipunji*) are almost exclusively arboreal while the remainder are predominantly terrestrial (Chalmers, 1968; Jones and Sabater-Pi, 1968; Struhsaker, 1971; Happold, 1973; Quris, 1975; Waser, 1977, 1984; Homewood, 1978; Wallis, 1983; Harding, 1984; Horn, 1987; Mori, 1988; Mitani, 1989, 1991; Olupot *et al.*, 1994, 1997; Shah, 1996; McGraw, 1998).

For many years, all mangabeys were subsumed under one genus—*Cercocebus*—with two species groups separated on the basis of habitat preference and a handful of cranio-dental characters (Schwartz, 1928; Dobroruka and Badalec, 1966; Thorington and Groves, 1970; Napier, 1981). The first indication that there were differences beyond those of mere support use or gross morphology came from an analysis of hemoglobin which revealed substantial incongruence in the α and β chains of *albigena* and *torquatus/atys* samples, respectively (Barnicot and Hewett-Emmett, 1972). The integrity of the mangabey clade was more formally questioned by Cronin and Sarich (1976) who showed that blood proteins of the arboreal (*albigena/aterrimus*) and terrestrial (*galeritus/torquatus*) species were quite distinct from each other and that the arboreal mangabeys clustered with baboons and geladas while the terrestrial mangabey species and mandrills were not a part of that clade. Additional hemoglobin data from Hewett-Emmett and Cook (1978) bolstered the conclusion that mangabeys were not a natural group and mandrills were not closely related to *Papio* and *Theropithecus*.

Groves (1978) made a significant contribution to the taxonomic literature when he identified numerous cranio-dental differences between the arboreal and terrestrial mangabey species. In addition to detailed discussions of subspecies validities and allocations, he proposed that mangabeys be divided into two genera based on the cranio-dental and habitus differences; the two arboreal species were placed in the genus *Lophocebus* (Palmer, 1903) while the remaining terrestrial species were retained in *Cercocebus*.

Nakatsukasa (1994 a, b, 1996) examined cercopithecine post-crania in attempting to identify the morphotype of the ancestral cercopithecid and infer its habitat. He demonstrated that many of the locomotor differences between arboreal and terrestrial mangabey species were reflected in their limb morphologies. Among other things, Nakatsukasa noted that the humerus of the arboreal *albigena* has a reduced greater tuberosity, weak muscular insertions on the shaft, less pronounced trochlear keels, a thinner diaphysis, and less retroflexed medial epicondyle. Compared to the terrestrial *torquatus*, the femur of *albigena* has a shorter greater trochanter, a thinner shaft, and a wider patellar groove. Nakatsukasa argued that although the generally less robust limbs seen in *albigena* were not consistent with the pattern seen in other arboreal mammals, most features of *albigena*—including more mobile joints—represent the derived condition relative to the terrestrial papionin ancestor (Nakatsukasa, 1996).

In recent years, there has been a flurry of molecular studies confirming mangabey diphyly and strongly supporting the notion that members of the genus *Cercocebus* are more closely allied with mandrills and drills than they are to arboreal mangabeys of the genus *Lophocebus* (Disotell, 1994, 1996; Disotell *et al.*, 1992; Harris & Disotell, 1998; Harris, 2000; Page and Goodman, 2001; Page *et al.*, 1999). In a recent analysis of a large morphological data set, Groves (2000) found strong evidence that *Cercocebus* and *Mandrillus* are sister taxa but was unable to resolve relationships among other papionins. Other recent morphological studies have provided additional, but not totally concordant evidence in support of the molecular phylogeny of African papionins. In a geometric morphometric study of ontogeny of the face in papionins, Collard and O'Higgins (2001) found that *Lophocebus* and *Cercocebus* shows facial trajectories similar to *Macaca* and thus, presumably retained the primitive condition. Patterns of facial growth in *Papio* and *Mandrillus* were distinct from those found in any of these taxa and thus their adult similarities in facial length were interpreted as homoplasies. In another geometric morphometric study of the papionin cranium, Singleton (2002) demonstrated that the pattern of shape variation exhibited by *Lophocebus* was clearly distinguishable from all other papionins and was most divergent from *Cercocebus*.

In the course of a study of mangabey comparative anatomy, we identified a suite of characters in the limbs and teeth of *Cercocebus* and *Mandrillus* directly related to a shared foraging regime that is absent in the limbs and teeth of *Lophocebus, Papio,* and *Theropithecus* (Fleagle and McGraw, 1999, 2002). Mandrills spend large portions of their day rummaging through the leaf litter on the forest floor in search of fallen nuts and frequently using their forelimbs to rip apart rotting logs (Hoshino, 1985; Lahm, 1986; Harrison, 1988; Rogers *et al.*, 1996; Caldecott *et al.*, 1996). Many items uncovered are hard object foods that are resistant to decomposition and require high bite forces to open. Hoshino (1985) noted that it is their ability to process foods other sympatric species cannot open that allows mandrills to maintain large group sizes during seasonal periods of food shortage. The reliance on hard object foods and habitual, aggressive use of forelimbs during foraging is clearly reflected in the upper limb bones and teeth of both drills and mandrills. These monkeys have expanded, heavily worn premolars and their forelimb bones show many features indicative of powerful wrist and elbow flexion and rotation (Fleagle and McGraw, 1999, 2002). The large muscle markings we observed are consistent with the findings of Jolly (1967, 1970, 1972) who noted that mandrills possess relatively larger forearm muscles than do baboons. We would therefore expect to find larger forearm flexors and wrist rotators in *Cercocebus* compared to *Lophocebus*.

The foraging behavior reported for mandrills is quite similar to that observed in at least one terrestrial mangabey species. In the Ivory Coast's Tai Forest, sooty mangabeys (*Cercocebus atys*) forage predominantly on the ground where they habitually paw through debris on the forest floor in search of fallen foods in a fashion similar to that observed in mandrills (McGraw, 1996; Bergmueller, 1998). Processing of certain food items requires high bite forces and the cracking of nuts by groups of sooty mangabeys can be heard throughout the Tai Forest. The features related to this foraging behavior that are present in mandrills including molarized and heavily worn premolars, expanded deltoid tuberosities, and proximally extended and laterally widened supinator crests are readily observed in *Cercocebus* spp. limbs and teeth (see Figure 5 in Fleagle and McGraw, 2002). In short, there are a number of striking features shared by *Cercocebus* and *Mandrillus*—to the exclusion of *Lophocebus*, *Papio*, and *Theropithecus*—that support the molecular phylogeny and confirm the diphyletic nature of mangabeys.

The goal of this paper is to examine diversity in the skeletal morphology among *Cercocebus* mangabeys that might provide clues to the biogeographic history of the group and to the phylogenetic and biogeographic origin of mandrills and drills. We believe that distribution information and morphology can be valuable tools in reconstructing evolutionary relationships and divergence scenarios. Unfortunately, samples of postcranial elements are too small and unevenly distributed among age and sex classes to permit broad intrageneric comparisons. Therefore, in this study, we examine variation of a few prominent facial characters in mangabey and mandrill skulls and correlate this variation with the biogeography of individual taxa in order to test hypotheses about the interrelationships of the *Cercocebus-Mandrillus* clade. We first review the known distributions of *Cercocebus* and *Mandrillus* species and then examine the size and distribution of a few diagnostic facial characters. In so doing, we hope to shed light on the evolution and radiation of this clade.

Distribution of *Cercocebus* and *Mandrillus* spp.

Terrestrial mangabeys of the genus *Cercocebus* are taxonomically diverse and widespread. Following Kingdon (1997) and Groves (2001) we recognize six allopatric species within the genus (Figure 1). *Cercocebus atys,* the sooty mangabey is the westernmost species ranging from Ivory Coast to Guinea. Napier (1981) gives a western range extension to Fouta Djallon, Guinea (11°30′N, 12°30′W) (Monard 1938). *Cercocebus atys* is the only *Cercocebus* taxon with delineated subspecies; the two subspecies—*atys* to the west and *lunulatus* to the east—are separated by the Nzo-Sassandra River system in the Ivory Coast. *Cercocebus torquatus,* the collared or red-capped mangabey, ranges from western Nigeria (Cross River) through Rio Muni to western Gabon. *Cercocebus agilis,* the agile mangabey, is endemic to a region in central Africa north of the

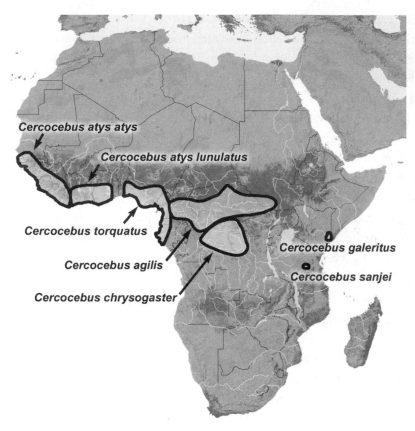

Figure 1. Distribution of six *Cercocebus* species.

Zaire River. It is found in Rio Muni, Equatorial Guinea, Gabon, Cameroon, and Democratic Republic of Congo. *Cercocebus chrysogaster*, the golden-bellied mangabey, is found in the lower Congo River basin. *Cercocebus galeritus*, the Tana River mangabey, is restricted to gallery forest along Kenya's Tana River. *Cercocebus sanjei*, the Sanje mangabey, has the most limited distribution and is found on the eastern slopes of the Uzungwa Mountains, Tanzania (Figure 1).

The distribution of drills and mandrills is more limited (Figure 2). Drills, *Mandrillus leucophaeus*, are restricted to southeastern Nigeria, northwestern Cameroon, and Bioko Island (Gartlan, 1970; Napier, 1981; Groves, 2001). Groves (2001) argues that the Bioko Island form should be considered a separate subspecies (*M. leucophaeus poensis*) based on its smaller skull and more grayish coat compared to the mainland form (*M. leucophaeus leucophaeus*). Mandrills, *Mandrillus sphinx*, are found in Cameroon, Rio Muni, Gabon, and

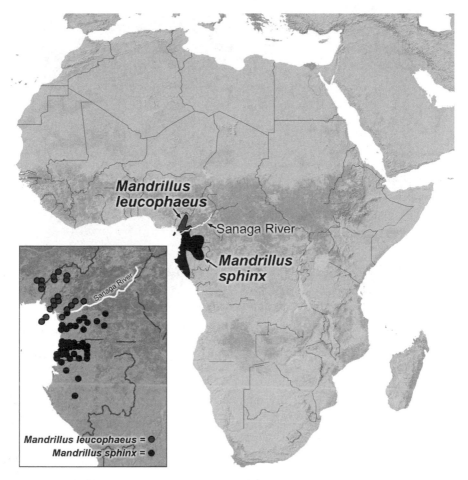

Figure 2.   Distribution of mandrills (*Mandrillus sphinx*) and drills (*M. leucophaeus*).

Congo as far as the Kouilou River (Napier, 1981; Groves, 2001). Although the precise geographic barrier separating drills and mandrills has been questioned, most authorities believe these taxa are allopatric (Grubb, 1973; Kingdon, 1997) (Figure 2).

METHODS

The most striking feature of the *Mandrillus* skull is the massive muzzle dominated by dramatic nasal ridges (Figure 3). These ridges are greatly enlarged in males and correspond to prominent, cobalt-blue paranasal swellings flanking a scarlet midline stripe. The size and color of the swellings is positively correlated

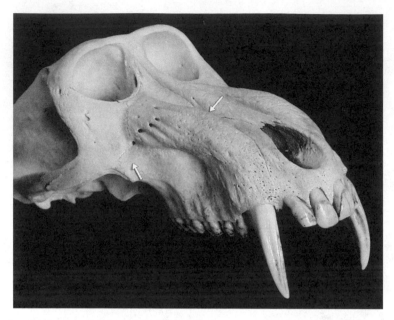

Figure 3.   Male mandrill (*Mandrillus sphinx*) skull. Notice the massive paranasal ridges and the lack of a suborbital fossa.

with testosterone levels and dominance rank in free-ranging mandrills males (Wickings and Dixson, 1992). Hylander and Johnson (2002) note that these ridges consisting of highly vascularized bone are examples of non-mechanical features of facial morphology used for attracting mates. This is almost certainly a derived feature and we hypothesized that the *Cercocebus* species most closely related to mandrills would display incipient ridging in its face.

Mangabeys—both *Cercocebus* and *Lophocebus*—are described as possessing deep fossae below the orbits (Napier and Napier, 1985; Szalay and Delson, 1979; Kingdon, 1997). Indeed, Groves (1978) remarked that the only skull feature found in all mangabey species is the suborbital fossa. Groves speculated that the fossa formed differently in the two groups, "and has probably developed independently in response, perhaps, to facial shortening from a long-faced ancestor to preserve a complex facial musculature." Kingdon (1997) has also argued that the fossa is a derived feature related to facial shortening from a long-snouted ancestor. We are less certain of the polarity of this feature among papionins. However, we were immediately struck by variation in the extent of maxillary excavation not only between *Lophocebus* and *Cercocebus*, but also within members of *Cercocebus* (Figure 4).

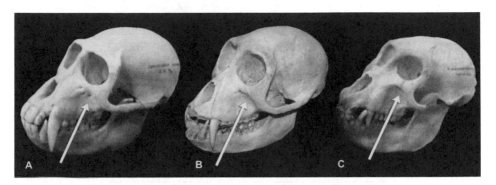

Figure 4. Skulls of male *Cercocebus torquatus* (A), *Cercocebus atys* (B), and *Lophocebus albigena* (C). Notice the significant difference in depth of the suborbital fossa in the two skulls as indicated by the arrows, and the paranasal ridges in *C. torquatus*. Note the intermediate morphology of *C. atys*.

We examined 68 *Cercocebus* and *Mandrillus* crania and 50 *Lophocebus* crania (Table 1). We were struck by the tremendous variation in the presence, size, and direction of the paranasal ridges in *Cercocebus* relative to the conditions observed in *Mandrillus* and *Lophocebus*. We used a scale of 0–4 to record ridge size: 0 (none discernible), 1 (small), 2 (moderate), 3 (large), and 4 (massive). We noted the direction of the ridge along the muzzle as running towards the canine root (C), towards the incisors (I), or intermediate (CI). In addition, we evaluated the extent of suborbital (maxillary) excavation in each taxon. We coded the size of the maxillary depression as 0 (no depression), 1 (small), 2 (moderate), (3) large, and 4 (massive). In view of this morphological diversity in facial morphology, we hypothesized that the differences in facial morphology would reflect relationships among mangabey subspecies and that the taxon most closely related to *Mandrillus* would show the greatest similarity in these derived facial features.

Table 1. Sample size

	Male	Female
Cercocebus torquatus	25	6
Cercocebus agilis	4	3
Cercocebus atys	4	2
Cercocebus galeritus	2	1
Mandrillus sphinx	6	2
Mandrillus leucophaeus	7	6
Lophocebus albigena	27	20
Lophocebus aterrimus	1	2

Table 2.   Maxillary excavation

	Male	Female
Mandrillus sphinx	0	0
Mandrillus leucophaeus	.9 (.6)	.5 (.4)
Cercocebus torquatus	.9 (.4)	.7 (.3)
Cercocebus agilis	2 (1.1)	1.7 (1.2)
Cercocebus atys	1.4 (.3)	2 (0)
Cercocebus galeritus	1.5 (.5)	1
Lophocebus albigena	2.8 (.4)	2.6 (.4)
Lophocebus aterrimus	3	3

[means and SD of suborbital fossa size scored 0–4]

RESULTS

Results from our analysis of maxillary excavation are presented in Table 2. As noted by many previous authors, both male and female *Lophocebus albigena* and *L. aterrimus* possess very deep suborbital fossae. In contrast, maxillary excavation is significantly less pronounced and varies considerably among members of *Cercocebus* (Figure 4). Most significantly, the collared mangabey, *Cercocebus torquatus,* is characterized by only slight suborbital depressions. Skulls from the remaining *Cercocebus* species, particularly *C. agilis*, exhibit deeper suborbital fossa; however none approach the condition seen in *Lophocebus* spp. The suborbital fossa of *C. atys* represents the intermediate condition. Mandrills and drills exhibit little maxillary excavation (e.g. Figures 3). In fact, there is no suborbital fossa in male or female *Mandrillus sphinx* skulls while *M. leucophaeus* males are characterized by a shallow depression below the orbits. It is notable that the extent of suborbital excavation in drills and *C. torquatus* is identical (Figure 5).

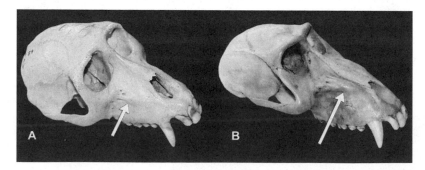

Figure 5.   Skull of male *Cercocebus torquatus* (A) and a drill *Mandrillus leucophaeus* (B). Notice the similarity in ridging indicated by arrows.

Table 3. Nasal ridge development and direction

	Ridge size	
	Male	Female
Mandrillus sphinx	4	1.7 (.5)
Mandrillus leucophaeus	3.3 (.7)	1.4 (.5)
Cercocebus torquatus	1.4 (.4)	.7 (.4)
Cercocebus agilis	.4 (.4)	0
Cercocebus atys	.5 (0)	.5 (0)
Cercocebus galeritus	.5 (0)	0
Lophocebus albigena	.8	.3

[means and SD of paranasal ridge size scored 0–4]

Results from our analysis of nasal ridge development are presented in Table 3. Mandrills have the most pronounced paranasal ridges. All male mandrills we examined exhibited the maximum (4) amount of ridge development. This feature is also well developed in drills although the extent of ridging is slightly reduced compared to mandrills. The extent of ridge development in female mandrills and drills is much less than in males. Collared mangabeys, *Cercocebus torquatus*, show the most pronounced ridge morphology of any *Cercocebus* species (Figure 6). Males of the remaining *Cercocebus* spp. examined showed

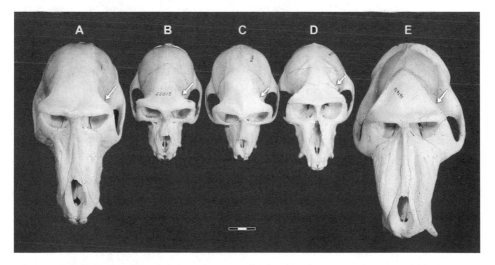

Figure 6. Superior view of the cranium of African papionins, showing differences in the trajectory of the temporal lines. (A) *Papio ursinus*; (B) *Lophocebus aterrimus*; (C) *Cercocebus agilis*; (D) *Cercocebus torquatus*; and (E) *Mandrillus sphinx*.

only minor ridge development and females of these species exhibit little, if any ridge development.

The prominence of paranasal ridging is not the only similarity in the mid-face of *Mandrillus* and *Cercocebus* spp. In these taxa, the paranasal ridges run medially towards the incisors while in the *Lophocebus albigena* specimens ridges, when discernible, run towards the canines.

In addition to differences in the faces of African papionins, we observed differences in the development of temporal lines among the taxa, a feature also noted by other observers of these primates (Groves, 2000; Chris Gilbert, pers. comm.). In *Papio*, *Theropithecus*, *Lophocebus*, some *Cercocebus*, and most other primates, the temporal lines follow the contour of the superior aspect of the cranium. They originate at the lateral rim of the orbit, diverge medially along the infratemporal fossa and then extend posteriorly to the occiput (Figure 6). However, in *Mandrillus* and *Cercocebus torquatus*, the temporal lines extend posteromedially in almost a straight line from the lateral walls of the orbit to the occiput, with little or no medial deflection along the infratemporal fossa. Although we did not quantify this feature, the similarity between *Cercocebus torquatus* and *Mandrillus* is striking and almost certainly a derived feature.

DISCUSSION

There is now substantial evidence that *Cercocebus* and *Lophocebus* skulls differ in a number of important ways (Groves, 1978; Singleton, 2002; Collard and O'Higgins, 2001). Our results suggest that it is misleading to characterize all mangabeys as having deep suborbital fossae. Arboreal mangabeys (*Lophocebus* spp.) certainly do, however this is not the case for most *Cercocebus* spp. Further, while most *Cercocebus* and all *Lophocebus* species possess only minor paranasal ridging, in *Lophocebus* the ridges run towards the canines whereas in *Cercocebus* they run medially towards the incisors. The distinctive morphology of the temporal lines seems limited to *Mandrillus* and *C. torquatus*.

There is certainly variation in these features both between species and across sexes within the *Cercocebus-Mandrillus* clade, however a general morphocline is recognizable. *Cercocebus torquatus* is most similar to *Mandrillus* in possessing the most pronounced paranasal ridges, the shallowest suborbital fossae, and the straight temporal lines. In contrast, *C. agilis* approaches the condition seen in *Lophocebus* characterized by a deep maxillary fossa and weakly developed—if any—nasal ridges. *C. atys* occupies a position intermediate to that displayed by

C. torquatus and *C. agilis*. Our limited sample of *C. galeritus* exhibits a mixture of features: maxillary excavation in males is similar to that seen in *C. atys* while ridge development falls with *C. agilis* and *C. atys*.

Taken together, these data suggest that in terms of maxillary fossa depth, nasal ridge development, and the morphology of the temporal lines, *C. torquatus* skulls are quite distinct from those of other *Cercocebus spp.* males. We believe that the features seen in *C. torquatus* and *Mandrillus spp.* represent the derived condition and that these species are sister taxa. We posit that the deep cheeks and absence of ridging seen in *C. agilis* represents the primitive condition since these features more closely approximate the condition seen in *Lophocebus*. In this scenario, the suborbital fossa in *Lophocebus* and some *Cercocebus* is not a derived feature, at least within extant African papionins, but is primitive for the group. However, the absence of suborbital fossae in macaques, the most likely outgroup to African papionins would argue against this view and suggest that suborbital fossae are derived only for the common ancestor of *Cercocebus* and *Lophocebus*.

Grubb (1978, 1982) proposed a hypothesis for the radiation of the *Cercocebus* clade. He identified five faunal regions (refugia) based on the total number of mammalian species and the number of endemic species in each region (Figure 7). His hypothetical dispersal route reconstructs the path of the ancestral *Cercocebus* population and the subsequent differentiation of its descendents (Figure 8). "At one stage in their history ... the mangabey (*Cercocebus agilis*) dispersed from Central to West Africa, forming presumably continuous populations across the continent ... *Cercocebus atys* differentiated when these continuous distributions were interrupted. Subsequently ... the westernmost species gave rise to eastward dispersing animals, *Cercocebus torquatus* (dispersing down the coast to Gabon). The discreteness of these species was emphasized once again by a break in distributions, the Volta River and Dahomey gap for *Cercocebus*. Intermediate populations .... between *Cercocebus atys* and *C. torquatus*, assigned subspecifically to the more western and ancestral species ... as *C. a. lunulatus* ... replace the nominate races in Ivory Coast and Ghana." (1978:544–545)

Grubb (1978, 1982) argues that the ancestral *Cercocebus* population dispersed from a central Africa refuge during wet periods and that radiating populations became isolated from parent populations during dry periods throughout the Quaternary (Kukla 1977) (Figure 8). According to his analysis, *Cercocebus agilis* best represents the ancestral species based on its presence in

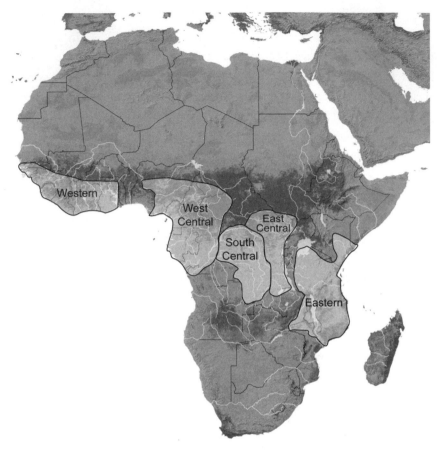

Figure 7. African refugia identified by Grubb (1978, 1982).

the hypothesized East Central refuge and the polarity of cranial characters established for *Cercocebus* by Groves.[1] During the approximately 17 glacial-interglacial events throughout the Quaternary, members of the original *Cercocebus* population migrated east, west, and south subsequently becoming isolated during arid periods. One population migrated as far as the western refuge where its modern descendent—*Cercocebus atys atys*—is found today in Sierra Leone, Liberia, and western Ivory Coast. To the south, *C. chrysogaster* became isolated in the Congo basin below the great bend in the Zaire River while *C. galeritus* is descended from a population that became isolated to the east in present day Kenya. Finally, Grubb hypothesized that *C. torquatus* was derived from a

---

[1] We are unable to determine how Groves (1978) established the character polarity.

Figure 8. Hypothetical dispersal route of *Cercocebus* spp. posited by Grubb (1978, 1982).

population of mangabeys (*C. atys lunulatus*) that had migrated back in an easterly direction. In this scheme, *C. atys* and *C. chrysogaster* represent populations of modern monkeys descended directly from the ancestral *C. agilis* population. *C. torquatus* is the product of a more complex branching event having been descended from *C. atys lunulatus* and it is therefore regarded as more derived.

Independent analysis of other mammalian species-groups with distributions similar to those of modern *Cercocebus* species have shown that this proposed radiation is plausible. Our confidence in the proposed dispersal route for terrestrial mangabeys beginning in Central Africa is strengthened by the fact that a similar route was used to explain the distribution, divergence, and relationships of members of the Mona (*Cercopithecus*) superspecies (Booth, 1955), red colobus monkeys (Struhsaker, 1981) and black and white colobus monkeys (Oates and Trocco, 1983; Oates *et al.*, 2000). If the dispersal scenario posited by Grubb (1978, 1982) is true, then the biogeography and pattern of facial morphology discussed in this analysis can be used to speculate about the affinities of individual species within the *Cercocebus-Mandrillus* clade.

Lophocebus is our outgroup and represents the primitive condition characterized by a deep suborbital fossa and little to no paranasal ridging. The deep maxillary fossa and lack of ridging seen in *C. agilis* is expected for the ancestral *Cercocebus* species found in central African refuge identified by Grubb (1978, 1982). The most derived condition is that seen in *C. torquatus* which includes little maxillary excavation and pronounced ridging. The intermediate morphologies present in *C. atys* and to a lesser extent (based on small sample) in *C. chrysogaster* are consistent with the model that these taxa represent modern but direct branching events from a *C. agilis* population that became isolated in west and central Africa, respectively. *C. torquatus* is most similar to mandrills in the facial characters we examined. The current distribution of collared mangabeys lies within West Central refuge zone identified by Grubb (1978, 1982). Although the precise range of *C. torquatus* is disputed, it undoubtedly overlaps with portions of that occupied presently by drills and mandrills, as indicated by similarities in the SIV viruses of northern mandrills and *Cercocebus torquatus*. (Telfer *et al.*, 2003). Thus in terms of geographic proximity, these species are certainly well positioned to be sister taxa.

Little is known about the free-ranging behavior of *C. torquatus*, however if collared mangabeys are the sister taxon of mandrills, we could expect similarities in their social behavior[2]. Mandrills are the most sexually dimorphic terrestrial cercopithecid in terms of overall body size and sex-specific adornments (Setchell *et al.*, 2001). Much of the dramatic coloration and other secondary sex characteristics displayed by male mandrills are undoubtedly related to attracting mates

---

[2] Given their molecular and anatomical similarities as well as their adjacent and perhaps overlapping distributions, it is possible that *Mandrillus* and *Cercocebus torquatus* interbreed in the wild. There are no reports of this behavior from free-ranging populations. However, we did learn of a monkey in the Brookfield Zoo (Chicago) born to a mandrill male and female *C. torquatus*. Dr. Anne Baker, then curator of mammals, writes: 'At Brookfield Zoo, mandrills and mangabeys share a large exhibit . . . while they are on exhibit together the mangabey and mandrill youngsters interact frequently. Mandrill males mount mangabey females, and vice versa. At the time the hybrid was conceived, we were not aware that mangabeys and mandrills could hybridize" (pers. comm). It is not known if this individual was fertile.

and there intense intermale competition in mandrill groups. If the bony ridges corresponding to the bright blue paranasal swellings on the mandrill muzzle is indicative of particularly high levels of male-male competition in this clade, then we might expect populations of *C. torquatus* to exhibit this feature of social behavior in wild populations. Further, levels of canine dimorphism are strongly correlated with levels of mate competition. We are not aware of any metric data for *C. torquatus* canines, however data available for other mangabey species are known to vary significantly (Plavcan and van Schaik, 1992). If collared mangabeys social groups and mating systems are organized along principles similar to those operating in mandrill society, we would predict the level of canine dimorphism in *C. torquatus* to be the greatest of all *Cercocebus* species and approaching the level present in mandrills. Our impression is that *C. torquatus* males have massive canines (see Figure 4).

A critical species not included in this analysis is the golden-bellied mangabey, *C. chrysogaster*. Kingdon (1997) called all members of *Cercocebus* "drill-mangabeys" and argued that *C. chrysogaster* was likely the sister taxon of drills. According to Kingdon, the golden-bellied mangabey is, "the most drill-like of drill mangabeys in having a naked, violet rump, bright-colored fur and relatively robust build (including the muzzle of adult males) . . . The tapered tail is carried in a backward arch (unlike other mangabeys)." Although the golden-bellied mangabey occupies a range immediately south of *C. agilis*, Kingdon hypothesizes that the two species may have been separated for a considerable length of time owing to the great width and flow rate of the Zaire River between them (Kingdon, 1997). It would be extremely interesting to examine the cranial anatomy of this rare monkey in light of Kingdon's argument that it shares so many external characteristics with drills. Unfortunately, we are unaware of any adult skeletal material for *C. chrysogaster* in museums. Testing this hypothesis that *C. chrysogaster* is the sister taxon of *Mandrillus* must await the acquisition of comparative material.

It is also worrisome that our morphological analysis is in conflict with all known molecular studies of African papionins which show no evidence of a unique genetic relationship between *Mandrillus* and any particular species of *Cercocebus* (e.g. Disotell, 1994, 1996; Disotell *et al.*, 1992, Harris and Disotell, 1998; Harris, 2000; Page and Goodman, 2001; Page *et al.*, 1999; Telfer *et al.*, 2003). This would suggest that the morphological similarities between *C. torquatus* and *Mandrillus* are parallelisms, probably associated with large size rather than synapomorphies. In this case, the similarities would be additional evidence for similar developmental patterns in the cranial anatomy of *Cercocebus* and *Mandrillus*. Alternatively, they might reflect some recent gene flow between *C. torquatus* and *Mandrillus* (which is also suggested as a possibility by the SIV viruses and the hybrids in captivity) that occurred subsequent to their divergence. We look forward to specific genetic studies aimed at testing these hypotheses.

CONCLUDING REMARKS

Papionin crania have been the subject of a number of recent studies (Collard and O'Higgins, 2001; Singleton, 2002; O'Higgins and Jones, 1998; O'Higgins and Collard, 2002; Leigh *et al.*, 2003). These authors have demonstrated that various cranial similarities between *Mandrillus* and *Cercocebus* on the one hand and *Papio* and *Lophocebus* on the other are due to a combination of complex evolutionary processes including sexual dimorphism, ontogeny, scaling factors, and homoplasy. Unfortunately, there appear to be an equal number of size-related characters that are shared by mangabeys only or by baboons and mandrills which only complicates matters further. Although it is apparent that homoplasy is rampant within the papionin group (Lockwood and Fleagle, 1999; Collard and Wood, 2001), there are strong indications from these studies that *Cercocebus* and *Mandrillus* skulls are most similar to each other while *Lophocebus* is the most divergent.

Our modest analysis of cranial features within the *Cercocebus-Mandrillus* clade has led to a number of conclusions that we offer as hypotheses to be tested with larger data sets. Arboreal mangabeys *Lophocebus* represent the primitive papionin facial morphology characterized by deep maxillary fossa and little to no paranasal ridging. Terrestrial mangabeys of the genus *Cercocebus* display variation in these characters, however the agile mangabey *Cercocebus agilis* appears to most closely approximate the primitive condition of minor ridging and deeper maxillary fossae. In contrast, the collared mangabey *Cercocebus torquatus* is distinct from the rest of its conspecifics in possessing only minor maxillary excavation, pronounced paranasal ridging, and a distinctive pattern of cranial temporal lines. In these respects, *C. torquatus* is most similar to mandrills and drills. We conclude that the conditions shared by *C. torquatus* and *Mandrillus* spp. represent the derived condition. *Cercocebus atys* exhibits an intermediate morphology. The distribution of characters within *Cercocebus* supports the hypothetical dispersal route for the clade proposed by Grubb (1978, 1982). Based

on their current distributions and shared cranial morphologies, it is most parsimonious to conclude that *C. torquatus* is the sister species to *Mandrillus*.
We predict that when details of *C. torquatus* social behavior become known,
this species will show marked affinities in mating strategies, including levels of
male-male competition, with mandrills and drills.

ACKNOWLEDGMENTS

We gratefully acknowledge the support of the L.S.B. Leakey Foundation. We
thank the curators of the following museums for access to specimens in their
care: American Museum of Natural History, British Museum of Natural History, Powell-Cotton Museum, Royal Museum of Central Africa, Museum of
Comparative Zoology (Harvard). We thank Dr Anne Baker, director of the
Rosamond Gifford Zoo, for providing information on the mangabey-mandrill
hybrid. Finally, we thank Luci Betti-Nash for preparing the maps, and Chris
Gilbert for insightful comments and suggestions.

REFERENCES

Barnicot, N. A. and Hewett-Emmett. D. 1972, Red cell and serum proteins of *Cercocebus, Presbytis, Colobus* and certain'n other species. *Folia primatol.* 17:442–457.

Bergmueller, R. 1998, *Nahrungsoekologie und Konkurrenz um Nahrungressourcen bei-
der Raunchgrauen Mangabe (Cercocebus atys)*. Diploma, University of Regensburg,
Germany.

Booth, A. H. 1955, Speciation in the mona monkey. *J. Mammal.* 36:434–449.

Caldecott, J. O., Feistner, A. T. C., and Gadsby, E. L. 1996, A comparison of ecological
strategies of pig-tailed macaques, mandrills and drills, in: Fa, J. E., and Lindburg,
D. G. eds., *Evolution and Ecology of Macaque Societies*, Cambridge University Press,
Cambridge, pp. 73–94.

Chalmers, N. R. 1968, Group composition, ecology and daily activity of free living
mangabeys in Uganda. *Folia Primatol.* 8:247–262.

Collard, M. and O'Higgins, P. 2001, Ontogeny and homoplasy in the papionin monkey
face. *Evol. Deve.* (35): 322–331.

Collard, M. and Wood, B. 2001, Homoplasy and the early hominid masticatory system:
inferences from analyses of extant hominoids and papionins. *J. Hum. Evol.* 41:167–
194.

Cronin, J. E. and Sarich, V.M. 1976, Molecular evidence for dual origin of mangabeys
among Old World monkeys. *Nature* 260:700–702.

Disotell, T. R. 1994, Generic-level relationships of the Papionini (Cercopithecoidea). *Am. J. Phys. Anthropol.* 94:47–57.

Disotell, T. R. 1996, The phylogeny of Old World monkeys. *Evol. Anthropol.* 5(1):18–24.

Disotell, T. R., Honeycutt, R. L., and Ruvolo, M. 1992, Mitochondrial DNA phylogeny of the Old World monkey tribe Papionini. *Mol. Biol. Evol.* 9:1–13.

Dobroruka, L. J. and Badalec, J. 1966, Zur artbildung der mangaben gattung *Cercocebus* (Cercopithecidae, Primates). *Rev. Zool. Bot. Afr.* LXXIII (3–4):345–350.

Fleagle, J. G. and McGraw, W. S. 1999, Skeletal and dental morphology supports diphyletic origin of baboons and mandrills. *Proc. Nat. Acad. Sci.* 96:1157–1161.

Fleagle, J. G. and McGraw, W. S. 2002, Skeletal and dental morphology of African papionins: unmasking a cryptic clade. *J. Hum. Evol.* 42:267–292.

Gartlan, J. S. 1970, Preliminary notes on the ecology and behavior of the drill *Mandrillus leucophaeus* Ritgen, 1824, in: Napier, J. R., and Napier, P. H., eds., *Old World Monkeys: Evolution, Systematics and Behavior*, Academic Press, New York.

Gebo, D. L. and Sargis, E. J. 1994, Terrestrial adaptations in the postcranial skeletons of guenons. *Am. J. Phys. Anthropol.* 93:341–371.

Gilbert, C. 2005, pers. comm.

Groves, C. P. 1978, Phylogenetic and population systematics of the mangabeys (Primates: Cercopithecoidea). *Primates* 19(1):1–34.

Groves, C. P. 2000, The phylogeny of the Cercopithecoidea, in: Whitehead, P. F., and Jolly, C. J., eds., *Old World Monkeys*, Cambridge University Press, Cambridge.

Groves, C. P. 2001, *Primate Taxonomy*, Smithsonian Institute Press, Washington, DC.

Grubb, P. 1973, Distribution, divergence and speciation of the drill and mandrill. *Folia Primatol.* 20:161–177.

Grubb, P. 1978, Patterns of speciation in African mammals. *Bull. Carn. Mus. Nat. Hist.* 6:152–167.

Grubb, P. 1982, Refuges and dispersal in the speciation of African mammals, in: Prance, G. T., ed., *Biological Diversification in the Tropics*, Columbia University Press, New York, pp. 537–553.

Happold, D. C. D. 1973, The Red Crowned mangabey, Cercocebus torquatus, in western Nigeria. *Folia primatol.* 20:423–428.

Harding, R. S. O. 1984, Primates of the Kilimi area, northwest Sierra Leone. *Folia primatol.* 42:96–114.

Harris, E. E. 2000, Molecular systematics of the Old World monkey tribe Paionini: Analysis of the total available genetic sequences. *J. Hum. Evol.* 38:235–256.

Harris, E. E. and Disotell, T. R. 1998, Nuclear gene trees and the phylogenetic relationship of the mangabeys (Primates:Papionini). *Mole. Biol. Evol.* 15:892–900.

Harrison, M. J. 1988, The mandrill in Gabon's rain forest–ecology, distribution and status. *Oryx* 22(4):218–228.

Hewett-Emmett, D. and Cook, C. N. 1978, Atypical evolution of papionin α-haemoglobins and indication that *Cercocebus* may not be a monophyletic genus, in: Chivers, D. J., and Joysey, K. A., eds., *Recent Advances in Primatology*, Academic Press, London, pp. 291–294.

Homewood, K. M. 1978, Feeding strategy of the Tana mangabey (*Cercocebus galeritus galeritus*) (Mammalia:Primates). *J. Zool. Lond.* 186:375–391.

Horn, A. D. 1987, The socioecology of the Black mangabey (*Cercocebus aterrimus*) near Lake Tumba, Zaire. *Am. J. Primatol.* 12:165–180.

Hoshino, J. 1985, Feeding ecology of mandrills (*Mandrillus sphinx*) in Campo Animal Reserve, Cameroon. *Primates* 26:248–273.

Hylander, W. L. and Johnson, K. R. 2002, Functional morphology and in vivo bone strain in the craniofacial region of primates: Beware of biomechanical stories about fossil bones, in: Plavcan, J. M., Kay, R. F., Jungers, W. L., and van Schaik, C. P., eds., *Reconstructing Behavior in the Primate Fossil Record*, Kluwer Academic, New York, pp. 43–72.

Jolly, C. J. 1967, The evolution of the baboons, in: Vagtborg, H., ed., *The Baboon in Medical Research, II*, University of Texas Press, Austin, pp. 23–50.

Jolly, C. J. 1970, The large African monkeys as an adaptive array, in: Napier, J. R., and Napier, P. H., eds., *Old World Monkeys: Evolution, Systematics and Behavior*, Academic Press, New York, pp. 139–174.

Jolly, C. J. 1972, The classification and natural history of *Theropithecus* (*Simopithecus*) (Andrews, 1916), baboons of the African Plio-Pleistocene. *Bulletin of the British Museum (Natural History), Geology.* 22:1–123.

Jones, C. and Sabater–Pi, J. 1968, Comparative ecology of *Cercocebus albigena* (Gray) and *Cercocebus torquatus* (Kerr) in Rio Muni, West Africa. *Folia primat.* 9:99–113.

Jones. T., Ehardt, C. L., Butynski, T. M., Davenport, T. R. B., Mpunga, N. E., Machaga, S. J., and De Luca D.W. 2005, The highland mangabey *Lophocebus kipunji*: A new species of African monkey. *Science* 308(20):1161–1164.

Kingdon, J. 1997, *The Kingdon Field Guide to African Mammals*, Academic Press, New York.

Lahm, S. A. 1986, Diet and habitat preference of Mandrillus sphinx in Gabon: Implications of foraging strategy. *Am. J. Primatol.* 11:9–26.

Leigh S. R., Shah, E., and Buchanan L. S. 2003, Ontogeny and phylogeny in papionin primates. *J. Hum. Evol.* 45: 285–316.

Lockwood, C. A. and Fleagle, J. G. 1999, The recognition and evaluation of homoplasy in primate and human evolution. *Ybk. Phys. Anthropol.* 42:189–232.

McGraw, W. S. 1996, *The positional behavior and habitat use of six sympatric monkeys in the Tai forest, Ivory Coast.* PhD thesis. SUNY at Stony Brook.

McGraw, W. S. 1998, Comparative locomotion and habitat use of six monkeys in the Tai Forest, Ivory Coast. *Am. J. Phys. Anthro.* 105:493–510.

Mitani, M. 1989, *Cercocebus torquatus*: Adaptive feeding and ranging behaviors related to seasonal fluctuations of food resources in the tropical rain forest of south-western Cameroon. *Primates* 30(3):307–323.

Mitani, M. 1991, Niche overlap and polyspecific associations among sympatric cercopithecids in the Campo Animal Reserve, southwestern Cameroon. *Primates* 32:137–151.

Monard, A. 1938, Resultants de la mission scientifique du Dr. Monard en Guinee Portugaise 1937–1938. *Arquivos do Museu Bocage* 9:121–150.

Mori, A. 1988, Utilization of fruiting trees by monkeys as analyzed from feeding traces under fruiting trees in the tropical rain forest of Cameroon. *Primates* 29:21–40.

Nakatsukasa, M. 1994a, Intrageneric variation of limb bones and implications for positional behavior in Old World monkeys. *Z. Morph. Anthrop.* 80(1):125–136.

Nakatsukasa, M. 1994b, Morphology of the humerus and femur in African mangabeys and guenons: functional adaptations and implications for the evolution of positional behavior. *Afr. Study Monogr.* 21:1–61.

Nakatsukasa, M. 1996, Locomotor differentiation and different skeletal morphologies in mangabeys (*Lophocebus* and *Cercocebus*). *Folia Primatol.* 66:15–24.

Napier, P. H. 1981, *Catalogue of Primates in the British Museum (Natural History) and elsewhere in the British Isles. Part II: Family Cercopithecidae, Subfamily Cercopithecinae*, British Museum (Natural History), London.

Napier, J.R. and Napier, P. H. 1985, *The Natural History of the Primates*, MIT Press, Cambridge.

Oates, J. F. and Trocco, T. F. 1983, Taxonomy and phylogeny of black and white colobus monkeys: Inferences from the analysis of loud call variation. *Folia Primatol.* 40:83–113.

Oates, J. F., Bocian, C. M., and Terranova, C. J. 2000, The loud calls of black and white colobus monkeys: their adaptive and taxonomic significance in light of new data, in: Whitehead, P. F., and Jolly, C. J., eds., *Old World Monkeys*, Cambridge University Press, Cambridge, pp. 431–452.

O'Higgins, P. and Jones, N. 1998, Facial growth in Cercocebus torquatus: an application of three-dimensional geometric morphometric techniques to the study of morphological variation. *J. Anat.* 193:251–272.

O'Higgins, P. and Collard, M. 2002, Sexual dimorphism and facial growth in papionin monkeys. *J. Zool., Lond.* 257:255–272.

Olupot, W., Chapman, C. A., Brown, C. H, and Waser, P. M. 1994, Mangabey (*Cercocebus albigena*) population density, group size, and ranging: A twenty year comparison. *Am. J. Primatol.* 32:197–205.

Olupot, W., Chapman, C. A., Waser, P. M., and Isabirye-Basuta, G. 1997, Mangabey (*Cercocebus albigena*) ranging patterns in relation to fruit availability and the risk of parasite infection in Kibale National park, Uganda. *Am. J. Primatol.* 43:65–78.

Page, S. L. and Goodman, M. 2001, Catarrhine phylogeny: noncoding DNA evidence for a diphyletic origin of the mangabeys and for a human-chimpanzee clade. *Mol. Phylog. Evol.* 18(1):14–25.

Page, S. L., Chiu, C., and Goodman, M. 1999, Molecular phylogeny of Old World monkeys (Cercopithecidae) as inferred from γ-globin DNA sequences. *Mol. Phylog. Evol.* 13(2):348–359.

Palmer, T. S. 1903, Some new generic names of mammals. *Science* 17:873.

Plavcan, J. M. and van Schaik, C. 1992, Intra-sexual competition and canine dimorphism in anthropoid primates. *Am. J. Phys. Anthropol.* 87:461–477.

Quiris, R. 1975, Ecologie et organisation sociale de *Cercocebus galeritus agilis* dans le Nord-Est du Gabon. *Terre et Vie*, 29:337–398.

Rogers, M. E., Abernathy, K. A., Fontaine, B., Wickings, E. J., White, L. J., and Tutin, CEG 1996, Ten days in the life of a mandrill horde in the Lope Reserve, Gabon. *Am. J. Primatol.* 40:297–313.

Schaaf, C. D., Butynski, T. M., and Hearn, G. W. 1990, The drill (*Mandrillus leucophaeus*) and other primates in the Gran Caldera Volcanica de Luba: Results of a survey conducted March 7–22, 1990. Unpublished report to the Government of the Republic of Equatorial Guinea.

Schwartz, E. 1928, The species of the genus *Cercocebus* E. Geoffroy. *Ann. Mag. Nat. Hist.* 10(1): 664–670.

Setchell, J. M., Lee, P. C., Wickings, E. J., and Dixson, A. F. 2001, Growth and ontogeny of sexual size dimorphism in the mandrill (*Mandrillus sphinx*). *Am. J. Phys. Anthropol.* 115:349–360.

Shah, N. F. 1996, Preliminary study of two sympatric mangabey species (*Cercocebus galeritus agilis* and *Cercocebus albigena*) in the Dzanga-Sangha Reserve, Central African Republic. *XVIth Congress of the International Primatological Society*.

Singleton, M. 2002, Patterns of cranial shape variation in the Papionini (Primates: Cercopithecinae). *J. Hum. Evol.* 42:547–578.

Struhsaker, T. T. 1971, Notes on *Cercocebus a. atys* in Senegal, West Africa. *Mammalia* 35:343–344.

Struhsaker, T. T. 1981. Vocalizations, phylogeny and paleogeography of red colobus monkeys (*Procolobus badius*). *African Journal of Ecology* 1:265–283.

Szalay, F. S. and Delson, E. 1979, *Evolutionary History of the Primates*, Academic Press, New York.

Telfer, P. T., Souquiere, S., Clifford, S. L., Abernethy, K. A., Bruford, S. M. W., Disotell, T. R., Sterner, K. N., Roques, P., Marx, P. A., and Wickings, E. J. 2003, Molecular evidence for deep phylogenetic divergence in *Mandrillus sphinx Mol. Ecol.* 12:2019–2024.

Thorington, R. W. and Groves, C. P. 1970, An annotated classification of the cercopithecoidea, in: Napier, J. R., and Napier, P. H., eds., *Old World Monkeys: Evolution, Systematics and Behavior*, Academic Press, New York.

Wallis, S. J. 1983, Sexual behavior and reproduction of *Cercocebus albigena johnstonii* in Kibale Forest, western Uganda. *Int. J. Primatol* 4:153–166.

Waser, P. 1977, Feeding, ranging and group size in the mangabey *C. albigena*, in: Clutton-Brock, T. H., ed., *Primate Ecology: Studies of Feeding and Ranging Behavior in Lemurs, Monkeys and Apes*, Academic Press, New York.

Waser, P. 1984, Ecological differences and behavioral contrasts between two mangabey species, in: Rodman, P.S., and Cant, J. G. H., eds., *Adaptations for Foraging in Nonhuman Primates*, Columbia University Press, New York.

Wickings, E. J. and Dixon, F. 1992, Testicular function, secondary sexual development and social status in male mandrills (*Mandrillus sphinx*). *Physiol. Behavior* 52:909–916.

Madagascar

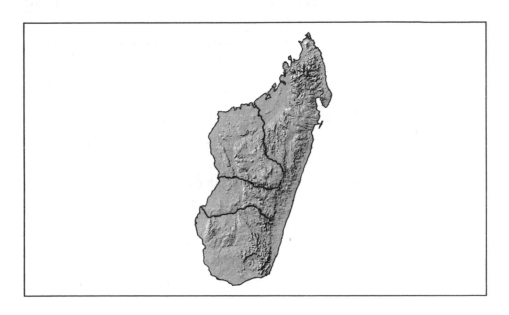

Madagascar has a surface area of 594,000 km,^{2} making it the fourth largest island on Earth. Madagascar has traditionally been divided into two main biogeographic zones: eastern humid forests and western dry forests. Abiotic factors (e.g., rainfall, temperature, humidity, and altitude) and biotic factors (e.g., plant composition, primate diversity, and abundance) vary considerably between these zones. Moreover, Madagascar has been subject to extreme levels of deforestation due to human activities such as agriculture and logging. This island nation has lost approximately 90% of its original forest cover since human colonization of the island 2000 years ago. The remaining forests have become increasingly fragmented, resulting in a patch-work of forest fragments; particularly in western dry region where few fragments exceed 800 ha in size (Ganzhorn *et al.*, 2001). Madagascar is considered one of the world's highest conservation priorities in terms of species diversity and endemism. For example, Madagascar

Primate Biogeography, edited by Shawn M. Lehman and John G. Fleagle.
Springer, New York, 2006.

is characterized by high levels of endemism for plants (ca. 2984 species, 83% endemic) and vertebrates (ca. 879 species, 84% endemic).

Lemurs in particular are of interest because they are found only on Madagascar. Extant lemurs are represented by five endemic families with 14 genera, approximately 33 species, and 71 subspecies. New species and subspecies are being described almost yearly as the result of primate surveys in unexplored forests and genetic studies. There is also a growing record of extinct taxa (Godfrey et al., 1999). Body sizes for extant lemurs range from a low of 30 g for *Microcebus myoxinus* to 6.43 kg for *Indri indri*. Lemurs feed upon a variety of plant parts and insects, and some taxa (*Hapalemur* sp.) are specialists on bamboo.

The papers in this section address questions concerning the historical and ecological biogeography of lemurs. In "Mouse Lemur Phylogeography Revises a Model of Ecogeographic Constraint in Madagascar," Anne Yoder and Kellie Heckman review genetic data to determine if mouse lemurs support biogeographic distinctions between eastern wet and western dry habitats. They find that biogeographic and ecological separation between western and eastern habitats has not been particularly restrictive to interpopulation gene flow among mouse lemur populations. Their results will have important implications for other lemur taxa given preconceived notions that small-bodied mammals have reduced dispersal abilities.

In "Lemur Biogeography," Jörg Ganzhorn, Steven Goodman, Stephen Nash, and Urs Thalmann seek to address questions about the effects of habitat zones and rivers/gallery forests on the dispersal of extant lemurs. They document a biogeographic history for lemurs that includes repeated dispersal between eastern wet forests and western dry forests, but not between northern and southern lemur populations. They also note a complex pattern of riverine barriers, including river orientation that may have limited the distribution of many—but not all—lemur taxa in eastern and western Madagascar. They conclude that further exploration of unstudied forest areas are needed, and that researchers should seek to integrate data from different fields of research (ecology, genetics, and morphology) to derive a clearer understanding of lemur biogeography.

In "Abiotic and Biotic Factors as Predictors of Species Richness on Madagascar," Nancy Stevens and Patrick O'Connor compare and contrast the distribution of mammals, birds, reptiles, and amphibians in protected areas to biogeographic patterns of geography, topography, climate, and vegetation. They find that lemur diversity is positively influenced by area and plant diversity but negatively influenced by altitude. These biogeographic patterns are most congruent with birds and nonprimate mammals. Stevens and O'Connor note the sensitivity of metanalyses to research effort, indicating the need to balance biological research and conservation action across protected areas at the landscape level.

REFERENCES

Ganzhorn, J. U., Lowry, P. P., Schatz, G. E., and Sommer, S. 2001, The biodiversity of Madagascar: one of the world's hottest hotspots on its way out. *Oryx* 35:346–348.

Godfrey, L. R., Jungers, W. L., Simons, E. L., Chatrath, P. S., and Rakotosamimanana, B. 1999, Past and present distributions of lemurs in Madagascar. In: Rakotosamimanana, B., Rasamimanana, H., Ganzhorn, J.U., and Goodman, S. M. eds., New Directions in Lemur Studies. Kluwer Academic/Plenum, New York, pp. 19–53.

CHAPTER EIGHT

Lemur Biogeography

Jörg U. Ganzhorn, Steven M. Goodman,
Stephen Nash, and Urs Thalmann

ABSTRACT

Madagascar is the fourth largest island in the world and is inhabited by a rich fauna of endemic lemuriform primates. Recent morphological and molecular, studies indicate that the biogeography of Malagasy lemurs is far more complex than previously thought. Small scale patterns of vegetation and river barriers seem to have been more important than large present-day barriers. Rivers can act as both barriers and corridors. Most significantly, there is increasing evidence that the major phylogenetic distinctions are between northern and southern taxa rather than between eastern and western forms. Lemur systematics is still in a state of ongoing revision, and additional surveys of poorly-known regions are needed to understand the complex biogeography of Malagasy primates.

Key Words: phylogeography, ecological biogeography, river barriers, river corridors

Madagascar, "La Grande Ile" off the coast of southeast Africa, is the fourth largest island on earth. Its 587,000 km^2 are only surpassed by the islands of Greenland, New Guinea, and Borneo. The island broke off from Africa some

Jörg U. Ganzhorn • Department Animal Ecology and Conservation, Biozentrum Grindel, Martin-Luther-King-Platz 3, 20146 Hamburg, Germany **Steven M. Goodman** • Field Museum of Natural History, 1400 South Lake Shore Drive, Chicago, Illinois 60605, USA, and WWF, BP 738, Antananarivo (101), Madagascar **Stephen Nash** • Conservation International, 1919 M Street NW, Washington D.C. 20036, USA, and Department of Anatomical Sciences, Health Sciences Center, State University of New York, Stony Brook NY 11794-8081, USA **Urs Thalmann** • Anthropological Institute and Museum, University of Zürich, Winterthurerstr. 190, 8057 Zürich, Switzerland

Primate Biogeography, edited by Shawn M. Lehman and John G. Fleagle.
Springer, New York, 2006.

150–160 million and from India some 88–95 million years ago (Rabinowitz *et al.*, 1983; Storey *et al.*, 1995). Due to the island's long isolation and low rates of colonization, the flora and fauna of Madagascar underwent numerous adaptive radiations, resulting in one of the world's most diverse biotas with remarkable levels of endemism (Myers *et al.*, 2000). Though Madagascar is separated from Africa only by the Mozambique Channel, 300–450 km wide, the prevailing winds and ocean currents were in recent geological history and still are unfavorable for repeated African source colonization of the island (Krause *et al.*, 1997). As a consequence, the lemurs of Madagascar are derived probably from a single colonization event from Africa during the Eocene (Yoder *et al.*, 1996). From there they radiated into their present diversity of at least 38 extant species and another ≥17 taxa that went extinct during the last 2000 years (reviewed e.g. by Godfrey *et al.*, 1997, 1999; Simons, 1997; Goodman *et al.*, 2003).

Lemur taxonomy is in the process of revision due to a surge in discoveries of new lemur species, additional information about distributions, and new methods in molecular genetics (Zimmermann *et al.*, 1998, 2000; Rasoloarison *et al.*, 2000; Rumpler, 2000; Thalmann and Geismann, 2000; Yoder *et al.*, 2000; Pastorini *et al.*, 2000, 2001a,b; Fausser *et al.*, 2002). On the basis of classical museum studies that had provided the basis for former lemur taxonomies, Groves (2001a) suggested further splitting of taxa, elevation of various subspecies to the species level, and resurrection of previously synonymized genera. Even though several of his conclusions are likely to be confirmed by future surveys and associated taxonomic work, we will follow herein the previously established taxonomy. We therefore follow the "pre-Groves" arrangement as summarized by Goodman *et al.* (2003; Table 1).

Since the arrival of humans on Madagascar some 2300 years ago (Burney *et al.*, 2004) Madagascar's ecosystems underwent very significant changes (Richard and Dewar, 1991; reviewed by Burney, 1997). This raises the question whether biogeographic analyses based on extant distribution patterns might provide a solid foundation for conclusions to be drawn about the biogeographic evolution of lemur distributions and the underlying constraints. Reconstruction of certain life history traits, particularly locomotion and diet based on anatomical characters, of the subfossil lemur species and their distribution in the recent past also showed that these lemur communities underwent significant changes in their functional composition (Godfrey *et al.*, 1997; Jernvall and Wright, 1998), but that the principle assignment of these communities to vegetation formations has apparently not changed over the last 2000 years

Table 1. Lemurs of Madagascar and some of their life history characteristics; supplemented from Ganzhorn *et al.* (1999) and Goodman *et al.* (2003)

Species	Body mass [kg]	Activity	Diet	Evergreen wet	Marsh	Deciduous dry	Riverine	Spiny forest
Microcebus berthae	0.03	N	OM			+++		
M. griseorufus	0.06	N	OM					+++
M. murinus	0.06	N	OM	+		+++	+++	?
M. myoxinus	0.05	N	OM			+++		
M. ravelobensis	0.06	N	OM			+++		
M. rufus	0.04	N	OM	+++				
M. sambiranensis	0.04	N	OM	+++				
M. tavaratra	0.06	N	OM			+++	+++	
Allocebus trichotis	0.09	N	OM	+++				
Cheirogaleus medius	0.28	N	OM	+		+++	+++	
C. major	0.4	N	OM	+++				
Mirza coquereli	0.3	N	OM	+		+++	+++	
Phaner furcifer[a]	0.46	N	OM	+		+++	+++	
Lepilemur dorsalis	0.5	N	FO	+++				
L. edwardsi	0.91	N	FO			+	+++	+++
L. leucopus	0.61	N	FO			+	+++	
L. microdon	0.97	N	FO	+++				
L. mustelinus	0.78	N	FO	+++				
L. ruficaudatus	0.77	N	FO			+++		
L. septentrionalis	0.75	N	FO			+++	+++	
Eulemur coronatus	1.2	C	FR	+++		+++	+++	

(Continued)

Table 1. (*Continued*)

Species	Body mass [kg]	Activity	Diet	Evergreen wet	Marsh	Deciduous dry	Riverine	Spiny forest
E. fulvus[b]	2	C	FR	+++		+++	+++	
E. macaco[c]	1.8	C	FR	+++		+++	+++	
E. mongoz	1.5	C	FR	+		+++	+++	
E. rubriventer	2.0	C	FR	+++				
Hapalemur aureus	1.5	D	FO	+++				
H. griseus[d]	0.7–1.4	D	FO	+++			+++	
H. simus[e]	2.6	C	FO	+++	+			
Lemur catta	2.2	D	FR			+++	+++	+++
Varecia variegata[f]	3.5	D	FR	+++			+++	
Avahi laniger	1.2	N	FO	+++			+++	
Avahi occidentalis	0.79	N	FO			+++	+++	
Avahi unicolor		N		+				
Indri indri	6.3	D	FO	+++				
Propithecus diadema[g]	6.1	D	FR	+++		+++		
P. tattersalli	3.5	D	FR			+++	+++	
P. verreauxi[b]	3.0–4.3	D	FR	+++		+++	+++	+++
Daubentonia madagascariensis	2.6–3.5	N	OM	+++		+++	+++	+++

N = nocturnal, C = cathemeral, D = diurnal; OM = omnivorous, FO = folivorous, FR = frugivorous

[a] *Phaner furcifer* includes *P. f. electromontis*, *P. f. parienti*, and *P. f. pallescens*.

[b] *Eulemur fulvus* occurs in dry and moist forests. It includes the taxa: *E. (f.) collaris*, *E. (f.) albocollaris*, *E. f. fulvus*, *E. f. rufus*, *E. f. albifrons*, and *E. f. sanfordi*.

[c] *Eulemur macaco* includes *E. m. macaco* and *E. m. flavifrons*.

[d] *Hapalemur griseus* includes the eastern forms *H. g. meridionalis*, *H. g. ssp.*, *H. g. griseus*, and *H. g. alaotrensis*. The western form is *H. g. occidentalis*.

[e] *H. simus* has been changed to *Prolemur simus* by Groves (2001a).

[f] *Varecia variegata* includes *V. v. variegata* and *V. v. rubra*.

[g] *Propithecus diadema* is restricted to moist forest. It includes the taxa (from south to north): *P. d. edwardsi*, *P. d. diadema*, *P. d. candidus*, and *P. d. perrieri*.

[b] *Propithecus verreauxi* is restricted to spiny and dry forests. It includes the taxa (from south to north): *P. v. verreauxi*, *P. v. deckeni*, *P. v. coronatus*, and *P. v. coquereli*; *P. v. deckeni* and *P. v. coronatus* have been elevated and combined by Groves (2001a) into a new species as *P. deckeni* and *P. deckeni coronatus* respectively. This taxonomy is not followed here.

(Godfrey *et al.*, 1999). This relative stability, at least over the last few hundred years, might allow to combine data on extant and subfossil lemur distributions as relevant units for biogeographic analyses.

The geographical distribution of organisms can be analyzed on various time scales (Myers and Giller, 1988). *Historical biogeography* reconstructs evolutionary processes over millions of years. In the case of Madagascar this approach would include continental drift and the analyses of the origin of taxa on higher taxonomic levels, including the question of the lemur origin and the colonization of Madagascar (Yoder *et al.*, 1996; Krause *et al.*, 1997; Martin, 2000; Marivaux *et al.*, 2001; Murphy *et al.*, 2001; Tavaré *et al.*, 2002). This aspect of lemur biogeography is subject of another review (Goodman and Ganzhorn, 2004b) and will not be addressed here. On more recent time scales climatic vicissitudes during the Pleistocene and associated changes in vegetation had profound impacts on the distribution of organisms that are still visible today (*Pleistocene biogeography*). These Pleistocene processes interact with the physiological capacity of organisms to cope with shifting environmental conditions and changes in interactions on the community level (*ecological biogeography*).

The goal of this paper is to review several hypotheses on the biogeographic patterns of lemur distributions on a more recent time scale. We focus on the questions:

(1) to what extent are lemur species limited to specific vegetation formations and their underlying climatic conditions?

(2) what is the evidence that rivers represent barriers to the dispersal and gene flow?

(3) which role might have been played by gallery forests along rivers or other corridors of vegetation to facilitate dispersal between east and west Madagascar?

These hypotheses will be summarized, discussed, and extended in the light of recent discoveries and of new information accumulated over the last few years.

HYPOTHESES TO EXPLAIN LEMUR BIOGEOGRAPHY

Vegetation Formations as the Basis for Distinct Lemur Communities (Ecological Biogeography)

Based on vegetation types, floristic formations, and geology the moist evergreen forests of eastern Madagascar are distinguished from the dry deciduous

(a)

(b)

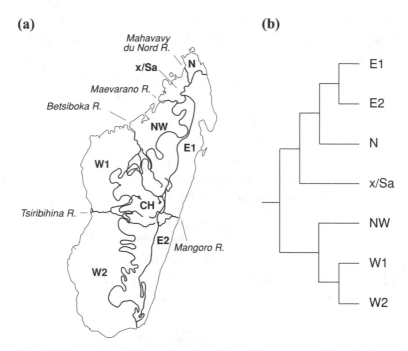

Figure 1. (a) Zoogeographic regions based on phytogeographic criteria and similarities of lemur communities as proposed by Martin (1972a, 1995; from Thalmann, 2000b). (b) Simplified general area cladogram based on species similarities of lemur communities.

formations of the west and south (Humbert, 1955; Koechlin *et al.*, 1974; Phillipson, 1994; Du Puy and Moat, 1996; Lowry *et al.*, 1997; Schatz, 2000; Figure 1). The evergreen forests receive from 1500 to more than 6000 mm of rain per year. The deciduous forests of the west and extreme north of Madagascar are subject to a distinct dry season of four to eight months without rain and annual precipitation of 500 to 2000 mm. On both the eastern and western sides of Madagascar, annual rainfall decreases from north to south. Parts of the south and southwest of the island, with its dry deciduous, riverine and spiny forest, characterized by Didiereaceae and other succulent plant, receive less than 500 mm of rain per year at irregular intervals with an extended dry season of more than eight months.

For some time biogeographers have noticed a division amongst lemur species that could be assigned to the eastern rain forest and other, sometimes congeneric, species of the western dry deciduous forests. This indicated that lemur species have specific adaptations inhibiting that certain taxa currently living and presumably adapted to the moist forest conditions can survive in the dry forest and *vice versa*. These ecogeographic zones include the eastern wet forest, an extension of the eastern forest towards the west, the Sambirano (not the Sambirano Domain *sensu* Humbert [1955]), a transition zone in the north, dry deciduous forest of the west, the spiny forest of the south, and the central highland with some forest patches. This basic classification provided the basis for subsequent analyses (Martin, 1972a; 1995; Petter *et al.*, 1977; Tattersall, 1982; Pollock, 1986; Richard and Dewar, 1991; Ganzhorn *et al.*, 1999; Wright, 1999; Thalmann, 2000b; Figure 1).

Rivers as Barriers

On the basis of the six ecogeographic zones listed above, Martin (1972a) proposed eight major zoogeographic regions elucidating lemur biogeography. Given the information available at the time, Martin's biogeographic description was based mainly on the most conspicuous diurnal lemurs. As noted above, the first and most obvious distinction seemed to be represented by lemur communities of the evergreen moist forests of the east and the Sambirano region *versus* the dry deciduous forests of the west. Except for the north and extreme south, these types of vegetation are separated by several hundred kilometers of savanna that is unsuitable for arboreal lemurs.

In addition to a floristic component that separated communities of the evergreen moist from the dry forest types, Martin's classification included information on varying lemur distributions within the eastern and western forests. Based on shifts in species composition of lemur communities he subdivided the eastern belt of rain forest and the western dry forests at about 20° southern latitude (Figure 1a). According to this scenario rivers represented barriers to dispersal and gene flow resulting in biogeographic subdivisions within the eastern and western domains. These rivers were proposed to be the Tsiribihina, Betsiboka, Maevarano, and Mahavavy du Nord in the west and northwest, and the Mangoro River in the east.

This subdivision can best be illustrated by the distribution of taxa within the genus *Propithecus*. In the west, the Tsiribihina separates *Propithecus v. verreauxi* from *P. v. deckeni* to the north. *P. v. deckeni* is then separated from *P. v. coronatus* by the Mahavavy du Sud River (Figure 2) and the Betsiboka represents the boundary for *P. v. coronatus* and *P. v. coquereli*. However, it should be noted that *P. v. coronatus*-like sifaka occur on the north side of the Manambolo River

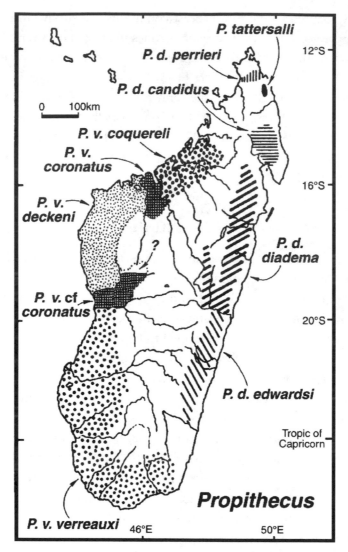

Figure 2. Distribution of *Propithecus* sspp. in relation to river systems.

(1 group only, 6 out of 7 animals showed *P. v. coronatus* coloration, 1 *P. v. deckeni* coloration; labeled *P. v. cf. coronatus* in Figure 2). Thus, it could be that the range of *P. v. coronatus* extends in a crescent-like fashion around the range of *P. v. deckeni*. Interviews with inhabitants on the north side of the Tsiribihina during a descent of the gorge seemed to confirm the presence of *P. v. coronatus* between the Tsiribihina and the Manambolo. The local name of the animals was said to be *"sifaka maintiloha"* which translates to black-headed sifaka. Sifaka are said to be very rare between the Manambolo and the Tsiribihina (Thalmann and

Rakotoarison, 1994, unpubl. data). This distribution could have come about by separate colonization of the west by *P. v. coronatus* originating from the east or central highland (e.g., from the region around Ambohitantely and the Bongalava mountains, moving west along the Tsiribihina River and between the Mahavavy du Sud and the Betsiboka River.

In the east the Mangoro River marks the southern boundary for *Indri indri* and separates *P. diadema edwardsi* from the more northern *P. d. diadema*. Other rivers represented the southern limit for *Varecia v. variegata* (Mananara River) and the northern boundary between *V. v. variegata* and *V. v. rubra* (Antainambalana River). Later, Martin's (1972a) view was refined for eastern Madagascar where lemur distributions seemed to indicate additional river systems as major zoogeographic barriers, including the Fanambana River in northeastern Madagascar (Martin 1995). Differences in vegetation characteristics together with rivers as barriers might then have created isolated geographical regions that allowed specialization and subsequent speciation on rather small scales.

Recent survey results support the principles of Martin's (1972a, 1995) analyses. A simplified cladogram based on similarities of the species composition of lemur communities at different sites as known in 1999 indicated that the north and the Sambirano region are linked to a "eastern clade." The western clade consists of the region "NW", "W1" and "W2" with "W1" and "W2" being more similar to one another than either one to "NW" (Figure 1b). According to this result, the Betsiboka River presented a more effective barrier to dispersal than the Tsiribihina, though this does not apply to all lemur taxa.

Rivers as Corridors

Apart from the east-west dichotomy described above and the subdivision of the eastern and western domain by rivers, there is a north-south zonation where taxa of the east and the west are more similar and probably closer related to each other than to neighboring populations to the north or south. An example of this is the occurrence of seemingly the same subspecies of brown lemur, *Eulemur fulvus rufus* and *E. f. fulvus*, in eastern and western Madagascar (Petter *et al.*, 1977; Tattersall, 1982; Figure 3).

Originally, this east-west connectivity had been attributed to the direction of various drainage systems. Several rivers originate on the eastern side of the central highlands and drain into the Mozambique Channel (Figure 4).

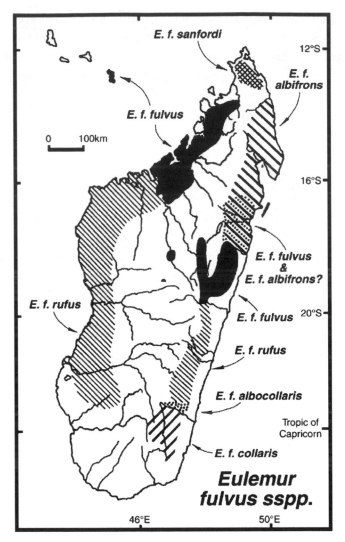

Figure 3. North-south zonation as illustrated by the occurrence of *Eulemur fulvus* subspecies. The distributional range of *E. f. rufus* and *E. f. fulvus* have been extended based on data by Irwin *et al.* (2000) and Lehman and Wright (2000).

At least prior to human intervention, forests along these rivers could have represented not only refugia for rain forest species during Pleistocene drought (sensu Maeve and Kellman, 1994) but also corridors for the exchange of forest species across the island. In numerous cases the sources of these eastern and western flowing watersheds are within close distance of one another (tens of meters to 1 km).

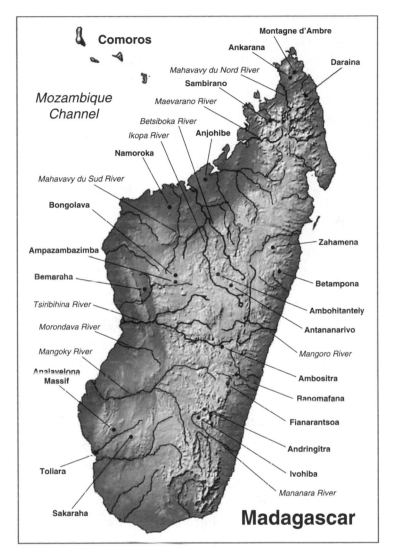

Figure 4.   River systems; rivers as possible corridors between east and west.

Possible candidates for riverine forests that bridged the central highlands are the Betsiboka River that originates east of Antananarivo and the Tsiribihina and the Mangoky rivers with their tributaries extending east beyond Ambositra and Fianarantsoa in the eastern rain forest. In principle, *Eulemur fulvus rufus* could have moved between eastern and western portions of the island along the Mangoky River in the south and along the Tsiribihina River in central Madagascar. Similarly, the Betsiboka River and its tributaries could have linked the western population of *Propithecus verreauxi coronatus* (occurring between the Mahavavy du Sud River and the Betsiboka River) with a remnant (and now extinct) population on the central highlands in the vicinity of Ambohitantely (Petter and Andriatsarafara, 1987). As outlined above, a second portion of the central population of *P. v. coronatus* could have moved west between the Tsiribihina and the Manambolo rivers. The link between east and west is further illustrated by now separated eastern and western subpopulations of *E. f. fulvus* in the zoogeographic region "E1" and "NW" with a remnant "stepping stone population" at Ambohitantely (Figure 3).

NEW INSIGHTS IN LEMUR DISTRIBUTIONS

East-West-Connectivity

Today the central highland seems to represent an unbridgeable barrier between the moist forests of the east and the dry forests of the west. Even though the central highland was apparently not entirely forested during the Pleistocene and Holocene (reviewed by Burney, 1997) accumulated evidence suggests that the forested regions of the east and the west were not as isolated as had been assumed by the earlier biogeographic analyses. Therefore the link between the east and the west do not seem to have been limited to forests along river systems. Rather there seem to have been belts of more humid forest extending from the east far into the west. These connections are indicated by mixed elements of eastern humid and western deciduous floristic and faunistic communities at certain sites. An excellent example of this are the upper portions of the Analavelona Massif, close to Toliara (Carleton *et al.*, 2001). This massif is within the most arid portion of the island, but the moist forest in the summital zone of this mountain contains numerous eastern species, although in this case the lemurs are all "western" species. On the basis of orographic factors, this massif seemingly maintains relict conditions of a previous more mesic conditions and could be considered a "Pleistocene refuge."

For lemur species this east-west connectivity is illustrated by isolated occurrences of the "western species" *Phaner furcifer* in the center of the eastern rain forest at Zahamena and Betampona or at Montagne d'Ambre (reviewed by Ganzhorn *et al.*, 1996/97) and *Cheirogaleus medius* in the northeast at Daraina (Mittermeier *et al.*, 1994). On the other hand "eastern" lemur species found in the forests of the west include *Cheirogaleus* cf. *major* in Bongolava and *C.* cf. *major* and *Hapalemur griseus occidentalis* in Bemaraha (Thalmann and

Rakotoarison, 1994; Ausilio and Raveloanrinoro, 1998; Thalmann, 2000a). The latter has also been reported from Namoroka (Thalmann *et al.*, 1999).

A somewhat different example is provided by *Lemur catta*. This species of the spiny and dry forests of the southwest also occurs on the west side of Andringitra south to Ivohibe (Goodman and Langrand, 1996). Its distribution might be continuous from the southwest to Andringitra and its limit in the Andringitra–Ivohibe region where it reaches its eastern limits due to habitat characteristics.

Vegetation types and specific habitat characteristics also do not seem to play the dominant role limiting lemur species to either the dry or the moist forest. A remarkable example is the Aye-aye (*Daubentonia madagascariensis*). Once thought to be a moist forest specialist, it has now been reported from most regions of Madagascar except for the very dry southwest (Sterling, 1998; Simons and Meyers, 2001).

Similarly, apart from the "dry forest species" *Phaner furcifer* in the moist forests of the north and east (see above), the dry forest species *Microcebus murinus* and *Cheirogaleus medius* seem to do well in some localized populations within the evergreen moist littoral forests of the south (Martin, 1972b; Ganzhorn, 1999; Ramanamanjato and Ganzhorn, 2001; Lahann *et al.*, in press). Thus, the physiologcal and ecological constraints of at least some lemurs towards environmental variation and the links between lemur species and phytogeographic units may not be as strong as assumed so far.

Apart from lemurs there are also other firm evidences of east-west distributions of extinct and extant vertebrates indicating that the region or a belt south of about 20° latitude might well have been a contiguous and suitable habitat for vertebrates adapted to more humid conditions (e.g., Goodman and Rakotondravony, 1996; Goodman and Rakotozafy, 1997).

North-South Zonation

Within the phytogeographic units of the eastern moist and the western dry forest, Martin (1972a) had already noticed a discontinuity of lemur species composition into northern and southern communities. But the different species were still considered to belong to "eastern" or "western" lemur species and the communities were still characterized as moist or dry forest communities (references listed above). This view has to be changed in favor of a much more complex biogeographic history including phylogenetic aspects and repeated dispersal between the east and the west. From a phylogenetic perspective, the north-south zonation has received support form studies in molecular genetics. According to these studies, the genus *Microcebus* apparently consists of a northern and a southern clade of species rather than an eastern and a western clade (Yoder *et al.*, 2000; Yoder this volume).

The Role of Rivers

The hypotheses on lemur biogeography outlined above emphasize the limiting role of barriers and the facilitating role of corridors in dispersal. The role of major rivers as barriers to lemur dispersal needs revision. In the east, *Hapalemur simus*, now confined to a small region between Ranomafana and Andringitra, was once widespread ranging up to Ankarana in the north, Anjohibe in the northwest, and Ampazambazimba in the center, thus bridging several river systems associated with the eastern and western watershed divides. Similarly, *Indri indri* occurred well beyond the Antainambalana River, and north to at least Ankarana and perhaps as far southwest as the Sakaraha area (reviewed by Godfrey *et al.*, 1997, 1999). Other smaller lemur species (if they actually represent only one taxon) do not seem to be limited by river systems in the east. These include *Lepilemur mustelinus*, *L. microdon*, *Eulemur rubriventer*, and *Avahi laniger*.

Rather, distributional boundaries seem to occur between major river systems. The ranges of *Indri indri* and *Eulemur rubriventer* stop between major river systems. What is currently known as *Hapalemur g. griseus* from north to south might actually represent two types that replace each other in the region of Ranomafana–Ifanadiana, apparently without being separated by any river system (Fausser *et al.*, 2002). The two southern subspecies (or species) of the *Eulemur fulvus* group (*E. (f.) albocollaris* and *E. (f.) collaris*) seem genetically distinct from the other *E. fulvus* group to warrant species status (Djlelati *et al.*, 1997; Wyner *et al.*, 1999). While *E. (f.) albocollaris* and *E. (f.) collaris* are separated by the Mananara River the biogeographic "break" between these two types and *E. fulvus* occurs north of the Mananara River where *E. (f.) albocollaris* and *E. f. rufus* form a zone of hybridization (Johnson and Wyner, 2000; Wyner *et al.*, 2002).

In the west, biogeographic boundaries seem even more complex. River systems seem clear boundaries for *Propithecus* subspp. (references quoted above; Thalmann *et al.*, 2002). But vicariant species turnover of other taxa occurs between river systems. This applies to *Lepilemur ruficaudatus* that extends north of the Tsiribihina River where it is replaced by *L. edwardsi* or lives in sympatry with some undescribed species of *Lepilemur* (Thalmann and Rakotoarison, 1994; Tomiuk *et al.*, 1997; Bachmann *et al.*, 2000). *Lemur catta* vanishes in the middle of a continuous forest between the rivers of Mangoky and Morondava (Petter *et al.*, 1977; Tattersall, 1982). *Eulemur fulvus rufus* occurs as far south as the Forêt des Mikeas in the west but seems to be restricted by the spiny forest formations further south rather than by river barriers. Also, the range of the southern population of *Mirza coquereli* does not seem to be limited by major river systems.

An important aspect that has not been taken into account in this hypothesis is the question whether certain species can or can not "cross-over" these river barriers towards their inland source as has been illustrated by Ayres and Clutton-Brock (1992) for South American primate species. Clearly the long lowland rivers of the west that move across considerable stretches of similar habitat are very different from the eastern rivers that to some extent drop off the central highlands and then within a few kilometers empty into the sea (Goodman and Ganzhorn 2004a). *Mirza coquereli* from western Madagascar might be a case where its upper elevational range is far below the level of where river sources are found and thus it is not possible that it skirts around rivers by passing over the level of the source. Further, orientation of the rivers will have a major influence here. If a western river is aligned east-west as opposed to north-south (e.g., Mahavavy River) this could have important influence in its role as a barrier.

OPEN QUESTIONS AND RECONSTRUCTION OF BIOGEOGRAPHIC PROCESSES THROUGH GENETIC AND COMMUNITY ANALYSES

Recent genetic work has revealed new perspectives and biogeographic implications for the distributions of extant lemur species and in turn has also created new questions. There is evidence that Malagasy lemurs are monophyletic and colonized Madagascar via rafting from Africa (Yoder *et al.*, 1996; reviewed by Martin, 2000). The question remains as to where the ancestral lemur landed and how they spread over the island. One has the impression that previous geographic analyses assumed that the colonization direction was from the east to the west. The eastern humid forests have more species than the west and many species of the east were assumed to be monotypic with wide distributions while ranges of extant taxa were smaller in the west. Therefore it seemed logical that colonization occurred from east to west with subsequent speciation in the west. The emerging view of eastern lemur taxonomy with many more distinct taxa makes this assumption obsolete, though possible colonization from India or even origination of lemurs on the Malagasy/Indian plate is still an option (reviewed by Martin, 2000; Marivaux *et al.*, 2001).

Biogeographic analyses needs to be based on evolutionary significant units that can be identified and, ideally, be put in a phylogenetic context. Both requirements are still incomplete for Malagasy lemurs at lower taxonomic levels. For eastern lemur taxa evidence is accumulating that some previously assumed monotypic species actually consist of several genetically and/or morphologically distinct forms. This applies to *Microcebus* cf. *rufus* (Yoder *et al.*, 2000; Pastorini *et al.*, 2001b), *Hapalemur* cf. *griseus* (Fausser *et al.*, 2002), *Varecia* cf. *variegata variegata* (Ross unpubl.), *Cheirogaleus* cf. *major* (Thalmann, 2000a; Irwin, unpubl.; Hapke, unpubl.), and *Avahi* cf. *laniger* (Rumpler, 2000). For western lemur communities it has been demonstrated already that species richness is by far greater than has been assumed for a long time, but forms that cannot easily be assigned to recognized taxa are reported from almost any survey.

Another concept that might need revision is the idea of major rivers acting as barriers, particularly in the west. While they are likely to act as barriers for some taxa, such as various species of *Propithecus*, it might be worthwhile to overlay vegetation changes during the Pleistocene with lemur distributions. During the Pleistocene the belt of eastern rain forest that has been continuous until very recently might have been disrupted by montane heath vegetation (Humbert's [1955] "High Mountain Domain") that might have occurred at elevations some 1000 m lower than at present (Carleton and Goodman, 1996; Burney, 1997). This type of analysis seems only useful once the pending taxonomic revision of eastern rain forest lemurs have advanced beyond their present stage.

Other analyses of molecular genetic data yielded unexpected results that are difficult to reconcile with the present ideas of lemur biogeography and their evolution. Based on mtDNA *Propithecus tattersalli* groups with *P. verreauxi coquereli*. If these genetic data will prove to be relevant in evolutionary terms these two taxa would have to be pooled and contrasted with the other subspecies/species of *P. verreauxi* (Pastorini *et al.*, 2001a). Similarly, the present taxonomy of *Eulemur fulvus* needs revision. The western form of what is currently considered as *E. f. rufus* might have to be considered as a species group, one of which has to be considered as a taxon that is clearly separate from the eastern *E. f. rufus* while the other is closely related to the eastern form (Pastorini *et al.*, 2000; C. Ross, pers. comm.).

UNSOLVED QUESTIONS AND DISCREPANCIES

The need for taxonomic revision of lemurs has now been stated and illustrated repeatedly. This will not be discussed in more detail here (Martin, 2000; Rasoloarison *et al.*, 2000; Thalmann and Geismann, 2000; Yoder *et al.*, 2000; Pastorini *et al.*, 2000; Zimmermann *et al.*, 2000; Groves, 2001b). We need to wait for the new results before any further analyses seem useful. Nevertheless we would like to illustrate some of the emerging problems below.

In several cases the genetic reconstruction of phylogenetic divergence does not match similarities in the species composition of extant lemur communities. This suggests repeated dispersal of different taxa (such as postulated for *Propithecus v. coronatus*; see Figure 2) and independent speciation events. The need for considering rather complex biogeographic scenarios is nicely illustrated by the reconstruction of the present forms of *Lepilemur* spp. (Ishak *et al.*, 1992; Rumpler, 2000; Figure 5).

According to this reconstruction based on chromosomal rearrangements *Lepilemur septentrionalis* and *L. mustelinus/L. microdon* belong to an eastern clade while the other forms, including *L. dorsalis* originated in the west. However, this reconstruction of phylogenetic evolution does not match zoogeographic similarities of lemur communities if it is overlaid over a map of the present distribution of species. Thalmann's (2000b) analysis of community similarities assign *L. septentrionalis* and *L. dorsalis* to eastern lemur communities that are distinguished from communities that contain *L. edwardsi*, *L. ruficaudatus*, and *L. leucopus* (Figure 6). If these genetic and biogeographic analyses are indeed correct, they indicate that phylogenetic and ecological biogeographic evolution do not need to coincide.

For the time being there is not a sufficient number of genetic studies available for comparisons across the island. Nevertheless, based on chromosomal and molecular data the south of Madagascar seems to be a region where genetic differentiation occurred in several genera. Similarly to the high and disruptive species turnover in the north of Madagascar, many taxa of the south seem separated genetically from their relatives. This has been postulated for *Eulemur* (*fulvus*) *collaris* and *E. (f.) albocollaris* which might have to be raised to species level (Rumpler, 1975; Wyner *et al.*, 1999; for an opposite view see Pastorini

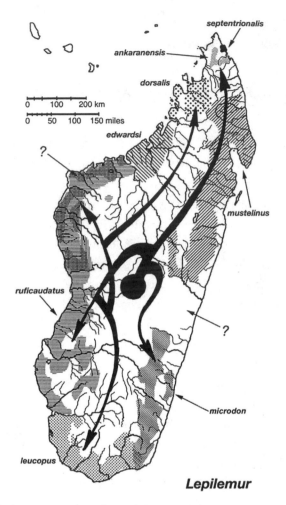

Figure 5. Evolutionary tree based on chromosomal arrangements within the genus *Lepilemur* superimposed over a distribution map of *Lepilemur* spp. (modified from Mittermeier *et al.*, 1994; Rumpler, 2000; Zaramody *et al.*, 2005).

et al., 2000) despite the hybridization of *E.* (*f.*) *albocollaris* with *E. f. rufus* at Andringitra (Wyner *et al.*, 2002). In the same region, *Hapalemur* cf. *griseus meridionalis* might also represent a taxon that is distinct from the northern *Hapalemur* cf. *griseus* (Fausser *et al.*, 2002). Similar situations are likely to be present also in the cheirogaleids (Hapke, pers. comm.).

Up to now distributional data, morphological traits, chromosomal characteristics, and molecular genetics have been used to reconstruct lemur biogeography. In some cases these different types of data yield conflicting information (Yoder, 1997). Reconciliation of the different aspects based on objective criteria

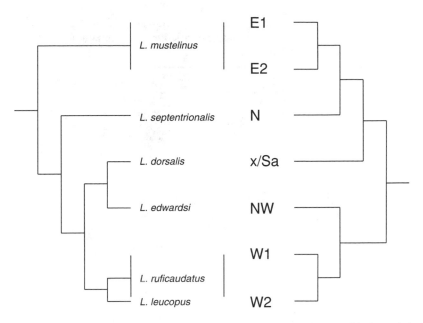

Figure 6. Phylogenetic relationships of *Lepilemur* spp. according to Ishak *et al.* (1992) (left) and the occurence of the different *Lepilemur* species in zoogeographic areas as depicted in Figure 1. "*L. septentrionalis*" is composed of two species: *L. septentrionalis* and *L. ankuranensis* (Ravoarimanana *et al.* (2004); see Fig. 5). This split does not affect the comparison shown here.

seems difficult for the time being as morphological traits are a result of several interacting coding genes and they are being compared with sequences of non-coding DNA, coding DNA, and chromosomal rearrangements of unknown consequences (for more detailed reviews see Yoder, 1997; Thalmann, 2003). Some of these discrepancies are summarized in Table 2.

CONCLUSIONS

During the last few decades there have been outstanding advances in the studies of lemur biology that eventually will lead into a comprehensive analysis of lemur biogeography that reflects reality. For the time being there is still a long way to go. But we now know that the eastern taxa might be as diverse and have distributions that might be as small as their western counterparts. Certainly, there is no uniform explanation that can account for all biogeographic phenomena observed. Regional and small-scale processes make analyses complicated (Thalmann and Rakotoarison, 1994; Thalmann *et al.*, 2002), though new

Table 2. Some examples of discrepancies between morphological characteristics, chromosomal data, and molecular genetics; extended from Thalmann (2003), supplemented with data from Fausser *et al.* (2002)

Distinction between	Morphology	Karyotype	mtDNA
Propithecus verreauxi coquereli *P. tattersalli*	yes	yes	no
P. v. deckeni *P. v. coronatus*	yes	no	no
Lepilemur edwardsi *L. ruficaudatus*	no	yes	yes
Hapalemur g. griseus *H. g. alaotrensis*	yes	no/yes [a]	no
H. g. griseus *H. g. meridionalis* *H. g.* sspp.	no	yes	yes
Cheirogaleus spp. *Microcebus* spp.	yes	no	yes

[a] *H. g. alaotrensis* with same number of chromosomes but more juxtacentromeric heterochromatin (Fausser *et al.*, 2002).

methodological approaches add new tools for biogeographic analyses (Haydon *et al.*, 1994; Thalmann, 2000b). The recent and ongoing anthropogenic de-struction of forests makes reconstruction of former biogeographic boundaries of taxa even more difficult. Therefore some of the issues presented here have to be considered preliminary and will certainly need revision in the future. First, it is certain that lemur taxonomy will undergo substantial revision. New taxa will be added to the present species and replace some of the educated guesses put forward in the literature at the moment. Second, most field surveys in poorly known regions of the island yield new insights and unexpected occur-rences of lemur species. Third, there is an urgent need for a better integration of molecular and morphological traits and for a better understanding of the biological relevance of molecular data that will allow biologically meaningful interpretations.

ACKNOWLEDGEMENTS

We thank S. M. Lehman and J. G. Fleagle for their invitation to contribute to this volume. Fieldwork associated with this review has been generously sup-ported by the DFG, Ellen Thorne Smith Fund of the Field Museum of Natural History, John D. and Catherine T. MacArthur Foundation, National Geo-graphic Society, Volkswagen Foundation, and World Wide Fund for Nature.

REFERENCES

Ausilio, E. and Raveloanrinoro, G. 1998, Les lémuriens de Bemaraha: Forêts de Tsimembo, de l'Antsingy et de la région de Tsiandro. *Lemur News* 3:4–7.

Ayres, J. M. and Clutton-Brock, T. H. 1992, River boundaries and species range size in Amazonian primates. *Am. Nat.* 140:531–537.

Bachmann, L., Rumpler, Y., Ganzhorn, J. U., and Tomiuk, J. 2000, Genetic differentiation among natural populations of *Lepilemur ruficaudatus*. *Int. J. Primatol.* 21: 853–864.

Burney, D. A. 1997, Theories and facts regarding Holocene environmental change before and after human colonization, in: Goodman, S. M. and Patterson, B. D., eds., *Natural Change and Human Impact in Madagascar*, Smithsonian Institution Press, Washington, DC, 75–89.

Burney, D. A., Burney, L. P., Godfrey, L. R., Jungers, W. L., Goodman, S. M., Wright, H. T., and Jull, A. J. T. 2004, A chronology for late Prehistoric Madagascar. *J. Hum. Evol.* 47:25–63.

Carleton, M. D. and Goodman, S. M. 1996, Systematic studies of Madagascar's endemic rodents (Muroidea: Nesomyinae): A new genus and species from the Central Highlands. In *A Floral and Faunal Inventory of the Eastern Slopes of the Réserve Naturelle Intégrale d'Andringitra, Madagascar: With Refrerence to Elevational Variation*: 231–256. Goodman, S. M., ed., *Fieldiana: Zoology,* Vol. 85, new series. Chicago: Field Museum of Natural History.

Carleton, M. D., Goodman, S. M., and Rakotondravony, D. 2001, A new species of tufted-tailed rat, genus *Eliurus* (Muridae: Nesomyinae), from western Madagascar, with notes on the distribution of *E. myoxinus. Proc. Biol. Soc. Washington* 114: 977–987.

Djlelati, R., Brun, B., and Rumpler, Y. 1997, Meiotic study of hybrids in the genus *Eulemur* and taxonomic considerations. *Am. J. Primatol.* 42: 235–245.

Du Puy, D. J. and Moat, J. 1996, A refined classification of the primary vegetation of Madagascar based on the underlying geology: Using GIS to map its distribution and to assess its conservation status, in: Lourenço, W. R., ed., *Biogeography of Madagascar*, ORSTOM, Paris, pp. 205–218.

Fausser, J.-L., Prosper, P., Donati, G., Ramanamanjato, J.-B., and Rumpler, Y. 2002, Phylogenetic relationships between *Hapalemur* species and subspecies based on mitochondrial DNA sequences. *BMC Evol. Biol.* 2: 4 (http://www.biomedcentral.com/1471-2148/2/4).

Ganzhorn, J. U. 1999, Progress report on the QMM faunal studies: Lemurs in the littoral forest of southeast Madagascar. *Lemur News* 3: 22–23.

Ganzhorn, J. U., Langrand, O., Wright, P. C., O'Connor, S., Rakotosamimanana, B., Feistner, A. T. C., and Rumpler, Y. 1996/1997, The state of lemur conservation in Madagascar. *Primate Cons.* 17: 70–86.

Ganzhorn, J. U., Wright, P. C., and Ratsimbazafy, J. 1999, Primate communities: Madagascar, in: Fleagle, J. G., Janson, C., and Reed, K. E., eds., *Primate Communities*, Cambridge University Press, Cambridge, pp. 75–89.

Godfrey, L. R., Jungers, W. L., Reed, K. E., Simons, E. L., and Chatrath, P. S. 1997, Inferences about past and present primate communities in Madagascar, in: Goodman, S. M. and Patterson, B. D., eds., *Natural Change and Human Impact in Madagascar*, Smithsonian Institution Press, Washington, DC, pp. 218–256.

Godfrey, L. R., Jungers, W. L., Simons, E. L., Chatrath, P. S., and Rakotosamimanana, B. 1999, Past and present distributions of lemurs in Madagascar, in: Rakotosamimanana, B., Rasamimanana, H., Ganzhorn, J. U., and Goodman, S. M., eds., *New Directions in Lemur Studies*, Kluwer Academic/Plenum Publishers, New York, pp. 19–53.

Goodman, S. M. and Ganzhorn, J. U. 2004a, Biogeography of lemurs in the humid forests of Madagascar: The role of geological history, elevational distribution and rivers. *J. Biogeog.* 31: 47–55.

Goodman, S. M. and Ganzhorn, J. U. 2004b, Elevational ranges of lemurs in the humid forests of Madagascar. *Int. J. Primatol.* 25: 331–350.

Goodman, S. M. and Langrand, O. 1996, A high mountain population of the ring-tailed lemur *Lemur catta* on the Andringitra Masif, Madagascar. *Oryx* 30: 259–268.

Goodman, S. M. and Rakotondravony, D. 1996, The Holocene distribution of *Hypogeomys* (Rodentia: Muridae: Nesomyinae) on Madagascar, in: Lourenço, W. R., ed., *Biogéographie de Madagascar*, ORSTOM, Paris, pp. 283–293.

Goodman, S. M. and Rakotozafy, L. M. A. 1997, Subfossil birds from coastal sites in western and southwestern Madagascar: A paleoenvironmental reconstruction, in: Goodman, S. M. and Patterson, B. D., eds., *Natural Change and Human Impact in Madagascar*, Smithsonian Institution Press, Washington, DC, pp. 257–279.

Goodman, S. M., Ganzhorn, J. U., and Rakotondravony, D. 2003, Mammals, in: Goodman, S. M. and Benstead, J. P., eds., *The Natural History of Madagascar*, The University of Chicago Press, Chicago, pp. 1159–1186.

Groves, C. P. 2001a, *Primate Taxonomy*, Smithsonian Institution Press, Washington, DC.

Groves, C. 2001b, Why taxonomic stability is a bad idea, or why are there so few species of primates (or are there?). *Evol. Anthropol.* 10: 192–198.

Hapke, A., Fietz, J., Nash, S., Rakotondravony, D., Rakotosamimanana, B., Ramanamanjato, J.-B., Randria, G., and Zischler, H. 2005, Biogeography of dwarf lemurs (*Cheirogaleus* spp.): Genetic evidence for unexpected patterns in southeastern Madagascar. *Int. J. Primatol.* 26: 873–901.

Haydon, D. T., Crother, B. I., and Pianka, E. R. 1994, New directions in biogeography? *Trends Ecol. Evol.* 9: 403–406.

Humbert, H. 1955, Les territoires phytogéographiques de Madagascar. Leur cartographie. *Année Biologique*, 3ème série 31: 195–204.

Irwin, M. T., Smith, T. M., and Wright, P. C. 2000, Census of three eastern rainforest sites north of Ranomafana National Park: Preliminary results and implications for lemur conservation. *Lemur News* 5: 20–22.

Ishak, B., Warter, S., Dutrillaux, B., and Rumpler, Y. 1992, Chromosomal rearrangements and speciation of sportive lemurs (*Lepilemur* species). *Folia Primatol.* 58: 121–130.

Jernvall, J. and Wright, P. C. 1998, Diversity components of impending primate extinctions. *Proc. Natl. Acad. Sci. USA* 95: 11279–11283.

Johnson, S. E. and Wyner, Y. 2000, Notes on the biogeography of *Eulemur fulvus albocollaris. Lemur News* 5: 25–28.

Koechlin, J., Guillaumet, J.-L., and Morat, P. 1974, *Flore et Végétation de Madagascar.* Vaduz: J. Cramer.

Krause, D. W., Hartmann, J. H., and Wells, N. A. 1997, Late Cretaceous vertebrates from Madagascar: Implications for biotic change in deep time, in: Goodman, S. M. and Patterson, B. D., eds., *Natural Change and Human Impact in Madagascar*, Smithsonian Institution Press, Washington, DC, pp. 3–43.

Lahann, P., Schmid, J., and Ganzhorn, J. U. (in press). Geographic variation in life history traits of *Microcebus murinus* in Madagascar. *Int. J. Primatol.*

Lehman, S. M. and Wright, P. C. 2000, Preliminary study of the conservation status of lemur communities in the Betsakafandrika region of eastern Madagascar. *Lemur News* 5: 23–25.

Lowry, P. P. II., Schatz, G. E., and Phillipson, P. B. 1997, The classification of natural and anthropogenic vegetation in Madagascar, in: Goodman, S. M. and Patterson, B. D., eds., *Natural Change and Human Impact in Madagascar*, Smithsonian Institution Press, Washington, DC, pp. 93–123.

Marivaux, L., Welcomme, J.-L., Antoine, P.-O, Métails, G., Baloch, I. M., Benammi, M., Chaimanee, Y., Ducrocq, S., and Jaeger, J.-J. 2001, A fossil lemur from the Oligocene of Pakistan. *Science* 294: 587–590.

Martin, R. D. 1972a, Adaptive radiation and behaviour of the Malagasy lemurs. *Phil. Trans. Roy. Soc. Lond. B* 264: 295–352.

Martin, R. D. 1972b, A preliminary study of the lesser mouse lemur (*Microcebus murinus* J.F. Miller 1777). *Z. Tierpsychologie* 9: 43–89.

Martin, R. D. 1995, Prosimians: From obscurity to extinction? in: Alterman, L., Izard, K., and Doyle, G. A., eds., *Creatures of the Dark*, Plenum Press, New York, pp. 535–563.

Martin, R. D. 2000, Origins, diversity and relationships of lemurs. *Int. J. Primatol.* 21: 1021–1049.

Meave, J., and Kellman, M. 1994, Maintenance of rain forest diversity in riparian forests of tropical savannas: Implications for species conservation during Pleistocene drought. *J. Biogeogr.* 21: 121–135.

Mittermeier, R. A., Tattersall, I., Konstant, W. R., Meyers, D. M., and Mast, R. B. 1994, *Lemurs of Madagascar*, Conservation International, Washington, DC.

Murphy, W. J., Eizirik, E., O'Brien, S. J., Madsen, O., Scally, M., Douady, C. J., Teeling, E., Ryder, O. A., Stanhope, M. J., de Jong, W. W., and Springer, M. S. 2001, Resolution of the early placental mammalian radiation using Bayesian phylogenetics. *Science* 294: 2348–2351.

Myers, A. A. and Giller, P. S. 1988, *Analytical Biogeography*, Chapman and Hall, London.

Myers, N., Mittermeier, R. A., Mittermeier, C. G., da Fonseca, G. A. B., and Kents, J. 2000, Biodiversity hotspots for conservation priorities. *Nature* 403: 853–858.

Pastorini, J., Forstner, M. R. J., and Martin, R. D. 2000, Relationships among Brown Lemurs (*Eulemur fulvus*) based on mitochondrial DNA sequences. *Mol. Phyl. Evol.* 16: 418–429.

Pastorini, J., Forstner, M. R. J., and Martin, R. D. 2001a, Phylogenetic history of sifakas (*Propithecus*: Lemuriformes) derived from mtDNA sequences. *Am. J. Primatol.* 53: 1–17.

Pastorini, J., Martin, R. D., Ehresmann, P., Zimmermann, E., and Forstner, M. R. J. 2001b, Molecular phylogeny of the lemur family Cheirogaleidae (Primates) based on mitochondrial DNA sequences. *Mol. Phyl. Evol.* 19: 45–56.

Petter, J.-J. and Andriatsarafara, S. 1987, Les lémuriens de l'ouest de Madagascar, in: Mittermeier, R. A., Rakotovao, L. H., Randrianasolo, V., Sterling, E. J., and Devitre, D., eds., *Priorités en Matière de Conservation des Espèces à Madagascar*, Gland: IUCN, 71–73.

Petter, J. J., Albignac, R., and Rumpler, Y. (eds.). 1977, *Faune de Madagascar: Mammifères Lémuriens*, Vol. 44, ORSTOM CNRS, Paris.

Phillipson, P. B. 1994, Madagascar, in: Davis, S. D., Heywood, V. H., and Hamilton, A. C., eds., *Centres of Plant Diversity*, Vol. 1, WWF and IUCN, Gland, pp. 271–281.

Pollock, J. I. 1986, Towards a conservation policy for Madagascar's eastern rainforests. *Primate Cons.* 7: 82–86.

Rabinowitz, P. D., Coffin, M. F., and Falvey, D. 1983, The separation of Madagascar and Africa. *Science* 220: 67–69.

Ramanamanjato, J.-B. and Ganzhorn, J. U. 2001, Effects of forest fragmentation, introduced *Rattus rattus* and the role of exotic tree plantations and secondary vegetation for the conservation of an endemic rodent and a small lemur in littoral forests of southeastern Madagascar. *Anim. Cons.* 4: 175–183.

Rasoloarison, R. M., Goodman, S. M., and Ganzhorn, J. U. 2000, A taxonomic revision of mouse lemurs (*Microcebus*) occurring in the western portions of Madagascar. *Int. J. Primatol.* 21: 963–1019.

Ravaoarimanana, I. B., Tiedemann, R., Montagnon, D., and Rumpler, Y. 2004, Molecular and cytogenetic evidence for cryptic speciation within a rare endemic Malagasy lemur, the northern sportive lemur (*Lepilemur septentrionalis*). *Mol. Phyl. Evol.* 31: 440–448.

Richard, A. F. and Dewar, R. E. 1991, Lemur ecology. *Ann. Rev. Ecol. Syst.* 22: 145–175.

Rumpler, Y. 1975, Chromosomal studies in the systematics of Malagasy lemurs, in: Tattersall, I. and Sussman, R. W., eds., *Lemur Biology*. Plenum Press, New York.

Rumpler, Y. 2000, What cytogenetic studies may tell us about species diversity and speciation in lemurs. *Int. J. Primatol.* 21: 865–881.

Schatz, G. E. 2000, Endemism in the Malagasy tree flora, in: Lourenço, W. and Goodman, S. M., eds., *Diversité et Endemisme à Madagascar*. Paris: Mémoires de la Société de Biogéographie, 1–9.

Simons, E. L. 1997, Lemurs: Old and new, in: Goodman, S. M. and Patterson, B. D., eds., *Natural Change and Human Impact in Madagascar*, Smithsonian Institution Press, Washington, DC, pp. 142–166.

Simons, E. L. and Meyers, D. 2001, Folklore and beliefs about the aye aye (*Daubentonia madagascariensis*). *Lemur News* 6: 11–16.

Sterling, E. J. 1998, Preliminary report on a survey for *Daubentonia madagascariensis* and other primate species in the west of Madagascar. *Lemur News* 3: 7–8.

Storey, M., Mahoney, J. J., Saunders, A. D., Duncan, R. A., Kelley, S. P., and Coffin, M. F., 1995, Timing of hot spot-related volcanism and the breakup of Madagascar and India. *Science* 267: 852–855.

Tattersall, I. 1982, *The Primates of Madagascar*, Columbia University Press, New York.

Tavaré, S., Marshall, C. R., Will, O., Soligo, C., and Martin, R. D. 2002, Using the fossil record to estimate the age of the last common ancestor of extant primates. *Nature* 416: 726–729.

Thalmann, U. 2000a, Greater dwarf lemurs from the Bongalava (Central Western Madagascar). *Lemur News* 5: 33–35.

Thalmann, U. 2000b, Lemur diversity and distribution in western Madagascar - inferences and predictions using a cladistic approach, in: Lourenço, W. R. and Goodman, S. M., eds., *Diversité et Endemisme à Madagascar*, Paris: Mémoires de la Société de Biogéographie, 191–202.

Thalmann, U. 2003, An integrative approach to the study of diversity and regional endemism in lemurs (Primates, Mammalia), and their conservation in western Madagascar, in: Legakis, A., Sfenthourakis, S., Polymeni, R. and Thessalou-Legaki, M., eds., *The New Panorama of Animal Evolution. Proceedings of the XVIIIthe (New) International Congress of Zoology. Hellenic Zoological Society (Athens, August 28— September 2, 2000)*. Sofia: Pensoft Publ, 393–402.

Thalmann, U. and Geissmann, T. 2000, Distribution and geographic variation in the Western Woolly Lemur (*Avahi occidentalis*) with description of a new species (*A. unicolor*). *Int. J. Primatol.* 21: 915–941.

Thalmann, U. and Rakotoarison, N. 1994, Distribution of lemurs in central western Madagascar, with a regional distribution hypothesis. *Folia Primatol.* 63: 156–161.

Thalmann, U., Müller, A. E., Kerloc'h, P., and Zaramody, A. 1999, A visit to the Strict Nature Reserve Tsingy de Namoroka (NW Madagascar). *Lemur News* 4: 16–19.

Thalmann, U., Kümmerli, R., and Zaramody, A. 2002, Why *Propithecus verreauxi deckeni* and *P. v. coronatus* are valid taxa—quantitative and qualitative arguments. *Lemur News* 7: 11–16.

Tomiuk, J., Bachmann, L., Leipoldt, M., Ganzhorn, J. U., Ries, R., Weis, M., and Loeschcke, V. 1997, Genetic diversity of *Lepilemur mustelinus ruficaudatus*, a nocturnal lemur of Madagascar. *Cons. Biol.* 11: 491–497.

Wright, P. C. 1999, Lemur traits and Madagascar ecology: Coping with an island environment. *Yearb. Phys. Anthro.* 42: 31–72.

Wyner, Y., Absher, R., Amato, G., Sterling, E., Stumpf, R., Rumpler, Y., and DeSalle, R. 1999, Species concepts and the determination of historic gene flow patterns in the *Eulemur fulvus* (brown lemur) complex. *Biol. J. Linn. Soc.* 66: 39–56.

Wyner, Y. M., Johnson, S. E., Stumpf, R. M., and DeSalle, R. 2002, Genetic assessment of a white-collared x red-fronted lemur hybrid zone at Andringitra, Madagascar. *Am. J. Primatol.* 67: 51–66.

Yoder, A. D. 1997, Back to the future: A synthesis of strepsirrhine systematics. *Evol. Anthropol.* 6: 11–22.

Yoder, A. D., Cartmill, M., Ruvolo, M., Smith, K., and Vigalys, R. 1996, Ancient single origin for Malagasy primates. *Proc. Natl. Acad. Sci. USA* 93: 5122–5126.

Yoder, A. D., Rasoloarison, R. M., Goodman, S. M., Irwin, J. A., Atsalis, S., Ravosa, S., and Ganzhorn, J. U. 2000, Remarkable species diversity in Malagasy mouse lemurs (Primates, *Microcebus*). *Proc. Natl. Acad. Sci. USA* 97: 11325–11330.

Zaramody, A., Andriaholinirina, N., Rousset, D., and Rabarivola, C., 2005, Nouvelle répartition respective de *Lepilemur microdon* et *L. mustelinus*, et de *L. ruficaudatus* et *L. edwardsi. Lemur News* 10: 19–20.

Zimmermann, E., Cepok, S., Rakotoarison, N., Zietemann, V., and Radespiel, U. 1998, Sympatric mouse lemurs in north-west Madagascar: A new rufous mouse lemur species (*Microcebus ravelobensis*). *Folia Primatol.* 69: 106–114.

Zimmermann, E., Masters, J., and Rumpler, Y. 2000, Introduction to diversity and speciation in the Prosimii. *Int. J. Primatal* 21: 789–791.

CHAPTER NINE

Mouse Lemur Phylogeography Revises a Model of Ecogeographic Constraint in Madagascar

Anne D. Yoder and Kellie L. Heckman

ABSTRACT

Mouse lemurs (genus *Microcebus*) are small nocturnal primates that are ubiquitously distributed throughout Madagascar. Until the past decade or so, it was believed that there were only two species, one that occupied the eastern regions of Madagascar (*M. rufus*) and one that occupied the western regions of Madagascar (*M. murinus*). Intensive field studies, accompanied by genetic analysis, have revealed that the two species taxonomy vastly underestimates the actual species diversity, however, with eight species now recognized. There are numerous indicators that even the eight species taxonomy is an insufficient representation of their actual evolutionary diversity. Our chapter reviews some of the evidence both for the presently acknowledged species diversity, and clarifies the evidence for supposing that there are other species yet to be identified. Primarily, the chapter focuses on the unexpected phylogeographic patterns revealed by mitochondrial DNA analysis. These data show that mouse lemur species do not form western and

Anne D. Yoder • Departments of Biology & BAA, Duke University, Box 90338, Durham, NC 27708   Kellie L. Heckman • Department of Ecology and Evolutionary Biology, Yale University, P.O. Box 208105, 21 Sachem Street, New Haven, CT 06520

Primate Biogeography, edited by Shawn M. Lehman and John G. Fleagle.
Springer, New York, 2006.

eastern clades, as ecogeographic evidence might suggest. Rather, there appears to be a historical separation of species into northern and southern clades. We emphasize the point, however, that this latter pattern is based on incomplete species and geographic sampling. Only complete sampling of populations from all regions of Madagascar will reveal the true historical patterns of mouse lemur evolution.

Key Words: molecular systematics, Strepsirhini, cheirogaleids

INTRODUCTION

Phylogeographic methods involve determining the historical relationships among gene lineages with attention to the contemporary spatial distribution of those lineages. In other words, the methodology entails deriving a gene tree for an array of individuals and/or taxa and then mapping that tree onto the geographic localities whence the individual DNA samples were collected. By doing so, one can potentially examine the effects of putative geographic barriers to gene flow. If geographic samples are reciprocally monophyletic with respect to a putative geographic barrier (Figure 1a), then that barrier is often inferred to have inhibited the dispersal of individuals and their genes. If, on the other hand, the geographic samples are not reciprocally monophyletic (Figure 1b), then we can infer either that the barrier is not a barrier at all, or that the barrier arose too recently yet to be recorded in the genetic data. Because these methods have a concern both for spatial patterning of individual alleles and for their historical relationship to other homologous alleles, phylogeography has emerged as the theoretical bridge that unites the traditionally distinct fields of phylogenetics and population genetics (Avise *et al.*, 1987; Avise, 1989).

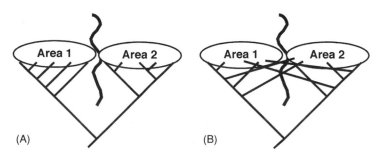

Figure 1. Schematic diagrams comparing models in which haplotype relationships show: (a) reciprocal monophyly with respect to a putative biogeographic barrier (wavy line) between two localities; versus (b) those that do not show reciprocal monophyly, indicating gene flow across barrier.

To date, phylogeographic studies have focused primarily on mitochondrial DNA (mtDNA) (e.g., Avise *et al.*, 1992; Da Silva and Patton, 1993; Taberlet and Bouvet, 1994; Ward, 1997; Avise and Walker, 1998; Eizirik *et al.*, 1998; Lucchini and Randi, 1998; Bensch and Hasselquist, 1999). Advantages of mtDNA for phylogeographic studies were highlighted by Avise *et al.* (1987) in their seminal paper: mtDNA is distinctive yet ubiquitously distributed, is easy to isolate, amplify, and sequence, has a simple genetic structure, is nonrecombining and uniparentally inherited, and evolves rapidly enough to provide information at even intra-populational levels. Yet, there are regions in the mtDNA genome conserved enough to be informative at much higher taxonomic levels. Despite all these advantages, reliance on mtDNA alone has been criticized (Pamilo and Nei, 1988; Hare, 2001; Ballard and Whitlock, 2004). Because the mtDNA genome is nonrecombining, one is examining a single locus, no matter how many mitochondrial genes one chooses to sequence. And, because it is maternally inherited, it is possible that organismal mating patterns (e.g., strong female philopatry) can potentially skew the results (Hoelzer, 1997). Several empirical and theoretical studies have indicated, however, that this may not be as problematic as has been suggested (Pamilo and Nei, 1988; Avise, 1992; Moore, 1995). Also, the problems associated with nuclear DNA (nDNA) markers are not trivial. Avise (1998) summarizes these as two: (1) it is difficult to isolate single alleles from a diploid organism; and (2) it is difficult to find markers that are accumulating mutations rapidly enough for fine-scale resolution, but are free of recombination. Consequently, even though nDNA markers are increasingly more common in phylogeographic studies (Hammer *et al.*, 1998), their use is still limited and largely in conjunction with mtDNA markers.

Given the various issues described above, one is typically faced with a single gene tree (usually mtDNA) for a taxonomically focused group of organisms. Even though striking patterns might present themselves in such an analysis, one might well ask, "how much about geographic structuring can safely be inferred from such limited data?" Clearly, inferences must be limited, especially given that stochastic processes such as isolation by distance can readily create similar patterns (Irwin, 2002). As a means for extending the power of such analyses, the newly emerging field of comparative phylogeography has been offered as a method for the investigation of "landscape evolution," including patterns of gene flow and genealogical vicariance, even in the absence of an a priori hypothesis of localized barriers to gene flow (Bermingham and Moritz, 1998). Da Silva and Patton (1998) detail the logic of the approach, which essentially involves the comparison of phylogeographic patterns for multiple co-distributed species. First, the observation of reciprocally monophyletic groups offers support for long-term spatial barriers to gene flow. Second, geographically concordant phylogenetic gaps for different taxa can identify common boundaries and/or historical events. Thus, these multiple vicariant biogeographic histories can be used as evidence for interpreting the biogeographic history of a region. Studies employing this method are growing in number, and show mixed results. While some have shown congruent patterns among the organisms, thereby indicating strong biogeographic patterns (Da Silva and Patton, 1998; Moritz and Faith, 1998), others have not (Zink, 1996; Taberlet *et al.*, 1998).

Landscape Evolution in Madagascar

Madagascar has long been recognized as an island of rare floral and faunal diversity. At present, it lies approximately 350 km to the east of Africa at the narrowest point of the Mozambique Channel and is otherwise completely isolated from other significant landmasses. The complex relationship between geological history and geographic isolation has conspired to create its unique assemblage of organisms. The Malagasy flora and fauna are a fascinating mix of singularity and diversity, singular due to the island's ancient isolation, and diverse due to the complexity of its topography and ecology. Due to its large surface area, and its varied assortment of microclimates and habitats, it is often referred to as a mini-continent (de Wit, 2003). Much of Madagascar's ecological variation relates to its sharply asymmetrical topography. The eastern edge, where it was once conjoined with India, is ruggedly mountainous, abruptly rising from the Indian Ocean to attain elevations of 2000 m, and is characterized by moist evergreen rainforest. Altitudes gradually diminish to sea level in the west, where the vegetation is predominated by dry deciduous forest. There, rainfall is sharply lower, with the extreme southwest receiving less than 35 cm/yr of rainfall. The intervening central plateau is composed primarily of depauperate grassland. Preliminary analyses indicate that the inherent dissimilarities in topography between eastern and western Madagascar have important bearing on the biogeography of these two portions of the island (Goodman and Ganzhorn, 2004b, 2004a).

Early in the 20th century, the prevailing view of Madagascar's presettlement landscape was that human habitation was entirely responsible for the abrupt disjunction between east and west, and that prior to the arrival of humans, Madagascar was entirely covered by closed forest formations with wildfire virtually absent (Humbert, 1927; Perrier de la Bâthie, 1927). Analyses of palyno-logical and geological data show, however, that much of Madagascar's central plateau has long been characterized by a mosaic of woodlands, shrublands, and grasslands (Burney, 1997). Moreover, analysis of Madagascar's topography, in concert with climatological and paleogeographic data, indicate that the "eastern edge" watershed has concentrated rain on the east coast and desiccated the west since the late Cretaceous (Wells, pers. com.). Superimposed on this east/west asymmetry is a north/south climatic gradient, most obvious in the west, wherein the island tends to become progressively drier in a north to south progression. In sum, these data suggest that disparate eastern and western ecological communities might have provided a separate suite of ecogeographic characteristics in which terrestrial vertebrates were able to diversify into the variety of forms that we observe today.

Inferring Big Patterns with Small Primates

The model of ecogeographic constraint summarized above presents a number of questions that can potentially be addressed with phylogeographic data. Most obviously, the long-term separation and ecological distinction between eastern and western habitats would suggest that organisms with limited means of dispersal (e.g., terrestrial vertebrates) that are distributed along both coasts might show strong historical roots within their respective geographic locales. Intuitively, congeneric species distributed along one coast would be more closely related to each other than to congeners distributed along the opposite coast and the range of individual species within a genus would not extend to both coasts. This pattern of east/west distribution is found commonly in range distributions for the majority of lemur species (Mittermeier et al., 1994). In fact, there are few examples of lemurs that do not exhibit a disjunct distribution (see Simons, 1993; Mittermeier et al., 1994; Sterling, 1994). It is important to note, however, that with the inclusion of subfossil specimens, many exceptions to the general pattern of east/west species disjunction can be observed, also with evidence for a potential dispersal corridor between east and west across the central highlands (Godfrey et al., 1999).

Until the late 1970s, *Microcebus* was considered monotypic by most authorities, containing only the species *murinus* (Schwarz, 1931). Upon broader geographic sampling and increased research activity, researchers who were studying mouse lemur populations reached the conclusion that there were actually two species in the genus (Martin, 1972, 1973; Petter *et al.*, 1977). One was a dry-adapted form that was restricted to the western portions of Madagascar, and the other was a wet-adapted form found in the eastern forests. The first, which retained the name *murinus*, was typified as a long-eared gray form, and the second, that was given the name *M. rufus*, was typified as a short-eared reddish form. Thus, the two species taxonomy emphasized both ecogeographic and morphological distinctions between the two mouse lemur types. Martin (1973) made particular note of the differing habitats and ecological constraints that appeared to define the two species, with *M. murinus* inhabiting dry deciduous and spiny desert forest and specializing on insectivory, and *M. rufus* inhabiting humid rain forest and showing dietary tendencies towards omnivory. Thus, the idea that both ecological and biogeographic mechanisms maintain species separation was an implicit assumption of the two-species taxonomy.

The two-species classification remained stable until relatively recently. Within the part decade, that classification has been radically altered. The revision began with the discovery that two distinct mouse lemur forms occur in non interbreeeding sympatry at the western locality of the Kirindy Forest (Kirindy/CFPF). One is the typical *M. murinus* of dry forests and the other is a distinctly smaller rufus-colored animal. The authors of that study (Schmid and Kappeler, 1994) concluded that the second form fit with the original diagnosis of *M. myoxinus*. Subsequently, a much larger mouse lemur type, also sympatric with *M. murinus*, was described from the northwest and designated as *M. ravelobensis*. More recently still, the number of mouse lemur species has been doubled by a morphological study that sampled broadly throughout western localities (Rasoloarison *et al.*, 2000).

In collaboration with the authors of that most recent study, Yoder and colleagues undertook a mtDNA phylogeographic study of the same western populations along with several eastern populations (Yoder *et al.*, 2000). The purpose of the study was both to test the species designations proposed by Rasoloarison *et al.* (2000) and to identify the historical relationships among the various populations sampled by that study. To accomplish these goals, we initially adopted a null hypothesis of species homogeneity and accordingly selected a rapidly evolving mtDNA marker likely to show variation at the intraspecific level. We sequenced an approximately 500 base pair segment of the mtDNA control region, homologous with the hypervariable region one region in humans (HV1), for all 118 individuals sampled by our study. Samples of *M. rufus* from two eastern localities were included and originally intended to serve as outgroups to the western populations. HV1 showed surprisingly high levels of sequence variation and yielded a tree wherein there were well-resolved clades that are perfectly congruent with the morphological species designations of Rasoloarison *et al.* (2000) but whose interrelationship could not be determined due to poor internal resolution (Figure 2). The latter result is presumably due to saturation, and in fact, was our first indication that the null hypothesis of a single species might be incorrect. The HV1 tree was also surprising in that the two populations of *M. rufus* do not form a clade, and instead, are both nested within clades that contain western populations. To attempt better resolution of deeper nodes, we sub-sampled individuals from each of the well supported HV1 clades and sequenced them for the more conserved cytochrome oxidase subunit II (684 bp) and cytochrome *b* (1140 bp) genes.

The combined analysis of the three mtDNA markers yields a tree in which the nine terminal clades from the HV1-only analysis are identically resolved (Figure 3) and has the additional strength of resolving two deep clades with strong support, thereby allowing a test of the east/west biogeographic constraint hypothesis. If this hypothesis held, we would expect a phylogeographic scenario much like the one illustrated in Figure 4 wherein individuals sampled from eastern localities would form one clade, and those from western localities would form another. An entirely different pattern emerged, however. As had been suggested by the HV1 analysis, there is no clear grouping into eastern and western clades. Rather, the populations sampled appear to form northern and southern clades (Figure 5). Aside from the departure from expectation, this result is surprising in that it is difficult to surmise what is, or could have been, the biogeographic barrier separating northern and southern mouse lemur communities. One possibility is highlighted by Pastorini *et al.*, (2003) who demonstrated the importance of rivers as barriers to gene flow for multiple populations of lemurs along the west coast. In that study, the authors found that both the Tsiribihina and the Betsiboka rivers were significant isolating mechanism for a number of lemur species and subspecies along the west coast of Madagascar.

Testing the Reality of the Pattern

To summarize, mouse lemurs show historical relationships that indicate close connections between eastern and western populations across similar latitudes, thereby falsifying any notion that the ecogeographic disjunction between

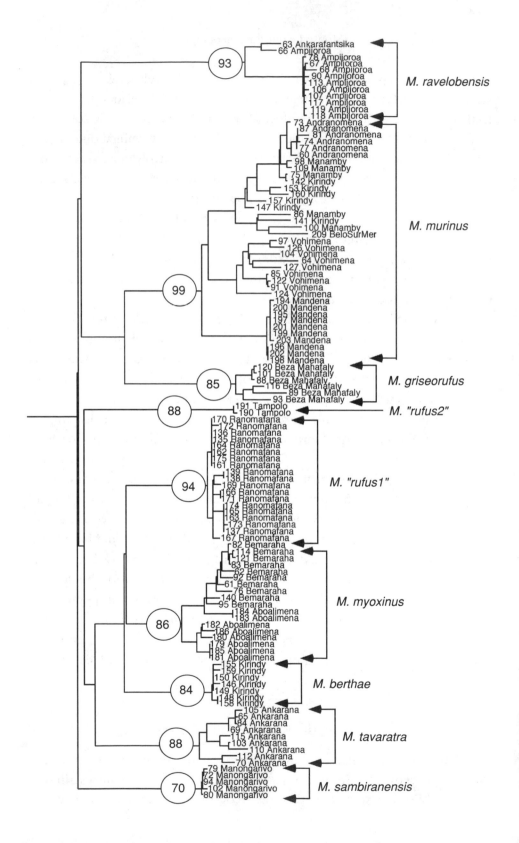

63 Ankarafantsika
66 Ampijoroa
78 Ampijoroa
67 Ampijoroa
68 Ampijoroa
90 Ampijoroa
113 Ampijoroa
106 Ampijoroa
107 Ampijoroa
117 Ampijoroa
119 Ampijoroa
118 Ampijoroa

93

M. ravelobensis

73 Andranomena
87 Andranomena
81 Andranomena
74 Andranomena
77 Andranomena
60 Andranomena
98 Manamby
109 Manamby
75 Manamby
142 Kirindy
153 Kirindy
160 Kirindy
157 Kirindy
147 Kirindy
86 Manamby
141 Kirindy
100 Manamby
209 BeloSurMer
97 Vohimena
126 Vohimena
104 Vohimena
64 Vohimena
127 Vohimena
85 Vohimena
122 Vohimena
91 Vohimena
124 Vohimena
194 Mandena
200 Mandena
195 Mandena
197 Mandena
201 Mandena
199 Mandena
203 Mandena
196 Mandena
202 Mandena
198 Mandena

M. murinus

99

120 Beza Mahafaly
101 Beza Mahafaly
88 Beza Mahafaly
116 Beza Mahafaly
89 Beza Mahafaly
93 Beza Mahafaly

85

M. griseorufus

191 Tampolo
190 Tampolo

88

M. "rufus2"

170 Ranomafana
172 Ranomafana
136 Ranomafana
135 Ranomafana
164 Ranomafana
162 Ranomafana
75 Ranomafana
161 Ranomafana
139 Ranomafana
138 Ranomafana
169 Ranomafana
166 Ranomafana
171 Ranomafana
174 Ranomafana
165 Ranomafana
163 Ranomafana
173 Ranomafana
137 Ranomafana
167 Ranomafana

94

M. "rufus1"

82 Bemaraha
114 Bemaraha
121 Bemaraha
83 Bemaraha
62 Bemaraha
92 Bemaraha
61 Bemaraha
76 Bemaraha
140 Bemaraha
95 Bemaraha
184 Aboalimena
183 Aboalimena
182 Aboalimena
186 Aboalimena
180 Aboalimena
179 Aboalimena
185 Aboalimena
181 Aboalimena

M. myoxinus

86

155 Kirindy
159 Kirindy
150 Kirindy
146 Kirindy
149 Kirindy
148 Kirindy
158 Kirindy

84

M. berthae

105 Ankarana
65 Ankarana
84 Ankarana
69 Ankarana
115 Ankarana
103 Ankarana
110 Ankarana
112 Ankarana
70 Ankarana

88

M. tavaratra

79 Manongarivo
72 Manongarivo
94 Manongarivo
102 Manongarivo
80 Manongarivo

70

M. sambiranensis

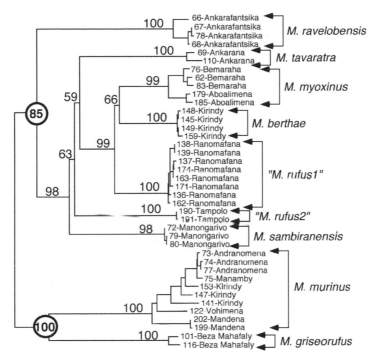

Figure 3. Distance tree of 2404 bp of combined HV1, cytochrome *b* and COII. Tree was generated in PAUP* 4.0b4a (PPC) (Swofford, 1998) by using HKY85 correction and weighted least squares (power = 2) algorithm. Location of midpoint root was confirmed by multiple outgroup rootings. Numbers on branches indicate statistical support from 100 bootstrap replicates with one random addition per replicate. Circled numbers highlight bootstrap support for two primary clades.

eastern and western habitats serves as an insurmountable barrier to dispersal. Moreover, an unexpected pattern of northern and southern clades emerged. Our results are mirrored to some extent by a study by Pastorini *et al.*, (2000) in which she also found that putative *M. rufus* are in a clade with *M. ravelobensis*

Figure 2. Distance tree of 118 mouse lemur mtDNA haplotypes derived from 580 bp alignment of control region sequence homologous with HV1 in humans. Individuals are identified by unique laboratory extraction number (Yoder Lab Extraction; YLE) and by locality. Tree was generated in PAUP* 4.0b4a (PPC) (Swofford, 1998) by using HKY85 correction and weighted least squares (power = 2) algorithm. Branches are proportional to expected number of changes per site. Numbers on branches indicate statistical support from 100 bootstrap replicates of the "fast" stepwise-addition algorithm for species-level clades. Tree was rooted with *Propithecus* and *Varecia* (not shown).

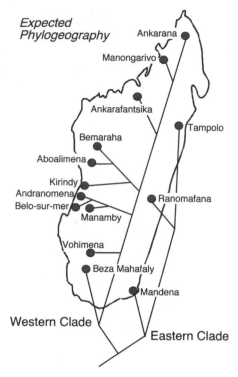

Figure 4. Phylogeographic model of mtDNA haplotypes that would be consistent with predicted east/west biogeographic disjunction.

from the northwest that excludes *M. murinus*, also from the west. That study did not address the north/south pattern observed in our study, however, due to its limited geographic sampling. On the other hand, the pattern of a north/south disjunction observed in the Yoder *et al.* (2000) study may simply be a consequence of limited sampling of *M. murinus* from north of the Tsiribihina River. To sufficiently test these biogeographic patterns in *Microcebus*, we need to expand our sampling of mouse lemurs from the northwest and from all regions of the east (work in progress) and include subfossil specimens from the central plateau (Godrey *et al.*, 1997). Furthermore, we need to expand our genetic sampling beyond the confines of the mitochondrial locus. To that end, we are assembling sequence data from a suite of independently segregating nuclear loci. Thus far, analysis of these data further indicate the lack of any east/west structuring of mouse lemur populations or species (Heckman *et al.*, in prep.).

The indication of a primary north/south biogeographic division in Madagascar is suggested by recent studies of other Malagasy lemurs. Within the species

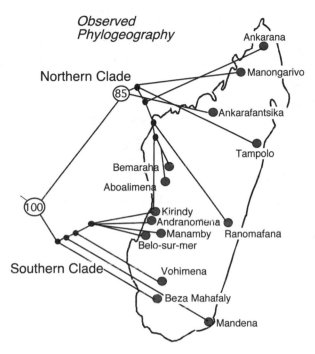

Figure 5. Actual phylogeographic structure of mtDNA haplotypes sampled by the Yoder *et al.* (2000) study. Note strong statistical support for primary division into northern and southern clades. Pattern is subject to further testing with additional geographic and haplotype sampling.

Eulemur fulvus, a north/south split is seen between subspecies marked in the west by the Betsiboka River (Pastorini *et al.*, 2000), rather than the Tsiribihina River as in the Yoder *et al.* (2000) study. In addition, in a study of the Indridae Pastorini *et al.* (2001) show an essential split of *P. verreauxi* into northern and southern clades also separated by the Betsiboka River. Looking more broadly, an example of biogeographic pattern outside of lemurs is found in the chameleon genus *Calumna*, where strong evidence for regional structuring of southern populations from western and northern populations is found. In the *Calumna* study, however, sampling locations are limited and results may represent a pattern of isolation by distance (Russell *et al.*, in prep.). This work is ongoing, and as with the mouse lemur study, additional intervening populations are being sampled.

In the meantime, the mouse lemur data confirm the fact that the biogeographic and ecological separation between western and eastern habitats has not been particularly restrictive to interpopulation gene flow among mouse lemur populations. Future studies will focus on increased sampling of taxa, localities, and ecosystems throughout Madagascar. Such multi layered analysis is the key to revealing patterns that are of universal impact, versus those that are merely idiosyncratic.

ACKNOWLEDGMENTS

This research has been supported by the Margot Marsh Biodiversity Foundation, a Northwestern University Research Grants Committee award, and by National Science Foundation grant DEB-9985205. We are grateful to David Burney and Neil Wells for their perseverance and insights into the complexities of Madagascar's climatological and geological past. We most especially thank Neil for sharing unpublished work of his in which this body of knowledge is masterfully summarized.

REFERENCES

Avise, J. C. 1989, Gene trees and organismal histories: A phylogenetic approach to population biology. *Evolution* 43:1192–1208.

Avise, J. C. 1992, Molecular population structure and the biogeographic history of a regional fauna: A case history with lessons for conservation biology. *Oikos* 63:62–76.

Avise, J. C. 1998, The history and purview of phylogeography: A personal reflection. *Mol. Ecol.* 7:371–379.

Avise, J. C., Alisauskas, R. T., Nelson, W. S., and Ankney, C. D. 1992, Matriarchal population genetic structure in an avian species with female natal philopatry. *Evolution* 46:1084–1096.

Avise, J. C., Arnold, J., Ball, R. M., Bermingham, T., Lamb, T., Neigel, J. E., Reeb, C. A., and Saunders, N. C. 1987, Intraspecific phylogeography: The mitochondrial DNA bridge between population genetics and systematics. *Ann. Rev. Ecol. Syst.* 18:489–522.

Avise, J. C. and Walker, D. 1998, Pleistocene phylogeographic effects on avian populations and the speciation process. *Proc. Roy. Soc. of Lond., Ser. B* 265:457–463.

Bensch, S. and Hasselquist, D. 1999, Phylogeographic population structure of great reed warblers: An analysis of mtDNA control region sequences. *Biol. J. Linn. Soc.* 66:171–185.

Bermingham, E. and Moritz, C. 1998, Comparative phylogeography: Concepts and applications. *Mol. Ecol.* 7:367–369.

Burney, D. A. 1997, Theories and facts regarding Holocene environmental change before and after human colonization, in: Goodman, S. M., and Patterson, B. D., eds., *Natural Change and Human Impact in Madagascar*, Washington, DC, Smithsonian Institution Press, pp. 75–89.

Da Silva, M. N. F. and Patton, J. L. 1993, Amazonian phylogeography: mtDNA sequence variation in arboreal echimyid rodents (Caviomorpha). *Mol. Phylogene. Evol.* 2:243–255.

Da Silva, M. N. F. and Patton, J. L. 1998, Molecular phylogeography and the evolution and conservation of Amazonian mammals. *Mol. Ecol.* 7:475–486.

de Wit, M. 2003, Madagascar: Heads it's a continent, tails its an island. *Ann. Rev. Earth Planet. Sci.* 31:213–248.

Eizirik, E., Bonatto, S. L., Johnson, W. E., Crawshaw, P. G., Vié, J. C., Brousset, D. M., O'Brien, S. J., and Salzano, F. M. 1998, Phylogeographic patterns and evolution of the mitochondrial DNA control region in two neotropical cats (Mammalia, Felidae). *J. Mol. Evol.* 47:613–624.

Goodman, S. M. and Ganzhorn, J. 2004a, Biogeography of lemurs in the humid forests of Madagascar: The role of elevational distribution and rivers. *J. Biogeogr.* 31:47–55.

Goodman, S. M. and Ganzhorn, J. 2004b, Elevational ranges of lemurs in the humid forests of Madagascar. *Int. J. Primatol.* 25:331–350.

Godfrey, L. R., Jungers, W. L., Reed, K. E., Simons, E. L., and Chatrath, P. S. 1997, Subfossil lemurs, in: Goodman S. M., Patterson B. D., eds., *Natural Change and Human Impact in Madagascar*, Smithsonian Institution Press, Washington, D.C., pp. 218–256.

Hammer, M. F., Karafet, T., Rasanayagam, A., Wood, E. T., Altheide, T. K., Jenkins, T., Griffiths, R. C., Templeton, A. R., and Zegura, S. L. 1998, Out of Africa and back again: A nested cladistic analysis of human Y chromosome variation. *Mol. Biol. Evol.* 15:427–441.

Humbert, H. 1927, Destruction d'une flore insulaire par le feu: Principaux aspects de la végétation à Madagascar. *Mémoires de l'Académie Malgache* 5:1–80.

Irwin, D. E. 2002, Phylogeographic breaks without geographic barriers to gene flow. *Evol. Int. J. Org. Evol.* 56:2383–2394.

Lucchini, V. and Randi, E. 1998, Mitochondrial DNA sequence variation and phylogeographical structure of rock partridge (*Alectoris graeca*) populations. *Heredity* 81:528–536.

Martin, R. D. 1972, A preliminary field-study of the lesser mouse lemur (Microcebus murinus J.F. Miller, 1777). *Zeitschrift für Tierpychologie* (9) Suppl:43–89.

Martin, R. D. 1973, A review of the behaviour and ecology of the lesser mouse lemur (*Microcebus murinus* J.F. Miller, 1777), in: Michael, R. P., and Crook, J. H., eds., *Comparative Ecology and Behaviour of Primates*, London, Academic Press, pp. 1–68.

Moore, W. S. 1995, Inferring phylogenies from mtDNA variation: Mitochondrial-gene trees versus nuclear-gene trees. *Evolution* 49:718–726.

Moritz, C. and Faith, D. P. 1998, Comparative phylogeography and the identification of genetically divergent areas for conservation. *Mol. Ecol.* 7:419–429.

Pamilo, P. and Nei, M. 1988, Relationships between gene trees and species trees. *Mol. Biol. Evol.* 5:568–583.

Pastorini, J. 2000, Molecular Systematics of Lemurs. Ph.D., Universität Zürich, Zurich.

Pastorini, J., Forstner, M. R. J., and Martin, R. D. 2001, Phylogenetic history of sifakas (*Propithecus: Lemuriformes*) derived from mtDNA sequences. *Am. J. Primatol.* 53:1–17.

Pastorini, J., Thalmann, U., and Martin, R. D. 2003, A molecular approach to comparative phylogeography of extant Malagasy lemurs. *Proceedings of the National Academy of Sciences* 100:5879–5884.

Perrier de la Bâthie, H. 1927, La végétation malgache. *Annals du Muséum Colonial, Marseille* 9:1–266.

Petter, J.-J., Albignac, R., and Rumpler, Y. (eds.) 1977, *Mammifères Lémuriens (Primates Prosimiens)*, ORSTOM and CNRS, Paris.

Rasoloarison, R. M., Goodman, S. M., and Ganzhorn, J. U. 2000, Taxonomic revision of mouse lemurs (*Microcebus*) in the western portions of Madagascar. *Int. J. Primatol.* 21:963–1019.

Schmid, J. and Kappeler, P. M. 1994, Sympatric mouse lemurs (*Microcebus* spp.) in western Madagascar. *Folia Primatol.* 63:162–170.

Schwarz, E. 1931, A revision of the genera and species of Madagascar Lemuridae. *Proc. Zoolog. Soc. Lond.* 1931:399–428.

Taberlet, P. and Bouvet, J. 1994, Mitochondrial DNA polymorphism, phylogeography, and conservation genetics of the brown bear *Ursus arctus* in Europe. *Proc. Roy. Acad. Lond., B* 255:195–200.

Taberlet, P., Fumagalli, L., and Wust-Saucy, A. G. 1998, Comparative phylogeography and postglacial colonization routes in Europe. *Mol. Ecol.* 7:453–464.

Ward, R. H. 1997, Phylogeography of human mtDNA: An Amerindian perspective, in: Donnelly, P. and Tavare, S., eds., *Progress in Population Genetics and Human Evolution*, New York, Springer-Verlag, pp. 33–53.

Yoder, A. D., Rasoloarison, R. M., Goodman, S. M., Irwin, J. A., Atsalis, S., Ravosa, M. J., and Ganzhorn, J. U. 2000, Remarkable species diversity in Malagasy mouse lemurs (Primates, *Microcebus*). *Proc. Natl. Acad. Sci. USA.* 97:11325–11330.

Zink, R. M. 1996, Comparative phylogeography in North American birds. *Evolution* 50:308–317.

CHAPTER TEN

Abiotic and Biotic Factors as Predictors of Species Richness on Madagascar

Nancy J. Stevens and Patrick M. O'Connor

ABSTRACT

Madagascar contains a diversity of endemic species. Yet levels of species richness vary among the isolated habitats scattered across the island. A number of ecological factors have been advanced to account for patterns of species richness. In particular, abiotic factors such as habitat area, latitude, altitude, temperature, and rainfall have been suggested to account for ultimate differences in the number of species a habitat may support. Biotic variables such as vegetation type have been suggested as more proximate factors in determining the diversity of habitats available for animals to occupy. Several studies have included Malagasy locales in evaluating large-scale relationships between ecological variables and species richness (e.g., Reed and Fleagle, 1995, Fleagle and Reed, 1996; Eeley and Lawes, 1999; Emmons, 1999; Ganzhorn *et al.*, 1999). This study combines data on geography, topography, climate, and vegetation with species lists from 27 national parks, reserves, and other protected areas to specifically address biogeographic patterns of species richness on Madagascar. Ecological variables are considered individually in order to determine which biotic and abiotic factors may best predict primate, mammal, bird, reptile, and amphibian richness.

Nancy J. Stevens and Patrick M. O'Connor • Department of Biomedical Sciences, College of Osteopathic Medicine, Ohio University, Athens, Ohio 45701

Primate Biogeography, edited by Shawn M. Lehman and John G. Fleagle.
Springer, New York, 2006.

Key Words: Primate biogeography, species-area relationships, ecological biogeography

INTRODUCTION

Madagascar is home to a wide range of endemic forms (e.g., Harcourt and Thornback, 1990; Wright, 1997). Over 53% of its birds, 90% of its known plant and invertebrate species, 95% of its reptiles, and nearly all of its fresh-water fishes, amphibians, tenrec insectivorans, cricetid rodents, carnivorans, and primates are unique to the island (Perrier de la Bathie, 1936; Albignac, 1972; Koechlin, 1972; Tattersall, 1982; Blommers-Schlosser and Blommers, 1984; Wilmé, 1996; Fisher, 1997; Raxworthy *et al.*, 2002; de Wit, 2003). Such endemicity has made Madagascar the darling of biogeographers for some time (Wallace, 1876), but the consensus among researchers is increasingly grim: many of these species are in rapid decline, as a combination of habi-tat destruction and hunting threatens much of the island's flora and fauna (e.g., Ratsimbazafy, 2002). Indeed, a great deal of research has documented habitat loss (e.g., Green and Sussman, 1990; Myers *et al.*, 2000) and even postulated the characteristics that make some species more vulnerable to ex-tinction than others (Jernvall and Wright, 1998; Wright and Jernvall, 1999; Wright, 1999a). The effects of recent extinctions upon biodiversity in Mada-gascar have been explored (Godfrey *et al.*, 1997; Simons, 1997), while at the same time a surge in conservation efforts has helped to create national parks and reserves to preserve as much of the remaining habitat as possible (e.g., Wright, 1997). In order to develop such protected areas, many habitat frag-ments have been censused (Goodman *et al.*, 1996, 1997, 1999; Johnson and Overdorff, 1999; Irwin *et al.*, 2000; Goodman and Benstead, 2003). As a re-sult, a good deal is known about species compositions in many of the habitat islands.

One thing is clear: levels of species richness are quite different among the isolated habitats scattered across Madagascar. Some of these differences may be related to vicariant and dispersal patterns by which animals came to live in dif-ferent areas on Madagascar (de Wit, 2003; Goodman *et al.*, 2003; Raxworthy, 2003; Yoder, 2003). Species richness likely reflects other historical or anthro-pogenic factors, such as climate change (Burney, 1997) and/or human activities (Dewar, 1997). Indeed, much research on Malagasy biogeography has focused on these issues (e.g., Goodman and Patterson, 1997). But the fact remains that some habitats may have ecological properties that allow them to support more species than do other habitats (Ceballos and Brown, 1995). Efforts have been made to understand the intrinsic ecological characteristics of certain species in order to provide for them on a case-by-case basis (e.g., Ratsimbazafy, 2002). Fewer recent studies have concentrated on identifying the types of habitats that maximize species richness on Madagascar.

Ecological approaches may lend insight by characterizing environments based on differing biotic and abiotic factors. Such factors are suggested to account for patterns in the number of species present in a given locale. Abiotic factors thought to account for ultimate differences in species richness include habitat area (Darlington, 1957; Terborgh and Winter, 1980; Pianka, 1994; Reed and Fleagle, 1995), latitude (Stevens, 1989; Eeley and Lawes, 1999), altitude (Donque, 1972), temperature (Pianka, 1994), and rainfall (Fleagle and Reed, 1996; Ganzhorn, 1997). Biotic variables such as vegetation type have been advanced as more proximate factors in determining the diversity of habitats available for animals to occupy (Paulian, 1961; Humbert, 1927, 1965; White, 1983; Pianka, 1994). The next section discusses the relationships between these factors and species richness individually.

ABIOTIC FACTORS

The topography of Madagascar provides for a range of habitat types (de Wit, 2003). The island rises in elevation from its coastlines to 2876 m at the highest peak in the Tsaratanana massif (Battistini and Rechard-Vindard, 1972), encompassing a number of altitudinal belts with different habitat characteristics (Donque, 1972). Many natural barriers exist on the landscape, including elevational clines (Battistini and Rechard-Vindard, 1972) and hydrological barriers (Aldegheri, 1972). In particular, some river systems are thought to play a role in separating primate habitats from one another (Martin, 1972). Topography also influences climate (Brenon, 1972; Donque, 1972), and is largely determined by geological features. This study considers these five related abiotic factors: *habitat area, latitude, altitude, temperature,* and *rainfall.*

Area

Habitat area has long been suggested to be a leading determinant of species richness. Two explanations for species-area relationships have been put forth. One mechanism to explain higher numbers of species in larger habitats is that large areas may support larger population sizes, which in turn could have a lower vulnerability to stochastic extinction events (Terborgh and Winter, 1980).

An alternate view is that habitat area is merely a proxy for habitat diversity (e.g., topographical diversity), and that area, itself, has no direct bearing upon species richness (e.g., Brown, 1988). This view predicts higher levels of habitat heterogeneity in larger habitats. As such, it views habitat area as a proxy for other ecological factors that ultimately result in species richness. More diverse habitats may support a greater number of species (Terborgh and Winter, 1980). This study predicts both more habitat diversity and greater species richness in larger locales.

Latitude

Studies examining large-scale biogeographic patterns often find relationships between the number of species present and the latitude of a given locale (e.g., Rapoport, 1982; Stevens, 1989; Ruggiero, 1994; Eeley and Foley, 1999; Eeley and Lawes, 1999; Harcourt, 2000). In general, species richness tends to increase towards the equator. Whereas there is likely no *direct* relationship between species richness and latitude, a number of explanations have been offered to explain latitudinal patterns. For example, species richness at lower latitudes may reflect habitat area; Peres and Janson (1999) observe that latitude relationships in South America may be confounded by the fact that it is wider near the equator and tapers off further south. The same is true for continental Africa. From the standpoint of climate, ascending in latitude has also been likened to moving toward higher altitudes (Pianka, 1994). Moreover, Madagascar has a north-south component in both its temperature and rainfall patterns (Donque, 1972). For these reasons, latitude relationships tend to sample richness at one of the coarsest scales. This study predicts that if there is a relationship between species richness and latitude, more species will be found at lower latitudes.

Altitude

Altitude can affect species richness in different ways. Goodman *et al.* (1996) have observed higher rodent and insectivore species richness at montane altitudes in Andringitra, whereas Hawkins *et al.* (1998) have demonstrated greater bird richness at lower altitudes at Anjanaharibe-Sud.

The island of Madagascar slopes upward steeply from the eastern coast toward the central plateau, and more gently down toward the western coast. The plateau has an additional downward slope in the northern direction, and accommodates the Tsaratanana, Andringitra, and Ankarana massifs, all of which are higher than 2500 m (Donque, 1972). This study examines whether mountainous or sea level environments support greater species richness in Madagascar. It predicts that locales with larger altitudinal ranges should include more habitat zones and higher species richness. In addition, the topography of Madagascar may help to explain climate patterns found in individual locales (Donque, 1972). High altitude can affect climate both by lowering temperatures, and by altering rainfall patterns (Donque, 1972). The individual effects of temperature and rainfall are discussed next.

Temperature

Both mean annual temperature and annual temperature range have the potential to affect patterns of vertebrate species richness. The bulk of the Madagascan landmass resides in the intertropical zone (Koechlin, 1972), which results in a largely tropical climate on the island (Figure 1). Yet Madagascar extends from the subequatorial region in the north, to the subtropical zone in the south (Donque, 1972).

Temperatures in the far south have a greater range of variability in comparison to those in the subequatorial region (Donque, 1972; Koechlin, 1972). For example, along the east coast, northern Diego Suarez experiences as little as 3°C differences in monthly average temperature, whereas southern Fort Dauphin experiences a nearly 7°C range (Donque 1972). Western localities show a similar pattern of increasing annual range of temperature with increased distance from the equator (Donque, 1972). In studies of global biogeographic patterns, higher species richness is often found in warmer, less seasonal climates (Wright *et al.*, 1993). This study examines patterns of species richness with mean annual temperatures in each locale. It predicts higher species richness in subequatorial environments with fewer annual fluctuations in temperature.

Rainfall

Rainfall has been suggested to exert an influence upon species richness patterns (Reed and Fleagle, 1995; Peres and Janson, 1999; Emmons, 1999). Studies looking at more global patterns of primate species richness have sampled well-studied Malagasy locales that represent extremes in rainfall patterns, such as the eastern rain forest, Ranomafana and the western dry forest, Morandava (e.g., Reed and Fleagle, 1995; Fleagle and Reed, 1996). Areas with higher rainfall

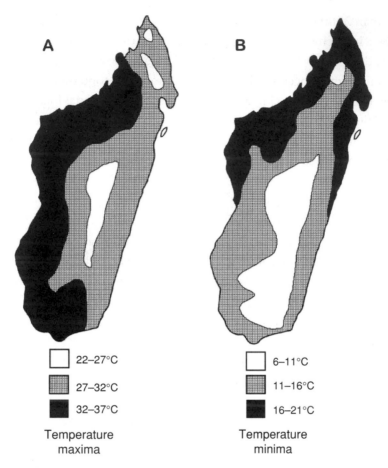

Figure 1. Temperature maxima and minima across Madagascar. Modified from Donque (1972).

generally possess greater species richness than do areas with lower rainfall. Examining these patterns over 27 locales within Madagascar should facilitate a more robust investigation to see if these patterns of species richness are maintained at a smaller spatial scale.

The eastern border of Madagascar experiences climatic influences generated by the warm, humid southeast Trade Winds (Donque, 1972; Koechlin, 1972). These winds are responsible for heavy rainfall (>1500 mm) along the eastern coast and along the eastern slopes of the high plateau (Koechlin, 1972). In contrast, western slopes receive very little humidity from the Trade Winds (Figure 2). Instead, the western coast receives its seasonal moisture from the monsoon regime that originates from the north (Koechlin, 1972). In general,

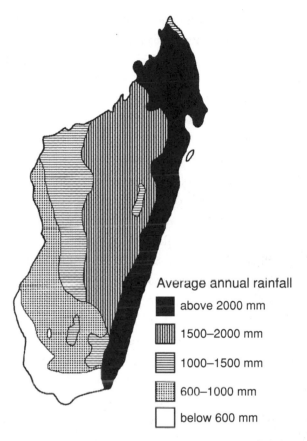

Figure 2. Average annual rainfall on Madagascar, based on Humbert (1965). Modified from Donque (1972).

the dry season is brief along the east coast and almost continuous in the far south. Perhaps the combination of Trade Winds and monsoons helps to create diverse climate zones that support the higher primate species richness in Madagascar relative to South America, Africa, and Asia observed by Reed and Fleagle (1995). This study predicts greater levels of species richness in areas with higher annual rainfall.

BIOTIC FACTORS

Biotic variables such as vegetation diversity may have a more proximate influence on species richness than the abiotic factors discussed above. Different habitats possess different plant types (Paulian, 1984). For example, tropical rain forests possess vines, epiphytes, and broad-leafed plants, whereas evergreens and smaller forms are common in colder areas of high latitude or altitude (Pianka, 1994). Deciduous trees are generally found in seasonal temperate zones that have moderate rainfall (Pianka, 1994). Various forms of sclerophyllous (thick-leaved) evergreen plants are often found in areas that experience a pronounced dry season of long duration (Pianka, 1994).

Most classifications of the vegetation of Madagascar are based upon phyto-geographic regions (Humbert, 1965). Humbert divided Madagascar's vegeta-tion into an eastern and a western component, based on coarsely defined plant types. White (1983) refined the phytogeographic regions into subregions, or domains. He divided the eastern region into eastern, central, high mountain, and sambirano domains, and the Western region into western and southern do-mains, in order to more accurately reflect the distribution of plant types in each region (Figure 3). Plant types are more easily identified from a distance than are individual individual species, and may be evaluated by large-scale analyses, (e.g., remote sensing by Humbert, 1965; Green and Sussman, 1990; Smith *et al.*, 1997). Because many floral habitats of Madagascar occur in remote areas, most plant species compositions are yet to be evaluated (DuPuy and Moat, 1996). An understanding of individual plant species distributions and dispersal patterns may eventually assist in fine-tuning biogeographical patterns (Lowry *et al.*, 1997). For the purposes of this study, habitat types will follow White's (1983) classification, with greater vertebrate species diversity predicted in locales with more habitat domains.

Habitat fragmentation as a result of human activities such as mining, log-ging, and agriculture has also exerted a profound impact upon many Malagasy species (Tattersall, 1982; Jolly and Jolly, 1984; Green and Sussman, 1990; Wright, 1997), a reminder that not all patterns of species richness are likely to be explained by the non-human agents considered in this study.

It is clear that a number of ecological variables differ among locales in Madagascar. These abiotic and biotic factors may play a role in maintaining lev-els of species richness. Although some studies have included the better-known Malagasy national parks in evaluating large-scale relationships between eco-logical variables and species richness (e.g., Reed and Fleagle, 1995; Ceballos and Brown, 1995; Fleagle and Reed, 1996; Ganzhorn *et al.*, 1999; Eeley and Lawes, 1999; Emmons, 1999), few have included numerous locales in different regions of Madagascar to examine patterns of vertebrate richness on the island. This study combines data on geography, topography, climate, and vegetation with species lists from 27 national parks, reserves, and other protected areas in

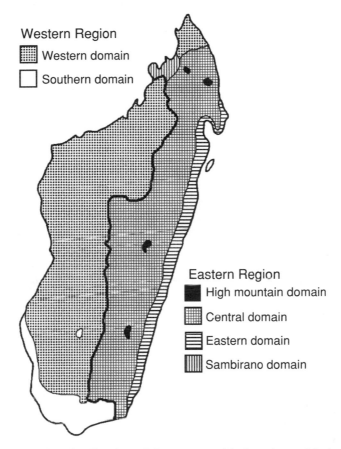

Figure 3. Vegetation classification of phytogeographic domains on Madagascar. Modified from White (1983).

Madagascar in order to determine which factors best predict vertebrate species richness.

METHODS

Data on ecological variables and species richness in national parks, special reserves, and other protected areas were compiled from the literature (Table 1 and Figure 4). Raw data were obtained from Humbert (1965); Donque (1972); Koechlin (1972); White (1983); Nicoll and Langrand (1989); Evans *et al.* (1994); Mittermeier *et al.* (1994); Raxworthy and Nussbaum (1994, 1996); Stephenson *et al.* (1994); Jenkins *et al.* (1996); DuPuy and Moat (1996); Goodman (1996, 1998); Sterling and Ramaroson (1996); Goodman and

Table 1. Raw data included in analysis

Reserve	Type	Area	Latitude	Min Alt	Max Alt	Alt Range	Rainfall	HiTemp	LoTemp
Isalo[1,2,3,4,5,6,7]	PN	81540	22	514	1268	754	850	30	14
Ranomafana[1,2,3,4,5,6,7,8,10]	PN	42000	21.1	800	1200	400	2600	30	14
Montage d'Ambre[1,2,3,4,5,6,7,18]	PN	18200	12.3	850	1475	625	2378	25	19
Bemaraha[1,2,3,4,5,6,7]	RNI	2E+05	18.5	150	750	600		35	19
Andohahela[1,2,3,4,5,6,7,9,11]	RNI	76020	24.5	90	1972	1882	1700	35	9
Zahamena[1,2,3,4,5,6,7]	RNI	73160	17	750	1512	762	1750	30	14
Ankarafantsika[1,2]	RNI	60520	16.1	80	333	253	1250	35	19
Marojejy[1,2,3,4,5,6,7,10]	RNI	60150	14	75	2133	2058	3000	30	19
Tsaratanana[1,2,3,4,5,6,7,10]	RNI	48622	13.75	227	2876	2649		25	9
Tsimanampetsotsa[1,2,3,4,5,6,7]	RNI	43200	24.1	38	114	76	350	35	14
Andringitra[1,2,3,4,5,6,7,8,9,10,14]	RNI	31160	22	650	2008	1958		30	5
Namoroka[1,2,3,4,5,6,7]	RNI	21742	16.2	71	227	156	1160	35	19
Lokobe[1,2,3,4,5,6,7]	RNI	740	13.2	0	430	430	2250	35	15
Manongarivo[1,2,3,4,5,6,7]	RS	35250	13.75	155	1876	1721		35	19
Analamera[1,2,3,4,5,6,7]	RS	34700	12.4	10	608	598	1250	30	19
Anjanaharibe-Sud[1,2,3,4,5,6,7,9,10]	RS	32100	14.4	500	2064	1564	2000	30	14
Kalambatritra[1,2,3,4,5,6,7,12]	RS	28250	23.2	1300	1500	200		30	9
Ankarana[1,2,3,4,5,6,7]	RS	18220	12.5	50	409	359	1890	30	19
Andranomena[1,2,3,4,5,6,7]	RS	6420	20.1	0	100	100	1200	35	14
Ambohitantely[1,2,3,4,5,6,7,15,16,17]	RS	5600	18.1	1448	1662	214	1678	30	9
Manombo[1,2,3,4,5,6,7,13]	RS	5020	23	0	137	137	2500	30	14
Bora[1,2,3,4,5,6,7]	RS	4780	14.75	115	411	296	1800	35	19
Cap Sainte Marie[1,2,3,4,5,6,7]	RS	1750	25.3	110	199	89	350	35	14
Beza-Mahafaly[1,2,3,4,5,6,7,9]	RS	580	23.4	100	200	100	550	35	14
Nosy Mangabe[1,2,3,4,5,6,7]	RS	520	15.3	0	332	332	3680	30	19
Pic d'Ivohibe[1,2,3,4,5,6,7]	RS	3453	22.3	775	2060	1285	3000	28	9
Analamazaotra[1,2,3,4,5,6,7,9,10]	RS	810	18.2	930	1040	110	1700	24	14

Patterson (1997); Schmid and Smolker (1998); Feistner and Schmid (1999); Goodman and Pidgeon (1999) and references therein, Andrianarimisa et al. (2000); Goodman and Rakotondravony (2000); Irwin et al. (2000); Goodman and Rasolonandrasana (2001); and Ratsimbazafy (2002). Species richness was defined as the number of primate, mammal, reptile, bird, and amphibian species recorded from each locale. Some authors have suggested that species richness patterns are potentially unstable because of disagreements on the degrees and types of attributes that constitute species-level distinctions (Harcourt, 2000). In this study, patterns of vertebrate richness were also compared at the generic level to assess how well ecological variables predict richness at different taxonomic scales.

Potential sources of error in species lists may arise from different methodologies used to survey individual locales (Ricklefs and Schluter, 1993; Johnson and Overdorff, 1999). Additionally, many lists are based on short-term surveys that may not reliably reflect inhabitants due to seasonal or yearly fluctuations in population size, or as a result of sampling only a portion of a given habitat (Irwin et al., 2000). Perhaps most importantly, short-term surveys can potentially

Table 1.   (*Continued*)

TempRan	# Dom	D Pri	Primates	Mammals	Reptiles	Amphib	Birds	DPr	Gen	Prim G	Mam G	Rept G	Amp G	Bird G
16	1	3	3	7	4		55	3	3	9	3			54
16	1	7	12	29	16	33	98	4	9	24	9	10		79
6	1	2	7	16	46	24	73	1	6	15	15			63
16	1	3	7	15	8		53	3	7	15	7			47
26	4	7	17	51	62	49	127	5	12	33	0			57
16	2	6	12	12	9		72	5	11	11	8			60
16	1	3	7	15	38	5	102	2	6	13	30	5		79
11	3	4	9	22	26	34	103	3	8	19	10	9		80
16	2	4	7	13	9	24	54	2	5	10	5	6		48
21	1	2	3	11	12		72	2	3	11	11			60
25	3	10	15	49	49	78	106	5	10	27	16	10		77
16	1	2	4	9	6	3	56	2	4	9	6	3		51
20	1	1	3	3	30		17	1	3	3	20			17
16	1	3	8	20	33	15	60	2	7	17	21	5		49
11	1	3	7	12	5		59	2	6	11	5			55
16	2	5	11	38	40	53	94	4	10	26	20	9		72
21	1	2	5	12	6		52	2	5	12	5			43
11	1	4	11	16	34		87	3	9	16	22			75
21	1	2	7	17	23		45	2	7	17	19			41
21	1	1	3	12	17	17	51	1	3	10	9	7		50
16	1	3	5	11	5		54	3	5	11	4			51
16	1	2	2	3	2		50	2	2	3	2			46
21	1	0	1	4	5		14	0	1	4	5			12
21	1	2	5	13	25		62	2	5	13	20			52
11	1	2	5	9	16		39	2	5	9	13			36
19	2	2	2	5			39	2	2	5				37
10	1	6	11	34	29	24	111	5	10	25	17	7		88

Raw data used in this study were obtained from the following sources:
[1]Nicoll and Langrand (1989), [2]Mittermeier et al. (1994), [3]Donque (1972), [4]Koechlin (1972), [5]Humbert (1965), [6]White (1983), [7]DuPuy and Moat (1996), [8]Goodman (1996) and references therein, [9]Goodman et al. (1997) and references therein, [10]Goodman (1998) and references therein, [11]Goodman (1999a, b) and references therein, [12]Irwin et al. (2000), [13]Ratsimbazafy (2002), [14]Goodman et al. (2001), [15]Andrianarimisa et al. (2000), [16]Goodman et al. (2000), [17]Stephenson et al. (1994), [18]Raxworthy and Nussbaum (1994).

under represent small, nocturnal, or otherwise cryptic taxa (Emmons, 1999). Often it was apparent that no survey had been conducted for one or more of the vertebrate groups included in this study. Most frequently, a locale would have census data only for primates or birds.

In order to reduce the effects of missing data on the analysis, it was necessary to identify a group less likely to be underrepresented in short-term surveys. Diurnal primates are among the most well studied vertebrates on Madagascar. They are larger in average body size and some have louder vocalizations than do other taxa. As such, they may have a higher probability of being accurately documented in remote, less thoroughly surveyed areas. For this reason, diurnal primates were extracted from the data set, and analyzed both separately and in combination with other taxa (following Emmons, 1999).

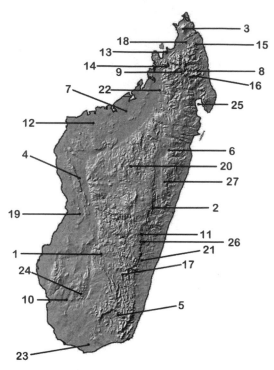

Figure 4. Locations of national parks, reserves-integrales, and special reserves included
in this study. Modified from Nicoll and Landgrand 1989. Numbers refer to the following localities: 1. Isalo, 2. Ranomafana, 3. Montage d'Ambre, 4. Bemaraha, 5. Ando-
hahla, 6. Zahamena, 7. Ankarafantsika, 8. Marojejy, 9. Tsaratanana, 10. Tsimanampet-
sotsa, 11. Andringitra, 12. Namoroka, 13. Lokobe, 14. Manongarivo, 15. Analamera,
16. Anjanaharibe-Sud, 17. Kalambatritra, 18. Ankarana, 19. Andranomena, 20. Ambohitantely, 21. Manombo, 22. Bora, 23. Cap Sainte Marie, 24. Beza-Mahafaly, 25. Nosy
Mangabe, 26. Pic d'Ivohibe, 27. Analamazaotra.

Abiotic variables were defined as habitat area (the total number of square
hectares in a locale); latitude; altitude (minimum, maximum, and range); temperature (minimum, maximum, and range); and annual rainfall. In addition,
White's (1983) classification of Humbert's original phytogeographic regions,
and the forest types reported for each locale (Nicoll and Landgrand, 1989) were
used to approximate vegetation diversity. Raw data compiled for this study are
summarized in Table 1.

Data were log-transformed and regression analyses were conducted (SPSS
ver. 8.0) to examine relationships between ecological variables and primate,
mammal, reptile, bird, and amphibian species richness. Significance levels and r^2

values for regressions are summarized in Table 2. Correlations among ecological variables are shown in Table 3, and correlations among species richness in the different vertebrate groups are depicted in Table 4.

RESULTS

Area Effects

This study included locales that range in size between 500 and 152,000 h. Regression analysis indicates that habitat area serves as a good predictor of primate, mammal, and bird species and generic richness (Table 2). This trend was strongest in the data set limited to diurnal primates ($p = 0.008$, Figure 5)

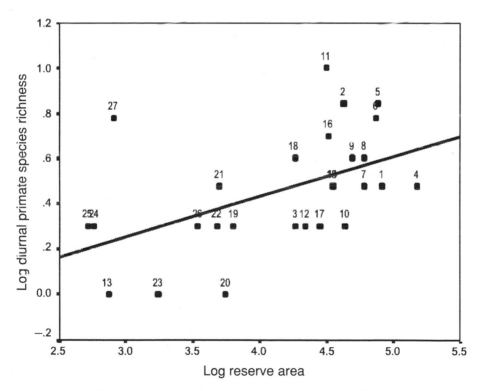

Figure 5. Log-log plot of diurnal primate species richness against locale area. Numbers correspond with the order of locales listed in Table 1. Note that locales that are well studied and/or that have heterogeneous habitats such as Ranomafana (2), Andringitra (11), Anjanaharibe-Sud (16), and Andohahela (5) plot above the regression line. Other locales such as Lokobe (13) and Ambohitantely (20) have lower than expected richness, perhaps due to either habitat disturbance or lack of long-term thorough studies.

Table 2. Regression values for species and generic richness vs. ecological variables

	Diurnal Primates		Primates		Mammals		Birds		Reptiles		Amphibians	
	Sig.	R sq	Sig.	R sq	Sig.	R sq	Sig.	R sq	Sig.	R sq	Sig.	R sq
Species richness												
Area	0.008	0.250	0.022	0.193	0.032	0.413	0.005	0.271	0.915	0.000	0.831	0.005
Latitude	0.936	0.000	0.267	0.049	0.793	0.053	0.754	0.004	0.639	0.009	0.290	0.123
Min Alt	0.117	0.096	0.511	0.017	0.144	0.289	0.050	0.145	0.919	0.000	0.175	0.194
Max Alt	0.016	0.212	0.039	0.160	0.043	0.393	0.064	0.130	0.221	0.062	0.001	0.720
Alt Range	0.006	0.261	0.021	0.196	0.089	0.334	0.124	0.092	0.145	0.086	0.042	0.384
RF	0.160	0.096	0.058	0.168	0.405	0.035	0.308	0.052	0.291	0.058	0.027	0.583
HiTemp	0.102	0.103	0.177	0.072	0.208	0.250	0.169	0.074	0.943	0.000	0.241	0.149
LoTemp	0.217	0.060	0.631	0.009	0.172	0.073	0.729	0.005	0.280	0.048	0.045	0.376
TemRan	0.697	0.006	0.314	0.041	0.790	0.003	0.279	0.047	0.718	0.006	0.484	0.056
No. of domains	0.002	0.328	0.015	0.215	0.008	0.250	0.029	0.177	0.022	0.200	0.031	0.419
Generic richness												
Area	0.092	0.114	0.035	0.165	0.022	0.192	0.010	0.238	0.418	0.029	0.958	0.000
Latitude	0.089	0.116	0.267	0.049	0.726	0.005	0.440	0.024	0.288	0.049	0.280	0.144
Min Alt	0.242	0.057	0.642	0.009	0.194	0.067	0.058	0.137	0.667	0.008	0.074	0.346
Max Alt	0.247	0.056	0.077	0.120	0.097	0.106	0.105	0.102	0.937	0.000	0.025	0.486
Alt Range	0.202	0.067	0.053	0.142	0.178	0.071	0.217	0.060	0.890	0.001	0.255	0.158
RF	0.788	0.004	0.057	0.170	0.497	0.023	0.234	0.070	0.631	0.013	0.009	0.774
HiTemp	0.588	0.012	0.225	0.058	0.242	0.054	0.067	0.128	0.842	0.002	0.192	0.203
LoTemp	0.276	0.049	0.887	0.001	0.382	0.031	0.995	0.000	0.810	0.003	0.173	0.219
TemRan	0.521	0.017	0.263	0.050	0.591	0.012	0.101	0.104	0.693	0.007	0.859	0.004
No. of domains	0.008	0.260	0.038	0.160	0.033	0.169	0.222	0.059	0.681	0.007	0.103	0.297

Significance levels and r^2 values for regressions of species richness against abiotic and biotic variables. (Rainfall estimates for parcel 1 of Andohahela were generously provided by S. Goodman.)

Table 3. Correlations among ecological variables

	Area	Latitude	Min Alt	Max Alt	Alt Range	RF	HiTemp	LoTemp	TemRan	No. of Domains
Area	.									
Latitude	0.723	.								
Min Alt	0.063	0.488	.							
Max Alt	0.036	0.285	0.000	.						
Alt Range	0.007	0.066	0.142	0.000	.					
RF	0.916	0.021	0.803	0.014	0.005	.				
HiTemp	0.871	0.231	0.046	0.004	0.157	0.017				
LoTemp	0.875	0.004	0.060	0.038	0.221	0.817	0.237	.		
TemRan	0.939	0.000	0.970	0.563	0.675	0.048	0.002	0.001	.	
No. of Domains	0.076	0.505	0.315	0.004	0.000	0.228	0.658	0.006	0.182	.

Pearson correlation significance levels in comparisons among ecological variables.

with larger locales generally accommodating more taxa. Low r^2 values suggest that area alone does not account for the majority of the variance observed in the data set (Table 2).

Species-area relationships in Madagascar may be explained in part by habitat heterogeneity, as defined by the number of vegetation domains present in a locale (Table 3). Locales with more vegetation domains tended to be larger (n.s., $p = 0.076$), although not all large locales possess greater heterogeneity (Figure 6).

Latitude Effects

Locales included in this study range in latitude from 12 to 25 degrees. Despite patterns observed at larger spatial scales (e.g., Harcourt, 2000), latitude fails to explain observed levels of species or generic richness for the vertebrate groups in the locales examined in this study (Table 2). Because differences in species and generic richness were not significant, they are not figured.

Altitude Effects

A significant relationship between minimum altitude and species richness was observed only for birds ($p = 0.05$). However, this relationship disappeared when examined at the generic level (Table 2).

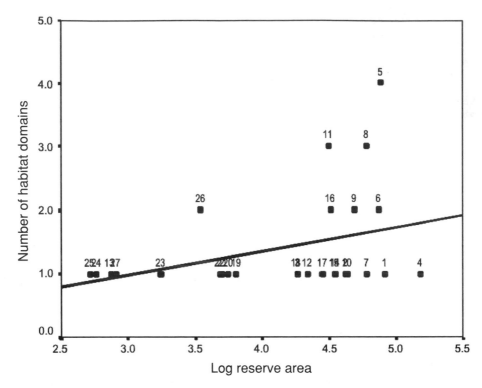

Figure 6. Plot of number of habitat domains against log-area for locales in this study. Whereas many locales of different sizes have just one habitat domain, only the larger locales have three or four domains.

Significant relationships were found between maximum altitude and primate species richness (total primate $p = 0.039$; diurnal primate $p = 0.016$). Similar to birds, significant relationships were not observed at the generic level in primates. Maximum altitude was also a strong predictor of amphibian species and generic richness, with more amphibians observed at higher altitudes (Figure 7).

Altitude range was highly correlated with habitat heterogeneity, as defined by the number of habitat domains in a given locale ($p < 0.001$, Table 3). As with habitat area, with the most heterogeneous locales exhibited greater altitude ranges, but not all locales with large altitude ranges possessed more than one habitat domain (Figure 8). Altitude range was significantly related to amphibian species richness ($p = 0.042$), overall primate species richness ($p = 0.021$), and diurnal primate species richness ($p = 0.006$, Figure 9). Again, these relationships were not significant at the generic level (Table 2). Although many altitude-related regression lines exhibit slopes significantly different from zero,

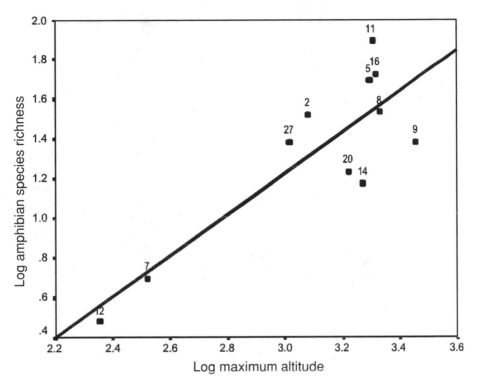

Figure 7. Log-log plot of amphibian species richness against maximum altitude. Primates exhibit a similar pattern. Again, notice that well-surveyed Andringitra (11), Anjanaharibe-Sud (16), and Ranomafana (2) fall above the regression line whereas Manongarivo (14) and Ambohitantely (20) do not.

low R squared values in most cases indicate that altitude, alone, does not explain the majority of variance in species richness (Table 2).

Temperature Effects

Temperatures range from lows under 5°C in locales such as Andringitra, and highs over 35°C in locales such as Beza-Mahafaly, with annual temperature ranges exceeding 25°C in some places. Low temperatures generally seem to predict more amphibian species ($p = 0.045$), yet it is unclear why this may be the case, and significant relationships were not observed in other vertebrate groups. Further, regression analyses indicate no significant relationships between species or generic richness and temperature maxima or range with any of the taxa examined in this study (Table 2).

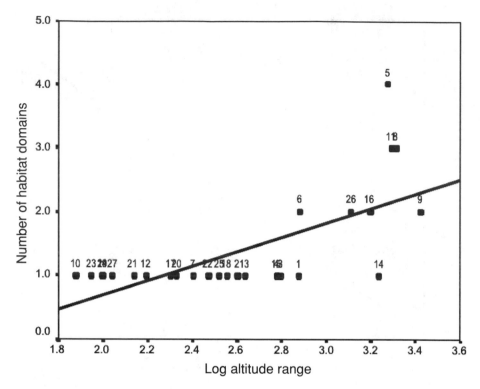

Figure 8. Plot to show the relationship between log-altitude range and the number of habitat domains. Whereas locales with differing altitude ranges may have just one domain, those with more domains tend to have greater topographic relief.

Rainfall Effects

Previous studies have revealed a strong positive correlation between annual rainfall and species richness (e.g., Reed and Fleagle, 1995). Annual rainfall in Madagascar ranges from over 2000 mm along the northern and eastern rain forests to less than 300 mm in the south and southwest (Koechlin, 1972). Surprisingly upon first examination, no significant relationships were found between rainfall and vertebrate richness in this study. A closer look revealed that low annual rainfalls recorded from Andohahela (Figure 10) represent only the dry forest parcel in that locale (Nicoll and Landgrand, 1989), whereas more species hail from the humid forest portion (Feistner and Schmid, 1999; Goodman *et al.*, 1999a and b; Hawkins and Goodman, 1999; Nussbaum *et al.*, 1999). When rainfall was corrected to reflect estimates for the humid parcel 1, primate species and generic richness approach significance ($p = 0.058$, $p = 0.057$ respectively). In

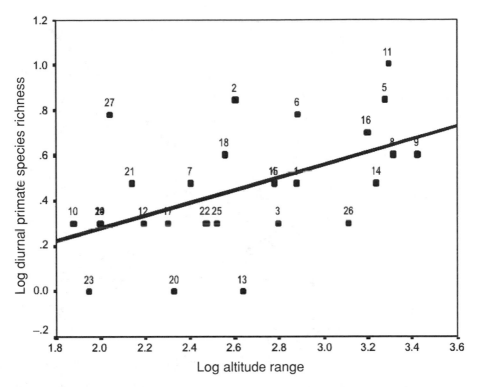

Figure 9. Log-log plot of diurnal primate species richness against altitude range. Although locales with greater topographic relief tended to have more primates, altitude range did not explain all of the variation in the data. For example, Lokobe (13) and Ambohitantely (20) continued to fall below the regression line, whereas Andringitra (11), Anjanaharibe-Sud (16), and Ranomafana (2) maintained more primates than expected.

addition, amphibian species and generic richness are well predicted by annual rainfall ($p = 0.027$, $p = 0.009$ respectively), with some of the highest r^2 values in the study (Table 2), indicating that much of the variance in the amphibian data set might be explained by rainfall alone (Figure 10).

Vegetation Effects

This study used phytogeographic domains (White, 1983) to provide coarse-grained information on vegetation type. Locales with more habitat domains have higher primate, mammal, bird, reptile, and amphibian species richness (Figure 11). This pattern holds at the generic level for primates and mammals

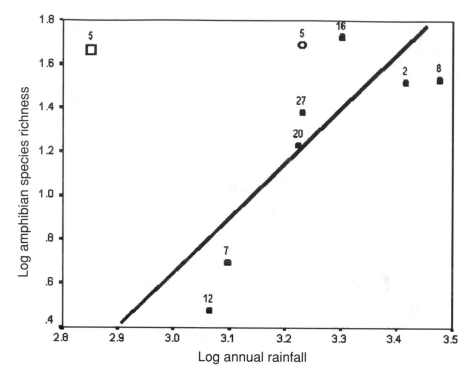

Figure 10. Log-log plot of amphibian species richness against annual rainfall. Published annual rainfall recorded for Andohahela in Nicoll and Landgrand (1989) likely were taken in the dry forest portion of the locale (open square). Most species are found in the humid forest. When rainfall estimates for the humid parcel were used (open circle), a significant trend with amphibian species richness is observed.

(Table 2). Future work should focus on the utility of more fine-grained data on plant diversity and distribution for predicting species richness in individual locales.

Relationships among Ecological Variables

A number of studies have observed correlations among abiotic and biotic factors (e.g., Donque, 1972; Brown, 1988; Peres and Janson, 1999). Indeed, significant relationships were observed between area and altitude maximum and range, between altitude and temperature, between latitude and rainfall and temperature, and between habitat heterogeneity and altitude, some of which have already been pictured (Figure 6 and 8). Significance levels for these comparisons can be found in Table 3.

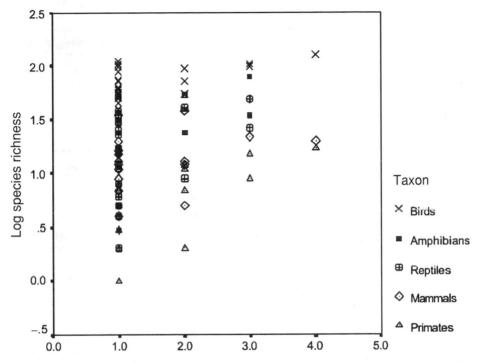

Figure 11. Plot of log-species richness for the different vertebrate groups examined in this study against number of habitat domains. Whereas locales with one habitat domain can have different levels of species richness, more heterogeneous locales tend to have more vertebrate species.

Relationships in Species Richness in Different Vertebrate Groups

Finally, vertebrate richness tends to be closely correlated among groups at both the species and generic level (Table 4). Notable exceptions were that amphibian and reptile richness were not highly correlated, perhaps because most amphibians require more humid habitats, whereas many reptiles prefer drier areas (Raxworthy *et al.*, 1998). In addition, no close relationships were observed between species or generic richness in birds and amphibians.

DISCUSSION

Analyses indicate trends between vertebrate species richness and factors such as habitat area, altitude maximum and range, and phytogeography. Other variables more commonly held to be associated with species richness, such as latitude and

Table 4.   Correlations among species richness in vertebrate groups

	Diurnal Primates	Primates	Mammals	Reptiles	Amphibians	Birds
Diurnal Primates	.	0.000	0.001	0.879	0.101	0.002
Primates	0.000	.	0.000	0.011	0.059	0.000
Mammals	0.000	0.000	.	0.015	0.013	0.000
Reptiles	0.065	0.001	0.000	.	0.659	0.277
Amphibians	0.023	0.013	0.002	0.075	.	0.080
Birds	0.000	0.000	0.000	0.025	0.144	.

Pearson correlation significance levels in comparisons among generic richness above the diagonal, and species richness below the diagonal.

climate, did not show significant relationships with overall primate, mammalian, and vertebrate richness.

As predicted, larger habitat areas tend to be associated with higher levels of species richness. The mechanism for this pattern is unclear. On a smaller scale, Goodman and Rakotondravony (2000) demonstrated that small mammal species richness declines with decreasing forest fragment size in different parts of Ambohitantely Special Reserve. Some have suggested that species-area relationships result from smaller fragments supporting smaller and less diverse populations that may more easily succumb to stochastic extinction events (Terborgh and Winter, 1980). Interestingly, this notion is not supported by multilocus genetic fingerprint studies of understory bird populations in fragments in Ambohitantely (Andrianarimisa *et al.*, 2000), where no differences were observed in genetic diversity in fragments of differing size. Similar work on non-volant vertebrates will be useful in further examining this issue.

Species-area relationships may also be related to increased habitat heterogeneity in some of the larger locales. Trends between habitat diversity and all vertebrate groups were significant at the coarse level of phytogeographic domain. Other sources of habitat heterogeneity such as mild levels of forest disturbance have been demonstrated to result in higher bird and primate species diversity on Madagascar (e.g., Goodman and Putnam, 1996; Ganzhorn, 1997). Habitat disturbance can act to increase the presence of understory plants, and thereby enhance overall habitat diversity.

In terms of topography, locales with greater maximum altitudes and altitude ranges also had higher primate richness. Perhaps this too relates to habitat heterogeneity, as many of these locales also had higher numbers of phytogeographic regions. Alternatively or in addition, it is possible that areas of high topographic relief are simply less desirable locales for human habitat disturbances such as hunting and tavy agriculture. Locales with greater maximum altitudes or altitude ranges also tended to have higher amphibian richness. This is surprising in that more fine-grained analyses have found the opposite pattern (e.g., Raxworthy *et al.*, 1998), suggesting that at this spatial scale amphibian richness may be tracking some other factor related to high elevation, such as rainfall.

Relationships between climate and species richness were not pronounced for most taxa. Rainfall and temperature did not exhibit significant relationships with primate, mammal, bird, or reptile richness. However, rainfall did predict amphibian richness when annual rainfall estimates were used for the humid portion of Andohahela Special Reserve. When Andohahela was excluded from the analysis, the relationship between amphibian richness and rainfall was even stronger, and no significant relationships were found between temperature minimum, maximum, or range for any of the vertebrate groups. Including this locale in minimum temperature analyses yielded a puzzling inverse correlation with amphibian species richness. Such drastic changes in results underscore a few points. First, even within locales, more fine-grained climatic data may reveal stronger patterns with vertebrate species richness. Second, until more reliable survey data are available for all locales, results should be viewed with caution, as a single datum point can vastly change significance levels in vertebrate richness patterns. Third, some ecological variables may be more informative than others at different spatial scales. Whereas amphibians tend to live at lower altitudes within a given locale (e.g., Raxworthy and Nussbaum, 1996), there may be greater overall species richness in locales with varied topographies that increase mean annual rainfall. Finally, it is important to note that rainfall and temperature are highly variable from season to season, and from year to year, thus longer-term studies are necessary to unravel relationships between climate and vertebrate richness.

Certain locales consistently exhibit greater species richness than predicted by ecological variables, whereas others consistently plot below the regression lines. Sites with lower than predicted levels of species richness tend to contain just one habitat type, whereas locales that consistently plot higher often have more (Figure 5, Table 1). For example, Andringitra samples three different phytogeographic domains, and Andohahela samples four, and both consistently show greater than predicted species richness (Figures 5, 6, 7, 8, and 9). Other locales may have higher known vertebrate richness simply because they are well studied, thus fewer taxa are likely to have escaped detection. For example,

Ranomafana has had more long-term studies conducted within its boundaries than have most other locales in this study (e.g., Overdorff, 1996; Tan, 1999; Wright, 1999b; Grassi, 2001). It is likely that further research in other locales will increase known vertebrate richness. In particular, locales such as Lokobe or Manongarivo that repeatedly plot below the regression line might be predicted to have higher richness than reported in the literature. This is supported by the observation that an earlier version of this study also predicted higher levels of primate species richness than were reported at that time for Kalambatritra Special Reserve (O'Connor and Stevens, 2000). Subsequent census in that locale (Irwin *et al.*, 2000) revealed four additional primate species, placing it much closer to regression lines for a locale of its size and vegetational diversity (Figures 6 and 9). Alternatively, other factors such as high human hunting pressure or habitat fragmentation may be associated with lower than predicted species richness, as has been suggested in areas such as Manombo Special Reserve (Ratsimbazafy, 2002) and Ambohitantely (Stephenson *et al.*, 1994).

Therefore, a mosaic of factors likely influences the patterns of species richness on Madagascar. Not only is this type of analysis sensitive to missing data due to incomplete censuses of the different habitats, but human hunting pressures, domestic introductions, predator presence/absence, habitat alteration due to mining and logging, and other factors may each contribute to differential preservation of species richness. Abiotic and biotic factors can reflect vertebrate species richness differently at different spatial scales (Bohning-Gaese, 1997, Fleagle *et al.*, 1999). Thus, more fine-grained analyses of vegetation, seasonality in temperature and rainfall, and other biotic and abiotic factors are likely to reflect new patterns with vertebrate richness.

Therefore, trends seen in this study should be viewed as preliminary until more fine-grained studies such as those of Goodman and colleagues (1996, 1997, 1998, 1999, 2000, 2001) supplement our knowledge about these and other habitat fragments around Madagascar. Although nothing can replace such baseline studies, it is nonetheless useful to identify ecological variables associated with the highest levels of species richness at intermediate scales in order to compliment conservation efforts associated with establishing new protected areas.

SUMMARY

Species richness is highly variable within the habitat isolates scattered across Madagascar. This study addresses the importance of a number of ecological variables for predicting species richness by combining data on topography, climate, and vegetation with species lists from 27 national parks, reserves, and other protected areas in Madagascar.

(1) Primate, mammal, and bird species and generic richness were significantly higher in larger locales. Reptiles and amphibians did not show this pattern.

(2) Primate species richness was greater at high altitudes, and in locales with greater altitude ranges. Greater range in altitude is related to habitat heterogeneity, and may also play a role in making habitats more difficult to traverse, causing them to be less attractive to hunters.

(3) Only amphibians showed significant trends with annual rainfall patterns, and none of the vertebrate groups examined in this study showed strong relationships between species richness and temperature or latitude.

(4) Vertebrate species richness was higher in locales containing more vegetational diversity, as defined by phytogeographic domains. This was true for primates, mammals, birds, reptiles, and amphibians. Higher vegetation diversity may increase the number of niches for vertebrates to occupy.

(5) Locales that have been intensively studied also tend to have greater species richness than less well researched locales. Based on all ecological variables considered in this study, locales such as Lokobe and Ambohitantely have fewer than expected vertebrate taxa.

More detailed information on species compositions of other locales may clarify patterns of richness with respect to these and other ecological variables. More fine-grained approaches will likely refine relationships among geological, climatic, and vegetational diversity and vertebrate species richness on Madagascar. Such inferences are useful, in they may assist conservationists in identifying habitat types that maximize vertebrate richness, an important consideration in light of the fact that many of Madagascar's endemic species face extinction.

ACKNOWLEDGMENTS

This work was completed during our graduate studies at Stony Brook University. We would like to thank J.G. Fleagle, P.C. Wright, and W.L. Jungers for helpful discussions on this project. Special thanks also to J. Ratsimbazafy, K. Samonds, M. Irwin, S. Goodman, S. Nash, L. Betti, C. Heesy, L. Jolley, R. Fajardo, and E. Stone and our anonymous reviewers for many useful comments along the way. We appreciate the efforts of S. Lehman and J.G. Fleagle in putting together this volume.

REFERENCES

Albignac, R. 1972, The carnivora of Madagascar, in: Battistini, R. and Richard-Vindard, G., eds., *Biogeography and Ecology in Madagascar*, W. Junk, The Hague, pp. 667–682.

Aldegheri, M. 1972, Rivers and Streams on Madagascar, in: Battistini, R. and Richard-Vindard, G., eds., *Biogeography and Ecology in Madagascar*, W. Junk, The Hague, pp. 261–310.

Andrianarimisa, A., Bachman, L., Ganzhorn, J. U., Goodman, S. M., and Tomiuk, J. 2000, Effects of forest fragmentation on genetic variation in endemic understory forest birds in central Madagascar. *J. Ornithol.* 141:152–159.

Battistini, R. and Richard-Vindard, G. 1972, *Biogeography and Ecology in Madagascar*, W. Junk, The Hague, pp. 765.

Blommers-Schlosser, R. M. A. and Blommers, L. H. M. 1984, The amphibians, in: Jolly, A., Oberle, P., and Albignac, R., eds., *Key Environments: Madagascar*, Pergamon Press, Oxford, pp. 89–104.

Bohning-Gaese, K. 1997, Determinants of avian species richness at different spatial scales. *J. Biogeogr.* 24:49–60.

Brenon, P. 1972, The geology of Madagascar, in: Battistini, R. and Richard-Vindard, G., eds., *Biogeography and Ecology in Madagascar*, W. Junk, The Hague, pp. 27–86.

Brown, J. H. 1988, Species diversity, in: Myers, A. A. and Giller, P. S., eds., *Analytical Biogeography: An Integrated Approach to the Study of Animal and Plant Distributions*, Chapman and Hall, London, pp. 57–89.

Burney, D. 1997, Theories and Facts regarding Holocene environmental change on Madagascar before and after human colonization, in: Goodman, S. M. and Patterson, B. D., eds., *Natural Change and Human Impact in Madagascar*. Smithsonian Institution Press, Washington DC, pp. 75–89.

Ceballos, G. and Brown, J. H. 1995, Global patterns of mammalian diversity, endemism and endangerment. *Cons. Biol.* 9:559–568.

Darlington, P. J. 1957, *Zoogeography: The Geographical Distribution of Animals*, Wiley Press, New York, p. 675.

Dewar, R. E. 1997, Were people responsible for the extinction of Madagascar's sub-fossils, and how will we ever know? in: Goodman, S. M. and Patterson, B. D., eds., *Natural Change and Human Impact in Madagascar*, Smithsonian Institution Press, Washington, DC, pp. 364–377.

de Wit, M. J. 2003, Madagascar: Heads it's a continent, tails it's and island. *Ann. Rev. Earth Planet. Sci.* 31:213–248.

Donque, G. 1972, The climatology of Madagascar. in: Battistini, R., and Richard-Vindard, G., eds., *Biogeography and Ecology in Madagascar*, W. Junk, The Hague, pp. 87–144.

Duckworth, J. W., Evans, M. I., and Hawkins, A. F. A. 1995, The lemurs of Marojejy Strict Nature Reserve, Madagascar—A status overview with notes on ecology and threats. *Int. J. Primatol.* 16:545–559.

Dumetz, N. 1999, High plant diversity of lowland rainforest vestiges in eastern Madagascar. *Bio. and Cons.* 8:273–315.

DuPuy, D. J. and Moat, J. 1996, A refined classification of the primary vegetation of Madagascar based on the underlying geology: using GIS to map its distribution and to assess its conservation status, in: Lourenco, W. R., ed., *Biogeographie de Madagascar*, Museum-Orstom, Paris, pp. 205–218.

Eeley, H. A. C. and Foley, R. A. 1999, Species richness, species range size and ecological specialization among African primates: Geographical patterns and conservation implications. *Bio. and Cons.* 8:1033–1056.

Eeley H. A. C. and Lawes, M. 1999, Large-scale patterns of species richness and species range size in anthropoid primates, in: Fleagle, J. G., Janson, C. H., and Reed, K. E., eds., *Primate Communities*, Cambridge University Press, Cambridge, pp. 191–219.

Emmons, L. H. 1999, Of mice and monkeys: Primates as predictors of mammal community richness. in: Fleagle, J. G., Janson, C. H., and Reed, K. E., eds., *Primate Communities*, Cambridge University Press, Cambridge, pp. 171–188.

Evans, M. I., Thompson, P. M., and Wilson, A. 1994, A survey of the lemurs of Ambatovaky Special Reserve, Madagascar. *Pri. Cons.* 14–15:13–21.

Feistner, A. T. C. and Schmid, J. 1999, Lemurs of the of the Réserve Naturelle Intégrale d'Andohahela, Madagascar. *Field. Zool.* 94:269–283.

Fisher, B. L. 1997, Biogeography and ecology of the ant fauna of Madagascar (Hymenoptera: Formicidae, *J. Nat. Hist.* 31:269–302.

Fleagle, J. G. and Reed, K. E. 1996, Comparing primate communities: A multivariate approach. *J. Hum. Evol.* 30:489–510.

Fleagle, J. G., Janson, C. H., and Reed, K. E. 1999, *Primate Communities*, Cambridge University Press, Cambridge, p. 329.

Ganzhorn, J. U. 1997, Habitat characteristics and lemur species richness in Madagascar. *Biotropica* 29:331–343.

Ganzhorn, J. U., Wright, P. C., and Ratsimbazafy, J. 1999, Primate Communities in Madagascar, in: Fleagle, J. G., Janson, C. H., and Reed, K. E., eds., *Primate Communities*, Cambridge University Press, Cambridge, pp. 75–89.

Godfrey, L. R., Jungers, W. L., Reed, K. E., Simons, E. L., and Chatrath, P. S. 1997, Subfossil Lemurs: Inferences about past and present primate communities in Madagascar, in: Goodman, S. M., and Patterson, B. D., eds., *Natural Change and Human Impact in Madagascar*, Smithsonian Institution Press, Washington, DC, pp. 218–256.

Goodman, S. M. 1996, The carnivores of the Réserve Naturelle Intégrale d'Andringitra, Madagascar. *Field. Zool.* 85:289–292.

Goodman, S. M. 1998, A floral and faunal inventory of the Réserve Spéciale d'Anjanaharibe-Sud, Madagascar: With reference to elevational variation. *Field. Zool.* No. 90. Field Museum of Natural History, Chicago. p. 246.

Goodman, S. M. and Benstead, J. P. 2003, *The Natural History of Madagascar*, Chicago University Press, Chicago. p. 1709.

Goodman, S. M. and Carleton, M. D. 1996, The rodents of the Réserve Naturelle Intégrale d'Andringitra, Madagascar: A study of elevational distribution and regional endemicity. *Field. Zool.* 85:257–282.

Goodman, S. M. and Carleton, M. D. 1998, The rodents of the Réserve Spéciale d'Anjanaharibe-Sud, Madagascar. *Field. Zool.* 90:201–222.

Goodman, S. M., Carleton, M. D., and Pidgeon, M. 1999a, Rodents of the Réserve Naturelle Intégrale d'Andohahela, Madagascar. *Field. Zool.* 94:217–249.

Goodman, S. M., Ganzhorn, J. U., and Rakotondravony, D. 2003, Introduction to the Mammals, in: Goodman, S. M. and Benstead, J. P., eds., *The Natural History of Madagascar*, University of Chicago Press, Chicago. pp. 1159–1186.

Goodman, S. M. and Jenkins, P. D. 1998, The insectivores of the Réserve Spéciale d'Anjanaharibe-Sud, Madagascar. *Field. Zool.* 90:139–162.

Goodman, S. M., Jenkins, P. D., and Pidgeon, M. 1999b, Lipotyphla (Tenrecidae and Soricidae) of the Réserve Naturelle Intégrale d'Andohahela, Madagascar. *Field. Zool.* 94:187–216.

Goodman, S. M. and Lewis, B. A. 1996, Description of the Réserve Naturelle Intégrale d'Andringitra, Madagascar. *Field. Zool.* 85:7–19.

Goodman, S. M. and Lewis, B. A. 1998, Description of the Réserve Spéciale d'Anjanaharibe-Sud, Madagascar. *Field. Zool.* 90:9–16.

Goodman, S. M. and Patterson, B. D. 1997, Natural Change and Human Impact in Madagascar, Smithsonian Institution Press, Washington, DC, p. 432.

Goodman, S. M. and Pidgeon, M. 1999, Carnivora of the Réserve Naturelle Intégrale d'Andohahela, Madagascar. *Field. Zool.* 94:259–268.

Goodman, S. M., Pidgeon, M., Hawkins, A. F. A., and Schulenberg, T. S. 1997, The birds of southeastern Madagascar. *Field. Zool.* Field Museum of Natural History, Chicago. 87:132.

Goodman, S. M. and Putnam, M. S. 1996, The birds of the eastern slopes of the Réserve Naturelle Intégrale d'Andringitra, Madagascar. *Field. Zool.* 85:171–190.

Goodman, S. M. and Rakotondravony, D. *et al.*, 2000, The effects of forest fragmentation and isolation on insectivorous small mammals (Lipotyphla) on the Central High Plateau of Madagascar. *J. Zool., Lond.* 250:193–200.

Goodman, S. M. and Rasolonandrasana, B. P. N. 2001, Elevational zonation of birds, insectivores, rodents and primates on the slopes of the Andringitra Massif, Madagascar. *J. Nat. Hist.* 35:285–300.

Goodman, S. M., Raxworthy, C. J., and Jenkins, P. D. 1996, Insectivore ecology in the Réserve Naturelle Intégrale d'Andringitra, Madagascar. *Field. Zool.* 85:218–230.

Grassi, C. 2001, The Behavioral Ecology of *Hapalemur griseus griseus*: The Influences of Microhabitat and Population Density on this Small-bodied Prosimian Folivore. Ph.D. dissertation, University of Texas, Austin.

Green, G. M. and Sussman, R.W. 1990, Deforestation history of the eastern rain forests of Madagascar from satellite images. *Science* 243:212–215.

Harcourt, A. H. 2000, Latitude and latitudinal extent: a global analysis of the Rapoport effect in a tropical mammalian taxon: primates. *J. Biogeo.* 27:1169–1182.

Harcourt, C. S. and Thornback, J. 1990, *Lemurs of Madagascar and the Comoros*, IUCN—The World Conservation Union, Gland.

Hawkins, A. F. A., Thiollay, J., and Goodman, S. M. 1998a, The birds of the Réserve Spéciale d'Anjanaharibe-Sud, Madagascar. *Field. Zool.* 90:93–128.

Hawkins, A. F. A. 1999, Altitudinal and latitudinal distribution of east Malagasy forest bird communities. *J. Biogeo.* 26:447–458.

Hawkins, A. F. A. and Goodman, S. M. 1999, Bird community variation with elevation and habitat in parcels 1 and 2 of the Réserve Naturelle Intégrale d'Andohahela, Madagascar. *Fieldi. Zool.* 94:175–186.

Humbert, H. 1927, Principaux aspects de la vegetation a Madagascar. *Mem. Acad. Malagache.* 5:1 80.

Humbert, H. 1965, Description des types de vegetation, in: Humbert, H. and Cours Darne, G., eds, *Notice de la carte de Madagascar. Travaux de la Section Scientifique et Technique de l'Institute Francois de Pondichery* 6:46–78.

Irwin, M. T., Samonds, K. E., and Raharison, J. L. 2000, A biological inventory of Reserve Speciale de Kalambatritra, with special emphasis on lemurs and birds. Unpublished report to ANGAP

Janson, C. H. and Peres, C. A. 1999, Species coexistence, distribution, and environmental determinants of neotropical primate richness: A community-level zoogeographic analysis, in: Fleagle, J. G., Janson, C. H., and Reed, K. E., eds., *Primate Communities*, Cambridge University Press, Cambridge. pp. 55–74.

Jenkins, P. D., Goodman, S. M., and Raxworthy, C. J. 1996, The shrew Tenrecs (*Microgale*) (Insectivora: Tenrecidae) of the Réserve Naturelle Intégrale d'Andringitra, Madagascar. *Field. Zool.* 85:191–216.

Jernvall, J. and Wright, P. C. 1998, Diversity components of impending primate extinctions. *Proc. Nat. Acad. Sci.* 95:11279–11283.

Johnson, S. E. and Overdorff, D. J. 1999, Census of brown lemurs in southeastern Madagascar: Methods-testing and conservation implications. *Am. J. Primatol.* 47:51–60.

Jolly, A. and Jolly, R. 1984, Malagasy economics and conservation: A tragedy without villains, in: Jolly, A., Oberle, P., and Albignac, R., eds., *Key Environments: Madagascar*, Pergamon Press, Oxford. pp. 211–218.

Lowry, P. P., Schatz, G. E., and Phillipson, P. B. 1997, The classification of natural and anthropogenic vegetation in Madagascar, in: Goodman, S. M. and Patterson, B. D., eds., *Natural Change and Human Impact in Madagascar*. Smithsonian Institution Press, Washington, DC pp. 75–89.

Koechlin, J. 1972, Flora and vegetation of Madagascar, in: Battistini, R. and Richard-Vindard, G., eds., *Biogeography and Ecology in Madagascar*, W. Junk, The Hague, pp. 145–190.

Martin, R. D. 1972, Adaptive radiation and behaviour of the Malagasy lemurs. *Philos. Trans. R. Soc. Lond. B* 264:295–352.

Millot, J. 1972, In conclusion, in: Battistini, R. and Richard-Vindard, G., eds., *Biogeography and Ecology in Madagascar*, W. Junk, The Hague pp. 741–756.

Mittermeier, R. A., Tattersall, I., Konstant, B., Meyers, D. M., and Mast, R. B. 1994, *Lemurs of Madagascar*, Conservation International, Washington, D C.

Myers, N., Mittermeier, C. G., da Fonseca, G. A. B., and Kent, J. 2000, Biodiversity hotspots for conservation priorities. *Nature* 403:853–858.

Nicoll, M. E. and Langrand, O. 1989, *Madagascar: Revue de la Conservation et des Aires Protegees*, World Wildlife Fund, Gland.

Nussbaum, R. A., Raxworthy, C. J., Raselimanana, A. P., and Ramanamanjato, J. 1999, Amphibians and reptiles of the Réserve Naturelle Intégrale d'Andohahela, Madagascar. *Field. Zool.* 94:155–173.

O'Connor, P. M. and Stevens, N. J. 2000, Biotic and abiotic factors as predictors of species richness in Madagascar. *Am. J. Phys. Anthropol., Supp.* 30:241.

Overdorff, D. J. 1996, Ecological correlates to activity and habitat use of two prosimian primates: *Eulemur rubriventer* and *Eulemur fulvus rufus* in Madagascar. *Am. J. Primatol.* 40:327–342.

Paulian, R. 1961, La zoographie de Madagascar et des iles voisines. L'Institut de Recherche Scientifique Tananarive-Tsimbazaza. Antananarivo.

Paulian, R. 1984, Introduction to the mammals, in: Jolly, A., Oberle, P., and Albignac, R., eds., *Key Environments: Madagascar*, Pergamon Press, Oxford. pp. 151–154.

Peres, C. A. and Janson, C. H. 1999, Species coexistence, distribution, and environmental determinants of neotropical primate richness: A community-level zoogeographic analysis, in: Fleagle, J. G., Janson, C. H., and Reed, K. E., eds., *Primate Communities*, Cambridge University Press, Cambridge. pp. 55–74.

Perrier de la Bathie, H. 1936, *Biogéographie des Plantes à Madagascar*, Société d'Editions Géographiques, Maritimes et Coloniales, Paris.

Pianka, E. R. 1994, *Evolutionary Ecology*, Harper Collins College Publishers, New York. p. 486.

Rapoport, E. H. 1982, *Areography. Geographical Strategies of Species*, Pergamon Press, Oxford. p. 269.

Ratsimbazafy, J. H. 2002, On the Brink of Extinction and the Process of Recovery: Responses of Black and White Ruffed Lemurs (*Varecia variegata variegata*) to Disturbance in Manombo Forest, Madagascar. Ph.D. dissertation, SUNY Stony Brook, New York, p. 190.

Raxworthy, C. J. 2003, Introduction to the Reptiles, in: Goodman, S. M. and J. P. Benstead, eds., *The Natural History of Madagascar*, University of Chicago Press, Chicago, pp. 934–949.

Raxworthy, C. J. and Nussbaum, R. A. 1994, A rainforest survey of amphibians, reptiles and small mammals at Montagne D'Ambre, Madagascar. *Bio. Cons.* 69:65–73.

Raxworthy, C. J. and Nussbaum, R. A. 1996, Amphibians and reptiles of the Réserve Naturelle Intégrale d'Andringitra, Madagascar: A study of elevational distribution and regional endemicity. *Field. Zool.* 85:158–170.

Raxworthy, C. J., Andreone, F., Nussbaum, R. A., Rabibisoa, N., and Randriamahazo, H. 1998, Amphibians and reptiles of the Anjanaharibe-Sud Massif, Madagascar: Elevational distribution and regional endemicity. *Field. Zool.* 90:79–92.

Raxworthy, C. J., Forstner, M. R. J., and Nussbaum, R. A. 2002, Chameleon radiation by oceanic dispersal. *Nature* 415:784–787.

Reed, K. E. and Fleagle, J. G. 1995, Geographic and climatic control of primate diversity. *Proc. Nat. Acad. Sci.* 92:7874–7876.

Ricklefs, R. E. and Schluter, D. 1993, *Species Diversity in Ecological Communities*, University of Chicago Press, Chicago. p. 414.

Rosenzweig, M. L. 1995, *Species Diversity in Space and Time*, Cambridge University Press, Cambridge. p. 436.

Ruggiero, A. 1994, Latitudinal correlates of the sizes of mammalian geographical ranges in South America. *J. Biogeo.* 21:545–559.

Schmid, J. and Smolker, R. 1998, Lemurs of the Réserve Spéciale d'Anjanaharibe-Sud, Madagascar. *Field. Zool.* 90:227–238.

Simons, E. L. 1997, Lemurs: old and new, in: Goodman, S. M. and Patterson, B. D., eds., *Natural Change and Human Impact in Madagascar*. Smithsonian Institution Press, Washington, DC pp.142–166.

Smith, A. P., Horning, N., and Moore, D. 1997, Regional biodiversity planning and lemur conservation with GIS in western Madagascar. *Cons. Biol.* 11:498–512.

Stephenson, P. J., Randriamahazo, H., Rakotoarison, N., and Racey, P. A. 1994, Conservation of mammalian species diversity in Ambohitantely Special Réserve, Madagascar. *Biol. Cons.* 69:213–218.

Sterling, E. J. and Ramaroson, M. G. 1996, Rapid assessment of the primate fauna of the eastern slopes of the Réserve Naturelle Intégrale d'Andringitra, Madagascar. *Field. Zool.* 85:293–305.

Stevens, G. C. 1989, The latitudinal gradient in geographical range: how many species coexist in the tropics. *Am. Nat.* 13:240–256.

Tan, C. L. 1999, Group composition, home range size, and diet of three sympatric bamboo lemur species (Genus *Hapalemur*) in Ranomafana National Park, Madagascar. *Int. J. Primatol.* 20:547–566.

Tattersall, I. 1982, *The Primates of Madagascar*, Columbia University Press, New York. p. 382.

Terborgh, J., and Winter, B. 1980, Some causes of extinction, in: Soule, M. E., and Wilcox, B. A., eds., *Conservation biology, an evolutionary-ecological approach*, Sinauer Associates, Sunderland, Massachusetts, pp. 119–133.

Wallace, A. R. 1876, *The Geographical Distribution of Animals*, MacMillan, London. p. 607.

White, F. 1983, *The Vegetation of Africa*, UNESCO, Paris.

Wilmé, L. 1996, Composition and characteristics of bird communities in Madagascar, in: Lourenco, W. R., ed., *Biogeographie de Madagascar*, Museum Orstom, Paris. pp. 349–362.

Wright, D. H., Currie, D. J., and Maurer, B. A. 1993, Energy supply and patterns of species richness on local and regional scales, in: Ricklefs, R. E. and Schluter, D., eds., *Species Diversity in Ecological Communities,* University of Chicago Press, Chicago. pp. 66–74.

Wright, P. C. 1997, The Future of Biodiversity in Madagascar: A View from Ranomafana National Park, in: S. M. Goodman and Patterson, B. D., eds., *Natural Change and Human Impact in Madagascar*, Smithsonian Institution Press, Washington, DC pp. 381–405.

Wright, P. C. 1999a, Lemur traits and Madagascar ecology: Coping with an island environment. *Yearb. Phys. Anthropol.* 42:31–72.

Wright, P. C. 1999b, The future of primate communities: A reflection of the present? in: Fleagle, J. G., Janson, C. H., and Reed, K. E., eds., *Primate Communities*, Cambridge University Press, Cambridge. pp. 295–309.

Wright, P. C. and Jernvall, J. 1999, The future of primate communities: a reflection of the present? in: Fleagle, J. G., Janson, C., and Reed, K., eds., *Primate communities*, Cambridge University Press, Cambridge, pp. 295–309.

Yoder, A. D. 2003, Phylogeny of the Lemurs, in: Goodman, S. M. and Patterson, B. D., eds., *Natural Change and Human Impact in Madagascar*, Smithsonian Institution Press, Washington DC. pp. 1242–1247.

Asia

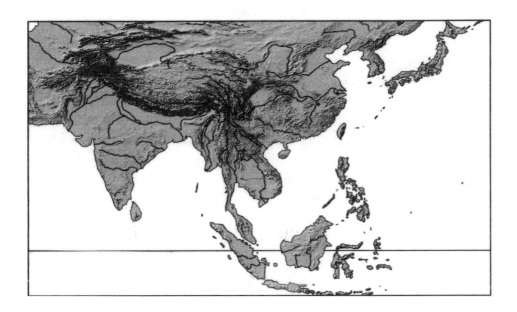

A sia is the largest continent on earth, but the area of tropical forest is less extensive than in Africa or South America, partly because most of Asia lies well above the equator, and partly because this continent has been greatly modified by human activity. The forests inhabited by primates in Asia vary considerably from the lush evergreen forests of Malay Peninsula, Borneo, and Sumatra to more deciduous forests in northern India, Bangladesh, and China (e.g. Gupta and Chivers, 1999). Compared with other tropical regions of the world, southeastern Asia has a much more complex biogeographic history because of the extensive recent and ongoing tectonic activity in the region which has modified the topography and patterns of river drainage, and because much of the region consists of islands that have been repeatedly connected to and separated from one another and the mainland during the past few million years (e.g. Hall and Holloway, 1998; Jablonski and Whitfort, 1999; Whitmore, 1987).

Primate Biogeography, edited by Shawn M. Lehman and John G. Fleagle.
Springer, New York, 2006.

The total number of primate species in Asia is much debated, but the most recent assessment identifies 77 species in 16 genera (Brandon-Jones *et al.*, 2004). There are two endemic families of primates in Asia—tarsiers and the lesser apes—each with numerous species. The other primates in Asia—lorises, colobine monkeys, cercopithecine monkeys, and a great ape, the orang-utan, are closely related to African primates. However, the diversity of the two cercopithecoid subfamilies is strikingly different on the two continents. In Asia, cercopithecines are limited to a single genus, *Macaca*, with many species (e.g. Abegg and Thierry, 2002). Colobines, in contrast, are far more diverse in Asia than in Africa (Davies and Oates, 1994).

The papers in this section address the complex biogeographic patterns among Southeast Asian primates from two very different perspectives. In "The Geography of Mammals and Rivers in Mainland Southeast Asia," Erik Meijaard and Colin Groves review the distribution of primates and other mammals in relation to the Brahmaputra, the Salween and the Mekong Rivers in an effort to explain the differences in mammalian richness and the relative distinctiveness of the faunas bounded by the rivers. They find that the present patterns of mammal distributions are probably due to a complex geological history of river capture in conjunction with the uplift of the Tibetan Plateau and Pleistocene glaciations.

In "Primate Biogeography and Ecology on the Sunda Shelf Islands: A Paleontological and Zooarcheological Perspective," Terry Harrison, John Krigbaum, and Jessica Manser use the evidence from paleontological sites and cave deposits to reconstruct the history of different primates on the islands of the Sunda Shelf during the Pleistocene. The combination of present distributions and evidence from the paleontological and zooarcheological records indicate differential patterns of initial colonization, speciation, and in some cases, of extinction, of individual species and genera on different islands. Most noticeably numerous widespread taxa have gone extinct on Java.

REFERENCES

Abbeg, C. and Thierry, B. 2002, Macaque evolution and dispersal in insular south-eat Asia. *Biolog. J. Linnean Soc.* 75:555–576.

Brandon-Jones, D., Eudey, A. A., Geissmann, T., Groves, C. P., Melnick, D. J., Morales, J. C., Shekelle, M., and Stewart, C.-B. 2004, Asian Primate Classification. *Int. J. Primatol.* 25:97–164.

Davies, A. G. and Oates, J. F. 1994, *Colobine Monkeys: Their Ecology, Behavior and Evolution*, Cambridge University Press, Cambridge.

Gupta, A. K. and Chivers, D. J. 1999, In: Fleagle, J. G., Janson, C. H., and Reed, K. E. eds., *Primate Communities*, Cambridge University Press, Cambridge, pp. 38–54.

Hall, R. and Holloway, J. D., eds., 1998, *Biogeography and Geological Evolution of Southesat Asia*, Backhuys, Leiden.

Jablonski, N. G. and Whitfort, M. J. 1999, Environmental change during the Quaternary in East Asia and its consequences for mammals. *Rec. Western Austral. Mus. Supp.* 57:307–315.

Whitmore, T. C. ed., 1987, *Biogeographical Evolution of the Malay Archipelago*, Clarendon Press, Oxford.

CHAPTER ELEVEN

The Geography of Mammals and Rivers in Mainland Southeast Asia

Erik Meijaard and Colin Peter Groves

ABSTRACT

This chapter describes the distribution of non-volant mammals of mainland Southeast Asia in relation to the region's main rivers, the Mekong, Salween, and Brahmaputra. We describe all species according to their general ecology and size to see whether the distribution ranges of the species can be characterized by these factors. The area east of the Mekong River appears to be relatively rich in mammal species compared to the area between the Mekong and Salween Rivers. The Mekong, however, does not seem to be an ecological barrier to the investigated species, unlike the Brahmaputra River that separates species of drier, open vegetation from forested adapted species. The species richness east of the Mekong River can possibly be explained by environmental changes in the Late Pliocene–Early Pleistocene, which may have isolated rain forest dependent species in the Vietnamese and Laotian mountains.

Key Words: Brahmaputra, divergence, evolution, Indochina, Mekong, palaeoenvironment, phylogeny, Salween, Quaternary

Erik Meijaard and Colin Peter Groves • School of Archaeology and Anthropology, Australian National University, Canberra, Australia

Primate Biogeography, edited by Shawn M. Lehman and John G. Fleagle.
Springer, New York, 2006.

INTRODUCTION

The distributions of the primates and other mammals of island Southeast Asia, and their implications for palaeogeographic and palaeovegetational reconstructions, have been the subject of numerous studies, if often in a rather piecemeal fashion, and are currently being extensively revised by one of us (EM). Less examined are primate and general mammalian distributions in mainland Southeast Asia, and much of the work consists of anecdotal writings by hunters (Duckworth, 1998). Also, from the 1960s until the 1990s, the political situation in several of the Indochinese countries restricted the access of zoologists to the field (MacKinnon, 2000). Still, mammalian distributions on the mainland too are non-uniform, and seem to require explanations, which, exactly as in the case of insular Southeast Asia, look back to a past when landforms and vegetation patterns were different from those of today.

The outstanding problem, which is especially noticeable in the case of the primates, is that of the Mekong. Entire genera and species-groups are confined to the east of this river: the genus *Pygathrix* (Cercopithecidae), the *Trachypithecus francoisi* group (Cercopithecidae), and the distinctive species *Nycticebus pygmaeus* (Lorisidae) are confined to the east of the Mekong (see Figure 1 for geographic locations), and so is the genus (or subgenus) *Nomascus* (Hylobatidae), with the exception of *N. concolor furvogaster*, which occurs in a small region to the west around $23°15'–40'N$, $99°05'–29'E$ (Ma and Wang, 1986), just below where the Mekong leaves the mountains and, now in flatter country, makes a large eastward meander, presumably cutting off this small population. On the other hand, there are no primates west of the Mekong that have a distribution limited by this river: *Trachypithecus germaini*, *T. phayrei*, *Nycticebus bengalensis* and the macaques are all distributed from as far west as the Bay of Bengal and then eastward across the Mekong without even subspecific differentiation. It is as if the Mekong has always been as wide and uncrossable as it evidently is today, but has very recently shifted its course in a way that captured portions of the ranges of the widespread species, while continuing to restrict the east-side species.

High species diversity of the eastern Mekong area is also indicated by the recent discovery of several mammal species new to science in Laos and Vietnam (e.g., Groves and Schaller, 2000) (see Table 1).

These new discoveries add to a considerable list of Annamite endemics (e.g., Corbet and Hill, 1992; MacKinnon, 2000), although it is yet unclear why this

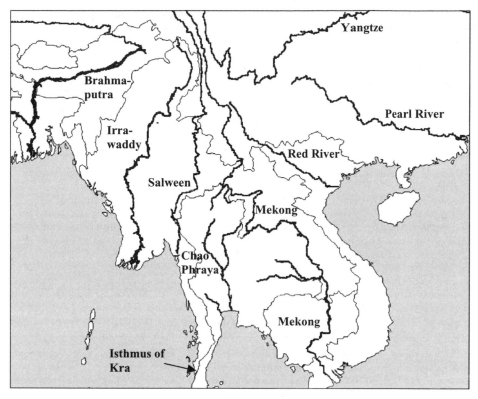

Figure 1. Map of Southeast Asia and the main rivers mentioned in the text.

area appears to be biogeographically distinct. This paper attempts to answer two questions. Firstly, it investigates whether there are significant spatial differences in species richness of the region, and whether Vietnam, Laos, and eastern Cambodia stand out as particularly rich. Secondly, we will investigate the factors that could have influenced the biogeographical patterns.

METHODS

To examine species that are and are not limited by the large rivers of South east Asia, we listed all non-volant mammal species in the Indochinese Faunal Division (see Corbet and Hill, 1992), using the Isthmus of Kra, the Brahmaputra River, and the Red River as the region's boundaries (see Figure 1); we also included Indian species for which the Brahmaputra River was the eastern limit. Species that only occur on the Thai and Burmese Peninsula and further south, such as *Trachypithecus obscurus* and *Presbytis femoralis* were excluded from the

Table 1. Recent discoveries of mammal species from Laos and Vietnam

Species	Remarks	Reference
Muntiacus vuquangensis (Artiodactyla: Cervidae)	new species	(Schaller and Vrba, 1996)
Muntiacus rooseveltorum (Artiodactyla: Cervidae)	rediscovered species	(Amato *et al.*, 1999)
Muntiacus truongsonensis (Artiodactyla: Cervidae)	new species	(Giao *et al.*, 1998)
Pseudonovibos spiralis (Artiodactyla: Bovidae)	new species[a]	(Peter and Feiler, 1994)
Pseudoryx nghetinhensis (Artiodactyla: Bovidae	new species	(Schaller and Rabinowitz, 1995)
Sus bucculentus (Artiodactyla: Suidae)	rediscovered species	(Groves *et al.*, 1997)
Nesolagus timminsi (Lagomorpha: Leporidae)	new species	(Averianov *et al.*, 2000)
Cynocephalus variegatus (Dermoptera: Cynocephalidae)	new subspecies[b]	(Ruggeri and Etterson, 1998)
Viverra tainguensis (Carnivora, Viverridae)	new species[c]	(Sokolov *et al.*, 1997)
Tragulus versicolor	rediscovered species	(Meijaard and Groves, 2004)

[a] The horns used to describe this new large bovid were recently reported to be a skilful forgery made by carving and distorting ordinary cow horns (Thomas *et al.*, 2001).
[b] Stafford and Szalay (2000) considered this population to be a dwarfed form of what they named *Galeopterus variegatus*, but, considering its isolation from other populations and its morphometric distinctiveness, this population may well be a new species.
[c] Walston and Véron (2001) investigated the evidence for this new species, and suggested that it was insufficient to distinguish it from *Viverra zibetha*.

analysis, as were those that have the largest part of their range in the Himalayan subregion and southern Chinese subdivision. Finally, we excluded the genera *Mus* and *Rattus* from this research because of continuing taxonomic difficulties in these groups, and the Lutrinae because of their aquatic habits.

We classified each species according to its ecological, phylogenetic, and morphological characteristics (see Appendix 1). For this we used the following general texts: Lekagul and McNeely (1977); Payne *et al.*, (1985); Schreiber *et al.*, (1989); Chapman and Flux (1990); Corbet and Hill (1992); Nisbett and Ciochon (1993); Alderton (1994); and several other publications that deal with particular species. Subsequently we analysed whether the biogeographical groups could be separated with statistical significance using these characteristics. Each species was assessed according to the following classes: habitat (forest, deciduous forest, bamboo, scrub, grass), altitudinal range (lowland, mountains), general lifestyle (terrestrial, arboreal), and male body weight (0–1 kg, 1–10 kg,

Table 2. Summary table of the distribution patterns of southeast Asian mammals in relation to the region's large rivers

Distribution patterns	nos. of species
Limited to the east of the Mekong	19[a]
On both sides of the Mekong, but not across the Salween	20
Across the Mekong and the Salween, but limited in the west by the Brahmaputra	25
Only northwest of the Brahmaputra	12
Only south and east of the Brahmaputra, but limited in the east by the Mekong	8
On both sides of the Brahmaputra, but limited in the east by the Salween	6
Not limited by any of these rivers	51

[a] This includes the *Muntiacus rooseveltorum* and *Trachypithecus francoisi* species groups for which the phylogeny has not yet been resolved, but which likely consist of several distinct species. It excludes *Pseudonovibos spiralis*, because its identity remains uncertain, and *Cynocephalus* cf. *variegatus* and *Viverra* tainguensis because it remains unclear whether these are distinct species.

10^+ kg). These characters were analysed by principal components analysis with SPSS 11.0 software, using the percentages of the total species per biogeographical unit that fell into the different class types; e.g., 79% of the species east of the Mekong were forest dependent, 5% occur in bamboo forest, 11% in scrubland, and 5% in grassland, and so forth. If species were included in two or more class types, for instance, a species occurring in forest, bamboo forest, and deciduous forest, the class types were each given an equal weight of 1 and added to the total count of class types occurrences. The biogeographical groups, as shown in Appendix 1, were plotted in graphs of the resulting principal components. These graphs were then compared to the component plot and matrix, which indicate the factors that strongly contribute to the variation of the individual components. Thus we could see which biogeographical groups were strongly correlated with the ecological and morphological characteristics.

RESULTS

Table 2 summarizes the distribution patterns of Southeast Asia's mammals in relation to the large rivers of the region. Neither the Red River nor the Chao Phraya clearly delimit the distribution of any mammal species.

The area east of the Mekong River has a total of ca. $115 (= 19 + 20 + 25 + 51)$ non-volant mammal species, of which 19 (17%) are endemic to that region; the area between the Salween and the Mekong has ca. $104 (=$

$20 + 25 + 8 + 51$) species, none of which are endemic to this area; the area between the Salween and the Brahmaputra has ca. $90 (= 25 + 8 + 6 + 51)$ species,
of which none are endemic to that region. There appears to be a decline in
species richness and endemic species richness from east to west. This relationship becomes even clearer when the number of species is compared to the size
of each of these areas. The Brahmaputra-Salween area is approximately 910,000
km^2, the area between the Salween and Mekong 870,000 km^2, and the area
east of the Mekong 570,000 km^2. The smallest area thus has the largest number
of species and endemic species.

The statistical analysis showed that ecological or morphological characters characterized several biogeographical groups. Three components with an
Eigenvalue > 1 were extracted, explaining 43%, 29%, and 11% of the total variance. The principal components analysis of the first and second components
(Figure 2a) shows that the species that occur northwest of the Brahmaputra are
distinct; the corresponding correlation matrix (Table 3) suggesting that this is
primarily because of the association of these species with scrub and grassland
habitats in the lowlands; also there are few small species and few arboreal species
in this group. The species further east correlate strongly with forest habitat. The
second component differentiates the species around the Salween River (SB/WS
and SB) from the others, primarily because of the high percentage of species
within the former groups that are found both in lowland and mountain habitats;
the high negative correlation of these species groups with a mixed arboreal and
terrestrial lifestyle is strongly biased by zero counts for species in these groups,
and it is unclear whether this has any ecological significance or whether it is the
result of small sample size. The species that occur on both sides of the Mekong,
and on both sides of the Salween, are mostly small to medium-sized, arboreal
forest species.

The diagram of the first and third components (Figure 2b), and its corresponding correlation matrix (Table 3), is mostly similar to Figure 2a, but it
differentiates the species in the area east of the Salween and across the Mekong;
the difference primarily arises from the relatively high number of species of
bamboo and deciduous forests.

The taxa limited to the east side of the Mekong do not really stand out; from
this we conclude that there is no evidence that any ecological factor limits their
distribution, so that (as far as we can tell) geographic barriers alone seem to be
responsible.

We also assessed which species were good swimmers, under the assumption
that most ungulates, larger carnivores, and larger murines are good swimmers,

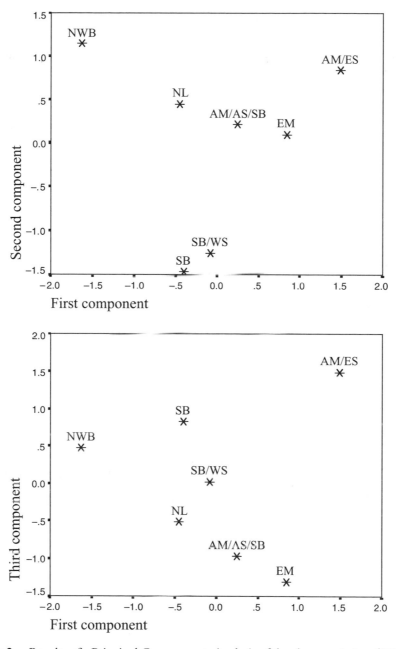

Figure 2. Results of a Principal Components Analysis of the characteristics of SE Asian mammals (Figure 2a = diagram of First vs. Second component; Figure 2b = diagram of First vs. Second component). The correlation table shows the correlation between the factors and the extracted components; in bold style the highest positive or negative correlations per component. Abbreviation stand for: NWB = northwest of Brahmaputra; SB/AM/AS = south of Brahmaputra, and across Mekong and Salween; SB/WS = south of Brahmaputra, and west of Salween; EM = east of Mekong; AM/ES = across Mekong, and east of Salween; SB/AS = south of Brahmaputra and across Salween; and NL = not limited by any of these rivers.

Table 3. Correlation matrix for principal component analysis in 2a and 2b. Numbers in bold show the highest correlation factor for each of the characters

Factor	Component		
	1	2	3
Rain forest	**0.85**	−0.35	−0.31
Bamboo forest	**0.81**	0.40	0.30
Deciduous forest	**0.55**	0.50	0.24
Scrubs and bushes	**−0.76**	0.63	0.02
Grasslands	**−0.89**	−0.23	0.35
Lowlands	−0.42	**0.55**	0.48
Lowlands and mountains	−0.17	**−0.72**	−0.20
Mountains	**0.82**	0.16	−0.42
Strictly arboreal	**0.87**	−0.23	0.36
Arboreal and terrestrial	0.03	**0.96**	−0.03
Strictly terrestrial	**−0.70**	−0.58	−0.28
Small-sized	**0.62**	−0.59	0.42
Medium-sized	−0.01	**0.78**	−0.52
Large-sized	**−0.71**	−0.14	0.07

and others are either poor swimmers or are reluctant to enter the water. We created a category for the flying squirrels, some of which are known to cross gaps of >100 m (see Payne *et al.*, 1985). Good swimmers and gliders were indicative of the following groups (graphs not shown): (1) species occurring on both sides of the Mekong, but that are limited in the west by the Brahmaputra; (2) species occurring west of the Salween, but that are limited in the west by the Brahmaputra; (3) species that are not limited by any of these rivers.

Overall, the data suggest that the Brahmaputra River is mostly an ecological barrier to many species, even to those that swim well, with drier, more open vegetation types dominating the northwest, and the forests in the east. The Mekong, on the other hand does not seem to separate different ecological groups, and most species that occur on either sides of this river are forest dependent, and often occur at higher altitudes. The species that are not limited by any of these rivers are often ecologically flexible (occurring both in lowlands and mountains, not limited to an arboreal lifestyle, and able to exist in a relatively wide range of vegetation types).

DISCUSSION

Our results indicate that the area east of the Mekong has a higher mammal diversity than the area between the Mekong and the Salween. Using slightly different biounit boundaries, MacKinnon (1997) found a similar trend of decreasing biodiversity from the Annam Mountains, to the Indochinese coast, coastal Burma, and central Indochina. He did not, however, investigate the main rivers as biogeographical breaks.

The species that occur in the area east of the Mekong and in the area between the Mekong and the Salween do not show any significant ecological differences, which would preclude an ecological explanation of the pattern. Below, we investigate other potential explanations, including the existence of Pleistocene refugia on mountains, changes in river courses, and catastrophic events.

Geomorphology: Mountain Building and River Courses

Three rivers of importance to the biogeography of non-volant, terrestrial species in Indochina and China—the Salween, Mekong, and Yangtze—originate on the Tibetan Plateau, then converge in the Three Rivers area where for 300 km they follow deeply incised and closely spaced parallel valleys only a few tens of kilometres apart (Hallet and Molnar, 2001). Hallet and Molnar considered the upper valleys of these three rivers to have existed since the Early Miocene or even earlier, and Lacassin et al., (1998) similarly found that they have occupied about the same upper course for at least several million years. Downstream of the gorges, however, the courses of these and other major rivers have been anything but stable. Métivier and Gaudemer (1999) suggested that the main Chinese rivers flowed southward rather than eastward until possibly the Early Pliocene. Their analysis of the sedimentary output of the Mekong and the Yangtze rivers showed that between the Late Miocene and Early Pliocene the former was reduced to a third of its mid-Miocene level while the latter increased fourfold. This may suggest that the Yangtze River originally occupied part of the Mekong catchment, and started to follow its present course only after the Early Pliocene. Métivier and Gaudemer (1999) suggested that the Mekong River discharged into the Gulf of Thailand or the Malay Basin before shifting its course to the present Mekong Basin.

Based on geological data by Hutchinson (1989) and mitochondrial DNA sequence data on freshwater snails, Attwood and Johnston (2001) provided a similar scenario for the development of the Southeast Asian drainage patterns. They suggested that the drainage systems of the Yangtze and the Mekong rivers were separated in the Late Miocene (seven to five million years ago (Myr)). During the Pliocene and Pleistocene, the Mekong and the Salween rivers flowed as one into the present Chao Phraya River Basin, with the Mekong following a course north of Chiang Rai before merging with the Salween. At 1.5 mya, the Mekong River still flowed into the Salween, although it now followed a more southern course along the present Ping River valley. Faulting then probably diverted the Mekong eastwards along its present course towards Vientiane, but later in the Pleistocene the Mekong once again flowed toward the present Chao Phraya delta, although this time via the Loei/Passac rivers.

After extending its course southwards to the location of the present Mul-Mekong river junction, the Mekong underwent further course changes. It appears that the channel now occupied by the Mekong south of Khong Island (the present lower Mekong River) originated not more than five thousand years ago (Kyr). This lower part of the Mekong River, which according to Attwood and Johnston (2001) remained separated from the upper Mekong until very recently, may have flowed westward in the Pliocene, just south of the mountains along the Cambodian-Thai border, or down the present Tonlé Sap and into the Gulf of Thailand near Kampot (Workman, 1977, cited in Attwood and Johnston, 2001). Mapping of sea-floor contours in the Gulf of Thailand supports such a course of the Mekong up until at least the Late Pleistocene (Sawamura and Laming, 1974).

There does not seem to be much support for a direct connection between the Salween and the Chao Phraya rivers, as suggested by Attwood and Johnston (see above). Such a connection is also unlikely because the fresh water fish fauna of the Salween is only remotely related to that of the Chao Phraya and Mekong, and has much stronger links to the Irrawaddy and Brahmaputra. The Mekong and Chao Phraya, on the other hand, have fish faunas very similar to each other (Kottelat, 1989).

In summary, the evidence points towards an eastward migration of the Mekong River during the Pleistocene of some 600 km.

Palaeoenvironments of the Region

Toward the latter part of the Early Miocene, rising global temperatures and sea levels corresponded with a change to predominantly moist forests in Indochina, and tropical and subtropical rain forests became established beyond the tropics. From this time, Dipterocarpaceae became prominent in Thailand (Watanasak, 1990, in Morley, 1999) and alternating wet and dry climates were characteristic of Vietnam (Dzanh, 1994, in Morley, 1999). This climate more or less lasted until the Late Miocene/Early Pliocene. Palaeobotanical data from India and Burma indicate that a rich tropical to subtropical vegetation still covered the region ca. five Myr under a prevailing warm humid climate (several authors in Poole and Davies, 2001). Likewise, the Late Miocene climate of North Vietnam was described as humid and subtropical by Dzanh (1996, cited in Covert *et al.*, 2001).

During the Pliocene, the increasingly arid climate, engendered by the rising Himalayas, caused a consequent change in the vegetation of this region (Poole and Davies, 2001). In mid-Pliocene times (ca. 3.5 Myr), the climate may have become warmer and wetter again, as indicated by a considerable expansion of evergreen forest in the north of this region and into China. Indochina was at that time mostly covered by rain forest, possibly with patches of deciduous forest in the area of present-day Burma (Dowsett *et al.*, 1994).

The Pliocene ended with a long cold period at app. 2.7 Myr. Another cold phase probably followed between 1.5 and 1.8 Myr (1.3 according to Singh and Srinivasan, 1993), and then a very long phase of several glaciations started around 800 Kyr and lasted until app. 10 Kyr (Beard *et al.*, 1982). No detailed data are presently available for vegetation types in the Middle Pleistocene, but Jablonski and Whitfort (1999) suggest that tropical forests still occurred in most Indochina and southern China. Between the Middle and Late Pleistocene this tropical forest zone further retracted southward, and disappeared from most of China, leaving behind mostly subtropical forests.

Late Pleistocene environments in Northeast Thailand (17°N, 103°E) were described by Penny (2001) and Kealhofer and Penny (1998). Between 40 and 10 Kyr the vegetation in this area, consisting of Fagaceous-Coniferous forest, was remarkably stable, with clear evidence of lower temperatures and possible geomorphic evidence of significant drying in that region. During the Late Pleistocene, a change in climatic conditions is also apparent on the Khorat Plateau in eastern Thailand; the pollen assemblages suggest xeric, species poor, and strongly seasonal vegetation, with seasonally inundated floodplains or stream margins (Dheeradilok, 1995).

These examples from Thailand and Cambodia resemble patterns described for Sichuan and, possibly, Yunnan (Sun *et al.*, 1986; Li and Liu, 1988; Jarvis, 1993, all referred to in Maxwell, 2001), although the dry late glacial period may have lasted longer in Indochina than in Southwest China. Data by Liu *et al.*, (1986) suggest that Yunnan, at a time approaching or including the last glacial maximum (LGM) (ca. 36–20 Kyr), experienced greater winter humidity and rainfall, while mean annual temperatures were only slightly, if at all, lower than present. Yunnan may therefore have represented a Chinese tropical refugium.

In summary, it can be said that up until ca. five Myr the climate was relatively stable, with predominantly humid and warm conditions. During the Pliocene, increasing aridity and increasing monsoonal activity probably led to greater habitat diversity, with wetter areas on and around mountains. During the Pleistocene, many glacial stages occurred, of which the most severe may have been around ca. 2.7 Myr, ca. 800 Kyr, and ca. 18 Kyr. These may have led to the creation of wet forest refugia in Yunnan, and possibly also in other mountain ranges such as the Annamite Mountains, and possibly the Burmese Mountains as well. Overall, the trend during the Pleistocene was a southward retraction of the tropical forest zone, and an expansion of the subtropical forest zone in Indochina (Jablonski and Whitfort, 1999).

Catastrophic Events

An event of great importance to the biogeography and evolution of species in Indochina occurred around 0.77 Myr (Bunopas *et al.*, 1999; recently revised to 0.793 Myr by Lee and Wei, 2000). At that time, a large comet slammed into the earth, probably with a centre of impact near Ubon, in eastern Thailand (Bunopas *et al.*, 1999), although other areas have been suggested: Tonlé Sap by Hartung and Koberl (1994, in Paine, 2001), eastern Cambodia by Lee and Wei (2000), or southern Laos by Dass and Glass (1999, in Paine, 2001). Bunopas *et al.*, (1999) described northeast and east of Khorat in eastern Thailand with completely burnt and petrified trees, apparently pushed down to the ground by tremendous force. Among these, *Stegolophodon* sp. teeth were found together with other mammal bones, and crocodile teeth and bones. Howard *et al.*, (2000, cited in Paine, 2001), who also worked in the Khorat region, found petrified trees with trunks up to two metres in diameter "shattered, branchless, snapped, uprooted and burnt to the core," together with evidence of catastrophic floods. Evidence of possible contemporaneous forest destruction was found in Bose, South China (Yamei *et al.*, 2000, in Paine, 2001). This and other evidence suggests sudden ground melting from an extraterrestrial impact, leading to continental forest fires, build-up of atmospheric sands, and loss of life on a global scale. J.T. Wasson (13 December 2001) suggests that the impacts destroyed all animals and plants that lived above ground over a large area, probably greater than 1 million km^2.

Phylogenetic Patterns in Relation to the Region's
Palaeoenvironments and Geomorphology

Phylogenetic analyses based on differences between taxa in their DNA structure often provide information on when two taxa started to diverge, using the assumption of a constant molecular clock. Table 4 sums up some of the relevant phylogenetic research in the region in relation to geomorphological events and palaeoenvironmental conditions.

The DNA-based phylogenies discussed above point to a high species divergence rate in the Late Pliocene–Early Pleistocene of the region, although clearly they only represent a small fraction of the total number of species divergences in the region. We infer from this that an east/west faunal division was intensifying in mainland Southeast Asia, and that, combined with glacially induced forest fragmentation, this was the genesis of the present marked endemicity east of the Mekong.

The Late Pliocene marked a considerable faunal turnover in China. Environmental fluctuations alternately favouring deciduous broad-leaved trees and herbaceous plants had already led to faunal changes between three and four Myr, but at the start of the Pleistocene faunal replacement entered a critical phase. By this time, ca. 1.7 Myr, many Pliocene taxa had vanished from China. Others were only able to survive by migrating to the south of China and into Sundaland (Ferguson, 1993). Ferguson remarked that, considering the lack of faunal change in southern China during the Pleistocene, the environment there has probably remained stable during the Quaternary. The significance of this Late Pliocene–Early Pleistocene period was also pointed out by Verneau *et al.*, (1998), who described the near simultaneous generation of at least five southeast Asian rat lineages at ca. 2.7 Myr, while Serizawa *et al.*, (2000) described the divergence within *Apodemus* field mice between two and four Myr. Elsewhere, a Late Pliocene–Early Pleistocene faunal turnover was found in the Caribbean (Budd and Klaus, 2001), and in eastern Africa (Behrensmeyer *et al.*, 1997). Finally, in tropical Australia many rain forest species evolved during the Pliocene to Early Pleistocene, and managed to survive, and in some species further split up, in refugia during Pleistocene glacial periods (Schneider *et al.*, 1998); a process very much like that found in the mountain endemics in our study area.

The Arakan Yomas, along the western coast of Burma, would have remained forested (if perhaps with deciduous or monsoon forest rather than strictly rain

Table 4. Selected species divergence events in relation to geomorphological and environmental changes

Geomorphological events	Palaeoenvironmental conditions	Taxon divergence events
	Late Miocene	
Yangtze (and Chiangjiang ?) flow via Red River course. Flow of Mekong reduced (cf. present day) by 50%. Mekong and Yangtze were separated **Wide river barrier between China and SE Asia**	Climate generally warm and humid, with tropical and subtropical vegetation throughout the region	The four gibbon subgenera *Nomascus* (Indochina), *Symphalangus* (Sundaland), *Hylobates* (Sundaland) and "*Bunopithecus*" (North Burma region)[a] **Sundaland/mainland division East/west division in mainland SE Asia**
	Early Pliocene	
Yangtze and Chiangjiang flow eastward; Yangtze flow increased by 4×. Mekong flows into Chao Phraya **China and Indochina regions now continuous East/ west division in mainland SE Asia**	Climate generally warm and humid, with tropical and subtropical vegetation throughout the region. Aridification slowly sets in	*Pangasius nasutus* (a freshwater fish species) from West and Central Kalimantan and East Sumatra split from *P. conchophilus* from Indochina[b] **Sundaland/mainland division**
	Middle Pliocene	
Mekong flows into Chao Phraya **East/west division in mainland SE Asia**	Development of savannah vegetation in the Irrawaddy area, but probably no grasslands in Indochina. In Yunnan, a mixed forest-bushland existed, which started to change to more evergreen forest. Rain forest in Indochina **Climatic east/west division**	The South Asian sloth bear (*Ursus ursinus*) diverged from the southeast Asian species[c] Divergence of species within *Hylobates*[a] **Brahmaputra barrier Divisions within Sundaland**
	Late Pliocene	
Mekong flows into Chao Phraya **East/west division in mainland SE Asia**	Several cold phases occurred leading to development of more open vegetation types during the coldest periods. In Yunnan, a grassland-bush environment existed in cool, subtropical	*Elaphodus* and *Muntiacus*[d] Rapid radiation occurred in the monophyletic pheasant genus *Lophura*[f] Divergence within the southeast and east Asian bear species[c]

Table 4. (*Continued*)

Geomorphological events	Palaeoenvironmental conditions	Taxon divergence events
	conditions. In Indochina, there is a marked increase in Gramineae. **Isolation of tropical or subtropical forest refugia**	Divergence between two assemblages of haplotypes in mainland Asian elephants (*Elephas maximus*)[f] *Nycticebus bengalensis* and *N. pygmaeus*, the species occurring only east of the Mekong[g] Divergence between *M. muntjak* and the other muntjaks[b] Divergence between the pheasants *Lophura diardi* (in Indochina) and *L. ignita* (in Peninsular and insular Malaysia)[e] **Sundaland/mainland division** **East/west division in mainland SE Asia**
	Early Pleistocene	
Mekong flows into Chao Phraya **East/west division in mainland SE Asia**	Generally open, mixed vegetation types **Climatic east/west division** **Isolation of tropical or sub-tropical forest refugia**	Divergence between *Muntiacus vuquangensis* group and *M. reevesi*[d] **East/west division in mainland SE Asia**
	Middle Pleistocene	
Mekong flows into Chao Phraya, via Ping River valley and later the Loei/Passac river valleys. **Present course of middle Mekong established**	Severe glacial period around 800 Kyr, probably leading to vegetation changes similar to those of the Late Pleistocene. This was preceded by the catastrophic impact of an extraterrestrial body. **Widespread extinctions in middle of mainland SE Asia.** **Isolation of tropical or sub-tropical forest refugia**	Divergence between the pheasants *Lophura leucomelanos* (in Himalayas and Burma) and *L. nycthemera* (in SE China, Indochina, Thailand)[e] Divergence between pheasant species (*Lophura* sp.) from Taiwan and from isolated rain forest patches in the Annamites[e] Divergence between *Muntiacus feae* and *M. muntjak*[d] Divergence between *Pangasius djambal* (a freshwater fish) from Indonesia and *P. bocourti* from Vietnam and Thailand[b] **Reinforcement of east/west division in mainland SE Asia**

(*Continued*)

Table 4. (*Continued*)

Geomorphological events	Palaeoenvironmental conditions	Taxon divergence events
	Late Pleistocene	
Mekong flows via the Mul River into the Chao Phraya; it is much more deeply incised and narrowed than at present. **Middle and part of present lower Mekong course established**	In NE Thailand, deciduous/ coniferous forest existed. Glacier in North Burma. Generally, strongly seasonal vegetation in the region. More open, deciduous vegetation in south China, although there may have been a tropical refugium in Yunnan. **Climatic east/west division. Isolation of tropical or subtropical forest refugia**	*Muntiacus crinifrons* and *M. gongshanensis*[d]
	Holocene	
Mekong diverted to its present course	Change of tropical vegetation in NE Thailand and Cambodia	

References:[a] Hayashi *et al.*, (1995);[b] Pouyaud *et al.*, (2000);[c] Waits *et al.*, (1999);[d] Wang and Lan (2000);[e] Randi *et al.*, (2000);[f] Fernando *et al.*, (2000);[g] Zhang *et al.*, (1993);[h] Lan *et al.*, (1995)
Bold: presumed implications for palaeogeography and faunal divisions.

forest) while the more central regions of mainland Southeast Asia were drier. Far to the east, the Vietnam/Laos border region has the southernmost high mountains of the southeast Asian mainland, which may have retained tropical rain forest during glacial times when lowland areas may have been covered in vegetation adapted to more seasonal rainfall patterns.

The possible Pleistocene refugia of Vietnam and Laos, and perhaps also Yunnan, may have enabled the survival of species adapted to tropical conditions, which may be an explanation of the species richness of these areas (also see Groves and Schaller, 2000). By the time the reestablishment of rain forest in the lowlands enabled these eastern refugial species to spread back into the lowlands, the upper and middle Mekong was already in its present valley, preventing their spread west into the central and western parts of mainland Southeast Asia.

We must comment on the detail of how some more widespread species, such as *Trachypithecus* and *Macaca* spp. and *Nycticebus bengalensis*, managed to get into the east-Mekong region. The migrations were evidently a very late event, because none of the species concerned show even subspecific differentiation east and west of the Mekong. We suppose that the final, late Holocene, phase of the lower Mekong's course changes were sufficiently meandering, across the Cambodian lowlands towards Tonlé Sap, to allow these species to disperse by passive transport as wide meanders were cut off. In this region, the native vegetation is not rain forest; and it is worth noting that these are all taxa of comparatively wide ecological tolerance.

CONCLUSIONS

Our results indicate that mammalian species richness east of the Mekong River is higher than in the area west of the Mekong. The Mekong does not seem to be an ecological barrier for the investigated species, unlike the Brahmaputra River that separates species of drier, open vegetation in the northwest from forested adapted species in the east. The Mekong has, however, been a much more important geographic barrier to mammals than the Salween and the Brahmaputra rivers.

We have described some of the main geomorphological and environmental events that took place during the Late Tertiary of the southeast Asian mainland region. The data are insufficient to support a clear cause-effect relationship between these events and local speciation processes, but there are many broad pointers, so we can hypothesize that Late Pliocene–Early Pleistocene environmental changes split up many tropical species leading to diversification, which was maintained during the Pleistocene by further glacial periods. During the last glacial maximum, this may have led to the isolation of rainforest-dependent species in several refugia, the most important of which was probably the Annamite Mountains range. An eastward shift of the Mekong River and a mid-Pleistocene catastrophic comet collision may have added to the selection pressures on species.

ACKNOWLEDGMENTS

We thank Will Duckworth and two anonymous reviewers for their constructive comments on an earlier draft.

APPENDIX 1

Species limited to the east of the Mekong River

Species	HAB	ALT	CO	SIZE
Callosciurus inornatus	F	M	A	s
Chrotogale owstoni	F	L	T/A	m
Cynogale lowei	F	L	T	m
Dremomys gularis	F	L	T	s
Hapalomys delacouri	BF	M	A	s
Hylobates concolor group	F	L	A	m
Maxomys moi	F/S	L/M	T/A	s
Megamuntiacus vuquangensis	F	M	T	l
Muntiacus rooseveltorum group	F	M	T	l
Nyctereutes procyonoides	F	L	T	m
Nycticebus pygmaeus	F	M	A	s
Nesolagus timminsi	F	L/M	T	m
Pygathrix spp.	F	L/M	A	m
Pseudoryx nghetinhensis	F	L/M	T	l
Rhinopithecus avunculus	F	M	A	l
Sciurotamias forrestii	S	M	T	s
Sus bucculentus	–	L	T	l
Tamiops maritimus	F	L/M	T	s
Trachypithecus francoisi group	F	M	A	m

Species that occur on both sides of the Mekong, but that are limited in the west by the Salween

Species	HAB	ALT	CO	SIZE
Acrocodia indica	F	L/M	T	l
Berylmys berdmorei	F/S	L	T	s
Bos sauveli	DF	L	T	l
Callosciurus finlaysoni	F	L	A	s
Callosciurus flavimanus	F	L/M	A	s
Chiromyscus chiropus	F	M	T	s
Dendrogale murina	F	L	T/A	s
Hylomys suillus	F	M	T	s
Hylopetes lepidus	F	L/M	A	s
Hylopetes spadiceus	F	L	A	s
Leopoldamys sabanus	F	L	T/A	s
Lepus peguensis	G	M	T	m
Menetes berdmorei	F/S	L	T	s
Rhizomys pruinosus	BF	M	T/A	m
Rhizomys sumatrensis	BF	L/M	T/A	m
Tamiops mcclellandi	F	M	A	s
Tamiops rodolphii	F	M	A	s
Trachypithecus germaini	F	L	T/A	m
Tragulus javanicus	F/S	L	T	m
Tragulus napu	F	L	T	m

Species that go across the Mekong and the Salween, but that are limited in the west by the Brahmaputra

Species	HAB	ALT	CO	SIZE
Arctogalidia trivirgata	F	L/M	A	m
Atherurus macrourus	F	L/M	T	m
Bandicota savilei	S/G	L	T	s
Berylmys bowersii	F	M	T	s
Bos javanicus	F/S	L	T	l
Cervus eldi	F	L	T	l
Chiropodomys gliroides	F	L/M	A	s
Crocidura fuliginosa	S/G	L/M	T	s
Dicerorhinus sumatrensis	F	L/M	T	l
Dremomys rufigenis	F	M	T	s
Hylopetes phayrei	F	M	A	s
Hystrix brachyura	F/S	L	T	m
Macaca arctoides	F	M	T	l
Macaca assamensis	F	M	A	m
Macaca fascicularis	F	L	T/A	m
Macaca leonina	F	L/M	A	m
Macaca mulatta	F/S/G	L/M	T/A	m
Maxomys surifer	F/S	L	T	s
Melogale moschata	F	L	T	m
Melogale personata	F	L	T	m
Naemorhedus sumatraensis	F	M	T	l
Niviventer confucianus	F	L/M	T/A	s
Niviventer langbianis	DF	L/M	T/A	s
Nycticebus bengalensis	F	L/M	A	m
Trachypithecus phayrei	F	M	A	m
Viverra megaspila	S	L	T	m

Species that only occur northwest of the Brahmaputra

Species	HAB	ALT	CO	SIZE
Antilope cervicapra	S/G	L	T	l
Axis axis	F/S/G	L	T	l
Biswamoyopterus biswasi	?	L	A	m
Caprolagus hispidus	G	L	T	m
Funambulus pennantii	F/S/G	L	T/A	s
Golunda ellioti	S/G	L	T	s
Mellivora capensis	S/G	L	T	m
Naemorhedus goral	F/S	M	T	l
Semnopithecus entellus	F/S/G	L/M	T/A	l
Sus salvanius	F	L	T	m
Ursus ursinus	F/S	L	T/A	l
Vulpes bengalensis	S	L/M	T	m

Species only occurring south and east of the Brahmaputra, across the Salween, but not across the Mekong River

Species	HAB	ALT	CO	SIZE
Berylmys mackenziei	F/S	L/M	T	s
Berylmys manipulus	F	L/M	T	s
Hadromys humei	S/G	L/M	T	s
Hylobates hoolock	F	L	A	m
Milliardia kathleenae	S/G	L	T	s
Petaurista sybilla	F	M	A	m
Petinomys setosus	F	L	T	s
Tapirus indicus	F	L/M	T	l

Species that occur on both sides of the Brahmaputra, but that are limited in the east by the Salween

Species	HAB	ALT	CO	SIZE
Callosciurus pygerythrus	F	L	A	s
Cervus duvaucelli	G	L	T	l
Diomys crumpi	F	L/M	T	s
Dremomys lokriah	F	M	T	s
Rhinoceros unicornis	S/G	L/M	T	l
Trachypithecus pileatus	F	L/M	A	l

Species not limited by any of the main rivers

Species	HAB	ALT	CO	SIZE
Arctictis binturong	F	L/M	A	l
Arctonyx collaris	F	L/M	T	l
Axis porcinus	G	L	T	l
Bandicota bengalensis	S/G	L	T	s
Bandicota indica	S/G	L	T	s
Bos gaurus	F/S/G	L	T	l
Bubalus arnee	DF/S/G	L	T	l
Callosciurus erythraeus	F/S/G	L/M	A	s
Canis aureus	S/G	L	T	m
Cannomys badius	BF	M	T	s
Catopuma temminckii	F/DF	L/M	T	l
Cervus unicolor	F	L/M	T	l
Crocidura attenuata	S/G	L/M	T	s
Cuon alpinus	F/S/G	L/M	T	l
Elephas maximus	F/S/G	L	T	l
Felis chaus	DF/G	L/M	T	m
Herpestes javanicus	S/G	L/M	T	s
Herpestes urva	S/G	L	T	m
Hylopetes alboniger	F	M	A	s
Leopoldamys edwardsi	F/S	M	T	s
Manis javanica	F/S	L/M	T/A	m

(*Continued*)

Species	HAB	ALT	CO	SIZE
Martes flavigula	F	L/M	T/A	m
Micromys minutus	G	L	T	s
Muntiacus muntjak	F/S/G	L	T	l
Mustela kathiah	F	L/M	T/A	m
Mustela sibirica	F/S/G	M	T/A	m
Mustela strigidorsa	F/S/G	M	T/A	m
Neofelis nebulosa	F	L/M	A	l
Niviventer fulvescens	F	L/M	T/A	s
Niviventer tenaster	F	M	T/A	s
Paguma larvata	F/S	L/M	A	m
Panthera pardus	F/S	L	T	l
Panthera tigris	F/S	L	T	l
Paradoxurus hermaphroditus	F/S/G	L/M	T/A	m
Pardofelis marmorata	F	L/M	T/A	m
Petaurista elegans	F/S	M	A	m
Petaurista petaurista	F	L/M	A	m
Petaurista philippensis	F	L	A	m
Prionailurus bengalensis	F/S/G	L/M	T	m
Prionailurus viverrinus	F/S	L/M	T	m
Prionodon pardicolor	F	M	T/A	m
Ratufa bicolor	F	L/M	A	m
Rhinoceros sondaicus	F	L/M	T	l
Sus scrofa	F/S/G	L/M	T	l
Trogopterus pearsonii	F	M	A	s
Tupaia belangeri	F/S	L/M	T/A	s
Ursus malayanus	F/S	L/M	T/A	l
Ursus thibetanus	F/S	M	T/A	l
Vandeleuria oleracea	F/S/G	L/M	T/A	s
Viverra zibetha	F/S/G	L/M	T	m
Viverricula indica	F/S/G	L	T	m

Appendix 1. Mammals of Southeast Asia, grouped according to their biogeographical limits. Codes used: F = forest; DF = deciduous forest; S = scrub; G = grass; L = lowland; M = mountains; T = terrestrial; A = arboreal; s = 0–1 kg; m = 1–10 kg; l = 10+ kg

REFERENCES

Alderton, D. 1994, *Foxes, Wolves and Wild Dogs of the World*, Blandford.

Amato, G., Egan, M. G., Schaller, G. S., Baker, R. H., Rosenbaum, H. C., Robichaud, W. G., and DeSalle, R. 1999, Rediscovery of Roosevelt's barking deer (*Muntiacus rooseveltorum*). *J. Mammal.* 80:639–643.

Attwood, S. W. and Johnston, D. A. 2001, Nucleotide sequence differences reveal genetic variation in *Neotricula aperta* (Gastropoda: Pomatiopsidae), the snail host of schistosomiasis in the lower Mekong Basin. *Biol. J. Linn. Soc.* 73:23–41.

Averianov, A. O., Abramov, A. V., and Tikhonov, A. N. 2000, A new species of *Nesolagus* (Lagomorpha, Leporidae) from Vietnam with osteological description. *Cont. Zool. Inst. St. Petersburg* 3:1–22.

Beard, J. H., Sangree, J. B., and Smith, L. A. 1982, Quaternary chronology, paleoclimate, depositional sequences, and eustatic cycles. *AAPG Bull.* 66:158–169.

Behrensmeyer, A. K., Todd, N. E., Potts, R., and Mcbrinn, G. E. 1997, Late pliocene faunal turnover in the Turkana Basin, Kenya and Ethiopia. *Science* 278:1589–1594.

Budd, A. F. and Klaus, J. S. 2001, The origin and early evolution of the *Montastraea* "*annularis*" species complex (Anthozoa: Scleractinia). *J. Palaeontol.* 75:527–545.

Bunopas, S., Wasson, J. T., Vella, P., Fontaine, H., Hada, S., Burrett, C., Suphajunya, T. and Khositanont, S. 1999, Catastrophic losses, mass mortality and forest fires suggest that a Pleistocene cometary impact in Thailand caused the Australasian tektite field. *J. Geol. Soc. Thailand* 1:1–17.

Chapman, J. A. and Flux, J. E. C. eds. 1990, *Rabbits, Hares and Pikas. Status Survey and Conservation Action Plan.* IUCN, Gland, Switzerland.

Corbet, G. B. and Hill, J. E. 1992, *The Mammals of the Indomalayan Region: A Systematic Review.* Oxford University Press, Oxford, UK.

Covert, H. H., Hamrick, M. W., Dzanh, T., and McKinney, K. C. 2001, Fossil mammals from the Late Miocene of Vietnam. *J. Vertebr. Palaeontol.* 21:633–636.

Dheeradilok, P. 1995, Quaternary coastal morphology and deposition in Thailand. *Quat. Int.* 26:49–54.

Dowsett, H., Thompson, R., Barron, J., Cronin, T., Fleming, F., Ishman, S., Poore, R., Willard, D., and Holtz, T. J. 1994, Joint investigations of the Middle Pliocene climate I: PRISM palaeoenvironmental reconstructions. *Global Planet. Change* 9:169–195.

Duckworth, J. W. 1998, A survey of large mammals in the central Annamite countains of Laos. *Z Säugetierk.* 6:239–250.

Ferguson, D. K. 1993, The impact of Late Cenozoic environmental changes in East Asia on the distribution of terrestrial plants and animals, in: Jablonski, N. G. and Chak-Lam, S., eds., *Evolving Landscapes and Evolving Biotas of East Asia Since the Mid-Tertiary. in: Proceedings of the Third Conference on the Evolution of the East Asian Environment*, Centre of Asian Studies. The University of Hong Kong, pp. 145–196.

Fernando, P., Pfrender, M. E., Encalada, S., and Lande, R. 2000, Mitochondrial DNA variation, phylogeography and population structure of the Asian elephant. *Heredity* 84:362–372.

Giao, P. M., Tuoc, D., Wikramanayake, E. D., Amato, G., Actander, P., and Mackinnon, J. 1998, Description of *Muntiacus truongsonensis*, a new species of muntjac (Artiodactyla: Muntiacidae) from central Vietnam, and implications for conservation. *Anim. Cons.* 1:61–68.

Groves, C. P. and Schaller, G. B. 2000, The phylogeny and biogeography of the newly discovered Annamite artiodactyls, in: Vrba, E. S. and Schaller, G. B., eds., *Antelopes, Deer, and Relatives. Fossil Record, Behavioral Ecology, Systematics, and Conservation*, Yale University Press, New Haven, USA, pp. 261–282.

Groves, C. P., Schaller, G. B., Amato, G., and Khounboline, K. 1997, Rediscovery of the wild *Sus bucculentus. Nature* 386:335.

Hallet B. and Molnar, P. 2001, Distorted drainage basins as markers of crustal strain east of the Himalayas. *J. Geophys. Res.* 106:13697–12709.

Hayashi, S., Hayasaka, K., Takenaka, O., and Horai, S. 1995, Molecular phylogeny of gibbons inferred from mitochondrial DNA sequences: Preliminary report. *J. Mol. Evol.* 41:359–365.

Hutchinson, C. S. 1989, *Geological Evolution of South-East Asia*, Clarendon Press, Oxford, UK.

Jablonski, N. G. and Whitfort, M. J. 1999, Environmental change during the Quaternary in East Asia and its consequences for mammals. *Rec. W. Aust. Mus. Suppl.* 57:307–315.

Kealhofer, L. and Penny, D. 1998, A combined pollen and phytolith record for fourteen thousand years of vegetation change in northeastern Thailand. *Rev. Palaeobot. Palynol.* 103:83–93.

Kottelat, M. 1989, Zoogeography of the fishes from Indochinese inland waters with an annotated check list. *Bull. Zoöl. Mus.* 12:1–54.

Lacassin, R., Replumaz, A., and Hervé Leloup, P. 1998, Hairpin river loops and slip-sense inversion on southeast Asian Strike-slip faults. *Geology* 26:703–706.

Lan, H., Wang, W., and Shi, L. 1995, Phylogeny of *Muntiacus* (Cervidae) based on mitochondrial DNA restriction maps. *Biochem. Gen.* 33:377–388.

Lee, M.-Y. and Wei, K.-Y. 2000, Australasian microtektites in the South China Sea and the West Philippine Sea: Implications for age, size, and location of the impact crater. *Meteorit. Planet. Sc.* 35:1151–1155.

Lekagul, B. and McNeely, J. A. 1977, *Mammals of Thailand*, Kurusapha Ladproa Press, Bangkok, Thailand.

Liu, J.-l., Tang, L., Qiao, Y., Head, M. J., and Walker, D. 1986, Late Quaternary vegetation history at Menghai, Yunnan Province, Southwest China. *J. Biog.* 13:399–418.

Ma, S.-L. and Wang, Y. X. 1986, The taxonomy and distribution of the gibbons in southern China and its adjacent region-with description of three new subspecies. *Zool. Res.* 7:393–410.

MacKinnon, J. 1997, *Protected Areas System Review of the Indo-Malayan Realm*, The Asian Bureau for Conservation Limited, Canterbury, UK.

MacKinnon, J. 2000, New mammals in the 21st century? *Ann. MI Bot. Gard.* 87:63–66.

Maxwell, A. L. 2001, Holocene monsoon changes inferred from lake sediment pollen and carbonate records, northeastern Cambodia. *Quat. Res.* 56:390–400.

Meijaard, E. and Groves, C. P. 2004, A taxonomic revision of the *Tragulus* mouse-deer (Artiodactyla). *Zoological Journal of the Linnean Society* 140(1):63–102.

Métivier, F. and Gaudemer, Y. 1999, Stability of output fluxes of large rivers in South and East Asia during the last 2 million years: implication on floodplain processes. *Basin Res.* 11:293–303.

Morley, R. J. 1999, Palynological evidence for Tertiary plant dispersals in the SE Asian region in relation to plate tectonics and climate, in: Hall, R. and Holloway, J. D., eds., *Biogeography and Geological Evolution of SE Asia*, Backhuys Publishers, Leiden, The Netherlands, pp. 211–234.

Nisbett, R. A. and Ciochon, R. L. 1993, Primates in northern Viet Nam: A review of the ecology and conservation status of extant species, with notes on Pleistocene localities. *Int. J. Primatol.* 14:765–795.

Paine, M. 2001, Source of the Australasian tektites. *Meteorite.*

Payne, J., Francis, C. M., and Phillipps, K. 1985, *A field guide to the mammals of Borneo*, The Sabah Society, Kota Kinabalu, Malaysia.

Penny, D. 2001, A 40,000 year palynological record from north-east Thailand; Implications for biogeography and palaeo-environmental reconstruction. *Palaeogeogr., Palaeoclimat., Palaeoecol.* 171:97–128.

Peter, W. P. and Feiler, A. 1994, Eine neue Bovidenart aus Vietnam und Cambodia (Mammalia: Ruminantia). *Faun. Abh. Staatl. Mus. Tierk. Dresden* 48:169–176.

Poole, I. and Davies, C. 2001, *Glutoxylon* Chowdhury (Anacardiaceae): The first record of fossil wood from Bangladesh. *Rev. Palaeobot. Palynol.* 113:261–272.

Pouyaud, L., Teugels, G. G., Gustiano, R., and Legendre, M. 2000, Contribution to the phylogeny of pangasiid catfishes based on allozymes and mitochondrial DNA. *J. Fish Biol.* 56:1509–1538.

Randi, E., Lucchini, V., Armijo-Prewitt, T., Kimball, R. T., Braun, E. L., and Ligon, J. D. 2000, Mitochondrial DNA phylogeny and speciation in the Tragopans. *The Auk* 117:1007–1019.

Ruggeri, N. and Etterson, M. 1998, The first record of Colugo (*Cynocaphalus variegatus*) from the LAO P.D.R. *Mammlogy* 62:450–451.

Sawamura, K. and Laming, D. J. C. 1974, Sea-floor valleys in the Gulf of Thailand and quaternary sea-level changes. *CCOP Newsletter* 1:23–27.

Schaller, G. B. and Rabinowitz, A. 1995, The Saola or spindlehorn bovid *Pseudoryx nghetinhensis* in Laos. *Oryx* 29:107–114.

Schaller, G. B. and Vrba, E. S. 1996, Description of the Giant Muntjac (*Megamuntiacus vuquangensis*) in Laos. *J. Mammal.* 77:675–683.

Schneider, C. J., Cunningham, M., and Moritz, C. 1998, Comparative phylogeography and the history of endemic vertebrates in the Wet Tropics rain forests of Australia. *Mol. Ecol.* 7:487–498.

Schreiber, A., Wirth, R., Riffel, M., and Van Rompaey, H. 1989, *Weasels, Civets, Mongooses, and their Relatives. An Action Plan for the Conservation of Mustelids and Viverrids*, IUCN, Gland, Switzerland.

Serizawa, K., Suzuki, H., and Tsuchiya, K. 2000, A phylogenetic view on species radiation in *Apodemus* inferred from variation of nuclear and mitochondrial genes. *Biochem. Gen.* 38:27–40.

Singh, A. D. and Srinivasan, M. S. 1993, Quaternary climatic changes indicated by planktonic foraminifera of Northern Indian Ocean. *Curr. Sci.* 64:908–915.

Sokolov, V. E., Rozhnov, V. V., and Pham Trong Anh 1997, New species of viverrids of the genus *Viverra* (Mammalia, Carnivora) from Vietnam. *Russ. J. Zool.* 1:204–207.

Stafford, B. J. and Szalay, F. S. 2000, Craniodental functional morphology and taxonomy of dermopterans. *J. Mamm.* 81:360–385.

Thomas, H., Seveau, A., and Hassanin, A. 2001, The enigmatic new Indochinese bovid, *Pseudonovibos spiralis*: An extraordinary forgery. *Comp. Rend. Acad. Sci., Series 3.* 24:81–86.

Verneau, O., Catzeflis, F., and Furano, A. V. 1998, Determining and dating recent rodent speciation events by using L1 (LINE-1) retrotransposons. *Proc. Nat. Acad. Sci. USA* 95:11284–11289.

Waits, L. P., Sullivan, J., O'Brién, S. J., and Ward, R. H. 1999, Rapid radiation events in the family Ursidae indicated by likelihood phylogenetic estimation from multiple fragments of mtDNA. *Mol. Phyl. Evol.* 13:82–92.

Walston, J. and Véron, G. 2001, Questionable status of the "Taynguyen civet", *Viverra tainguensis* Sokolov, Rozhnov and Pham Trong Anh, 1997 (Mammalia: Carnivora: Viverridae). *Mammal Biol.* 66:181–184.

Wang, W. and Lan, H. 2000, Rapid and parallel chromosomal number reductions in muntjac deer inferred from mitochondrial DNA phylogeny. *Mol. Biol. Evol.* 17:1326–1333.

Zhang, Y.-p., Chen, Z.-p., and Shi, L.-m. 1993, Phylogeny of the slow loris (genus *Nycticebus*): An approach using mitochondrial DNA restriction enzyme analysis. *Int. J. Primatol.* 14:167–175.

CHAPTER TWELVE

Primate Biogeography and Ecology on the Sunda Shelf Islands: A Paleontological and Zooarchaeological Perspective

Terry Harrison, John Krigbaum, and Jessica Manser

ABSTRACT

Sundaland, with its complicated history of island formation and landbridge connections with mainland Southeast Asia, has figured prominently in studies of primate biogeography. The non-human primates on Sundaland are taxonomically diverse (comprising

Terry Harrison • Center for the Study of Human Origins, Department of Anthropology, New York University, 25 Waverly Place, New York, NY 10003.   **John Krigbaum** • Department of Anthropology, University of Florida, 1112 Turlington Hall, Gainesville, FL 32611.   **Jessica Manser** • Department of Anthropology, New York University, 25 Waverly Place, New York, NY 10003

Primate Biogeography, edited by Shawn M. Lehman and John G. Fleagle.
Springer, New York, 2006.

27 species), and they exhibit relatively high levels of provinciality and endemism. By combining archaeological and paleontological evidence, with data from molecular, paleoclimatological and paleoecological studies, it is possible to reconstruct the major zoogeographic events that took place in the formation of the present-day catarrhine primate community on the Sunda Shelf islands. It can be inferred that by the Late Pliocene the main islands of the Sunda Shelf had a primate fauna that included *Pongo pygmaeus* (Sumatra, Java and Borneo), *Hylobates* spp. of the *lar*-group (Sumatra, Mentawai Islands, Borneo, and Java), *Macaca nemestrina* (Sumatra, Mentawai Islands, Borneo, and Java), the common ancestor of the *Trachypithecus auratus/cristatus* clade (Java and Sumatra), and *Presbytis* spp. (Sumatra, Mentawai Islands, Borneo, Sumatra, and Java). Most of these taxa probably arrived during the Pretiglian cold phase, starting at ~2.8 Ma, when sea levels fell by more than 100 m. It is also likely that *Nasalis larvatus* (Borneo) and *Simias concolor* (Mentawai Islands) were already present as endemic taxa in the Late Pliocene, and that their last common ancestor had arrived in the Sunda islands by the early Pliocene. Soon after this initial period of colonization, *Hylobates* and *Presbytis* underwent rapid speciation as a consequence of vicariance and relictual survivorship, giving rise to *P. thomasi* on Sumatra, *H. klossii* and *P. potenziani* on the Mentawai Islands, *H. albibarbis*, *H. muelleri*, *P. hosei*, *P. frontata*, and *P. rubicunda* on Borneo, and *H. moloch* and *P. comata* on Java. During the Late Pliocene and Early Pleistocene, probably associated with a cold climate maximum at ~1.8 Ma, *Presbytis melalophos* and *P. femoralis*, along with *Macaca fascicularis*, colonized Sumatra, the Natuna Islands and Borneo from the Malay Peninsula. At about the same time, the orang-utan populations on Sumatra, Java and Borneo began to differentiate from each other. *Hylobates lar*, *H. agilis* and *H. syndactylus* extended their range from the Malay Pensinsula into Sumatra (and Java), probably during the Middle to Late Pleistocene, coincident with the arrival of *Trachypithecus cristatus* on mainland Southeast Asia. Meanwhile, *Pongo pygmaeus*, *Hylobates syndactylus* and *Macaca nemestrina* were extirpated on Java, probably as a consequence of a combination of ecological changes and the impact of early hominin incursions.

Key Words: Sundaland, zooarchaeology, paleontology, biogeography, ecology, primates

INTRODUCTION

The Sunda Shelf, with an estimated area of ~1,850,000 km^2, lies partially submerged beneath the Java Sea and the southwestern part of the South China Sea (Tjia, 1980; Hanebuth *et al.*, 2000). Sundaland is the name given to that area of the Sunda Shelf that emerged during periods of low sea level, particularly during the Quaternary, when sea levels fell by at least 120 m below present-day levels (Figure 1). It includes the Malay Peninsula, Borneo, Sumatra, Java, Bali, Palawan, the Mentawai Islands, and the smaller intervening

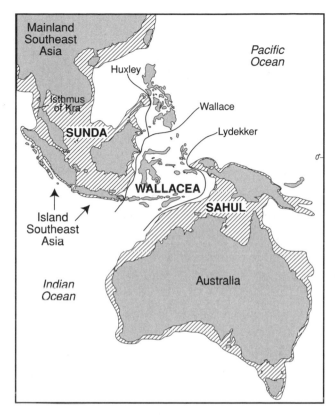

Figure 1. Map of Australasia showing extent of the Sunda and Sahul continental landmasses with Wallacea in between. Biogeographical lines of Wallace, Huxley, and Lydekker are indicated, as is the Isthmus of Kra.

islands. Periodically during the Pliocene and Pleistocene, Sundaland formed a continuous subcontinental landmass connecting Sumatra, Java, Borneo, and Palawan with the Asian mainland (Haile, 1973; Sartono, 1973; Verstappen, 1975; Tjia, 1980). The eastern boundary of Sundaland is delimited by Wallace's Line, as modified by Huxley (1868), which runs between Palawan and Luzon, Borneo and Sulawesi, and Bali and Lombok—this defines the boundary of the Oriental zoogeographic province (Huxley, 1868; Simpson, 1977). To the east is Sahulland, comprising New Guinea, Australia, and Tasmania. Its western boundary, represented by Lydekker's Line, delimits the Australasian zoogeographic province (Lydekker, 1896; Simpson, 1977).

During the Last Glacial Maximum (LGM) at ~21–18 ka, when sea levels were at their lowest, Sahulland and Sundaland continued to be separated by deep oceanic troughs. Wallacea, the region between Sahulland and Sundaland, consists of the Lesser Sunda Islands (i.e., Lombok, Sumbawa, Sumba, Flores, and Timor), Sulawesi, the Philippines (excluding Palawan), and the Moluccas (Dickerson, 1928; Scrivenor et al., 1941–42; Simpson, 1977; Darlington, 1980; Groves, 1985). During the later Pleistocene glacial periods the islands of Wallacea remained unconnected to Sahulland or Sundaland (Audley-Charles, 1981; Ollier, 1985). It is possible, however, that the Philippines proper and Sulawesi may have been connected to Borneo by landbridges during the Pliocene and Early Pleistocene, when sea levels were strongly influenced by cold-climate peaks (Sartono, 1973; Heaney, 1985, 1986; Prentice and Denton, 1988; Shackleton, 1995; Moss and Wilson, 1998).

Given the position of Sundaland at the junction between two major zoogeographic provinces that straddle the Equator, and its complicated history of island formation and landbridge connections with mainland Southeast Asia, it is not surprising that the Sunda Shelf islands have figured prominently in studies of mammalian biogeography (e.g., Simpson, 1977; Heaney, 1984, 1986; Groves, 1985; Han and Sheldon, 2000; Mercer and Roth, 2003). In particular, the systematics and zoogeographic relationships of the primates on Sundaland have been extensively studied, but it is evident from recent phylogenetic analyses of the primates in the region, using various lines of evidence, that the reconstructed biogeographic history is exceedingly complicated (e.g., Brandon- Jones, 1996, 1998, 2001; Rosenblum et al., 1997a, b; Morales and Melnick, 1998; Harcourt and Schwartz, 2001; Evans et al., 2003). There are currently 27 species of non-human primates with geographic distributions that encompass Sundaland, and since these comprise more than one-third of all large mammals from the region, they represent an important faunal component. Only two of these taxa are non-catarrhine primates, Nycticebus coucang (slow loris) and Tarsius bancanus (western tarsier). The catarrhines of Sundaland are diverse (comprising 41% of all Asian catarrhine species), and they exhibit relatively high levels of provinciality and endemism (76% of the 25 species are unique to Sundaland).

We present here a study of the catarrhine primate faunas from paleontological and archaeological sites in Borneo, Sumatra, and Java. Although the fauna from these sites represents a limited database, their identification and study introduces a unique diachronic perspective on the biogeography and ecology of the Sunda Shelf islands. This analysis is of particular interest because it provides insights into prehistoric human hunting strategies and dietary preferences, and offers clues to understanding regional paleoecological change—both important factors that likely influenced the zoogeographic distribution of primates on Sundaland.

Although there appear to have been few demonstrable local extinctions on the Sunda Shelf islands during the Quaternary (*Hylobates syndactylus*, *Pongo pygmaeus* and *Macaca nemestrina* became extinct on Java; Hooijer, 1948, 1960, 1962a), climatic perturbations and the arrival of modern humans during the later Pleistocene may have had a profound impact on the primate fauna. Zooarchaeological evidence indicates that primates were extensively exploited for food, but the presumed low density of humans on the Sunda Shelf islands, and their limited technologies for hunting arboreal mammals, suggest that they had little impact on primate distributions, except at the local level (Harrison, 1996). A more important factor influencing the distribution of primate species seems to have been ecological changes caused by cooler, more seasonal climatic conditions during the Pliocene and Pleistocene, and concomitant glacio-eustatic sea level fluctuations associated with northern hemisphere glacial phases (Verstappen, 1975; Heaney, 1991; Prentice and Denton, 1988; Shackleton, 1995). These ecological changes apparently had a much more significant impact on the structure and geographic distribution of the catarrhine primate community (see Harrison, 1996, 2000).

Finally, we attempt to combine the archaeological and paleontological evidence with data from molecular, paleoclimatological, and paleoecological studies in order to recreate the major zoogeographic events that took place in the formation of the present-day catarrhine primate community on the Sunda Shelf islands. Given the limitations of the evidence available, such a scenario involves a good deal of speculation, but we believe that the analysis represents a useful first step in the development of a broader synthesis, and one that can be used as a provisional model to be reassessed as new evidence comes to light on the zoogeographic relationships and evolutionary history of Southeast Asian primates.

PALEOENVIRONMENTAL CONTEXT

The geological history of Southeast Asia has been reviewed by Katili (1975), Audley-Charles (1981, 1987), Ollier (1985), McCabe and Cole (1989), Hutchison (1989), Hall (1998, 2001), and Metcalfe (1998). The major geological processes and events that led to the formation and present-day positions of the islands and landmasses of Sundaland were completed by the Early Pliocene at ~5 Ma (Audley-Charles, 1981; Moss and Wilson, 1998; Hall, 1998). Nevertheless, Southeast Asia has continued to experience volcanic and tectonic activity (Ashton, 1972; Aldiss and Ghazali, 1984; Hall, 1998). Danau Toba in northern Sumatra is associated with the most dramatic examples of Pleistocene volcanism in the region (dated at about 840 ka, 501 ka, and 74 ka; Diehl *et al.*, 1987; Ninkovich *et al.*, 1978; Chesner *et al.*, 1991), and it presumably had a significant impact on the local ecology. The largest and most recent erup-tion during the Late Pleistocene dispersed ash as far as India and the Strait of Malacca (Ninkovich *et al.*, 1978; Rose and Chesner, 1987; Chesner *et al.*, 1991; Bühring and Sarnthein, 2000), and it has been implicated in a volcanic winter that may have hastened, at least regionally, the global cooling trend that followed the Odderade interstadial (Rampino *et al.*, 1988; Chesner *et al.*, 1991; Rampino and Self, 1992; Zielinski *et al.*, 1996).

During the Pliocene, global temperatures were generally warmer, with estimated maximum temperatures 3.6°C higher than at present (Dowsett *et al.*, 1992, 1996; Crowley, 1991, 1996; Sloan *et al.*, 1996). Tropical sea surface temperature estimates, based on marine microfossils, are comparable to present-day (Dowsett *et al.*, 1996; Crowley, 1996), but there is evidence of cooler conditions towards the end of the Pliocene, presumably correlated with the Pretiglian cold phase of northern latitudes dated at ~2.75 Ma (King, 1996; Ravelo *et al.*, 2004). The main Pleistocene glacial spikes occur at ~1.8 Ma, (Eburonian), ~0.92 Ma (Menapian), ~630 ka (OIS 16), ~430 ka (OIS 12), ~350 ka (OIS 10), ~140 ka (OIS 6) and ~18 ka (OIS 2) (Chappell and Shackleton, 1986; Chappell *et al.*, 1996). Average regional temperature estimates during the Pleistocene, based largely on proxy data from deep-sea cores, indicate cooler conditions by ~3–5°C on land and ~2–4°C in the seas (Petersen, 1969; Verstappen, 1975; Hope *et al.*, 1976; Rind and Peteet, 1985; Flenley, 1985; Tan, 1985; Chappell *et al.*, 1996). In particular, oxygen isotope data confirm cooler seas during glacials and interstadials, and suggest that tropical seas were ~5–6°C cooler than today (e.g., Rind and Peteet, 1985; Chappell, 1994; Stute *et al.*, 1995; Broecker, 1996; Bush and Philander, 1998; McCulloch *et al.*, 1999; Lea *et al.*, 2000). Temperatures on land and sea rose by 1–3°C during Pleistocene interglacial periods and the Holocene climatic optimum (Verstappen, 1975, 1997; Flohn, 1981; Gagan *et al.*, 1998). The Early Holocene in low latitude regions was still marked by significantly colder sea surface temperatures (~6°C cooler than at present), followed by a rapid increase in temperature up to present-day values by ~4 ka (Guilderson *et al.*, 1994; McCulloch *et al.*, 1996; Beck *et al.*, 1997).

During Pleistocene glacial phases, climatic conditions in Southeast Asia were drier and cooler than at present, with longer dry seasons and shorter wet seasons, and more pronounced seasonality (Petersen, 1969; Morley, 1982; Debaveye *et al.*, 1986; Thomas, 1987; Heaney, 1991; van der Kaars and Dam,

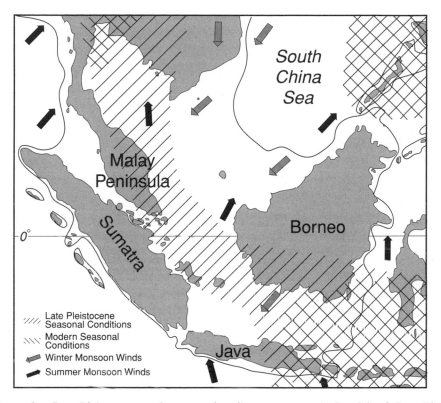

Figure 2. Late Pleistocene and present-day climate patterns in Sundaland. Late Pleis-tocene conditions (e.g., during the LGM) were considerably more seasonal than today. The extent of dual winter and summer monsoons (arrows) is more intense in present-day perhumid areas than in more seasonal ones. Figure adapted from Heaney (1991).

1995; Verstappen, 1997; Brandon-Jones, 1998). Despite a decrease in rain-fall, Pleistocene Southeast Asia may not have been as dry as other equatorial regions (Verstappen, 1975). Rainfall increased during warm interstadials and interglacials, and again during the terminal Pleistocene, followed by alternat-ing dry and moist periods in the Early and Middle Holocene (Flohn, 1981; Verstappen, 1997).

Palynological studies suggest that changes in vegetation in Southeast Asia during the Late Pleistocene and Holocene correlate with documented changes in temperature and precipitation. In general, during drier Pleistocene glacial periods grasslands and montane forests expanded, while tropical lowland rain-forests contracted (Figure 2). Marine and terrestrial core data from Sumatra, Java, eastern Indonesia, New Guinea, and northern Australia record a decrease in forest coverage, an increase in the spread of grasslands, and an altitudinal lowering of montane vegetation zones by about 300–500 m during the LGM

(Van Andel *et al.*, 1967; Hope *et al.*, 1976; Walker and Flenley, 1979; Flenley, 1979; Maloney, 1980, 1981; Morley, 1982; Stuijts, 1983/1984; Newsome and Flenley, 1988; Stuijts *et al.*, 1988; Hope and Tulip, 1994; Haberle, 1998; van der Kaars, 1998; Kershaw *et al.*, 2001). Van der Kaars (1998) and Kershaw *et al.* (2001) suggest that pollen spectra indicate a 30–50% decline in precipitation and a reduction in mean temperatures by as much as 6–7°C. Morley and Flenley (1987) infer a dry seasonal grassland belt running southwards along the Malay Peninsula and into Java (Flohn, 1981; van der Kaars, 1991; van der Kaars and Dam, 1995). Kershaw *et al.* (2001), by contrast, contend that perhumid rainforest was only replaced by grassland in regions already experiencing some degree of seasonality, and that reduced precipitation did not have a major impact on evergreen tropical forest in core areas, such as Borneo, during the LGM. Since the exposed continental shelves appear to have been covered by rainforest in more mesic areas and by grassland, woodland, and sedgeland in drier areas, there was a net increase in the availability of tropical forest habitats on Sundaland during the LGM, although it was probably more fragmented than at present. Kershaw *et al.* (2001) do acknowledge, however, that a dry corridor may have occurred during cold phases prior to the LGM. The warmer interglacials and terminal Pleistocene experienced expansion of the core rainforest areas, generally denser vegetation cover in drier areas, and the retreat of montane vegetation zones (Flenley, 1979; Verstappen, 1975, 1997; Flohn, 1981; van der Kaars, 1991; Hope and Tulip, 1994). Mangrove forests seem to have dominated the lowland coastal areas during these periods, as it does today (Biswas, 1973).

In conjunction with the generally warmer temperatures during the Pliocene, average global sea levels are estimated to have been between 20–60 m higher than at present (Crowley, 1996; Dowsett *et al.*, 1996; Haq *et al.*, 1987; Wardlaw and Quinn, 1991). Data from deep-sea cores provide a well documented chronology for global glacio-eustatic sea level changes over the past 5 Ma (Chappell and Shackleton, 1986; Shackleton, 1987, 1995; Prentice and Denton, 1988; Chappell, 1994; Chappell *et al.*, 1996). Figure 3 outlines Late Quaternary sea level fluctuations based on d^{18}O data from the Huon Peninsula and the Sulu Sea (Chappell and Shackleton, 1986; Martinson *et al.*, 1987; Linsley, 1996; McCulloch *et al.*, 1999), and these correspond to high-resolution global estimates of sea level changes (Fairbanks, 1989; Guilderson *et al.*, 1994).

The LGM witnessed a dramatic decrease in sea level of 120–135 m (Van Andel *et al.*, 1967; Van Andel and Veevers, 1967; Yokoyama *et al.*, 2000). Lower

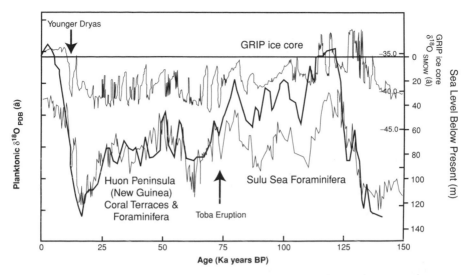

Figure 3. Late Quaternary sea level data based on Huon Peninsula (New Guinea) coral, d^{18}O (PDB) and Sulu Sea d^{18}O (SMOW). Oxygen isotope stages correspond to global climate trends. For example, OIS 2 corresponds to the Last Glacial Maximum (LGM) ~21–17 ka. Other climatic events (e.g., Toba eruption, Younger Dryas) are also noted. Figure adapted from Linsley (1996).

sea levels throughout the Plio-Pleistocene would have exposed substantial regions of the Sunda shelf (Molengraff and Weber, 1920; Fairbridge, 1953; Dobby, 1960; Petersen, 1969; Flint, 1971; Sawamura and Laming, 1974; Verstappen, 1975, 1997; Wang *et al.*, 1999; Hanebuth *et al.*, 2000; Sun *et al.*, 2000). Figure 4 depicts the exposed shelf during these periods, and shows major river systems that originated in the highlands of Borneo, Sumatra, Java, and the Malay Peninsula and transected the lowland areas (Molengraff and Weber, 1920; Umbgrove, 1938; Haile, 1973; Verstappen, 1975; Voris, 2000). In conjunction with tropical forest fragmentation, these drainage systems would probably have represented significant zoogeographic barriers to mammals, and may have contributed to increased population isolation and speciation.

The post-LGM period, up to and including the Middle Holocene, was characterized by a general trend of sea level increase (Tjia, 1970,1980; Emery *et al.*, 1971; Barham and Harris, 1983; Chappell and Shackleton, 1986; Shackleton, 1987; Edwards *et al.*, 1993; Chappell, 1994; Chappell *et al.*, 1996; Verstappen, 1997; Fleming *et al.*, 1998; Hanebuth *et al.*, 2000). A sea level curve based on sediment analysis from the Sunda Shelf documents an increase in sea level between 19–13 ka (from −114 to −64 m) (Hanebuth *et al.*, 2000).

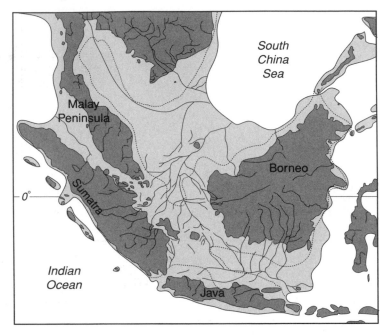

Figure 4. Map of Sundaland and adjoining islands showing paleo-river systems in relation to the 100 m isobath (solid line) and the 50 m isobath (dotted line). During low sea stands, these larger river systems were likely important zoogeographic barriers.

Average global sea levels rose to 10–20 m below present at the beginning of the Holocene (~10 ka) (Shackleton, 1987; Edwards *et al.*, 1993; Chappell, 1994; Linsley, 1996; Fleming *et al.*, 1998). During the mid-Holocene, average global sea level was at, or slightly above, present-day level (Tjia, 1970, 1977, 1984; Shackleton, 1987; Linsley, 1996; Edwards *et al.*, 1993). Sediment profiling in the Strait of Malacca confirms transgression of this area during the Holocene, reaching a maximum height of +5 m above present-day levels at ~5–4 ka (Kudrass and Schlüter, 1994). Similar estimates are inferred from cores taken off the coast of northeastern Johore on the Malay Peninsula (Nossin, 1962). Radiocarbon dating of shells and corals from cores taken from the north coast of Java indicate maximum sea level at 3,650 C[14] yrs BP +3 m above present (Thommeret and Thommeret, 1978). Although minor fluctuations occurred throughout the Holocene, sea level reached a maximum height at ~5 ka of about 4–5 m above present level (Tjia, 1970, 1977, 1984; Verstappen, 1997), and since the mid-Holocene they have decreased to present-day levels (Tjia, 1970; Fleming *et al.*, 1998).

In summary, glacial periods in the northern hemisphere were associated in Southeast Asia with cooler and drier conditions, in which tropical forests were fragmented and seasonal woodland and savanna habitats expanded. During these glacial periods, beginning in the Late Pliocene, a belt of dry woodland and savanna probably extended southwards from the eastern side of the Malay Peninsula, through southern and eastern Borneo, to eastern Java and the Lesser Sunda Islands (Morley and Flenley, 1987; Heaney, 1991; Gathorne-Hardy et al., 2002). This dry zone, with habitats largely unsuitable for arboreal primates, in conjunction with major river systems (now largely submerged under South China and Java seas), presumably provided a significant barrier to migration (Heaney, 1991; Voris, 2000). Four distinct subprovinces resulted: (1) central and northern Borneo; (2) Malay Peninsula including Sumatra; (3) Mentawai Islands; and (4) Western Java (see Figure 2). These areas probably retained tropical forest refugia with isolated and impoverished primate communities (see Brandon-Jones, 1998, 2001). A combination of relictual survivorship, vicariance, and differential recolonization resulted in distinct primate faunas in each subprovince (Table 1).

The Mentawai Islands have four catarrhine species, all of which are endemic. They are probably specialized insular members of a primate fauna that probably inhabited Sundaland during the Late Pliocene (when the Mentawai Islands were connected to Sumatra by a landbridge across the Mentawai Straits) (Samuel et al., 1997). Borneo is unusual in having a high level of endemicity given its size (Table 1), but this is presumably due to its degree of isolation from other Sunda subprovinces during glacial periods, starting in the Late Pliocene. Sumatra and Java, by contrast, were more readily recolonized from mainland Southeast Asia via the Malay Peninsula (we find this scenario more plausible than Brandon-Jones' [1996, 1998, 2001] suggestion that the Mentawai Islands represented the relictual source for subsequent colonizations of Sumatra and Borneo), although Java sustained an impoverished and endemic primate fauna due to partial isolation and habitat differences.

THE ARCHAEOLOGICAL AND PALEONTOLOGICAL RECORD

Of the relatively few archaeological and paleontological localities on the major islands of Sundaland many have yielded the remains of non-human catarrhine primates. These provide an important line of evidence that helps in reconstructing the ecological and zoogeographic history of the region. A list of key localities is presented in Table 2 (see Figure 5 for locations). Unfortunately, few of the sites have been radiometrically dated; most have been correlated using

Table 1. List of extant non-human catarrhine primates on Sundaland[a]

	Malay Peninsula	Borneo	Sumatra	Mentawai Island	Java
Cercopithecidae					
Colobinae					
Presbytis comata (Javan surili)					X
Presbytis femoralis (Banded surili)[b]	X	X	X		
Presbytis frontata (White-fronted langur)		X			
Presbytis hosei (Hose's langur)		X			
Presbytis melalophos (Sumatran surili)			X		
Presbytis potenziani (Mentawai langur)				X	
Presbytis rubicunda (Maroon leaf monkey)		X			
Presbytis thomasi (Thomas's langur)			X		
Trachypithecus cristatus (Silvery lutung)	X	X	X		
Trachypithecus auratus (Javan lutung)					X
Trachypithecus obscurus (Dusky leaf monkey)	X				
Nasalis larvatus (Proboscis monkey)		X			
Simias concolor (Pig-tailed langur)				X	
Cercopithecinae					
Macaca arctoides (Stump-tailed macaque)	X				
Macaca fascicularis (Long-tailed macaque)	X	X	X		X
Macaca nemestrina (Sunda Pig-tailed macaque)	X	X	X		
Macaca pagensis (Mentawai macaque)				X	
Hylobatidae					
Hylobates agilis (Agile gibbon)	X		X		
Hylobates albibarbis (Bornean white-bearded gibbon)		X			
Hylobates lar (White-handed gibbon)	X		X		
Hylobates klossii (Kloss gibbon)				X	
Hylobates moloch (Silvery gibbon)					X
Hylobates muelleri (Müller's Bornean gibbon)		X			
Hylobates syndactylus (Siamang)	X		X		
Hominidae					
Pongo pygmaeus (Orang-utan)		X	X		
Total catarrhine species	9	11	10	4	4
Number of endemic species	0	6	2	4	3

[a] Sources: Medway (1970); Oates *et al.* (1994); Fooden (1975, 1995); Rowe (1996); Groves (2001a); Brandon-Jones *et al.* (2004). Common names after Groves (2001a).
[b] Includes *P. chrysomelas* and *P. siamensis* recognized as distinct species by Groves (2001a).

Table 2. List of archaeological and paleontological sites on Sunda Shelf islands that have yielded non-human catarrhine primates

Island	Locality	Inferred age	Key references
Borneo	Niah Cave (Sarawak)	Late Pleistocene–Early Holocene	Harrisson, 1958, 1970; Hooijer, 1960, 1961, 1962a,1962b; Medway, 1977; Harrison, 1996, 2000
	Paku Flats (Sarawak)	Late Pleistocene–Early Holocene	Everett, 1880; Harrison, 2000
	Jambusan (Sarawak)	Late Pleistocene–Early Holocene	Everett, 1880; Harrison, 2000
	Gua Sireh (Sarawak)	Early Holocene	Ipoi, 1993; Bellwood, 1997; Harrison, 2000
	Madai (Sabah)	Holocene	Bellwood, 1988; Cranbrook, 1988; Harrison, 1998, 2000
Java	Sangiran	?Late Pliocene–Middle Pleistocene	Aimi, 1981; Jablonski and Tyler, 1999
	Djetis	?Early–Middle Pleistocene	Hooijer, 1948
	Trinil	Middle Pleistocene	Hooijer, 1948, 1962a; van den Bergh et al., 2001
	Bangle	Middle Pleistocene	Hooijer, 1962a
	Soember Kepoeh	Middle Pleistocene	Hooijer, 1962a
	Tegoean	Middle Pleistocene	Hooijer, 1962a
	Saradan	Middle Pleistocene	Hooijer, 1962a
	Glagahombo	Middle Pleistocene	Aimi and Aziz, 1985
	Ndangklampok	Middle Pleistocene	Aimi and Aziz, 1985
	Kali Brangkal	Middle Pleistocene	Aimi and Aziz, 1985
	Ngandong	?Late Pleistocene	Aziz, 1989; van den Bergh et al., 2001
	Punung Fissures	Late Pleistocene	Hooijer, 1948; Badoux, 1959; van den Bergh et al., 2001
	Wajak	Holocene	Hooijer, 1962a; van den Brink, 1982; Aziz and De Vos, 1989
	Gua Jimbe	Holocene	Hooijer, 1962a
	Gua Ketjil	Holocene	Hooijer, 1962a
	Sampung	Holocene	Dammerman, 1934
Sumatra	Lida Ayer	Late Pleistocene–Early Holocene	Dubois, 1891; Hooijer, 1948, 1962a; Harrison, 2000
	Sibrambang	Late Pleistocene–Early Holocene	Dubois, 1891; Hooijer, 1948, 1962a; Harrison, 2000
	Djambu	Late Pleistocene–Early Holocene	Dubois, 1891; Hooijer, 1948, 1962a; Harrison, 2000

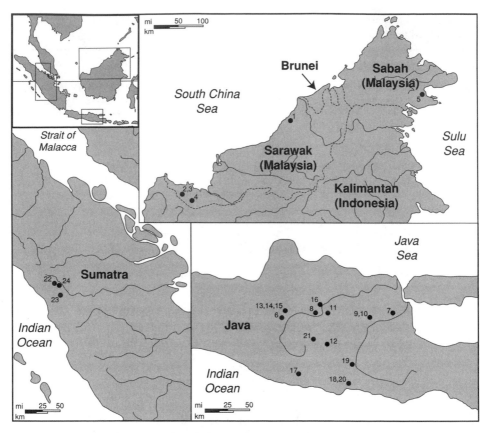

Figure 5. Map of principal Sundaic Islands showing location of sites discussed in text and listed in Table 2. Sites: 1. Niah Cave; 2. Paku Flats; 3. Jambusan; 4. Gua Sireh; 5. Madai; 6. Sangiran; 7. Djetis; 8. Trinil; 9. Bangle; 10. Soember Kepoeh; 11. Tegoean; 12. Saradan; 13. Glagahombo; 14. Ndangklampok; 15. Kali Brangkal; 16. Ngandong; 17. Punung; 18. Wajak; 19. Gua Jimbe; 20. Gua Ketjil; 21. Sampung; 22. Lida Ayer; 23. Sibrambang; 24. Djambu.

biostratigraphic data. The best series of dates is associated with Sangiran (Late Pliocene to Middle Pleistocene) and Niah Cave (Late Pleistocene to Holocene). ^{40}Ar/^{39}Ar dating of the Sangiran Dome succession has produced dates of ~0.8–1.6 Ma for the Kabuh Formation and ~1.6–2.0 Ma for the Pucangan Formation (Swisher *et al.*, 1994; Larick *et al.*, 2000). At Niah, a series of samples of charred bone and charcoal have yielded ^{14}C dates that indicate that the archaeological occupation of the site dates, more or less continuously, from older than 41,500 ± 1000 BP up to the present-day (Harrisson, 1959a; Harrison,

1996; Krigbaum, 2001). Radiocarbon dating at Gua Sireh and Madai Caves shows that occupation horizons with primates are younger than 10,500 BP (Bellwood, 1988; Ipoi, 1993; Harrison, 1998).

Of the nine genera of extant primates currently found on the Sunda Shelf islands, six are represented in the archaeological and paleontological record. Only *Tarsius, Nasalis,* and *Simias* are not recorded, while *Nycticebus* is known only from a few fragmentary specimens from Niah Cave in Sarawak (Medway, 1977) and from Holocene cave sites in Java (van den Bergh *et al.*, 2001). Each of the catarrhine genera known to occur—*Pongo, Hylobates, Macaca, Presbytis, and Trachypithecus*—are discussed in turn below.

Pongo

Two subspecies of orang-utan are generally distinguished—*Pongo pygmaeus pygmaeus* Linnaeus, 1760, and *Pongo pygmaeus abelii* Lesson, 1827, from Borneo and Sumatra respectively (von Koenigswald, 1982; Napier and Napier, 1985; Groves and Holthuis, 1985; Courtenay *et al.*, 1988; Groves, 1989). Recently, there has been some discussion concerning the possibility of recognizing additional subspecies or even separating the Sumatran and Bornean populations at the species level (Groves, 1986, 1989, 2001a; Courtenay *et al.*, 1988; Groves *et al.*, 1992; Rowe, 1996; Zhi *et al.*, 1996; Delgado and van Schaik, 2000; Brandon-Jones *et al.*, 2004). Certainly the two subspecies can be readily distinguished from each other based on hair color and length, distribution of facial hair, the size and shape of the throat pouches and cheek flanges in males, and various cranio-dental characteristics (MacKinnon, 1973; Groves, 1986, 2001a; Weitzel *et al.*, 1988; Courtenay *et al.*, 1988; Uchida, 1998). These morphological distinctions are supported by molecular and karyological data (Janczewski *et al.*, 1990; Ryder and Chemnick, 1993; Xu and Arnason, 1996; Zhi *et al.*, 1996; Karesh *et al.*, 1997; Rijksen and Meijaard, 1999). However, we conservatively subdivide extant orang-utans into two subspecies, and consider that the degree of morphological and molecular distinctiveness reflects the fact that these demes have been effectively isolated from one another for a considerable period of time, probably since the Early Pleistocene (Ryder and Chemnick, 1993; Zhi *et al.*, 1996; Karesh *et al.*, 1997).

The geographic range of the orang-utan is today limited to northern Sumatra and Borneo (von Koenigswald, 1982; Röhrer-Ertl, 1988; Groves, 1989, 2001a), but evidence from the paleontological and archaeological record shows that the species (or at least genus) was more widely distributed throughout Southeast Asia during the Late Quaternary. Fossil and subfossil material has been recovered from sites in southern China, Vietnam, Laos, Cambodia, Thailand, Sumatra, Java, and Borneo (Pei, 1935; Hooijer, 1948, 1961; Kahlke, 1972; Delson, 1977; Aigner, 1978; Han and Xu, 1985; Gu *et al.*, 1987; Olsen and Ciochon, 1990; Cuong, 1992; Nisbett and Ciochon, 1993; Schwartz *et al.*, 1994, 1995; Drawhorn, 1995; Tougard and Ducrocq, 1999; Bacon and Long, 2001). On mainland Southeast Asia, the remains of orang-utans have been found at Liucheng in southern China that date back to the Early Pleistocene (although de Vos [1984] has argued that the fauna also contains Middle to Late Pleistocene taxa). It also appears that the species survived until the latest Pleistocene in Vietnam (Tan, 1985; Schwartz *et al.*, 1995; Bacon and Long, 2001). The fossil record documenting the evolutionary history of orang-utans in the islands of Southeast Asia is much more scanty, however, and is restricted to occurrences on Java (the Late Pleistocene or Early Holocene Punung fissures of Gunung Kidul, and possibly also the Middle Pleistocene of Sangiran and Trinil), Sumatra (Early Holocene cave sites of Lida Ayer, Sibrambang and Djambu in the Padang Highlands) and Borneo (Late Pleistocene to Holocene of Niah Cave, Madai, Bau Caves and Gua Sireh) (Hooijer, 1948; Drawhorn, 1995; Harrison, 1996, 1998, 2000).

During the Early and Middle Pleistocene orang-utans were widely distributed throughout mainland Southeast Asia. Material from South China and Vietnam, assigned to *Pongo pygmaeus weidenreichi* (Hooijer, 1948), is ~20% larger on average in its dental dimensions than those of living orang-utans. Apparently, the geographic range of orang-utans extended onto the islands of Sundaland at this time. The oldest record of fossil orang-utans from Sundaland is probably from the Middle Pleistocene of Java. Two upper molars from Trinil, originally attributed to *Homo erectus* by Dubois (1896), have been considered by most subsequent workers to belong to *Pongo* (Hooijer, 1948). De Vos and Sondaar (1982) and de Vos (1983, 1984) argue, however, that the Trinil specimens are not orang-utans, and that this taxon is absent from the Middle Pleistocene faunas of Java, but one of us (TH) has examined the original specimens and agrees with Hooijer's (1948) assessment that the teeth most probably represent M^3 and M^4 of an orang-utan. Furthermore, von Koenigswald (1940), Widianto (1991), and Aziz and Saefudin (1996) have reported additional isolated teeth of fossil orang-utans from Middle Pleistocene sediments in the Sangiran Dome region of Java.

Dubois collected several thousand specimens of subfossil orang-utans from Early Holocene cave sites (i.e., Sibrambang, Lida Ayer, and Djambu) in the Padang Highlands of central Sumatra (Dubois, 1891; Hooijer, 1948). On the basis of their larger overall dental size (the teeth are ~15% on average larger in terms of occlusal area than those of the living subspecies [Harrison, 2000]), greater degree of canine, sexual dimorphism, and differences in the relative size of the teeth, Hooijer (1948) recognized a new subspecies—*Pongo pygmaeus palaeosumatrensis*. The small sample of orang-utan specimens from the Bau Cave sites in Sarawak (i.e., Paku Flats and Jambusan), of uncertain age, also have teeth that are comparable in size to *Pongo pygmaeus palaeosumatrensis*.

Orang-utans are found throughout the sequence at Niah Cave, and their remains are common (comprising more than 30% of non-human primate specimens) (Hooijer, 1961; Harrison, 1996, 2000). They have also been recovered from Madai Caves and Gua Sireh, but their remains are relatively much more rare than at Niah (only 9% and 3% of the primates respectively). These differences might reflect important ecological distinctions, but it is more likely that the very high frequency at Niah is due to the fact that the human occupants specialized in hunting orang-utans (Harrison, 2000). If this is so, it may account, at least in part, for the absence of orang-utans in the local environs of Niah today (apart from sporadic sightings the nearest modern day occurrence of orang-utans is more than 200 km from Niah) (Harrisson, 1959b; Reynolds, 1967; Rijksen and Meijaard, 1999).

During glacial periods, when the Sunda Shelf was exposed, orang-utans with a range that extended throughout mainland Southeast Asia as far north as southern China, were able to gain access into Sumatra, Borneo and Java. Despite intermittent land connections, however, a number of biogeographic barriers impeded free migration of orang-utans throughout the Sunda Shelf. These included major river systems that transected the present-day South China and Java seas, and a belt of drier more seasonal woodlands and grasslands that bordered the eastern edge of the Malay Peninsula, and continued onto the lowland areas between Sumatra and Borneo, through southern Kalimantan, and eastern Java and the Lesser Sunda Islands (see Figures 3 and 4; Morley and Flenley, 1987; Heaney, 1991). The population of orang-utans on the Malay Peninsula would have been easily able to colonize Sumatra during glacial times, but access to Borneo from Sumatra, presumably via the Bangka–Belitung–Karimata island chain, was more difficult because of the predominance of drier, more seasonal conditions. As a consequence of these biogeographical factors, Bornean orang-utans may have undergone a population bottleneck, as well as a greater degree of isolation and vicariance (Rijksen and Meijaard, 1999), and this has contributed to the degree of morphological and molecular separation between the two subspecies (as well as the differentiation between the individual populations on Borneo, i.e., *pygmaeus* in Sarawak and NW Kalimantan, *wurmbii* in SW Kalimantan, and *subsp. nov.* in Sabah and NE Kalimantan [see Groves, 2001a; Brandon-Jones *et al.*, 2004]). These factors also account for the unusually high level of endemism in the general primate (and broader mammalian) fauna (Table 1), despite the fact that Borneo is a relatively large island that has had frequent land connections with mainland Southeast Asia.

Hunting by humans may have been a contributing factor in the extirpation of orang-utans from south China, mainland Southeast Asia, and Java by the Early Holocene (Harrison, 1996, 1998, 2000; Delgado and van Schaik, 2000; Harrison *et al.*, 2002). The relictual populations of orang-utans survived in the perhumid tropical forests of Borneo and Sumatra until the present, probably because of several factors: (1) these populations were apparently smaller in body size and probably more committed to a fully arboreal habit than mainland orang-utans (Smith and Pilbeam, 1980; Harrison, 1996, 2000); (2) current evidence suggests that modern human hunter-gatherers were unable to exploit extensively tropical forests in Southeast Asia prior to the introduction of agriculture, because of the limited subsistence base that these ecosystems offer to obligate hunter-gatherers (Bailey *et al.*, 1989; Bailey and Headland, 1991). Thus, orang-utans were able to survive in Borneo and Sumatra because of their specialized arboreality and because of the low population densities of hominins on these islands (Harrison, 1996, 2000; Harrison *et al.*, 2002).

Hylobates

Seven species of hylobatids are currently recognized on Sundaland (Table 1). *Hylobates agilis, Hylobates lar,* and *Hylobates syndactylus* are found on both the Malay Peninsula and Sumatra, while *H. moloch* (Java), *H. klossii* (Mentawai Islands), *H. muelleri* (Borneo) and *H. albibarbis* (southwestern Borneo) are endemic species found on Sunda Shelf islands (Marshall and Sugardjito, 1986; Weitzel *et al.*, 1988; Rowe, 1996; Fleagle, 1999; Groves, 2001a; Brandon-Jones *et al.*, 2004).

Gibbons are rare at archaeological sites on the Sunda islands. At Niah, gibbons comprise only 0.5% of the total primate fauna, and they are known from just a few dental and postcranial specimens from other Early Holocene sites on Borneo (Harrison, 1998, 2000). Comparisons support Hooijer's suggestion that they are comparable in morphology and similar in size or slightly larger than extant *H. muelleri* (Hooijer, 1960, 1962a,b). Fossil gibbons are also scarce in the cave sites of similar age on Sumatra, and only a few isolated teeth have been recovered from Middle to Late Pleistocene localities in Java (Von Koenigswald, 1940; Badoux, 1959; Hooijer, 1960). Fossil and subfossil siamangs (*Hylobates syndactylus*) have been recovered from the Middle and Late Pleistocene of Java (where they are now locally extinct) and from Late Pleistocene sites on Sumatra (Von Koenigswald, 1940; Badoux, 1959; Hooijer, 1960). The specimens from Sumatra are somewhat larger than their modern counterparts, and have been assigned to a separate subspecies, *H. syndactylus subfossilis* (Hooijer, 1960).

Little about the zoogeography of gibbons and siamangs can be deduced from the current archaeological and paleontological evidence. Given the occurrence of *Hylobates* in Java during the Middle Pleistocene, they probably extended their range in the Sunda Shelf islands coincident with *Pongo* during the Pliocene, and speciated soon thereafter in their respective centers of endemicity (i.e., in Java, Borneo, and the Mentawai Islands). The underrepresentation of hylobatids in the archaeological samples is probably a reflection of the difficulty in hunting such fast-moving, upper canopy-dwelling primates, especially before the advent of bone projectile point technologies (Medway, 1959, 1977; Harrison, 1996). Given these findings, human hunting may be an important factor influencing the zoogeographic distribution of hominoids. During the Middle Pleistocene orang-utans and gibbons had similar geographic distributions in Southeast Asia, but only gibbons were able to maintain this range subsequent to the arrival of *Homo erectus* and *Homo sapiens* into the region. Humans may have been a contributing factor in the differential geographic distributions of gibbons and orang-utans, but other ecological variables may have been equally important (see Jablonski, 1998; Jablonski and Whitfort, 1999; Harcourt, 1999; Jablonski *et al.*, 2000; Harcourt and Schwartz, 2001).

Macaca

Four species of macaques occur in Sundaland, although *Macaca arctoides* is restricted to the northern limit of the region and *M. pagensis* is known only from Siberut, Sipora, and Pagai Islands in the Mentawai archipelago (Table 1).

The remaining two species, *M. nemestrina* and *M. fascicularis*, are widely distributed throughout the Sunda Shelf islands (Medway, 1970; Fooden, 1975, 1995, 2000; Brandon-Jones *et al.*, 2004). Unlike the hominoids and most colobines, *Macaca* has been able to colonize deep-water fringing islands beyond the Sunda Shelf, possibly by sweepstake dispersal, although human introductions cannot be discounted (Fooden, 1995; Heinsohn, 2001). *Macaca fascicularis* has a range that extends into the Nicobar Islands, north of the Mentawai Islands, the Lesser Sunda Islands, and the Philippines (Fooden and Albrecht, 1993; Fooden, 1995). *Macaca nemestrina* has been less successful in this regard, but a *nemestrina*-like ancestor was able to colonize Sulawesi, presumably from Borneo (Rosenblum *et al.*, 1997a; Morales and Melnick, 1998; Groves, 2001b).

Macaca fascicularis and *M. nemestrina* have been recovered from archaeological sites in Borneo dating back to the Late Pleistocene (Hooijer, 1962a,b; Harrison, 2000; Table 2). At Niah, *M. fascicularis* is much more common than *M. nemestrina* and among primates is second in abundance only to *Pongo*. Previous analyses have shown that the dental remains of long-tailed macaques are on average 13% larger than their modern conspecifics, with a gradual diminution in size through time (Harrison, 1996). A possible explanation is provided by the ecogeographic relationship between cooler climatic conditions during the Late Pleistocene and Bergmann's rule. The specimens from Niah correspond in dental size with those living in Thailand today, where annual temperatures are similar to those inferred for northern Borneo during the LGM ($\sim$5–6°C lower than at present) (Fooden and Albrecht, 1993; Harrison, 1996). All of the specimens of *Macaca nemestrina* from Niah fall within the range of modern Bornean pig-tailed macaques.

The Early Holocene sites in Sumatra have produced a large collection of macaques. By comparison to Niah and to the modern communities on Sumatra and Borneo, *M. nemestrina* is more common than *M. fascicularis*. This may be due to ecological differences, but more likely reflects differences in human hunting strategies (Harrison, 1998). Fossil macaques from Middle Pleistocene to Holocene sites on Java are all referable to *Macaca fascicularis mordax* that occurs today on Java and Bali, although the Middle Pleistocene material tends to be somewhat larger than their extant counterparts (Hooijer, 1962a; Aziz, 1989). *Macaca nemestrina* is absent from the modern primate fauna of Java, but a single specimen of this species has been recovered from the Middle Pleistocene of Sangiran (Aimi, 1981), which indicates that it did reach Java during glacial periods. Reis & Garong (2001) have recently reported the recovery of specimens of *Macaca fascicularis* from early Holocene cave sites on Palawan.

Presbytis

At least eight species of *Presbytis* occur today on the Sunda Shelf islands (Table 1), and reconstructions of their biogeographic history suggest a complex pattern of dispersal and speciation (Wilson and Wilson, 1975; Brandon-Jones, 1996, 1998, 2001). Unfortunately, the paleontological and archaeological record does not help clarify the problem, mainly because the fossil record for *Presbytis* is too sparse and the material is too fragmentary in most cases to be assigned to a particular species with any degree of confidence. Brandon-Jones (1996, 1998, 2001) has suggested that the glacial peaks during the later Pleistocene had a profound impact on the diversity and zoogeographic distribution of colobines in Sundaland. It seems more likely, however, given the morphological and taxonomic differentiation of members of this clade, and what we know about the timing of differentiation of other Southeast Asian primate species, that most of the speciation events leading to modern *Presbytis* species on the Sunda Shelf occurred deeper in time, perhaps as far back as the Pliocene or Early Pleistocene.

Presbytis is well represented at Niah, where it comprises 23% of the primate fauna. As noted above, it is difficult to assign these specimens to a species. Of the four extant species on Borneo, all but *P. frontata* occur today in the local environs of Niah (Payne *et al.*, 1985). *Presbytis rubicunda*, the maroon leaf-monkey, is the most common in the surrounding forests, but *P. hosei* and *P. femoralis* occur in the general area. Comparisons show that the material is most similar in dental morphology to *P. rubicunda* and *P. hosei*, particularly the former. The current distribution of these two species, and the relatively large size of the teeth from Niah, favor their attribution to *P. rubicunda*, although it is likely that the sample contains a mixed assemblage, including some specimens of *P. hosei*. Comparisons of dental size show that the Niah material is larger on average than all modern day species from Borneo. It would seem that *Presbytis*, like *Macaca fascicularis*, has undergone a diminution in size during the Late Pleistocene.

The colobines from Gua Sireh comprise 35% of the primate fauna. All specimens studied from that site appear to be assignable to a single species of *Presbytis*, although species identification is uncertain. The only *Presbytis* living today in western Sarawak is *P. femoralis*, and the Gua Sireh material is consistent in size and morphology with this species. *Presbytis* is also known from the Middle Pleistocene site of Soember Kepoeh in Central Java (Hooijer, 1962a), which shows that the taxon occurred east of its current distribution in the past. There are no examples of *Presbytis* from Late Pleistocene–Early Holocene sites in central Java, suggesting that local extinction had already occurred by this time. *Presbytis* is also recorded from Bau in Sarawak and from Early Holocene sites in central Sumatra, but again the specimens cannot be readily identified to species.

Trachypithecus

Two species of *Trachypithecus* are found today on the Sunda Shelf islands—*T. cristatus* on Borneo and Sumatra (as well as mainland Southeast Asia), and *T. auratus* on Java, Bali, and Lombok (Oates *et al.*, 1994; Rowe, 1996; Rosenblum *et al.*, 1997b). *Trachypithecus* is the only colobine that extends its geographic distribution beyond Wallace's Line, although there is the possibility that it was introduced on Lombok (Brandon-Jones, 1998; Heinsohn, 2001). Clearly, an important factor influencing the movement of primates between different subprovinces is their relative abilities to colonize new areas, particularly across stretches of ocean. Apart from modern humans, colobines and hominoids were apparently unable to extend their ranges beyond the Sunda Shelf, whereas macaques were able to establish themselves in the Philippines, Sulawesi, and the Lesser Sunda Islands, as far East as Timor and Flores.

Trachypithecus is well represented at Niah (comprising 13% of the primate fauna), but specimens are not found uniformly throughout the deposits. *Trachypithecus* is absent from depths greater than 60", whereas *Presbytis* is found throughout the sequence. This could indicate a change in hunting strategies or dietary preferences (Harrison, 1996). However, the appearance may coincide with the LGM when Niah was located ~200 km inland (presently, it is only ~17 km from the coast). Since *T. cristatus* prefers riverine forests, peat swamps, and mangrove, it is found mainly on the coastal plain of Borneo. It is likely that during the LGM, the human inhabitants of Niah were beyond the immediate range of *Trachypithecus*, but as sea levels rose at the end of the Pleistocene it became increasingly possible for hunters to obtain this species close to Niah.

Dental comparisons show that *Trachypithecus* from Niah is significantly larger than extant *Trachypithecus cristatus* from Borneo (Harrison, 2000). Nevertheless, the Niah material is still much smaller than fossil *Trachypithecus* from Java attributed to *T. auratus robustus* (Tegoean, Middle Pleistocene) and *T. auratus sangiranensis* (Sangiran, ? Late Pliocene—see Larick *et al.*, 2000 for a critical review of the dating), which are both larger than the modern Javan lutung (Hooijer, 1962a; Jablonski and Tyler, 1999). Given the early dates of these latter taxa, and consistent with the molecular evidence (Rosenblum *et al.*, 1997b), it can be inferred that the *T. auratus/cristatus* clade originated in Java, and later spread to Sumatra and Borneo, and eventually to the Malay Peninsula where it encountered *T. obscurus*.

SYNTHESIS AND DISCUSSION

The study of the non-human catarrhine fauna from archaeological and pale ontological sites on the major islands on the Sunda Shelf provides a unique diachronic perspective on the zoogeography and ecology of Sundaland. In particular, it provides valuable insight into the significance of two major extrinsic factors that may have had a profound impact on the distribution of primates in the region: (1) the arrival of hominins since the late Pliocene or early Pleistocene (who influenced primate distributions through hunting, habitat disturbance and destruction, and the translocation of primates as "ethnotramps" [Heinsohn, 2001]); and (2) climatic perturbations associated with the onset of the major phase of glacial cycles in the northern hemisphere from the Late Pliocene onwards.

Of the seven genera of catarrhines living today on Sundaland, five are found in the archaeological and paleontological record—only *Simias* and *Nasalis* are not represented. The geographic range of *Pongo* is restricted today to northern Sumatra and Borneo, but the paleontological and archaeological evidence shows that the species was widely distributed throughout the Sunda Shelf islands, as well as on the mainland of Southeast Asia during the Pleistocene. During glacial periods, when tropical forests were fragmented and the exposed Sunda Shelf was transected by major river systems, populations of orang-utans on Borneo, Sumatra, and Java were probably isolated and severely reduced in size. It is likely that intermittent land connections between Sumatra and the Malay Peninsula permitted gene flow between orang-utan populations across the Strait of Malacca. However, the combination of a belt of drier, more seasonal woodland and grassland and major rivers separating Sumatra from Borneo during glacial times, when the two landmasses were connected, represented a major impediment to gene flow between orang-utan populations. The bottleneck and isolation that occurred in the orang-utan populations on Borneo, probably during the Early Pleistocene, is reflected in the morphological and molecular distinctiveness of the endemic subspecies and the three major populations on Borneo (Ryder and Chemnick, 1993; Xu and Arnason, 1996; Zhi *et al.*, 1996; Karesh *et al.*, 1997; Rijksen and Meijaard, 1999; Groves, 2001a). Ryder and Chemnick (1993) and Zhi *et al.* (1996) have estimated that the separation of the Sumatran and Bornean populations occurred at ~1.5–1.7 Ma. This divergence may have been precipitated by ecological changes that accompanied the Eburonian cold phase at the start of the Early Pleistocene (at ~1.8 Ma). Orang-utans survived on Java only until the Late Pleistocene or Early Holocene. It is possible that hunting by humans was a contributing factor in the extirpation of orang-utans from mainland Southeast Asia and Java, and that relictual populations survived on Borneo and Sumatra only because of their specialized arboreality and the low population densities of humans on these islands.

Based on current evidence from paleontological, molecular, and paleoecological studies, the following scenario is proposed to explain the present-day zoogeographic distribution of *Pongo*. During the Pliocene, orang-utans were widely distributed throughout Southeast Asia, and they probably entered Sundaland during the Pretiglian cold phase (at ~2.7 Ma) in the Late Pliocene, when sea levels were at least 100 m below present-day levels. At the start of the Pleistocene, correlated with a major cold climate peak at ~1.8 Ma, and associated with increased seasonality in tropical Southeast Asia, orang-utan populations became fragmented and isolated. During this period, the population on Borneo (and probably that on Java) became differentiated genetically, morphologically, and behaviorally, from the orang-utan population on Sumatra and mainland Southeast Asia (this same biogeographical pattern is mirrored in the subspecies of *Nycticebus coucang*—with *N. c. coucang* on Sumatra, Natuna, and the Malay Peninsula, *N. c. javanicus* on Java, and *N. c. menagensis* on Borneo [Brandon-Jones *et al.*, 2004]). Gene flow between the Sumatran and mainland populations probably continued throughout the Pleistocene as a consequence of intermittent landbridge connections. By the end of the Pleistocene, orang-utans had become extinct throughout mainland Southeast Asia, leaving relictual and well differentiated populations on Sumatra, Borneo, and Java. At the beginning of the Holocene, extirpation of the Javan population resulted in the present-day distribution of orang-utans.

Gibbons and siamangs are rare at archaeological or paleontological sites on the Sunda islands. Their underrepresentation at archaeological sites may be associated with the difficulty that humans experienced in hunting them prior to the advent of projectile point technologies in the Late Pleistocene (see Harrison 1996, 1998, 2000). Unfortunately, little can be deduced about their zoogeography from the limited evidence available. Given the occurrence of *Hylobates* in Java during the Middle Pleistocene, it is likely that they extended their range into Sundaland at about the same time as *Pongo* (i.e., during the Late Pliocene). The common ancestor of the *lar*-group gibbons probably colonized the major islands at this time, including Sumatra, Java, Borneo, and the Mentawai Islands, and speciated soon thereafter in their respective centers of endemism. This is consistent with recent molecular studies (Hayashi *et al.*, 1995; Hall *et al.*, 1998) that indicate that the *lar*-group diverged during the Pliocene, probably around 3.5 Ma. They may have entered the Sunda Shelf islands during a short cold phase at ~3.4–3.6 Ma, or more likely at the start of the Pretiglian cold phase at ~2.8 Ma, when sea levels fell by more than 50 m (Prentice and Denton, 1988), and landbridge connections were formed between mainland Southeast Asia and the major Sunda islands. Molecular evidence suggests that the sub-genus *Symphalangus* diverged from the other hylobatids by ~5–6 Ma (Hayashi *et al.*, 1995; Hall *et al.*, 1998), but their limited distribution in the Malay Peninsula and Sumatra today, their occurrence in Java during the Pleistocene, and their absence from the Mentawai Islands and Borneo, suggests that their initial range expansion from mainland Southeast Asia occurred later than the *lar*-group gibbons.

The inferred timing and pattern of colonization of the Sunda islands by *Macaca* is complex (Fooden, 1975, 1995; Eudy, 1980; Delson, 1980; Abegg & Thierry, 2002; Evans *et al.*, 2003). Unlike the hominoids and most colobines, *Macaca* has been able to colonize deep-water fringing islands beyond the Sunda Shelf. The absence of *M. fascicularis* in Sulawesi suggests that the progenitor of *M. nemestrina* may have arrived in Borneo prior to *M. fascicularis*. The occurrence of *M. pagensis* on the Mentawai islands, derived from a *M. nemestrina*-like ancestor, in the absence of *M. fascicularis*, leads to a similar conclusion (see Delson, 1980; Abegg and Thierry, 2002 for more detailed reviews). Tosi *et al.* (2003) have inferred from molecular evidence that the Sulawesi macaques separated from *M. nemestrina* at 2.6–3.0 Ma. Similarly, Roos *et al.*, (2003) have estimated a divergence of *M. pagensis* from *M. nemestrina* at 2.2. Ma (with a second migratory event into Siberut at 1.1 Ma). This suggests that *Macaca nemestrina* colonized the Sunda islands during the mid-Pliocene, possibly associated with the onset of the Pretiglian cold phase at ~2.8 Ma (Prentice and

Denton, 1988), when sea levels dropped by more than 100 m. *Macaca fascicularis* followed, probably during the late Pliocene or early Pleistocene. Molecular data indicate that *M. fascicularis* diverged from the other members of the *fascicularis* species group at ~2.2–2.5 Ma (Tosi *et al.*, 2003), and this implies that *M. fascicularis* probably entered the Sunda islands from mainland Southeast Asia at the end of the Pretiglian cold phase at 2.3 Ma or during the Eburonian cold phase starting at ~2.0 Ma (Prentice and Denton, 1988). The later arrival of *M. fascicularis* into the region, at a time when sea level regressions were less dramatic, prevented it from colonizing deep-water fringing islands, such as Sulawesi and Mentawai Islands, even though its occurrence on numerous present-day islands shows that the species is a very successful island colonizer (Fooden, 1995; Abegg and Thierry, 2002).

Brandon-Jones (1996, 1998, 2001) has suggested that the glacial peaks during the later Pleistocene had a profound impact on the diversity and zoogeographic distribution of colobines on Sundaland. However, it seems more likely, given the morphological and taxonomic differentiation of the clade, and based on the fossil evidence available, that most of the speciation events leading to extant *Presbytis* and *Trachypithecus* species on the Sunda Shelf occurred more distant in time, probably as far back as the Pliocene and Early Pleistocene. It is suggested here that the *Trachypithecus auratus/cristatus* clade originated in Java, probably during the late Pliocene, and later spread to Sumatra, Borneo, the Natuna Islands, and the Malay Peninsula. The zoogeographic relationships of *Presbytis* spp. are complex, and much better paleontological and molecular data are needed to reconstruct their evolutionary history. Even so, the last common ancestor of *Presbytis* spp. was probably present on one of the larger Sunda Shelf islands, probably Borneo, during the Pliocene, and this gave rise to the various endemic species on Borneo, Java, Sumatra, and the Mentawai Islands, and the genus later extended its range from Sumatra onto the Malay Peninsula. Given that the evidence suggests that *Hylobates* and *Macaca* colonized the Mentawai islands during the Pretiglian cold phase at ~2.5–2.8 Ma, and that there were probably no subsequent influxes of catarrhine primates (but see Roos *et al.*, 2003 for an alternative scenario for *Macaca*), it seems that *Presbytis* (and *Simias*) were already present on the Sunda Shelf islands by mid-Pliocene times.

Hence, it can be inferred that by the Late Pliocene the main islands of the Sunda Shelf had a primate fauna that included *Pongo pygmaeus* (Sumatra, Java, and Borneo), *Hylobates* spp. of the *lar*-group (Sumatra, Mentawai Islands, Borneo, and Java), *Macaca nemestrina* (Sumatra, Mentawai Islands, Borneo, and Java), the common ancestor of the *Trachypithecus auratus/cristatus* clade (Java and Sumatra), and *Presbytis* spp. (Sumatra, Mentawai Islands, Borneo, Sumatra, and Java). Most of these taxa probably arrived during the Pretiglian cold phase, starting at ~2.8 Ma, when sea levels fell by more than 100 m, although it is conceivable that some may have arrived earlier, during a shorter cold phase at ~3.4–3.6 Ma, when sea levels declined by at least 50 m. Although supporting evidence is lacking, it is also highly probable that *Nasalis larvatus* (Borneo) and *Simias concolor* (Mentawai Islands) were already present as endemic taxa in the Late Pliocene, and that their last common ancestor had arrived in the Sunda islands by the early Pliocene. Soon after this initial period of colonization, *Hylobates* and *Presbytis* underwent rapid speciation as a consequence of vicariance and relictual survivorship, giving rise to *P. thomasi* on Sumatra, *H. klossii* and *P. potenziani* on the Mentawai Islands, *H. albibarbis, H. muelleri, P. hosei, P. frontata,* and *P. rubicunda* on Borneo, and *H. moloch* and *P. comata* on Java. During the Late Pliocene and Early Pleistocene, probably associated with the Eburonian cold climate maximum at ~1.8 Ma, *Presbytis melalophos* and *P. femoralis,* along with *Macaca fascicularis,* colonized Sumatra, the Natuna Islands, and Borneo from the Malay Peninsula. At about the same time the isolated orang-utan populations on Sumatra, Java, and Borneo were beginning to differentiate from each other. *Hylobates lar, H. agilis,* and *H. syndactylus* extended their range from the Malay Pensinsula into Sumatra (and Java), probably during the Middle to Late Pleistocene, coincident with the arrival of *Trachypithecus cristatus* on mainland Southeast Asia. Meanwhile, *Pongo pygmaeus, Hylobates syndactylus,* and *Macaca nemestrina* were extirpated on Java, probably as a consequence of a combination of ecological changes and early hominin incursions.

ACKNOWLEDGEMENTS

We thank the following individuals and institutions for permission to study material in their care: Peter Kedit and Ipoi Datan, Sarawak Museum, Kuching; Jon de Vos, Rijksmuseum van Natuurlijke Historie, Leiden; Jerry Hooker, Peter Andrews and Paula Jenkins, Natural History Museum, London; Guy Musser, Ross MacPhee and Nancy Simmons, American Museum of Natural History, New York; and C.M. Yang, Zoological Reference Collection, National University of Singapore. We are grateful to Peter Bellwood, Doug Brandon-Jones, Russell Ciochon, Ipoi Datan, Eric Delson, Jon de Vos, John Fleagle, Nina Jablonski,

Don Melnick, Denise Su, and Gu Yumin for discussions, ideas, and helpful suggestions. We appreciate the constructive comments and advice of three anonymous reviewers that helped improve the quality and substance of this chapter. Special thanks go to Shawn Lehman and John Fleagle for inviting us to participate in the symposium on primate biogeography at the AAPA meetings in San Antonio, and to write this contribution for the accompanying volume. This research was supported by travel grants from New York University and an NYU Research Challenge Fund grant (to TH).

REFERENCES

Abegg, C. and Thierry, B. 2002, Macaque evolution and dispersal in insular south-east Asia. *Biol. J. Linnean Soc.* 75: 555–576.

Aigner, J. S. 1978, Pleistocene faunal and cultural stations in South China, in: Ikawa-Smith, F., ed., *Early Palaeolithic in South and East Asia.* Mouton, The Hague, pp. 129–160.

Aimi, M. 1981, Fossil *Macaca nemestrina* (Linnaeus, 1766) from Java, Indonesia. *Primates* 22:409–413.

Aimi, M. and Aziz, F. 1985, Vertebrate fossils from the Sangiran Dome, Mojokerto, Trinil and Sambungmacan areas, in: Watanabe, N. and Kadar, D., eds., *Quaternary Geology of the Hominid Fossil Bearing Formations in Java.* Special Publication No. 4, Geological Research and Development Centre, Indonesia, pp. 155–168.

Aldiss, D. T. and Ghazali, S. A. 1984, The regional geology and evolution of the Toba volcanotectonic depression, Indonesia. *J. Geol. Soc. London* 141:487–500.

Ashton, P. S. 1972, The Quaternary geomorphological history of Western Malesia and lowland forest phytogeography. *Transactions Second Aberdeen-Hull Symposium on Malesian Ecology*, Misc. Ser. 13:35–62.

Audley-Charles, M. G. 1981, Geological history of the region of Wallace's line, in: Whitmore, T. C., ed., *Wallace's Line and Plate Tectonics.* Clarendon Press, Oxford, pp. 24–35.

Audley-Charles, M. G. 1987, Dispersal of Gondwanaland: Relevance to evolution of the angiosperms, in: Whitmore, T. C., ed., *Biogeographical Evolution of the Malay Archipelago.* Clarendon Press, Oxford, pp. 5–25.

Aziz, F. 1989, *Macaca fascicularis* (Raffles) from Ngandong, East Java. *Publication of the Geological Research and Development Centre, Bandung, Indonesia, Paleontology Series* 5:50–56.

Aziz, F. and de Vos, J. 1989, Rediscovery of the Wajak site (Java, Indonesia). *J. Anthropol. Soc. Nippon* 97:133–144.

Aziz, F. and Saefudin, I. 1996, An isolated tooth of orang utan (*Pongo* pygmaeus) from the Sangiran area, Central Java, Indonesia. *Publication of the Geological Research and Development Centre, Bandung, Indonesia, Paleontology Series* 8:47–50.

Bacon, A-M. and Long, V. T. 2001, The first discovery of a complete skeleton of a fossil orang-utan in a cave of the Hoa Binh Province, Vietnam. *J. Hum. Evol.* 41:227–241.

Badoux, D. M. 1959, *Fossil Mammals from two Fissure Deposits at Punung (Java)*. Kemink, Utrecht.

Bailey, R., Head, G., Jenike, M., Owen, B., Rechtman, R., and Zechenter, E. 1989, Hunting and gathering in tropical rain forest: Is it possible? *Am. Anthropol.* 91:59–82.

Bailey, R. C. and Headland, T. N. 1991, The tropical rain forest: Is it a productive environment for human foragers. *Hum. Ecol.* 19:261–285.

Barham, A. J. and Harris, D. R. 1983, Prehistory and paleoecology of Torres Strait, in: Masters, P. M. and Fleming, N. C., eds., *Quaternary Coastlines and Marine Archaeology*. Academic Press, London, pp. 529–557.

Beck, J. W., Recy, J., Taylor, F., Edwards, R. L., and Cabioch, G. 1997, Abrupt changes in early Holocene tropical sea surface temperature derived from coral records. *Nature* 385:705–707.

Bellwood, P. 1988, The Madai excavations: Sites MAD 1, MAD 2, and MAD 3, in: Bellwood, P., ed., *Archaeological Research in South-Eastern Sabah*. Sabah Museum Monograph, Vol 2, pp. 97–127.

Bellwood, P. 1997, *Prehistory of the Indo-Malaysian Archipelago*. Revised Edition. University of Hawai'i Press, Honolulu.

Biswas, B. 1973, Quaternary changes in sea-level in the South China Sea. *Bull. Geol. Soc. Malaysia* 6:229–256.

Brandon-Jones, D. 1996, The Asian Colobinae (Mammalia: Cercopithecidae) as indicators of Quaternary climatic change. *Biol. J. Linnean Soc.* 59:327–350.

Brandon-Jones, D. 1998, Pre-glacial Bornean primate impoverishment and Wallace's line, in: Hall, R. and Holloway, J. D., eds., *Biogeography and Geological Evolution of Southeast Asia*, Backhuys, Leiden, pp. 393–404.

Brandon-Jones, D. 2001, Borneo as a biogeographic barrier to Asian-Australasian migration, in: Metcalfe, I., Smith, J. M. B., Morwood, M., and Davidson, I., eds., *Faunal and Floral Migrations and Evolution in SE Asia-Australasia*, A. A. Balkema, Lisse, pp. 333–342.

Brandon-Jones, D., Eudey, A. A., Geissmann, T., Groves, C. P., Melnick, D. J., Morales, J. C., Shekelle, M., and Stewart, C.-B. 2004, Asian primate classification. *Int. J. Primatol.* 25:97–164.

Broecker, W. 1996, Glacial climate in the tropics. *Science* 272:1902–1904.

Bühring, C. and Sarnthein, M. 2000, Toba ash layers in the South China Sea: Evidence of contrasting wind directions during eruption ca. 74 ka. *Geology* 28:275–278.

Bush, A. B. G. and Philander, S. G. H. 1998, The role of ocean-atmosphere interactions in tropical cooling during the last glacial maximum. *Science* 279:1341–1344.

Chappell, J. and Shackleton, N. J. 1986, Oxygen isotopes and sea level. *Nature* 324:137–140.

Chappell, J. 1994, Upper Quaternary sea levels, coral terraces, oxygen isotopes and deep-sea temperatures. *J. Geogr.* 103:828–840.

Chappell, J., Omura, A., Esat, T., McCulloch, M., Pandolfi, J., Ota, Y., and Pillans, B. 1996, Reconciliation of late Quaternary sea levels derived from coral terraces at Huon Peninsula with deep sea oxygen isotope records. *Earth Planet. Sci. Let.* 141:227–236.

Chesner, C. A., Rose, W. I., Drake, A. D. R., and Westgate, J. A. 1991, Eruptive history of Earth's largest Quaternary caldera (Toba, Indonesia) clarified. *Geology* 19:200–203.

Courtenay, J., Groves, C., and Andrews, P. 1988, Inter- or intra-island variation? An assessment of the differences between Bornean and Sumatran orang-utans, in: Schwartz, J. H., ed., *Orang-utan Biology*. Oxford University Press, Oxford, pp. 19–29.

Cranbrook, Earl of 1988, Report on bones from the Madai and Baturong Cave excavations, in: Bellwood, P., ed., *Archaeological Research in South-Eastern Sabah*. Sabah Museum Monograph, Vol. 2, pp. 142–154.

Crowley, T. J. 1991, Modeling Pliocene warmth. *Quat. Sci. Rev.* 10:275–282.

Crowley, T. J. 1996, Pliocene climates: the nature of the problem. *Marine Micropaleontol.* 27:3–12.

Cuong, N. L. 1992, A reconsideration of the chronology of hominid fossils in Vietnam, in: Akazawa, T., Aoki, K., and Kimura, T., eds., *The Evolution and Dispersal of Modern Humans in Asia*. Hokusen-Sha, Tokyo, pp. 321–335.

Dammerman, K. W. 1934, On prehistoric mammals from the Sampoeng Cave, central Java. *Truebia* 14:477–486.

Darlington, P. J. 1980, *Zoogeography: The Geographical Distribution of Animals*. John Wiley, New York.

Delgado, R. A. and van Schaik, C. P. 2000, The behavioral ecology and conservation of the orangutan (*Pongo pygmaeus*): A tale of two islands. *Evol. Anthropol.* 9:201–218.

de Vos, J. 1983, The *Pongo* faunas from Java and Sumatra and their significance for biostratigraphical and paleo-ecological interpretations. *Proc. Koninkl. Ned. Akad. Wetensch.* [B] 86:417–425.

de Vos, J. 1984, Reconsideration of Pleistocene cave faunas from South China and their relation to the faunas from Java. *Cour. Forsch. Inst. Senckenberg* 69:259–266.

de Vos, J. and Sondaar, P. 1982, The importance of the Dubois collection reconsidered. *Mod. Quat. Res. SE Asia* 7:35–63.

Debaveye J., De Dapper, M., De Paepe, P., and Gybels, R. 1986, Quaternary volcanic ash deposits in the Padang Terap district, Kedah, peninsular Malaysia. *Bull. Geol. Soc. Malaysia* 19:533–549.

Delson, E. 1977, Vertebrate paleontology, especially of non-human primates, in China, in: Howells, W. W. and Tsuchitani, P. J., eds., *Paleoanthropology in the People's Republic of China*. National Academy of Sciences, Washington DC , pp. 40–65.

Delson, E. 1980, Fossil macaques, phyletic relationships and a scenario of deployment, in: Lindberg, D. G., ed., *The Macaques: Studies in Ecology, Behavior and Evolution*. Van Nostrand Reinhold, New York, pp. 10–30.

Dickerson, R. E. 1928, Distribution of life in the Philippines. *Monograph of the Bureau of Science (Manila)* 2, 1–322.

Diehl, J. F., Onstott, T. C., Chesner, C. A., and Knight, M. D. 1987, No short reversals of Brunhes age recorded in the Toba tuffs, north Sumatra, Indonesia. *Geophys. Res. Let.* 14:753–756.

Dobby, E. H. G. 1960, *Southeast Asia*. University of London Press, London.

Dowsett, H., Barron, J., and Poore, R. 1996, Middle Pliocene sea surface temperatures: A global reconstruction. *Marine Micropaleontol.* 27:13–25.

Dowsett, H. J., Cronin, T. M., Poore, R. Z., Thompson, R. S., Whatley, R. C., and Wood, A. M. 1992, Micropaleontological evidence for increased meridional heat transport in the North Atlantic Ocean during the Pliocene. *Science* 258:1133–1135.

Drawhorn, G. M. 1995, The Systematics and Paleodemography of Fossil Orangutans (Genus *Pongo*). Ph.D. Dissertation, University of California, Davis.

Dubois, E. 1891, Voorloopig bericht omtrent het onderzoek naar de pleistocene en tertiaire Vertebraten-Fauna van Sumatra en Java, gedurende het jaar 1890. *Natuurk. Tijdschr. Ned. Indië* 51:93–100.

Dubois, E. 1896, On *Pithecanthropus erectus*, a transitional form between man and the apes. *Scientific Transactions of the Royal Dublin Society, Series 2*, 6:1–18.

Edwards, R. L., Beck, J. W., Burr, G. S., Donahue, D. J., Chappell, J., Bloom, A. L., Druffel, E. R. M., and Taylor, F. W. 1993, A large drop in atmospheric $^{14}C/^{12}C$ and reduced melting in the Younger Dryas, documented with ^{230}Th ages of corals. *Science* 260:962–968.

Emery, K. O., Niino, Hiroshi, and Sullivan, B. 1971, Post-Pleistocene levels of the East China Sea, in: Turekian, K. K., ed., *The Late Cenozoic Glacial Ages*. Yale University Press, New Haven, pp. 381–390.

Evans, B. J., Supriatna, J., Andayani, N., and Melnick, D. J. 2003, Diversification of Sulawesi macaque monkeys: Decoupled evolution of mitochondrial and autosomal DNA. *Evolution* 57:1931–1946.

Everett, A. H. 1888, Report on the exploration of the caves of Borneo. *J. Straits Br. Roy. Asiatic Soc.* 6:274–284.

Eudey, A. A. 1980, Pleistocene glacial phenomena and the evolution of Asian macaques, in: Lindberg, D. G., ed., *The Macaques: Studies in Ecology, Behavior and Evolution.* Van Nostrand Reinhold, New York, pp. 52–83.

Fairbanks, R. G. 1989, A 17,000-year glacio-eustatic sea level record: Influence of glacial melting rates on the Younger Dryas event and deep-ocean circulation. *Nature* 342:637–642.

Fairbridge, R. W. 1953, The Sahul Shelf, northern Australia, its structure and geological relationships. *J. Roy. Soc. West. Australia* 37:1–33.

Fleagle, J. G. 1999, *Primate Adaptation and Evolution.* 2nd edn., Academic Press, San Diego.

Fleming, K., Johnston, P., Zwartz, D., Yokoyama, Y., Lambeck, K., and Chappell, J. 1998, Refining the eustatic sea-level curve since the Last Glacial Maximum using far- and intermediate-field sites. *Earth Planet. Sci. Let.* 163:327–342.

Flenley, J. R. 1979, *The Equatorial Rain Forest: A Geological History.* Butterworths, London.

Flenley, J. R. 1985, Quaternary vegetational and climatic history of island Southeast Asia. *Mod. Quat. Res. SE Asia* 9:55–63.

Flint, R. F. 1971, *Glacial and Quaternary Geology.* Wiley, New York.

Flohn, H. 1981, Tropical climate variations during late Pleistocene and early Holocene, in: Berger, A., ed., *Climatic Variations and Variability: Facts and Theories.* D. Reidel, Boston, pp. 233–242.

Fooden, J. 1975, Taxonomy and evolution of liontail and pigtail macaques (Primates: Cercopithecidae). *Fieldiana Zool.* 67:1–169.

Fooden, J. 1995, Systematic review of southeast Asian longtail macaques, *Macaca fascicularis* (Raffles, 1821). *Fieldiana Zool.* 81:1–206.

Fooden, J. 2000, Systematic review of the rhesus macaque, *Macaca mulatta* (Zimmermann, 1780). *Fieldiana Zool.* 96:1–180.

Fooden, J. and Albrecht, G. H. 1993, Latitudinal and insular variation of skull size in crabeating macaques (Primates, Cercopithecidae: *Macaca fascicularis*). *Am. J. Phys. Anthropol.* 92:521–538.

Gagan, M. K., Ayliffe, L. K., Hopley, D., Cali, J. A., Mortimer, G. E., Chappell, J., McCulloch, M. T., and Head, M. J. 1998, Temperature and surface-ocean water balance of the mid-Holocene tropical western Pacific. *Science* 279:1014–1018.

Gathorne-Hardy, F. J., Syaukani, Davies, R. G., Eggleton, P., and Jones, D. T. 2002, Quaternary rainforest refugia in south-east Asia: Using termites (Isoptera) as indicators. *Biol. J. Linnean Soc.* 75:453–466.

Groves, C. P. 1985, Plio-Pleistocene mammals in island Southeast Asia. *Mod. Quat. Res. SE Asia* 9:43–54.

Groves, C. P. 1986, Systematics of the great apes, in: Swindler, D. R. and Erwin, J., eds., *Comparative Primate Biology, Vol. 1: Systematics Evolution, and Anatomy.* Alan R. Liss, New York, pp. 187–217.

Groves, C. P. 1989, *A Theory of Human and Primate Evolution.* Clarendon Press, Oxford.

Groves, C. 2001a, *Primate Taxonomy.* Smithsonian Institution Press, Washington D C.

Groves, C. 2001b, Mammals in Sulawesi: Where did they come from and when, and what happened to them when they got there? in: Metcalfe, I., Smith, J. M. B., Morwood, M., and Davidson, I., eds., *Faunal and Floral Migrations and Evolution in SE Asia-Australasia,* A. A. Balkema, Lisse, pp. 333–342.

Groves, C. P., and Holthuis, L. B. 1985, The nomenclature of the orang-utan. *Zool. Meded. Leiden* 59:411–417.

Groves, C. P. Westwood, C., and Shea, B. T. 1992, Unfinished business: Mahalanobis and a clockwork orang. *J. Hum. Evol.* 22:327–340.

Gu Y., Huang, W., Song, F., Guo, X., and Chen, D. 1987, The study of some fossil orang-utan teeth from Guangdong and Guangxi. *Acta Anthropol. Sinica* 6:272–283.

Guilderson, T. P., Fairbanks, R. G., and Rubenstone, J. L. 1994, Tropical temperature variations since 20,000 years ago: Modulating interhemispheric climate change. *Science* 263:663–665.

Haberle, S. G. 1998, Late Quaternary vegetation change in the Tari Basin, Papua New Guinea. *Palaeogeogr., Palaeoclimatol., Palaeoecol.* 137:1–24.

Haile, N. S. 1973, The geomorphology and geology of the northern part of the Sunda Shelf and its place in the Sunda Mountain System. *Pacific Geol.* 6:73–89.

Hall, L. M., Jones, D. S., and Wood, B. A. 1998, Evolution of the gibbon subgenera inferred from Cytochrome b DNA sequence data. *Mol. Phylogenet. Evol.* 10:281–286.

Hall, R. 1998, The plate tectonics of Cenozoic SE Asia and the distribution of land and sea, in: Hall, R. and Holloway, J. D., eds., *Biogeography and Geological Evolution of Southeast Asia.* Backhuys, Leiden, pp. 99–131

Hall, R. 2001, Cenozoic reconstructions of SE Asia and the SW Pacific: Changing patterns of land and sea, in: Metcalfe, I., Smith, J. M. B., Morwood, M., and Davidson, I., eds., *Faunal and Floral Migrations and Evolution in SE Asia-Australasia,* A. A. Balkema, Lisse, pp. 35–56.

Han, D. and Xu, C. 1985, Pleistocene mammalian faunas of China, in: Rukang, W. and Olsen, J. W., eds., *Palaeoanthropology and Palaeolithic Archaeology in the People's Republic of China.* Academic Press, New York, pp. 267–289.

Han, K.-H., Sheldon, F. H., and Steubing, R. B. 2000, Interspecific relationships and biogeography of some Bornean tree shrews (Tupaiidae: *Tupaia*), based on DNA hybridization and morphometric comparisons. *Biol. J. Linnean Soc.* 70:1–14.

Hanebuth, T., Stattegger, K., and Grootes, P. M. 2000, Rapid flooding of the Sunda Shelf: A late-glacial sea-level record. *Science* 288:1033–1035.

Haq, B. U., Hardenbol, J., and Vail, P. R. 1987, Chronology of fluctuating sea levels since the Triassic. *Science* 235:1156–1167.

Harcourt, A. H. 1999, Biogeographic relationships of primates on South-East Asian islands. *Global Ecol. Biogeogr.* 8:55–61.

Harcourt, A. H., and Schwartz, M. W. 2001, Primate evolution: A biology of Holocene extinction and survival on the Southeast Asian Sunda Shelf islands. *Am. J. Phys. Anthropol.* 114:4–17.

Harrison, T. 1996, The palaeoecological context at Niah Cave, Sarawak: Evidence from the primate fauna. *Bull. Indo-Pacific Prehist. Assoc.* 14:90–100.

Harrison, T. 1998, Vertebrate faunal remains from Madai Caves (MAD 1/28), Sabah, East Malaysia. *Bull. Indo-Pacific Prehist. Assoc.* 17:85–92.

Harrison, T. 2000, Archaeological and ecological implications of the primate fauna from prehistoric sites in Borneo. *Bull. Indo-Pacific Prehist. Assoc.* 20:133–146.

Harrison, T., Ji, X., and Su, D. 2002, On the systematic status of the late Neogene hominoids from Yunnan Province, China. *J. Hum. Evol.* 43:207–227.

Harrisson, T. 1958, The caves of Niah: A history of prehistory. *Sarawak Mus. J.* 8: 549–595.

Harrisson, T. 1959a, Radiocarbon-^{14}C dating B.C. from Niah: A note. *Sarawak Mus. J.* 9:136–138.

Harrisson, T. 1959b, A remarkably remote orang-utan; 1958–60. *Sarawak Mus. J.* 9:448–451.

Harrisson, T. 1970, The prehistory of Borneo. *Asian Perspect.* 13:17–45.

Hayashi, S., Hayasaka, K., Takenaka, O., and Horai, S. 1995, Molecular phylogeny of gibbons inferred from mitochondrial DNA sequences: Preliminary report. *J. Mol. Evol.* 41:359–365.

Heaney, L. R. 1984, Mammalian species richness on islands on the Sunda Shelf, Southeast Asia. *Oecologia* 61:11–17.

Heaney, L. R. 1985, Zoogeographic evidence for middle and Late Pleistocene land bridges to the Philippine Islands. *Mod. Quat. Res. SE Asia* 9:127–143.

Heaney, L. R. 1986, Biogeography of mammals in Southeast Asia: Estimates of rates of colonization, extinction, and speciation. *Biol. J. Linnean Soc.* 28:127–165.

Heaney, L. R. 1991, A synopsis of climatic and vegetational change in southeast Asia. *Climat. Change* 19:53–61.

Heinsohn, T. E. 2001, Human influences on vertebrate zoogeography: animal translocation and biological invasions across and to the east of Wallace's Line, in: Metcalfe, I., Smith, J. M. B., Morwood, M., and Davidson, I., eds., *Faunal and Floral Migrations and Evolution in SE Asia-Australasia*, A. A. Balkema, Lisse, pp. 153–170.

Hooijer, D. A. 1948, Prehistoric teeth of man and of the orang-utan from central Sumatra, with notes on the fossil orang-utan from Java and southern China. *Zool. Meded. Leiden* 29:175–293.

Hooijer, D. A. 1960, Quaternary gibbons from the Malay Archipelago. *Zool. Verhand. Leiden* 46:1–42.

Hooijer, D. A. 1961, The orang-utan in Niah Cave pre-history. *Sarawak Mus. J.* 9:408–421.

Hooijer, D. A. 1962a, Quaternary langurs and macaques from the Malay Archipelago. *Zool. Verhand. Leiden* 55:1–64.

Hooijer, D. A. 1962b, Prehistoric bone: The gibbons and monkeys of Niah Great Cave. *Sarawak Mus. J.* 10:428–449.

Hope, G. and Tulip, J. 1994, A long vegetation history from lowland Irian Jaya, Indonesia. *Palaeogeogr. Palaeoclimatol. Palaeoecol.* 109:385–398.

Hope, G. S., Peterson, J. A., Radok, U., and Allison, I., eds., 1976, *The Equatorial Glaciers of New Guinea*. A.A. Balkema, Leiden.

Hutchison, C. S. 1989, Displaced terranes of the southwest Pacific, in: Ben-Avraham, Z., ed., *The Evolution of the Pacific Ocean Margins*, pp. 161–175. Clarendon Press, Oxford.

Huxley, T. H. 1868, On the classification and distribution of the Alectoromorphae and Heteromorphae. *Proc. Zool. Soc. London* 1868:294–319.

Ipoi, D. 1993, Archaeological excavations at Gua Sireh (Serian) and Lubang Angin (Gunung Mulu National Park), Sarawak, Malaysia. *Sarawak Mus. J.* 45, Special Monograph 6.

Jablonski, N. G. 1998, The response of catarrhine primates to Pleistocene environmental fluctuations in East Asia. *Primates* 39:29–37.

Jablonski, N. G. and Tyler, D. E. 1999, *Trachypithecus auratus sangiranensis*, a new fossil monkey from Sangiran, Central Java, Indonesia. *Inter. J. Primatol.* 20:319–326.

Jablonski, N. G. and Whitfort, M. J. 1999, Environmental change during the Quaternary in East Asia and its consequences for mammals. *Rec. Western Austral. Mus.* Supplement 57:307–315.

Jablonski, N. G., Whitfort, M. J., Roberts-Smith, N., and Xu, Q. 2000, The influence of life history and diet on the distribution of catarrhine primates during the Pleistocene in eastern Asia. *J. Hum. Evol.* 39:131–157.

Janczewski, D. N, Goldman, D., and O'Brien, S. J. 1990, Molecular divergence of orangutan (*Pongo pygmaeus*) subspecies based on isozyme and two-dimensional gel electrophoresis. *J. Hered.* 81:375–387.

Kahlke, H. D. 1972, A review of the Pleistocene history of the orang-utan (*Pongo* Lacépède 1799). *Asian Perspect.* 15:5–14.

Karesh, W. B., Frazier, H., Sahjuti, D., Andau, M., Gombek, F., Zhi, L., Janczewski, D., and O'Brien, S. J. 1997, Orangutan genetic diversity, in: Sodaro, C., ed., *Orangutan Species Survival Plan, Husbandry Manual*. Chicago Zoological Park, Chicago.

Katili, J. A. 1975, Volcanism and plate tectonics in the Indonesian island arcs. *Tectonophysics* 26:165–188.

Kershaw, A. P., Penny, D., va der Kaars, S., Anshari, G., and Thamotherampillai, A. 2001, Vegetation and climate in lowland southeast Asia at the Last Glacial Maximum, in: Metcalfe, I., Smith, J. M. B., Morwood, M., and Davidson, I., eds., *Faunal and Floral Migrations and Evolution in SE Asia-Australasia*, A.A. Balkema, Lisse, pp. 225–236.

King, T. 1996, Equatorial Pacific sea surface temperatures, faunal patterns, and carbonate burial during the Pliocene. *Marine Micropaleontol.* 27:63–84.

Krigbaum, J. S. 2001, Human Paleodiet in Tropical Southeast Asia: Isotopic Evidence from Niah Cave and Gua Cha. Ph.D. Dissertation, New York University, NY.

Kudrass, H. R., and Schlüter, H. U. 1994, Development of cassiterite-bearing sediments and their relation to Late Pleistocene sea-level changes in the Straits of Malacca. *Marine Geol.* 120:175–202.

Larick, R., Ciochon, R. L., Zaim, Y., Sudijono, Suminto, Rizal, Y., and Aziz, F. 2000, Lithostratigraphic context for kln-1993.05-SNJ, a fossil colobine maxilla from Jokotingkir, Sangiran Dome. *Int. J. Primatol.* 21:731–759.

Lea, D. W., Pak, D. K., and Spero, H. J. 2000, Climate impact of Late Quaternary equatorial Pacific sea surface temperature variations. *Science* 289:1719–1724.

Linsley, B. K. 1996, Oxygen-isotope record of sea level and climate variations in the Sulu Sea over the past 150,000 years. *Nature* 380:234–237.

Lydekker, R. 1896, *A Geographical History of Mammals*. Cambridge University Press, Cambridge.

MacKinnon, J. 1973, Orang-utans in Sumatra. *Oryx* 12:234–242.

Maloney, B. K. 1980, Pollen analytical evidence for early forest clearance in North Sumatra. *Nature* 287:324–326.

Maloney, B. K. 1981, A pollen diagram from Tao Sipinggan, a lake site in the Batak Highlands of North Sumatra, Indonesia. *Mod. Quat. Res. SE Asia* 6:57–76.

Marshall, J. and Sugardjito, J. 1986, Gibbon systematics, in: Swindler, D. R. and Erwin, J., eds., *Comparative Primate Biology, Vol. 1: Systematics Evolution, and Anatomy*. Alan R. Liss, New York, pp. 137–186.

Martinson, D. G., Pisias, N. G., Hays, J. D., Imbrie, J., Moore, T. C. J., and Shackleton, N. J. 1987, Age dating and the orbital theory of the ice ages: Development of a high-resolution 0 to 300,000-year chronostratigraphy. *Quat. Res.* 27:1–29.

McCabe, R. and Cole, J. 1989, Speculations on the Late Mesozoic and Cenozoic evolution of the southeast Asian margin, in: Ben-Avraham, Z., ed., *The Evolution of the Pacific Ocean Margins*. Clarendon Press, Oxford, pp. 143–160.

McCulloch, M. T., Mortimer, G. E., Esat, T., Xianhua, L., Pillans, B., and Chappell, J. 1996, High resolution windows into early Holocene climate: Sr/Ca coral records from the Huon Peninsula. *Earth Planet. Sci. Lett.* 138:169–178.

McCulloch, M. T., Tudhope, A. W., Esat, T. M., Mortimer, G. E., Chappell, J., Pillans, B., Chivas, A. R., and Omura, A. 1999, Coral record of equatorial sea-surface temperatures during the penultimate deglaciation at Huon Peninsula. *Science* 283:202–204.

Medway, L. 1959, Niah animal bone: II (1954–8). *Sarawak Mus. J.* 9:151–163.

Medway, L. 1970, The monkeys of Sundaland. Ecology and systematics of the cercopithecids of a humid equatorial environment, in: Napier, J. R. and Napier, P. H. eds., *Old World Monkeys: Evolution, Systematics, and Behavior.* Academic Press, New York, pp. 513–553.

Medway, L. 1977, The Niah excavations and an assessment of the impact of early man on mammals in Borneo. *Asian Perspect.* 20:51–69.

Mercer, J. M. and Roth, V. L. 2003, The effects of Cenozoic global change on squirrel phylogeny. *Science* 299:1568–1572.

Metcalfe, I. 1998, Palaeozoic and Mesozoic geological evolution of the SE Asian region: Multidisciplinary constraints and implications for biogeography, in: Hall, R. and Holloway, J. D., eds., *Biogeography and Geological Evolution of Southeast Asia.* Backhuys, Leiden, pp. 25–41.

Molengraff, G. A. F. and Weber, M. 1920, On the relation between Pleistocene glacial period and the origin of the Sunda Sea (Java- and South China Sea) and its influence on coral reefs and on the land- and freshwater fauna. *Proc. Roy. Acad. Amsterdam* 23:395–439.

Morales, J. C. and Melnick, D. J. 1998, Phylogenetic relationships of the macaques (Cercopithecidae: *Macaca*), as revealed by high resolution restriction site mapping of mitochondrial robosomal genes. *J. Hum. Evol.* 34:1–23.

Morley, R. J. 1982, A palaeoecological interpretation of a 10,000-year pollen record from Danau Padang, Central Sumatra, Indonesia. *J. Biogeogr.* 9:151–190.

Morley, R. J. and Flenley, J. R. 1987, Late Cainozoic vegetational and environmental changes in the Malay Archipelago, in: Whitmore, T. C., ed., *Biogeographical Evolution of the Malay Archipelago.* Clarendon Press, Oxford, pp. 50–59.

Moss, S. J. and Wilson, M. E. J. 1998, Biogeographic implications of the Tertiary palaeogeographic evolution of Sulawesi and Borneo, in: Hall, R. and Holloway, J. D., eds., *Biogeography and Geological Evolution of Southeast Asia*, Backhuys, Leiden, pp. 133–163.

Napier, J. R. and Napier, P. H. 1985, *The Natural History of the Primates.* MIT Press Cambridge, MA.

Newsome, J. and Flenley, J. R. 1988, Late Quaternary vegetational history of the Central Highlands of Sumatra. II. Palaeopalynology and vegetational history. *J. Biogeogr.* 15: 555–578.

Ninkovich, D., Shackleton, N. J., Abdel-Monem, A. A., Obradovich, J. D., and Izett, G. 1978, KAr age of the Late Pleistocene eruption of Toba, north Sumatra. *Nature* 276:574–577.

Nisbett, R. A. and Ciochon, R. L. 1993, Primates in northern Viet Nam: A review of the ecology and conservation status of extant species, with notes on Pleistocene localities. *Inter. J. Primatol.* 14:765–795.

Nossin, J. J. 1962, Coastal sedimentation in northeastern Johore (Malaya). *Z. Geomorphol.* 6:296–316.

Oates, J. F., Davies, A. G., and Delson, E. 1994, The diversity of living colobines, in: Davies, A. G., and Oates, J. F., eds., *Colobine Monkeys: Their Ecology, Behaviour and Evolution.* Cambridge University Press, Cambridge, pp. 45–73.

Ollier, C. D. 1985, The geological background to prehistory in island Southeast Asia. *Mod. Quat. Res. SE Asia* 9:25–42.

Olsen, J. W. and Ciochon. R. L. 1990, A review of evidence for postulated Middle Pleistocene occupations in Viet Nam. *J. Hum. Evol.* 19:761–788.

Payne, J., Francis, C. M., and Phillipps, K. 1985, *A Field Guide to the Mammals of Borneo.* The Sabah Society, Kota Kinabalu.

Pei, W. C. 1935, Fossil mammals from the Kwangsi caves. *Bull. Geol. Soc. China* 14:413–426.

Petersen, R. M. 1969, Würm II climate at Niah Cave. *Sarawak Mus. J.* 17:67–79.

Prentice, M. L. and Denton, G. H. 1988, The deep-sea oxygen isotope record, the global ice sheet system and hominid evolution, in: Grine, F. E., ed., *Evolutionary History of the "Robust" Australopithecines.* Aldine de Gruyter, New York, pp. 383–403.

Rampino, M. R. and Self, S. 1992, Volcanic winter and accelerated glaciation following the Toba super-eruption. *Nature* 359:50–52.

Rampino, M. R., Self, S., and Stothers, R. B. 1988, Volcanic winters. *Ann. Rev. Earth Planet. Sci.* 16:73–99.

Ravelo, A. C., Andreasen, D. H., Lyle, M., Lyle, A. O., and Wara, M. W. 2004, Regional climate shifts caused by gradual global cooling in the Pliocene epoch. *Nature* 429:263–267.

Reis, K. R. and Garong, A. M. 2001, Late Quaternary terrestrial vertebrates from Palawan Island, Philippines. *Palaeogeogr., Palaeoclimatol., Palaeoecol.* 171:409–421.

Reynolds, V. 1967, *The Apes.* Cassell, London.

Rijksen, H. D. and Meijaard, E. 1999, *Our Vanishing Relative: The Status of Wild Orangutans at the Close of the Twentieth Century.* Kluwer Academic Publishers, Dordrecht.

Rind, D. and Peteet, D. 1985, Terrestrial conditions at the Last Glacial Maximum and CLIMAP sea-surface temperature estimates: Are they consistent? *Quat. Res.* 24:1–22.

Röhrer-Ertl, O. 1988, Research history, nomenclature, and taxonomy of the orang-utan, in: Schwartz, J. H., ed., *Orang-utan Biology*. Oxford University Press, Oxford, pp. 7–18.

Roos, C., Ziegler, T., Hodges J. K., Zischler, H., and Abegg, C. 2003, Molecular phylogeny of Mentawai macaques: Taxonomic and biogeographic implications. *Mol. Phylogenet. Evol.* 29:139–150.

Rose, W. I. and Chesner, C. A. 1987, Dispersal of ash in the great Toba eruption, 75 ka. *Geology* 15:913–917.

Rosenblum, L. L., Supriatna, J., and Melnick, D. J. 1997a, Phylogeographic analysis of pigtail macaque populations (*Macaca nemestrina*) inferred from mitochondrial DNA. *Am. J. Phys. Anthropol.* 104:35–45.

Rosenblum, L. L., Supriatna, J., Hasan, M. N., and Melnick, D. J. 1997b, High mitochondrial DNA diversity with little structure within and among leaf monkey populations (*Trachypithecus cristatus* and *Trachypithecus auratus*). *Int. J. Primatol.* 18:1005–1028.

Rowe, N. 1996, *The Pictorial Guide to the Living Primates*. Pogonias Press, East Hampton, New York.

Ryder, O. A. and Chemnick, L. G. 1993, Chromosomal and mitochondrial DNA variation in orang utans. *J. Heredity* 84:405–409.

Samuel, M. A., Harbury, N. A., Bakri, A., Banner, F., and Hartono, L. 1997, A new stratigraphy for the islands of the Sumatran Forearc, Indonesia. *J. Asian Earth Sci.* 15:339–380.

Sartono, S. 1973, On Pleistocene migration routes of vertebrate fauna in Southeast Asia. *Bull. Geol. Soc. Malaysia* 6:273–286.

Sawamura, K. and Laming, D. J. C. 1974, Sea floor valleys in the Gulf of Thailand and Quaternary sea-level changes. *Comm. Coast. Offshore Prospect. Newsl.* 1:29–54.

Schwartz, J. H., Long, V. T., Cuong, N. L., Kha, L. T., and Tattersall, I. 1994, A diverse hominoid fauna from the late Middle Pleistocene breccia cave of Tham Khuyen, Socialist Republic of Vietnam. *Anthropol. Papers Am. Museum Nat. Hist.* 73:1–11.

Schwartz, J. H., Long, V. T., Cuong, N. L., Kha, L. T., and Tattersall, I. 1995, A review of the Pleistocene hominoid fauna of the Socialist Republic of Vietnam (excluding Hylobatidae). *Anthropol. Papers Am. Museum Nat. Hist.* 76:1–24.

Scrivenor, J. B., Burkhill, I. H., Smith, M. A., Corbet, A. S., Airy Shaw, H. K., Richards, P. W., and Zeuner, F. E. 1941–1942, A discussion on the biogeographic division of the Indo-Australian archipelago, with criticism of the Wallace and Weber lines and of any other dividing lines and with an attempt to obtain uniformity in the names used for the divisions. *Proc. Linnean Soc. London* 154:120–165.

Shackleton, N. J. 1987, Oxygen isotopes, ice volume and sea level. *Quat. Sci. Rev.* 6:183–190.

Shackleton, N. J. 1995, New data on the evolution of Pliocene climatic variability. in: Vrba, E. S., Denton, G. H., Partridge, T. C., and Burckle, L. H., eds., *Palaeoclimate and Evolution with Emphasis on Human Origins.* Yale University Press, New Haven, pp. 242–248.

Simpson, G. G. 1977, Too many lines; the limits of the Oriental and Australian zoogeographic regions. *Proc. Am. Phil. Soc.* 121:107–120.

Sloan, L. C., Crowley, T. J., and Pollard, D. 1996, Modeling of Middle Pliocene climate with the NCAR GENESIS general circulation model. *Marine Micropaleontol.* 27:51–61.

Smith, R. J., and Pilbeam, D. R. 1980, Evolution of the orang-utan. *Nature* 284:447–448.

Stuijts, I. 1983/1984, Palynological study of Situ Bayongbong, West Java. *Mod. Quat. Res. SE Asia* 8:17–27.

Stuijts, I., Newsome, J. C., and Flenley, J. R. 1988, Evidence for Late Quaternary vegetational change in the Sumatran and Javan highlands. *Rev. Palaeobot. Palynol.* 55:207–216.

Stute, M., Forster, M., Frischkorn, H., Serejo, A., Clark, J. F., Schlosser, P., Broecker, W. S., and Bonani, G. 1995, Cooling of tropical Brazil (5°C) during the last glacial maximum. *Science* 269:379–383.

Sun, X., Li, X., Luo, Y., and Chen, X. 2000, The vegetation and climate at the last glaciation on the emerged continental shelf of the South China Sea. *Palaeogeogr., Palaeoclimatol., Palaeoecol.* 160:301–316.

Swisher, C. C., Curtis, G. H., Jacob, T., Getty, A. G., Suprijo, A., and Widiasmoro 1994, Age of the earliest known hominids in Java, Indonesia. *Science* 263:1118–1121.

Tan, H. V. 1985, The Late Pleistocene climate in Southeast Asia: New data from Vietnam. *Mod. Quat. Res. SE Asia* 9:81–86.

Thomas, M. F. 1987, Quaternary sedimentation in West Kalimantan (Indonesian Borneo). *Trop. Geomorphol. News.* 4:23.

Thommeret, J. and Thommeret, Y. 1978, C14 datings of some Holocene sea levels on the North coast of the island of Java (Indonesia). *Mod. Quat. Res. SE Asia* 4:51–56.

Tjia, H. D. 1970, Quaternary shorelines of the Sunda Land Southeast Asia. *Geol. Mijnbouw* 49:135–144.

Tjia, H. D. 1977, Sea level variations during the last six thousand years in peninsular Malaysia. *Sains Malaysiana* 6:171–183.

Tjia, H. D. 1980, The Sunda Shelf, Southeast Asia. *Z. Geomorphol.* 24:405–427.

Tjia, H. D. 1984, Changes during the Holocene in the coastal areas of the Sunda shelf. *Ilmu Alam* 12/13:91–105.

Tosi, A. J., Disotell, T. R., Morales, J. C., and Melnick, D. J. 2003, Cercopithecine Y-chromosome data provide a test of competing morphological evolutionary hypotheses. *Mol. Phylogenet. Evol.* 27:510–521.

Tougard, C. and Ducrocq, S. 1999, Abnormal fossil upper molar of *Pongo* from Thailand: Quaternary climatic changes in Southeast Asia as a possible cause. *Int. J. Primatol.* 20:599–607.

Uchida, A. 1998, Variation in tooth morphology of *Pongo pygmaeus*. *J. Hum. Evol.* 34:71–79.

Umbgrove, J. H. F. 1938, On the time of origin of the submarine relief of the East Indies. *Comptes Rendus du Congrès International Geographie, Amsterdam* 2:150–159.

Van Andel, T. H. and Veevers, J. J. 1967, *Morphology and Sediments of the Timor Sea*. Bureau of Mineral Resources, Geology and Geophysics, Canberra.

Van Andel, T. H., Heath, R., Moore, T. C., and McGeary, D. F. R. 1967, Late Quaternary history, climate and oceanography of the Timor Sea, northwestern Australia. *Am. J. Sci.* 265:737–758.

Van den Bergh, G. D., de Vos, J., and Sondaar, P. Y. 2001, The Late Quaternary palaeogeography of mammal evolution in the Indonesian Archipelago. *Palaeogeogr., Palaeoclimatol., Palaeoecol.* 171:385–408.

Van den Brink, L. M. 1982, On the mammal fauna of the Wajak Cave, Java (Indonesia). *Mod. Quat. Res. SE Asia* 7:177–193.

Van der Kaars, S. 1998, Marine and terrestrial pollen records of the last glacial cycle from the Indonesian region: Bandung basin and Banda Sea. *Palaeoclimat: Data Model.* 3:209–219.

Van der Kaars, W. A. 1991, Palynology of eastern Indonesian marine piston-cores: A Late Quaternary vegetational and climatic record for Australasia. *Palaeogeogr., Palaeoclimatol., Palaeoecol.* 85:239–302.

Van der Kaars, W. A. and Dam, M. A. C. 1995, A 135,000-year record of vegetational and climatic change from the Bandung area, West-Java, Indonesia. *Palaeogeogr., Palaeoclimatol., Palaeoecol.* 117:55–72.

Verstappen, H. Th. 1975, On palaeo climates and landform development in Malesia. *Mod. Quat. Res. SE Asia* 1:3–35.

Verstappen, H. Th. 1997, The effect of climatic change on Southeast Asian geomorphology. *J. Quat. Sci.* 12:413–418.

Von Koenigswald, G. H. R. 1940, Neue Pithecanthropus-Funde 1936–1938. *Wet. Med. Dienst Mijnb. Ned. Indië* 28:1–205.

Von Koenigswald, G. H. R. 1982, Distribution and evolution of the orang utan, *Pongo pygmaeus* (Hoppius), in: de Boer, L. E. M., ed., *The Orang Utan. Its Biology and Conservation*. Junk, The Hague, pp. 1–15.

Voris, H. K. 2000, Maps of Pleistocene sea levels in Southeast Asia: Shorelines, river systems and time durations. *J. Biogeogr.* 27:1153–1167.

Walker, D. and Flenley, J. R. 1979, Late Quaternary vegetational history of the Enga Province of upland Papua New Guinea. *Phil. Trans. Roy. Soc.* [B] 286:265–344.

Wang. L., Sarnthein, M., Erlenkeuser, H., Grimalt, J., Grootes, P., Heilig, S., Ivanova, E., Kienast, M., Pelejero, C., and Pflaumann, U. 1999, East Asian monsoon climate during the Late Pleistocene: High-resolution sediment records from the South China Sea. *Mar. Geol.* 156:245–284.

Wardlaw, B. R. and Quinn, T. M. 1991, The record of Pliocene sea-level change at Enewetak Atoll. *Quat. Sci. Rev.* 10:247–258.

Weitzel, V., Yang, C. M., and Groves, C. P. 1988, A catalogue of Primates in the Singapore Zoological Reference Collection. *Raffles Bull. Zool.* 36:1–166.

Widianto, H. 1991, The hominid dental remains of Java: A metrical study. *Bull. Indo-Pacific Prehist. Assoc.* 11:23–35.

Wilson, W. L. and Wilson, C. C. 1975, Species-specific vocalizations and the determination of phylogenetic affinities of the *Presbytis aygula-melalophos* group in Sumatra. *Contemporary Primatology*, 5th International Congress of Primatology, Nagoya 1974. Karger, Basel, pp. 459–463.

Xu, X. and Arnason, U. 1996, The mitochondrial DNA molecule of Sumatran orangutan and a molecular proposal for two (Bornean and Sumatran) species of orangutan. *J. Mol. Evol.* 43:431–437.

Yokoyama, Y., Lambeck, K., De Deckker, P., Johnston, P., and Fifield, L. K. 2000, Timing of the last glacial maximum from observed sea-level minima. *Nature* 406:713–716.

Zhi, L., Karesh, W. B., Janczewski, D. N., Frazier-Taylor, H., Sajuthi, D., Gombek, F., Andau, M., Martenson, J. S., and O'Brien, S. J. 1996, Genomic differentiation among natural populations of orang-utan (*Pongo pygmaeus*). *Curr. Biol.* 6:1326–1336.

Zielinski, G. A., Mayewski, P. A., Meeker, L. D., Whitlow, S., Twickler, M. S., and Taylor, K. 1996, Potential atmospheric impact of the Toba mega-eruption ~71,000 years ago. *Geophys. Res. Lett.* 23:837–840.

Primate Biogeography
in Deep Time

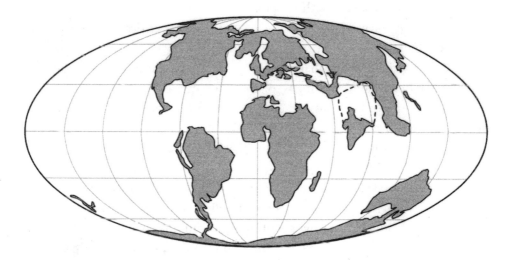

T he chapters in the previous sections have been concerned primarily
with understanding the present distribution of primates in different
parts of the world. On a small scale, primate biogeography seems
to be explicable in terms of a variety of factors, including, the geography of rivers, mountains, and oceans; the role of latitude, climate, and
vegetation structure; and the effects of human activity. The chapters in this section examine primate biogeography from a broader geographical and temporal
perspective and discuss the biogeography of primate evolution over the past 65
million years or so. In addition, these chapters examine the complex reciprocal
relationship between biogeography and phylogeny reconstruction.

In Chapter 13, "Primate Biogeography and the Fossil Record: The Role of
Plate Tectonics, Climate, and Chance," John Fleagle and Christopher Gilbert
provide a broad overview of the primate fossil record in the context of major

Primate Biogeography, edited by Shawn M. Lehman and John G. Fleagle.
Springer, New York, 2006.

changes in the positions and connections of the continents, changes in global climate, and what appear to be chance events, during the past 55 million years. They find that many of the most dramatic dispersal events in the primate fossil record can be related to major events in Earth history.

In Chapter 14, "Biogeographic Origins of Primate Major Taxa," Christopher Heesy, Nancy Stevens, and Karen Samonds evaluate alternative hypotheses regarding the geographic origin of major taxonomic groups (e.g. Primates, Anthropoidea) in the context of the phylogeny of living and fossil primates. They find that the phylogenetic positions of key fossil taxa are critical for understanding the biogeographic history of living groups.

In Chapter15, "Mammalian Biogeography and Anthropoid Origins," Christopher Beard addresses many of the same questions regarding the biogeographic origin of Primates and anthropoids in the context of the evolution of other orders of mammals. He argues that throughout its history, Africa has been a recipient of mammalian groups from other continents rather than a source.

In "Continental Paleobiogeography as Phylogenetic Evidence," James Rossie and Erik Seiffert examine, more formally, the role of biogeographic information in phylogenetic analysis. Following techniques previously developed for the inclusion of stratigraphic information into the phylogenetic analysis of fossil taxa, they create a methodology for incorporating the biogeographic relationships among continents and barriers to dispersal into parsimony analyses of primate relationships. They then apply this method in separate analyses of early anthropoids and Miocene catarrhines and find that the addition of biogeographic data helps clarify many relationships that could not be resolved using traditional approaches.

CHAPTER THIRTEEN

The Biogeography of Primate Evolution: The Role of Plate Tectonics, Climate and Chance

John G. Fleagle and Christopher C. Gilbert

ABSTRACT

The present biogeographic distribution of primates is the result of a complex series of events throughout the Cenozoic Era. In past, primates have been both widespread and diverse in North America and Europe and have had much broader distributions in South America, Madagascar and Asia. Many of the dramatic changes in primate biogeography can be related to global climatic events and to geological changes in the interconnections among continents. The evolutionary history of primate seems to have involved a wide range of traditional dispersal mechanisms, including land bridges, chance dispersal over open ocean, and intermediate island hopping. The assemblages of primates found in most of the world contain members of several different clades, each with a distinctive biogeographical history. Our understandings of primate phylogeny and primate biogeography are interwoven, and reciprocally illuminating.

John G. Fleagle • Department of Anatomical Sciences, Health Sciences Center, Stony Brook University, Stony Brook, New York 11794-8081 **Christopher C. Gilbert** • Interdepartmental Doctoral Program in Anthropological Sciences, SBS Building S-501, Stony Brook University, Stony Brook, New York 11794-4364

Primate Biogeography, edited by Shawn M. Lehman and John G. Fleagle.
Springer, New York, 2006.

Key Words: Fossil record, dispersal, anthropoids, phylogeny, continental drift, Tethys, latutude

INTRODUCTION

Perhaps the most commonly documented, but least appreciated, lesson from the study of paleontology is the fact that *the past was strikingly different from the present*. While the principle of uniformitarianism—that the processes underlying geological changes of the past are the same as those operating today—has been a cornerstone of geological sciences for over 150 years, it is nevertheless true that the animals that inhabited the earth have changed through geological time as have the geographical relationships of the continents and the patterns of global and regional climate. Not surprisingly, these aspects of the earth's history are closely interrelated. Climate is driven by the relative size, topography, and positions of the major bodies of land and water as well as by astronomical cycles and perturbations. Similarly, the availability of resources, the nature of available habitats, and the dispersal abilities of organisms (including both competitors and predators) are influenced by both continental connections and climate. In addition, although the principle of uniformitarianism remains the basic principle of all scientific endeavor, it is increasingly evident that over the long course of geological time, rare but dramatic events—either of extraterrestrial origin, such as impacts from asteroids, or indigenous such as the joining or separating of continental plates—have had profound effects on the nature of the evolution of life on earth.

The goal of this paper is to trace the biogeography of primates over the past 65 million years, comparing the distribution of the living taxa and clades with those of the past. In particular we will attempt to relate major aspects of primate biogeography to the changes in global geography and global climate during the Cenozoic Era. Thus, before discussing details of primate biogeography, it is necessary to first review the history of paleogeography and paleoclimate over the past 65 million years.

Tectonics and Paleogeography

Primate evolution during the past 65 million years has taken place in the context of a dynamic earth in which the positions and composition of continents, and especially their interconnections, have changed considerably (Figure 1). Perhaps

Cenozoic Continental Geography

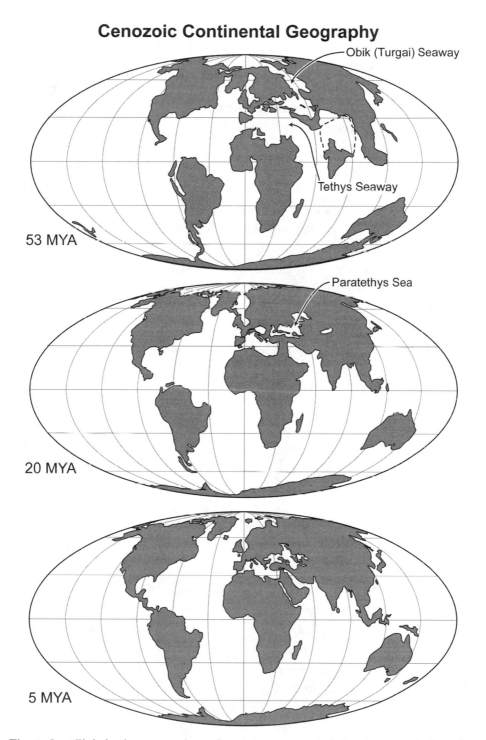

Figure 1.   Global paleogeography at three important periods in primate evolution: the Early Eocene (53 mya); the Early Miocene (20 mya) and the Early Pliocene (5 mya).

most dramatic is the history of South America. Throughout most the Cenozoic Era, South America was an island continent slowly moving apart from Africa by the expanding South Atlantic. No connections to North America existed until roughly 5 million years ago (mya) when Central America came into being, permitting a major faunal interchange between the two continents (Stehli and Webb, 1985). However, until approximately 30 mya, South America held a slender connection with Antarctica, which was, in turn, connected to Australia. The Atlantic rift, which separated South America from Africa during the Mesozoic Era, opened in a South to North direction, and continues today in Iceland. Thus while South America was well-separated from Africa by the early part of the Cenozoic, North America and Europe remained connected in the North until the beginning of the Eocene Epoch.

On the opposite side of the globe, India has been the great wanderer. Originally a part of the southern supercontinent of Gondwana along with Antarctica, Africa, Madagascar, Australia, and South America, the Indian subcontinent separated from Africa in the Cretaceous approximately 120–125 mya and slowly drifted northward towards Asia. The initial collision with the Asian mainland seems to have occurred in the Early Eocene and has continued, giving rise to the Himalayan mountains. As a result, the northernmost part of the subcontinent which made the initial contact with Asia has long been buried deep below the Himalayas. The geological history of eastern Asia is less clear, but it seems that the coherence of many parts of that continent is a relatively recent event. Throughout the Pleistocene, the islands of the Sunda Shelf have been alternately connected and separated as sea levels fell and rose in response to the glacial cycles.

The history of connections between Asia and North America via the Bering Straits seem to have been relatively stable throughout much of the Cenozoic, changing only with the rise and fall of sea level (Beard and Dawson, 1999).

Around North Africa and Europe, the Mediterranean Sea is a relatively recent phenomenon. From the beginning of the Cenozoic, African and Eurasia were separated by a continuous seaway, the Tethys Sea, extending from Gibraltar in the west to India in the east. There are, however, suggestions of some type of island arcs or some other filtered connection between Europe and Africa in the Paleocene. It was only in the Early Miocene that the northward movement and counter-clockwise rotation of Africa brought it into contact with western Asia to cut off the Tethys seaway and form the Mediterranean and also to provide a land bridge between Africa and Eurasia. The initial disconnection of the Tethys approximately 20 mya was not a permanent change in continental geography and was followed by a brief reopening in the Middle Miocene (Rogl, 1999; Whybrow Andrews, 2000). Throughout much of the early Cenozoic, Europe and Asia were separated by a large epicontinental seaway, the West Siberian or Obik Sea, which extended north from the Turgai Straits to the Arctic Sea. However, this barrier seems to have been breached from time to time in periods of low sea level (McKenna, 1975; Hooker and Dashzeveg, 2003). Much of southern Europe has been formed relatively recently by the coalescence of separate islands between the old Tethys Seaway and the more northern Paratethys Sea (Bernor, 1983).

Climate

The past 60 million years of Earth's history have also been associated with major climatic changes on both a long-term scale, which are well-documented, and short-term oscillations, which are less firmly documented before the last two million years or so. Most of these major climatic changes can be related to tectonic events such as changing connections among continents and oceans in conjunction with astronomical cycles (Zachos et al., 2001). Large scale climate change has been reconstructed from a variety of different types of data, including isotopic studies of deep sea cores and of terrestrial sediments, studies of marine invertebrates, and studies of leaf shape and distributions of fossil plants. However, the major patterns of global climate change during the past 65 million years are remarkably consistent (Figure 2).

The most significant climatic change in the early part of the Cenozoic was a dramatic rise in global temperatures during the Early Eocene, commonly called the Early Eocene Climatic Optimum. However, this event was preceded by a rapid temperature spike at the Paleocene-Eocene boundary, the Late Paleocene Thermal Maximum (Wing et al., 2005). The Early Eocene climates were the warmest of the whole Cenozoic and this time period was characterized by tropical faunas and floras well into high latitudes (Hickey et al., 1983). This Early Eocene rise is probably associated in some way with North Atlantic rifting and the collision of India with Asia but the mechanisms are not clear (Zachos et al., 2001).

Temperatures cooled in the Late Eocene, followed by a slight temperature rise at the beginning of the Oligocene and a dramatic drop in temperatures during the Early Oligocene (Prothero and Berggren, 1992). The major

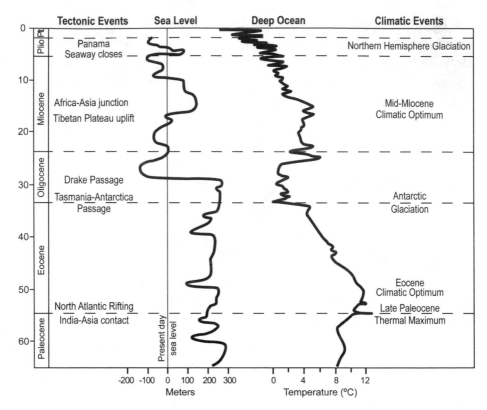

Figure 2. Major geological and climatic changes during the last 65 million years.

temperature drop in the Early Oligocene seems to be related to the formation of circum-Antarctic currents following the separation of Australia and South America from Antarctica and the opening of the Drake and the Tasmania-Antarctica passages although the exact timing of these individual events is still unclear (e.g., Zachos *et al.*, 2001). This Early Oligocene temperature drop also coincides with a major drop in sea levels associated with the formation of continental glaciers on Antarctica, and perhaps other continents as well (Zachos *et al.*, 2001).

Global climates rose slightly from the Late Oligocene reaching a peak in early Middle Miocene, the mid-Miocene Climatic Optimum (Zachos *et al.*, 2001). Global temperatures dropped after that and a drying trend continued into the Late Miocene with the Mediterranean Sea drying up completely on at least one occasion and maybe more (Hsu *et al.*, 1973).

At the beginning of the Pliocene, temperatures warmed and sea levels rose, refilling the Mediterranean Sea that had disappeared during the Late Miocene.

In the Early Pliocene, North America and South America became connected by the Central American land bridge (e.g., see Stehli and Webb, 1985). A major cooling event occurred near the end of the Pliocene, approximately 2.5 mya (e.g., Denton, 2001), which has been hypothesized to have had a large effect on the distribution and turnover rates of mammalian taxa, including primates (e.g., Wesselman, 1985; Vrba, 1992; Turner and Wood, 1993); however, other studies have found evidence of major faunal changes at different times (e.g., McKee, 1996; 2001; Behrensmeyer *et al.*, 1997; Frost, In Press).

The Pleistocene has been characterized by a series of glacial periods, perhaps set into motion by the formation of the Isthmus of Panama (Stanley, 1996). In addition to dramatic cycles of warmer and cooler temperatures, the formation of continental glaciers had the effect of lowering sea levels and exposing land bridges from time to time. For example, the exposed Bering Strait land bridge during the Pleistocene allowed for various mammalian groups, including humans, to disperse form Asia to North America. Currently, the earth is experiencing a warm interglacial period, and sea levels are relatively high.

PRIMATE BIOGEOGRAPHY

It is in the context of these major changes in paleogeography, climate change, and sea level that we can look at the fossil record of primate biogeography during the past 55 million years.

Overall Distribution of Primates

Today, nonhuman primates are associated with tropical or subtropical climates and most taxa are found between the latitudes of 30° North and 30° South. The most notable exceptions in the Northern Hemisphere are the Japanese macaques which live at a latitude of over 40° in snow and ice. There are no nonhuman primates in North America north of Central Mexico and none in Europe except the Barbary macaques on the island of Gibraltar. In the southern hemisphere, a few species extend beyond 30° to the tip of Africa and the temperate forests of northernmost Argentina. There are no primates except humans on Australia, New Guinea, or New Zealand. There is a growing literature on the extent to which primate distributions follow various biogeographic "rules" such as species-area relationships (Reed and Fleagle, 1995), or patterns of species richness and range size, and assemblage composition (e.g., Ganzhorn,

1998; Eeley and Foley, 1999; Eeley and Lawes, 1999; Harcourt, 2000; Lehman, 2004).

Primates as an order have relatively lower tolerance for colder climates than many other mammalian orders such as Carnivora, which extend from the Arctic to Tierra de Fuego, or bats or rodents which are quite successful in temperate climates. However, there is considerable variation in latitudinal range among primates related to habitat requirements. Individual genera and species of non-human primates may be more or less eurytrophic (Eeley and Lawes, 1999), but none can match the global distribution of *Homo sapiens*.

In earlier parts of the Cenozoic, primates had a much larger latitudinal distribution. During the climatic warming in the Eocene primates were abundant at high latitudes up to about 50° N in both North America and in Europe (Great Britain, France, and Germany). The plesiadapiform *Ignacius* has been found well into the Arctic Circle on Ellesmere Island (McKenna, 1980). Likewise in the Early-Middle Miocene warming peak, primates lived at high latitudes at the southernmost ends of South America as well as into northern Europe and Asia. As discussed later, in both of these instances, the expanded range of primate distributions corresponds to global climate changes.

PRIMATE ORIGINS AND EARLY PRIMATE EVOLUTION

Primate Origins

Primates that are clearly related to living members of the order first appear in the fossil record abruptly and more or less synchronously in Asia, Europe, and North America at the base of the Eocene Epoch along with the first members of several other "modern" mammalian orders, including perissodactyls and artiodactyls (e.g., Fleagle, 1999; Hooker, 2000; Beard, 2002). This dispersal (Figure 3) is coincident with the dramatic warming event at the Paleocene-Eocene boundary, and was made possible by the combination of a major rise in global temperatures and extension of tropical climates to high latitudes as well as the geographical connections between Asia, Europe, and North America at high latitudes (e.g., Hooker, 2004; Gingerich and Ting, 2004). On each of these continents these early primates appear to be immigrants with no clear ancestors in underlying deposits. Moreover, there is debate over the direction in which this earliest Eocene dispersal took place, with some favoring an Asia-Europe-North America direction (Hooker, 2004) and others an Asia-North

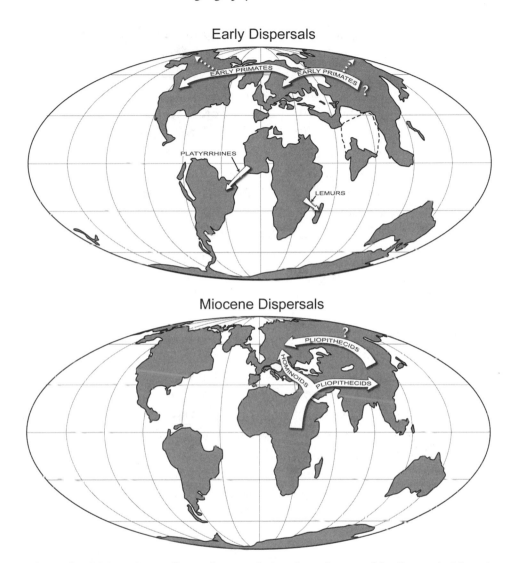

Figure 3. Major primate dispersal events during the early part of the Cenozoic (above) and during the Early Miocene (below).

America-Europe direction (Beard and Dawson, 1999). Although the biogeographical and phyletic origins of primates are unresolved, there have been many attempts to address this problem.

Over a decade ago, Krause and Maas (1990) noted that the abrupt appearance of primates in the fossil record of the northern continents was roughly coincident with the initial contact between the Indian subcontinent and Eurasia during the early Eocene, and hypothesized that primates (as well as artiodactyls

and perissodactyls) originated on the Indian subcontinent as it was adrift in
the Indian Ocean and subsequently dispersed into Eurasia upon contact (see
also Miller *et al.*, 2005). However, there are several difficulties with this scenario. One is timing. The Indian subcontinent appears to have separated from
other parts of Gondwana by 120 mya, well before there is any fossil evidence of
placental mammals anywhere. The last separation of India with any part of the
Gondwanan landmass was the separation from Madagascar at approximately 90
mya. There is no evidence from the Cretaceous fossil record of Madagascar for
any of the modern mammal groups and all of the extant mammals on that island
seem to be derived from Africa, via rafting (Krause, 2001). In a more direct
test of this hypothesis, Clyde *et al.*, (2003) studied the mammalian fossil record
from sediments adjacent to the suture zone between India and Asia in southern
Pakistan. They found that deposits prior to contact contained only endemic
primitive mammals and that the first evidence of modern groups—primates,
artiodactyls and perissodactyls—was well after initial contact and after these
mammals were known on other continents. This suggests that they dispersed
into the Indian subcontinent after its contact with Eurasia.

A phylogenetic and biogeographical approach, discussed in two other chapters of this volume (see Beard, this volume; Heesy *et al.*, this volume), involves
consideration of the biogeography of the taxa most commonly recognized as
the sister taxa of primates: plesiadapiforms, Dermoptera, and Scandentia. Two
of these, Dermoptera and Scandentia, have an Asian distribution for all living
and fossil species. For plesiadapiforms, the situation is more complicated. Plesiadapiforms are abundant during the Paleocene and Eocene of North America
and Europe, and they are present, but less common in the Eocene of Asia.
The fact that all three outgroups to primates have an Asian distribution is
strong evidence for an Asian origin for primates (Heesy *et al.*, this volume;
Beard, 1998; this volume; Beard and Dawson, 1999). Likewise, one of the fossil taxa most commonly cited as a basal primate, *Altanius*, is from Mongolia
(Dashzeveg and McKenna, 1977; Gingerich *et al.*, 1991). The major caveat to
this view comes from the presence in Africa of the Late Paleocene *Altiatlasias*.
This taxon is found earlier than any other known primate on any continent.
However, it is known only from isolated teeth and its affinities are unclear.
It has been suggested by some to be a plesiadapiform (Hooker *et al.*, 1999),
by others to be a basal primate (Gingerich, 1990; Sige *et al.*, 1990), and by
still others to be an anthropoid (Godinot, 1994; Beard, this volume; Seiffert
et al., 2005). The Early and Middle Eocene of Africa is a very poorly sample

time period, so *Altiatlasias* stands out, regardless of its affinities. Moreover, the presence of any primate or even a plesiadapiform in the Paleocene of Africa makes that continent a viable candidate for primate origins (Heesy *et al.*, this volume).

A more direct phylogenetic approach has come from the study of the earliest primates themselves. As discussed later, the first primate to appear in the fossil record of Asia and North America (and possibly Europe) is the omomyoid *Teilhardina*. Phylogenetic analyses of omomyoid primates by Ni *et al.*, (2004) and Hooker (2004) find that the Asian species is the most primitive suggesting that this continent is the origin for the taxon. In addition, the North American *Teilhardina americana* is considered the most derived species, suggesting that the earliest primates traveled from Asia, to Europe, to North America. However, it is worth noting that these analyses were of *Teilhardina americana*, not the earlier *Teilhardina brandti*.

In summary, most analyses point to Asia as the most likely source for primate origins (Beard, 2002), but the issue is far from resolved. Also unresolved is the nature of primate dispersal from their point of origin. If an Asian origin for primates is accepted, it is unclear whether the dispersal route of the group went first through Europe and then to North America or first across the Bering Strait into North America and then into Europe. It is also possible that primates dispersed from Asia in both directions to North America and Europe virtually simultaneously. Even more confusing is the position of *Altiatlasias* from the Paleocene of North Africa. If *Altiatlasius* is considered the earliest primate and Africa is regarded as the continent of primate origins, a northward migration of the group from Africa to Europe may have been possible during the Early Eocene (e.g., Gingerich, 1990), but the Tethys Seaway would still have likely been a barrier. Likewise, the Tethys Seaway would have made a northeastern migration route from Africa to Asia and then either to North America or Europe difficult. Further clarification of the geographic origins of primates and their subsequent pattern of dispersal awaits future research and fossil discoveries.

The Earliest Fossil Primates

The primates that appear in the earliest Eocene of North America and Europe belong to two distinct taxa—the adapoids and the omomyoids. Although the earlier members of these groups are strikingly similar in details of their dentition, they are more distinctive in cranial and postcranial anatomy. The adapoids show features that align them with living strepsirrhines (lemurs and lorises) (see Dagosto, 1988; Beard et al., 1988), while the omomyoids have features characteristic of haplorhines (tarsiers and anthropoids) (see Beard et al., 1988; Ross, 1994; Kay et al., 1997). They appear in the earliest Eocene of North America and Europe where they continue on both continents until the end of the Eocene. Omomyoids are also known from the Early Eocene of China and the Eocene of Pakistan. Adapoids are also found in the Eocene and Oligocene of Africa, the Middle Eocene of China, and the Miocene of southern Asia.

As mentioned above, two of the earliest and dentally most primitive taxa of both adapoids (*Cantius*) and omomyoids (*Teilhardina*) are found in the earliest Eocene of both Europe and North America, reflecting the Northern connection among these continents at the base of the Eocene as well as the Paleocene-Eocene boundary warming which enabled migration through the high latitude connections. However, for the remainder of that epoch, the radiations of the groups in Europe and North America are distinctive, despite the warm global climates indicating that by early Eocene the North Atlantic land connection had been breached. Hooker (2004) has further noted that the faunal identity between North America and Europe disappeared earliest among the primates but continued somewhat later for perissodactyls and artiodactyls indicating that there was some sort of filtering effect in place even before the connection disappeared.

In North America, the early primate fauna consists of a radiation of time-successive and closely related genera and species of notharctines derived from the basal *Cantius* that go extinct in the Middle Eocene when global temperature drop. Apart from *Cantius*, which is also found in Europe during the earliest Eocene, the notharctines are endemic to North America. Although North American Eocene faunas are known mostly from the western USA, the few Eocene primates from the Eastern USA in Georgia and Alabama are similar to those found in the West (Beard and Dawson, 2001). Coincident with the radiation of notharctines is a radiation of omomyoids placed in two subfamilies. The smaller and more primitive anaptomorphines are more common in the earliest part of the Eocene and the omomyines become more common in the Middle Eocene and last until the end of the epoch.

As discussed above, the earliest fossil prosimians of Europe are the same taxa as those of North America—the notharctine adapoid *Cantius* and the omomyoid *Teilhardina*. However, European primates from the rest of the epoch are distinct. The adapoids consist of the cercomoniines, a separate subfamily related to the notharctines, and the adapids. The omomyoids of Europe are less diverse than those of North America and belong to a distinct family, the Microchoeridae. It has frequently been argued that microchoerids are derived from anaptomorphines, and Hooker (2004) has recently suggested that this endemic European family is derived from a species of *Teilhardina*.

The sole exception to the subfamilial distinctiveness of the North American and European primate faunas throughout the Eocene is the presence of the cercomoniine genus *Mahgarita* in the latest Eocene of southern Texas. This taxon is a confusing biogeographical anomaly as the land connection between Europe and North America had long been breached and cercomoniines are not known from the Eocene of eastern Asia. Slightly younger deposits in the same region of Texas have yielded another anomalous taxon, *Rooneyia*, that has traditionally been regarded as an omomyoid, but not obviously related to other omomyoids from North America. More recent analyses have argued that *Rooneyia* is not an omomyoid, but some sort of unique haplorhine, perhaps more closely related to anthropoids (Ross *et al.*, 1998; Kay *et al.*, 2004).

The primate fauna of Europe changes dramatically in the Late Eocene with the arrival and subsequent radiation of a new family, the adapids. Beard *et al.*, (1994) have argued that adapids are immigrants from Asia based on the presence of the genus *Adapoides* in the Middle Eocene of Asia. At the end of the Eocene, most of the adapoids and microchoerids go extinct along with many other mammalian taxa in the major faunal turnover known as the Grand Coupure (e.g., Hooker, 1992).

The distribution of adapoids and omomyoids in Europe and North America clearly reflects the changes in continental relationships and climate during the Eocene. There are identical taxa shared between the continents in the earliest Eocene, when North America and Europe were connected, but the primate faunas subsequently diverged with no taxa in common and very different radiations by the end of the Eocene. At the end of the Eocene, with the decrease in global temperatures, primates disappeared from both continents, with perhaps a few exceptions. In Europe, the disappearance of primates is part of the major faunal transition called the Grand Coupure. This event has traditionally been related to the dramatic drop in global temperatures at the end of the Eocene, resulting from the opening of circum Antarctic currents with the separation of Tasmania and Australia from Antarctica (Zachos *et al.*, 2001).

Endemic Eocene faunas of Asia are less well-known than those of Europe and North America, but is has become increasingly evident in recent years that they have a very different composition. Despite the numerous arguments for Asia as the site of primate origins and the presence of *Teilhardina* (see above), both adapoids and omomyoids are relative scarce. Several omomyoids, from the Middle and Late Eocene of China are shared with North America suggesting repeated faunal connections through the Bering straits (Beard, 1998), and a few isolated teeth from India and Pakistan have been argued to demonstrate connections with Europe. Apart from the Middle Eocene *Adapoides*, which may be an Asian adapid, most of the adapoid primates from Asia seem to belong to an endemic group, the Sivaladapidae (Qi and Beard, 1998). In contrast with the adapoids of Europe and North America, which became extinct with the climatic deterioration at the end of the Eocene, the sivaladapids survived in southern Asia until the end of the Miocene. In addition, the Eocene faunas of Asia also contain two endemic groups whose affinities with other primates are extensively debated—the amphipithecids and eosimiids, and a third group that is still present today—the tarsiers.

Although Eocene faunas of Africa are poorly known, there are several adapoids from North Africa that show similarities to European cercomoniines (Hartenberger and Marandat, 1992; Rasmussen, 1994; Simons *et al.*, 1995; Simons, 1997). Isolated omomyoid teeth have also been described from the later Eocene, but the identity of these is unclear.

The Origins of Extant Prosimian Clades

There are two living groups of nonanthropoid primates: (1) strepsirrhines, the lemurs of Madagascar, the lorises and galagos of Africa, and the lorises of Asia; and (2) tarsiers, from the islands of the Sunda Shelf (Philippines, Borneo, Sulawesi, and Sumatra). The biogeography of the fossil primates most clearly related to these living taxa is generally similar to the present-day distributions with a couple of possible exceptions (Fleagle, 1999).

Today, lorises and galagos have a broad Old World distribution. The galagos are restricted to sub-Saharan Africa while two genera of lorises are found in Africa and two in Asia. Because the lorises of Africa and those of Asia each consist of a plump form and a very slender form, primatologists have debated for decades whether the African and Asian taxa were natural groups (e.g., Schwartz, 1992; Rasmussen and Nekaris, 1998). All studies of molecular systematics support the view that the geographic groups are clades and that the adaptive divergences in body form have taken place in parallel (Yoder, 1997). Several fossil lorises and galagos have been described from the Early to Middle Miocene and Pliocene of East Africa, within or near the range of the current taxa. The temporal and geographic ranges of lorises and galagos have recently been extended by the description of a fossil loris and a fossil galago from the Middle Eocene of Egypt (Seiffert *et al.*, 2003). These new taxa are farther north in Africa than any living lorisoid, and they also exist prior to the formation of the Sahara Desert. In addition, these new fossils from Egypt seem to push the origin of the living strepsirrhine clades in Africa at least back to the Middle Eocene (Seiffert *et al.*, 2003; Rossie and Seiffert, this volume). Although these taxa seem to have tooth combs like modern strepsirrhines (a very significant synapomorphy), the allocation to the families Lorisidae and Galagidae is still only based on inferred dental similarities. Additional remains are needed to confirm the allocation, but all available evidence indicates that the living lorisid clades have been in Africa since the Eocene.

In contrast, the present-day distribution of lorises in Asia seems to be a more recent event dating to the Middle Miocene (Jacobs, 1981; MacPhee and Jacobs, 1986; Flynn, pers. comm.). It seems likely that the migration of lorises to Asia was part of the same biogeographical event as the dispersal of the early apes (discussed below) since the fossil lorises are found in the same deposits as *Sivapithecus*.

Although the phylogenetic unity of the Malagasy primates has been questioned from time to time, all recent studies of molecular systematics indicate that they are a natural group, with the Aye-Aye as the sister taxon of all other families (e.g., Yoder, 1997). There are numerous extinct taxa also known from Madagascar. Although the extinct species are generally larger than extant species, and many are adaptively very distinctive, all seem closely related to living Malagasy taxa (e.g., Jungers *et al.*, 2002; Godfrey and Jungers, 2002, 2003). However, the biogeographic origin of lemurs on the Island of Madagascar has been topic of considerable debate (e.g., Krause, 1997; McCall, 1997; Yoder *et al.*, 2003). Because there is no fossil record from Madagascar between the end of the Late Cretaceous, more than 65 mya, and the end of the Pleistocene, 15,000 years ago, there is no direct evidence for the timing of the appearance of lemurs on Madagascar. However, the lack of any modern orders of mammals on the island in the Cretaceous and biomolecular studies showing that all modern clades diverged after Madagascar was separated from Africa (e.g., Yoder *et al.*, 1996;

2003; Yoder, 1997; Yoder and Yang, 2004), both indicate that the modern lemurs arrived via some sort of rafting from Africa (Figure 3).

The one possible exception to the cladistic and biogeographical unity of the Malagasy primates is the genus *Bugtilemur* from the Oligocene of Pakistan. *Bugtilemur* has been argued to be nested within the extant cheirogaleids (Marivaux *et al.*, 2001); however, this species is known from only isolated teeth and it lacks a tooth comb—a distinctive feature of all other living members of this clade—making the allocation very suspect (Seiffert *et al.*, 2003).

The other extant group of prosimians, the tarsiers, are found today on numerous islands of Southeast Asia: Borneo, Sumatra, and Sulawesi as well as several of the Philippines. The fossil record extends the range of the family Tarsiidae well onto the mainland of Asia, with several species and genera from the Eocene of China (Beard, 1998b) and one from the Miocene of Thailand (Ginsberg and Mein, 1986). This temporal distribution suggests a long record of tarsiids in mainland Asia. More contentious is *Afrotarsius* from the early Oligocene of Egypt which is considered by some to be a fossil tarsiid (e.g., Simons and Bown, 1985; Rasmussen *et al.*, 1998) and by others to be an early anthropoid (see Beard, this volume). There is considerable debate over the geographic origin of tarsiers. Most authorities probaabaly favor an Asia origin, but there is support for an African origin as well (e.g. Miller *et al.*, 2005). As Simons (2003) has emphasized, most of these fossil tarsiers are known only from dental remains, leaving open the question of the extent to which they share the many distinctive cranial and postcranial features of the extant *Tarsius*, and obscuring to some degree their true affinities. If *Afrotarsius* is a tarsier, it would extend the geographic range of the family tremendously.

ANTHROPOIDS

There are three major groups of anthropoids alive today. Platyrrhine monkeys are restricted to South and Central America. Old World monkeys (Cercopithecoidea) are found in Africa, Asia, and the island of Gibraltar. Apes are found in Africa and Asia, except for humans who are found all over the world. Most authorities believe that tarsiers or omomyoids are the sister taxa of living anthropoids (see Ross and Kay, 2004). However, in addition to these crown anthropoids, there are many taxa of stem anthropoids, and the biogeography of anthropoid origins is a major topic of debate (see Heesy *et al.*, this volume; Beard, this volume; Seiffert *et al.*, 2004; Rossie and Seiffert, this volume). As with the phylogenetic origin of anthropoids, many different approaches have been used to determine the biogeographic origin of anthropoids, using the distribution of extant taxa, the distribution of fossil taxa, and considerations of biogeographic barriers and corridors.

Anthropoid Origins

The earliest anthropoids that have all (or almost all) of the features characteristic of living anthropoids are from the Late Eocene and Early Oligocene of Africa, in the Fayum depression of Egypt. Somewhat older, more poorly known taxa from Algeria, *Algeripithecus*, *Tabelia*, and *Biretea*, appear morphologically similar to the Fayum taxa, but are much smaller and known only from isolated teeth. The Late Paleocene *Altiatlasias* from Morocco is also known only from isolated teeth and is considered by some as an anthropoid (Godinot, 1994; Beard, this volume; Seiffert *et al.*, 2005) but by others as a basal primate (Sige *et al.*, 1990; Gingerich, 1990) or a plesiadapiform (Hooker *et al.*, 1999). These fossils establish undoubted anthropoids in Africa by the Middle Eocene or earlier, but only the Late Eocene and Oligocene taxa are known from cranial and postcranial remains.

In Asia, "modern anthropoids" appear only in the Early-Middle Miocene, but there is an increasing diversity of Eocene primates from both China and Myanmar. However, there is also considerable debate concerning whether they are anthropoids or not (Beard, this volume; Gunnell and Miller, 2001; Ciochon and Gunnell, 2002, 2004; Kay *et al.*, 2004). Eosimiids are known from the Middle Eocene of China and Myanmar. They are known primarily from dental remains with some attributed ankle bones (Beard *et al.*, 1994, 1996; Gebo *et al.*, 2000, 2001; Beard and Wang, 2004;). The dentition of eosimiids is much more primitive than that of any of the African anthropoids except *Altiatlasius*, and eosimiids are considered stem anthropoids by most authorities (Kay *et al.*, 2004; Seiffert *et al.*, 2004; Rossie and Seiffert, this volume; Beard, this volume). The more primitive dental morphology and the presence in Asia of tarsiers, which are thought by many to be the sister taxon of anthropoids, make Asia a likely source for anthropoid origins (Beard, this volume; Heesy *et al.*, this volume). Beard (this volume) suggests that *Altiatlasius* is a Paleocene anthropoid migrant from Asia closely related to the Eocene Eosimiidae, where these Middle and Late Eocene taxa represent late-surviving members of the initial anthropoid radiation. If *Afrotarsius* is a basal anthropoid, it would also possibly represent a remnant of the African descendants of the eosimiids. However, if *Afrotarsius* is a tarsicr, the case for Africa as the source of anthropoids is equally strong (Heesy *et al.*, this volume). If *Altiatlasias* is an anthropoid related to eosimiids, any scenario must account for large amounts of lost time and intermediate taxa from the Paleocene of Africa to the Middle Eocene of China. As previously mentioned, the Tethys Seaway would have made a migration either from Africa to Asia or from Asia into Africa difficult during the Late Paleocene to Early Eocene. In the current situation, any assessment of the biogeography of anthropoid origins is heavily determined by the phylogenetic placement of these key taxa.

Compared with other major evolutionary events in the primate fossil record, the origin of anthropoids can not easily be related to any tectonic or climatic changes. Part of this lack of any correspondence may reflect uncertainty regarding the time and place of anthropoid origins. If the origin of anthropoids was in the late Cretaceous as many molecular clock estimates suggest (see Miller *et al.*, 2005), the possible biogeographic scenarios for anthropoid originsare virtually limitless because of vast amount of missing fossil information. Nevertheless, it is increasingly clear that anthropoids were present in Africa and probably Asia by the Middle Eocene (Seiffert *et al.*, 2005). It is conceivable that the early evolution of the group and the extensive radiation of tiny anthropoids suggested by limb bones in China (Gebo *et al.*, 2001) is related to the Eocene Climatic Optimum which would have greatly expanded tropical vegetation in higher latitudes. However, most known early anthropoids are from lower latitudes that are likely to have experienced less dramatic change. Similarly, it is difficult to account for any dispersals between Asia and Africa (Beard, this Volume) across the Tethys Seaway during the Eocene or Early Oligocene in the context of the paleogeography of the time (Figure 1). Indeed, apart from *Afrotarsius*, faunas of the Eocene and Early Oligocene of Africa and Asia show no evidence of a connection (Rossie and Seiffert, this volume).

More controversial are the amphipithecids, known from the Middle Eocene of Myanmar (e.g., Takai and Shigehara, 2004). They are considered by some to be stem anthropoids, others to be catarrhines, and still others to be adapoid prosimians (Ciochon and Gunnell, 2002). Like the eosimiids, they are known primarily from dental remains, with some attributed cranial material and a few attributed postcranial bones. In any case, they are apparently a group that is endemic to Southeast Asia.

Platyrrhines and Platyrhine Evolution

Living platyrrhines range through most of tropical South and Central America from northern Argentina in the South to Central Mexico in the North. They are currently divided into two families with five subfamilies: callitrichines, aotines, and cebines in the Cebidae, and pitheciines and atelines in the Atelidae. The fossil record of platyrrhines has a very scattered distribution in both time and space (Fleagle *et al.*, 1997). Except for the earliest genus, *Branisella*, most fossil taxa can be placed (with some disagreement) in the extant subfamilies and families. Most extant subfamilies are known at least from the Late Miocene of Colombia (Fleagle *et al.*, 1997; Hartwig and Meldrum, 2002).

Much of the platyrrhine fossil record is from geographical areas that lack primates today, documenting a greater geographic range of platyrrhines in the past. There is an extensive record (over seven genera) of fossil monkeys from Early and Middle Miocene of southernmost Argentina and Chile (Patagonia) at latitudes up to 52° S (Fleagle and Tejedor, 2002). This corresponds with the Early-Middle Miocene warming of global temperatures. In the north, there is increasing fossil evidence of a diverse fauna (3 genera) of very unusual platyrrhines from the islands of the Caribbean: Cuba, Jamaica, and Hispaniola (MacPhee and Horovitz, 2002). Most of the fossils are from recent Pleistocene time periods, but there is one fossil dated to 20 mya. MacPhee and colleagues have argued that the Caribbean taxa represent a single endemic radiation that seems to have survived into the late Pleistocene and maybe even into the last few hundred years (MacPhee *et al.*, 1995; Horovitz and MacPhee, 1999). Although the mammalian faunas of the Caribbean have traditionally been thought to be the result of rafting, MacPhee and colleagues have argued that primates may have originally colonized the region through land connections with northern South America in the Late Paleogene (Iturralde-Vinent and MacPhee, 1999; MacPhee and Horovitz, 2002).

The most biogeographically challenging aspect of platyrrhine evolution concerns the origin of the entire clade (e.g., Hartwig, 1994; Fleagle, 1999). South America was an island continent throughout most of the Tertiary, and most of the orders of mammals found in Paleocene through Miocene deposits are endemic families or orders almost exclusively restricted to that continent. Primates first appear in the Late Oligocene and become common only in the Early Miocene. Rodents also appear first in the Oligocene. Both groups are almost certainly immigrants from some other continent, and paleontologists have debated for much of this century how and from where primates reached South America.

For much of the twentieth century it was widely held that the higher primates of the New World, platyrrhines, and the higher primates of the Old World, catarrhines, evolved their higher primate features in parallel from different prosimian ancestors. Platyrrhines were thought to have been descended from the North American Eocene prosimians, either adapoids or possibly omomyoids (Gazin, 1958; Simons, 1961). However, with the increasing acceptance of phylogenetic methods and molecular systematics, the evidence has become overwhelming that platyrrhines and catarrhines are sister taxa and almost certainly share a common ancestry, a point appreciated by Scott (1913) early in the twentieth century. In addition, other than the platyrrhine monkeys of the Caribbean, there is no evidence of anthropoids in North America prior to the appearance of humans in the past 20,000 years.

In contrast to the absence of likely platyrrhine ancestors in North America, many of the early anthropoids from the Eocene and Oligocene of Africa share numerous, primitive, morphological features with extant and fossil South American platyrrhines. One genus, *Proteopithecus*, has absolutely no features that would distinguish it from a basal platyrrhine, and indeed, some authorities have argued that it is a part of the radiation of crown platyrrhines (Takai *et al.*, 2000). However, this raises a difficult biogeographical issue. South America is separated from Africa by a distance of at least 2600 km, making a phylogenetic and biogeographic link between the primate faunas of the two continents seem very unlikely. With the discovery of plate tectonics in the 1960s, however, Hoffstetter and Lavocat (1970) offered a novel solution, arguing that primates (and caviomorph rodents) rafted from Africa to South America in the early Tertiary when the South Atlantic was much narrower. Although their initial suggestion met with considerable resistance, it has become increasingly accepted in comparison to arguments for a North American or Antarctic origin of platyrrhines (Hartwig, 1994; Fleagle, 1999; Houle, 1999). Multiple factors such as the prevailing paleowinds and paleocurrents across the Atlantic Ocean from 50 to 30 mya would have made the journey from Africa to South America anywhere from 8 to 15 days (Houle, 1999), making rafting a much more plausible scenario than it first seems. Moreover, the fist appearance of rodents and monkeys in South America comes just after the major drop in sea level in the early Oligocene, a time when the distance between continents would have been reduced. The absence of any anthropoids from North America combined with the considerable morphological evidence of a South American-African connection within the rodent and primate faunas, the rafting hypothesis is the most likely scenario for the biogeographic origin of platyrrhines.

If rodents and anthropoid primates from the Late Eocene or Early Oligocene of Africa rafted to South America, why not lorises and galagos as well? After all, lorises and galagos are also documented in Africa at this time (Seiffert *et al.*, 2003), and aspects of strepsirrhine life history make them possibly better candidates for rafting than those of anthropoids. However, dispersal by rafting across the Atlantic Ocean is clearly a chance event, an example of "sweepstakes" dispersal. One can only speculate that by a stroke of good luck anthropoids were able to "win" the sweepstakes while lorises and galagos did not.

Early Catarrhines

By the Late Eocene or Early Oligocene of Africa, the primate fauna contained not just primitive anthropoids and potential platyrrhine ancestors, but also stem catarrhines: anthropoids that were more closely related to living catarrhines than to platyrrhines, but that lacked any of the distinctive features of either Old World monkeys or apes. It seems likely that basal catarrhines evolved in Africa and were restricted to that continent until the late Early Miocene when the Tethys Seaway first closed off to form the Mediteranean Sea (Whybrow and Andrews, 2000; Rogl, 1999; Figure 3). Among the first catarrhines in Eurasia during the Early Miocene were the pliopithecids, in Asia. This group probably originated in Africa and appears in the Early Miocene of China before they subsequently spread westward into Europe (Harrison and Gu, 1999; Harrison, 2002; in press; Rossie and Seiffert, this volume). This view is supported by the presence of incipient pliopithecid features in some of the stem catarrhines from the Miocene of East Africa. The dispersal of pliopithecids from Africa into Eurasia probably took place as a part of the same event as the hominoid dispersal from Africa to Eurasia during the Early Miocene (see below), as suggested by new dates for the earliest hominoids in Eurasia (Begun and Nargolwalla, 2004).

The extant Old World monkeys, Family Cercopithecidae, are among the most successful and taxonomically diverse living primates with species and genera in two major subfamilies: Cercopithecinae and Colobinae. Both subfamilies are found in Africa and Asia, but they have very different patterns of diversity on the two continents. In Africa, cercopithecines are the most diverse with 11 genera and roughly 40 species; in contrast, there are only six species of

African colobines placed in three genera. In Asia, there are between 20 and 30 species of colobines placed in up to eight genera. There is only a single Asian cercopithecine genus, *Macaca*, but it is very widespread and divided into 18 species, six of which are found on the island of Sulawesi.

The fossil record provides a very different view of the biogeography of Old World monkeys both within and between continents. The origin of the group appears to be clearly African. The single family of stem cercopithecoids, the Victoriapithecidae, is known only from Africa as are the most likely sister taxa among the basal catarrhines (Rossie and Seiffert, this volume). The divergence of the modern subfamilies is documented earliest in Africa by the presence of a colobine at 11 mya (Benefit and Pickford, 1986; Delson, 1994). In the latest Miocene, Pliocene, and Pleistocene of Africa there are numerous taxa of fossil colobines that are larger and very different adaptively from the extant taxa. In addition there are isolated teeth that are generally attributed to the extant genera.

Very soon after their first appearance in Africa during the Late Miocene, colobines are widespread in Europe (Delson, 1994) and they first appear in Asia (Northern India and Pakistan) by the Late Miocene, between 7 and 5 mya (Barry, 1987) and also in the latest Miocene of China (Delson, 1994). In the Pliocene, there are two colobines known from Europe, one of which seems to extend across to northern Asia. There are numerous fossils attributed to *Rhinopithecus* from the Pliocene and Pleistocene of China (Pan and Jablonski, 1987) and remains of both of the small living genera, *Presbytis* and *Trachypithecus*, from the islands of the Sunda Shelf. Thus, compared with the extant colobine diversity of Asia, there is surprisingly little material in the fossil record.

The biogeographic record of fossil cercopithecines is similar to that of colobines. The earliest cercopithecines are from the latest Miocene of Africa and Europe dated to approximately 7 mya (Delson, 1973, 1975; Szalay and Delson, 1979; Leakey *et al.*, 1996, 2003). In the Pliocene and Pleistocene of eastern, southern and northern Africa there is a moderate diversity of large baboon and gelada-like taxa, but a very limited record for *Cercopithecus* guenons. Outside of Africa, the record of cercopithecines in mostly limited to the genus *Macaca* and two large relatives (*Procynocephalus* and *Paradolichocebus*) which appear in Europe in the latest Miocene through Pleistocene and in Asia by the Early Pliocene into the Pleistocene. The most striking biogeographic anomaly among fossil cercopithecines is the genus *Theropithecus*, which is limited today to the highlands of Ethiopia but was more widespread and diverse in the Pliocene and Pleistocene of Africa. More strikingly, *Theropithecus* has been reported from early Pleistocene sites in Israel, India, and Spain, and possibly Italy (Gupta and Sahni, 1981; Gibert *et al.*, 1995; Belmaker, 2002; Rook *et al.*, 2004), a distribution that is only exceeded by *Homo* among anthropoid primate genera.

Most of the evolutionary history of the cercopithecids postdates major tectonic events in the Old World, and there are few obvious patterns linking their biogeography to major climate changes during the past 15 million years. However, it seems likely that the dispersal of both colobines and cercopithecines out of Africa in the latest Miocene was facilitated by the Messinian Crisis when the Mediterranean dried up. More broadly, several authors have related the expanding diversity of Old World monkeys in Africa during the past 15 million years to general climatic deterioration and reduction of rainforests across the continent (e.g., Delson, 1975; Fleagle, 1999). Similarly, the spread of cercopithecoids in Europe seems to be favored by drier open habitats, in contrast with the distribution of hominoids (Eronen and Rook, 2004). Pickford (1993) correlates the remarkable distribution of *Theropithecus* during the Pleistocene to warmer, interglacial climatic conditions in which the entire East African fauna seems to expand into higher latitudes, possibly explaining the occurrence of the genus in North Africa and the Iberian peninsula. Moreover, the present patterns of Old World monkey biogeography in the Sunda Shelf have also been driven by the fluctuating Plio-Pleistocene sea levels associated with cold periods and glacial cycles (e.g., Delson, 1980; Eudey, 1980; Brandon-Jones, 1996; Abegg and Thierry, 2002; Harrison *et al.*, this volume). Likewise it has been argued that the current biogeography of African Old World monkeys, specifically guenons, is related to Pleistocene refugia and other vegetation changes associated with glacial cycles (Hamilton, 1988; Oates, 1988; Colyn *et al.*, 1991).

Hominoid Biogeography

Like their sister taxon, the Cercopithecoidea, the hominoids had an African origin in the Late Oligocene or Early Miocene. There is increasing evidence that most of the fossil "apes" from the Early and Middle Miocene of East Africa are more probably stem hominoids that precede the divergence of living apes or even stem catarrhines that precede the Old World monkey-ape divergence (Harrison, 2002; Rossie and Seiffert, this volume). Only *Morotopithecus* and *Proconsul* have been argued to show synapomorphies with living hominoids (e.g., Rae, 1993; MacLatchy *et al.*, 2000).

The first evidence of African mammals dispersing from Africa to Eurasia occurs at about 18 mya, but evidence of hominoids in Eurasia is somewhat later. The earliest evidence of a "fossil ape" outside of Africa is between 16.5 and 17 mya from southern Germany and also from Pasalar in Turkey, roughly the same time that the earliest pliopithecids are found in China (Harrison and Gu, 1999; Begun 2003). From the Middle Miocene until the Pliocene, fossil hominoids are very rare in Africa and yet widespread in Eurasia. Begun and colleagues (Begun, 2001; Begun *et al.*, 2003) have argued for a series of migrations of hominoids between Africa and Eurasia in the Middle and Late Miocene. On the basis of the diversity of hominoids in Eurasia and the absence in Africa during this time, Begun and others (Begun *et al.* 1997; 2003; Stewart and Disotell, 1998) have argued that the clade consisting of African apes and humans (and our extinct relatives) originated in Eurasia and then reinvaded Africa just prior to the divergence of African apes and hominins. However, this scenario has been challenged by others for a variety of reasons (e.g., Andrews and Bernor, 1998; Cote 2004).

The initial Miocene dispersal of land mammals from Africa into Eurasia seems to correspond to the collision of Africa with Eurasia, an event which sealed off the Tethys Seaway and formed the Mediterranean Sea (Figure 3). However, this land bridge was apparently ephemeral (Rogl, 1999), and between roughly 16 and 14.5 mya it was breached for several million years, again isolating the two continental areas. This diversity of fossil apes (and pliopithecids) in Eurasia during the Middle Miocene also corresponds to a major period of global warming—the highest since the Eocene (e.g., Zachos *et al.*, 2001). The disappearance of fossil hominoids from Europe in the Late Miocene as well as their distribution throughout the epoch seems to reflect climatic deterioration (e.g., Eronen and Rook, 2004).

In the Pliocene, fossil hominoids are known primarily on the basis of abundant hominins from eastern and southern Africa and a few apes from the tropical regions of Asia. The present day diversity and biogeography of hylobatids in Southeast Asia is generally thought to be the result of the separation and coalescence of islands on the Sunda Shelf during Pleistocene sea level changes (e.g., Chivers, 1977; Marshall and Sugardito, 1986).

The biogeographic history of African hominins in the Pliocene is a debated issue (Turner and Wood, 1993; Strait and Wood 1999), as is the relationship between species turnovers and climatic change (Vrba, 1988; 1992; deMenocal, 2004; Frost, in press). The earliest fossil hominids appear in the latest Miocene

Hominid Dispersals

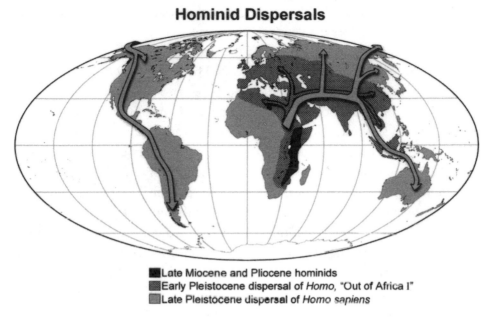

■Late Miocene and Pliocene hominids
▤Early Pleistocene dispersal of *Homo*, "Out of Africa I"
▢Late Pleistocene dispersal of *Homo sapiens*

Figure 4. Major patterns in the distribution and dispersal of hominids from the Late Miocene until the end of the Pleistocene.

of eastern and central Africa. By the Pliocene there was a considerable diversity of species known from these same parts of Africa (Figure 4). Between two and seven hominin dispersal events between East and South Africa are hypothesized during the Plio-Pleistocene, and while a hominin turnover-pulse is argued for during the 2.5 climatic cooling event, cercopithecoid monkey diversity at this time does not reflect a similar turnover event (Vrba, 1988; 1992; Turner and Wood, 1993; Bromage *et al.*, 1995; Strait and Wood, 1999; Frost, in press). In any case, the Pleistocene is characterized by at least two major dispersals (Figure 4) of hominins from Africa to Eurasia (e.g., Stringer, 2000)—first in the early Pleistocene by a taxon similar to *Homo erectus* (Anton and Swisher, 2004) and a second by *Homo sapiens* approximately 50,000 years ago (Lahr and Foley, 1994). There may well have been numerous other dispersals, but the record is open to differing interpretations. This final dispersal of *Homo sapiens* was unprecedented in primate history as it marked the first time any primate had reached Australia and the first time an anthropoid reached the temperate parts of North America. Although climate and geography have undoubtedly played important roles in guiding the spread of *Homo sapiens* over the world, the initial dispersal and subsequent history of our species is unusual in that it is a radiation that has used technological ability to occupy a greater diversity of climates and geographies than any other primate.

The initial appearance of *Homo* and the onset of Pleistocene glacial cycles with their tropical correlates have been related by Stanley (1996) to the rise of the Isthmus of Panama and closing off of the equatorial connection between the Atlantic and Pacific oceans. The timing and extent of the first dispersal of hominids outside of Africa in the early Pleistocene are contested issues and it is not clear how permanent initial dispersal events may have been (e.g., Dennell, 2003). Permanent range extension into the northern latitudes does not appear to have occurred until later in the Pleistocene, presumably when technology had advanced far enough to allow for adaptation to much cooler climates (e.g., Hoffecker, 2004). Range extension southward occurred by 50,000 in Australia despite the obvious water barrier (Stringer, 2000). As in other cases during the course of primate evolution, this barrier was most likely crossed by rafting. However, in stark contrast to the earlier primate rafting events, the rafting of the Pacific Ocean during Pleistocene human evolution was undoubtedly the result of technology and the physical building of sea-worthy vessels which subsequently permitted colonization of Pacific islands over immense distances. The migration of hominins into North America during the Late Pleistocene was probably facilitated by the exposure of the Bering Strait land bridge due to lower sea levels during a glacial period although many favor an overwater route for earliest dispersal (e.g., Dixon, 2001).

DISCUSSION

In the preceding sections we have reviewed the history of primate evolution on a clade by clade basis noting the role of tectonics and climate in the history of various primate radiations through time. In order to summarize the overall patterns, we can review this history in terms of the major tectonic and climatic changes over the past 65 million years. Climate and tectonics are obviously interrelated both causally and in their influence on primate evolution. Nevertheless, in the following sections we will make an attempt to identify their major effects on the course of primate evolution.

Tectonics and Primate Evolution

The relative positions and connection among continents have clearly had major effects on the history of primate evolution. The initial radiation of primates throughout the northern continents at the beginning of the Eocene was made possible by the lingering connection between North America and Europe. Likewise, the subsequent differentiation of the adapoids and omomyoids in Europe and North America reflects the breaching of that land bridge early in the Eocene. The differences between the primates of Africa and Asia throughout the Eocene and Oligocene are undoubtedly the result of the lack of any substantial connection between those continental regions during the Paleogene. The absence of primates in South America during the Eocene and the appearance of only as single endemic radiation resulting from a chance rafting dispersal event reflects the isolation of that continent throughout most of the Cenozoic. There is currently no evidence that the initial connection of the Indian subcontinent with Asia had any major role in primate evolution, contrary to hypotheses linking this event with initial primate dispersals (e.g., Krause and Maas, 1990; see also Miller *et al.*, 2005).

The dispersal of primitive catarrhines, hominoids, and presumably lorisoids from Africa into Eurasia in the late early Miocene was only made possible by the initial connection between Africa and western Asia, closing off the Tethys Seaway. Subsequent breaching and reestablishment of this connection has been implicated in the pattern of hominoid dispersals between Africa and Europe during the Middle and Late Miocene (e.g., Begun, 2003). The connection of South America to North America in the beginning of the Pliocene enabled platyrrhine monkeys to disperse into tropical parts of North America and much later this connection probably facilitated the spread of humans from North to South.

In addition to the direct contributions to primate evolution, tectonic events also have less dramatic but very significant longterm effects of primate evolution and biogeography as documented in other chapters of this volume. Mountain ranges and river drainages are major determinants of speciation events and primate distributions today. They have almost certainly been so in the past. More indirectly, but in many cases more profound, have been the effects of tectonic events in generating the major climatic changes throughout the Cenozoic (e.g., Zachos *et al.*, 2001) that have been important in guiding the course of primate evolution, as noted above and summarized below.

Climate and Primate Evolution

Climate and tectonics are obviously interrelated both causally and in their influence on primate evolution. This is most evident in the initial dispersal of

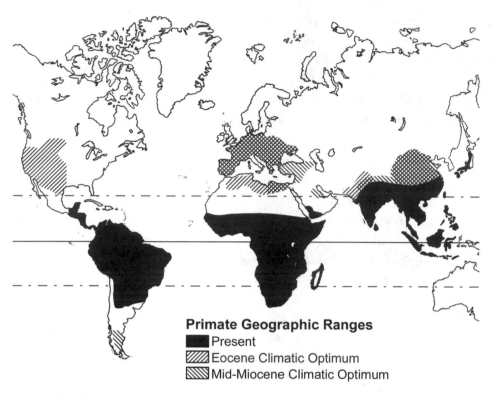

Figure 5. The geographic distribution of primates today compared with that during the two warmest times of the past 55 million years: the Eocene Climatic Optimum, and the mid-Miocene Climatic Optimum.

primates throughout the northern continents of North America, Europe, and Asia at the beginning of the Eocene Epoch. Although this dispersal was likely made possible by various land connections, a dispersal route at high latitudes was only feasible because of the dramatic temperature rise that took place at the Paleocene Eocene boundary (Gingerich and Ting, 2004). Probably the most significant pattern in primate biogeography over the past 65 million years is the diversification of primates at high latitudes that accompanied the two major temperature optima during the Cenozoic—one in the Eocene and one in the Middle Miocene (Figure 5).

In the Eocene, prosimian primates were both widespread and diverse in North America and Europe and almost totally disappeared from both continents by the end of the epoch with only a few exceptions. In the Middle Miocene, primates were again widespread and diverse in Europe. Hominoids and primitive catarrhines disappeared by the end of the epoch, but the more climatically tolerant Old World monkeys persisted into the Pliocene and Pleistocene. Similarly, in the Middle Miocene of South America, platyrrhine monkeys were found nearly to Tierra de Fuego, well beyond their known latitudinal range at any other time.

Less dramatic patterns in past primate distributions have also been related to climate. Changes in the primate faunas throughout the Early Eocene have been argued to reflect migrations from the south with increasing temperatures (Beard, 1988; Gunnell, 2002). In the Miocene and Pliocene of Europe, the distribution of fossil hominoids has been related to changing patterns of moisture and aridity (Eronen and Rook, 2004). In the Pliocene and Pleistocene of Africa, the appearance and disappearance of hominid taxa has been attributed to climate changes, but other taxa seem to show different patterns (e.g., Frost, in press). In a more recent framework, many explanations for the distribution of living primates in tropical area of the world today attribute them to Pleistocene climate changes and associated fragmentation of tropical forests (Hamilton, 1988; Oates, 1988; Colyn et al., 1991).

Mechanisms of Dispersal

In this review of the influence of tectonics and climate on primate biogeography we have mostly discussed the most dramatic events, usually dispersals of major primate clades from one continent to another, rather than distributions of individual taxa. It is evident from this review that primate evolution has involved almost all of the common mechanisms of dispersal (e.g., McKenna, 1975). Many dispersals have depended on standard land bridges, effected by tectonic change. The most striking example is the dispersal of hominoids (and probably lorisoids) out of Africa in the early Miocene. At the opposite extreme, it is also clear that sweepstakes dispersals through over-water rafting have also played important roles in the biogeographic history of primates. While rafting is a seemingly unlikely event in vertebrate dispersal patterns, especially among mammals, this mode of dispersal almost certainly accounts for the appearance of strepsirrhines of Madagascar and platyrrhines in South America, and possibly many more events (Rossie and Seiffert, this volume). The reasons for the prevalence of rafting during the course of primate evolution remain to be explained. However, the Malagasy strepsirrhine rafting event over the Mozambique Channel is more easily accounted for because of certain aspects of strepsirrhine life history, such as a low basal metabolic rate and extended periods of hibernation-like torpor. The case of the platyrrhines is more difficult to explain, as anthropoid primates have higher metabolic rates and do not have the ability for prolonged periods of torpor. A two-week rafting event across the Atlantic must have involved a floating island with an adequate food and water supply. However, such natural rafts are known to exist (Houle, 1999). The rafting dispersal by later hominins as the result of technology is a more specialized case, but again, the prevalence of over-water dispersal during primate evolution seems truly amazing for a mammalian order. Other dispersals such as that of the earliest primates at the Paleocene-Eocene boundary seem to have involved some sort of filter bridge since only a few taxa seem to have been involved. Other, less well-documented events such as dispersals between Africa and Eurasia during the Paleocene or Eocene may have involved some type of island hopping across the Tethys. One hypothetical method that does not seem to have taken place is McKenna's (1973) Noah's Ark mechanism in which a continent (India) transported a whole fauna to a new home (Asia). All available evidence indicates that when India docked on the Asian mainland, primates were not aboard.

Primate Communities

In our review of the fossil record of primate biogeography we have addressed the geographic history of individual clades largely in isolation. It is evident from this review that each clade seems to have had an idiosyncratic history of dispersal through geological time. Thus the biogeographical history of colobines as Old World monkeys is quite different from that of cercopithecines, or apes, or lorises, even though each seems to have an Africa origin and to have subsequently dispersed to other continents. An important corollary of these different histories is that the communities that we find in the forests of the world today are time-specific assemblages of individual units with very different histories. Consider, for example, the composition of many Asian communities which contain tarsiers, an endemic group dating back to the Eocene, apes which dispersed to Asia in the early Miocene, and numerous Old World monkeys which only entered Asia, again from Africa, perhaps via Europe, in the late Miocene or Pliocene. In contrast, Malagasy communities are all made up of endemic groups that seem to have radiated at approximately the same time. Platyrrhine communities similar to those seen today were already in place by the Middle Miocene of Columbia (Wheeler, 2003). Thus, the extant communities of different biogeographical regions are composed of elements with very different temporal and geographical histories (Fleagle and Reed, 1999); and also, the communities of previous epochs had very different phylogenetic and adaptive compositions from those of today (e.g., Fleagle and Kay, 1985; Fleagle, 1999). Study of the phylogenetic and biogeographical history of extant communities is a particular exciting approach in modern community ecology (Webb et al., 2002; Dohoghue and Moore, 2003).

Limitations and Biases of the Fossil Record

In any discussion of biogeography based on fossils, it is important to remember that the fossil record of primates, like that of other organisms, is subject to a tremendous series of limitations and biases that greatly affect the reliability of conclusions that can be drawn. The greatest of these is the general absence of any record for most of the earth's surface for most of geological time. The record that we do have is little more than an occasional window into particular time and particular places. Moreover the record is not randomly distributed. Madagascar has no fossil record at all between 65 mya and 10,000 years ago. Europe and North America have many fossils from this time period. The distribution of the known fossil record is due to many factors both past and present.

Thus western North America has an extensive and relatively continuous fossil record for much of the last 65 million years, although generally in different regions at different times. This is because western North America was a region of considerable geological activity and sedimentation for much of the Cenozoic and also the region is relatively arid and open today, exposing fossil at the surface for miles and miles where they can be recovered. In contrast, there is a very limited fossil record from eastern North America both because it had less geological activity during the past 65 million years and also is more covered with vegetation. Similarly eastern Africa has an extensive fossil record from the past 20 million years or so because the region was geologically active and is currently well-exposed, whereas the fossil record from most of the rest of the continent is very limited.

Certainly the lack of a fossil record from many parts of many continents limits our understanding of primate biogeography in the past, but it is probably impossible to determine what and how much we are missing in any particular case. For example, despite considerable distance and latitudinal differences, the primate fauna from one early Eocene site in Alabama is strikingly similar to early Eocene sites in Wyoming (Beard and Dawson, 2001). In contrast, *Mahgarita* and *Rooneyia*, the only two fossil primates from the Eocene-Oligocene boundary of southern Texas, are unlike any other primates known from North America, even though southern Texas is very close to extensive Eocene deposits in New Mexico. The major significance of the incompleteness of the fossil record for understanding primate biogeography is that we always have to keep in mind that paleontology is much more a record of presences than absences. Every important new find usually brings biogeographic surprises.

The Critical Role of Phylogeny

In is also necessary to keep in mind that all biogeographic scenarios or hypotheses are based on an hypothesis of phylogeny. In the case of fossils which are frequently known from a few teeth or jaws phylogenetic relationships and the implied biogeographic implications can be very tenuous. The most obvious example is *Altiatlasias* from the Paleocene of Morocco. As the only fossil primate from the Paleogene of Africa this is a critical taxon for primate biogeography. However its significance is very different depending upon whether it is a plesiadapiform, a basal primate, or an early anthropoid. This taxon is known only from isolated teeth and its phylogenetic affinities are not resolved. Likewise, the biogeographic significance of two other poorly known taxa, *Bugtilemur* and *Afrotarsius*, are dramatically different depending upon how their phylogenetic relationships are interpreted. It may seem ironic and frustrating that on the one hand biogeography is critical for understanding phylogeny (Rossie and Seiffert, this volume), while at the same time a correct understanding of phylogenetic relationships is critical for understanding biogeography. However, such is the case.

CONCLUSIONS

The present geographical distribution of primates is a single slice in time from a dynamic pattern of primate biogeography that has been continuously changing since the first appearance of the order over 55 mya. As we have tried to show in the previous pages, many of the origins, dispersals, and extinctions that have been responsible for the everchanging patterns of primate biogeography can be related to climatic changes on a global scale. Many of these global changes, in turn, can be related to changes in the configurations of continents and seas on the surface of the earth due to the processes of plate tectonics. Still others seem to have no clear explanation and may well reflect chance events resulting from the stochastic interactions of numerous factors. In this contribution we have focused on major geographic events such as dispersals between continents and on distribution patterns of major clades rather than on smaller scale geographic phenomena and taxa. In part this reflects the broad scope required by a chapter covering a whole order of mammals over 55 million years and partly because of the limitations of the sampling available in the fossil record. Nevertheless, even from this broad perspective, it is clear that individual clades have generally had very idiosyncratic biogeographic histories, presumably reflecting their differential characteristics of habitat use, dispersal abilities, demographic and life history parameters, as well as interactions with other taxa as competitors or predators. Our efforts at understanding the history of primate biogeography will be greatly advanced when we can integrate broad patterns of primate dispersal to the proximate factors involved.

REFERENCES

Abegg, C. and Thierry, B. 2002, Macaque evolution and dispersal in insular south-east Asia. *Biol. J. Linn. Soc.* 75:555–576.

Andrews, P. and Bernor, R. L. 1999, Vicariance biogeography and paleoecology of Eurasian Miocene hominoid primates, in: Agusti, J., Rook, L., and Andrews, P., eds., *Hominoid Evolution and Climatic Change in Europe. Vol 1: The Evoloution of Neogene Terrestrial Ecosystems in Europe*, Cambridge University Press, Cambridge, pp. 454–487.

Anton, S. C. and Swisher, C. C. 2004, Early dispersals of Homo from Africa. *Ann. Rev. Anthropol.* 33:271–296.

Barry, J. C. 1987, The history and chronology of Siwalik cercopithecids. *Hum. Evol.* 2:47–58.

Beard, K. C. 1988, New notharctine primate fossils from the Early Eocene of New Mexico and Southern Wyoming and the phylogeny of Notharctinae. *Am. J. Phys. Anthropol.* 75:439–469.

Beard, K. C. 1998, East of Eden: Asia as an important center of taxonomic origination in mammalian evolution. *Bull. Carnegie Mus. Nat. Hist.* 34:5–39.

Beard, K. C. 1998, A new genus of Tarsiidae (Mammalia: Primates) from the middle Eocene of Shanxi Province, China, with notes on the historical biogeography of tarsiers. *Bull. Carnegie Mus. Nat. Hist.* 34.

Beard, K. C. 2002, East of Eden at the Paleocene/Eocene boundary. *Science* 295:2028–2029.

Beard, K. C. 2005, Mammalian biogeography and anthropoid origins, in: Lehman, S. and Fleagle, J.G., eds., *Primate Biogeogr*, Kluwer Scientific Press.

Beard, K. C., Dagosto M., Gebo D. L., and Godinot M. 1988, Interrelationships among primate higher taxa. *Nature* 331:712–714.

Beard, K. C. and Dawson, M. R. 1999, Intercontinental dispersal of Holarctic land mammals near the Paleocene/Eocene boundary: Paleogeographic, paleoclimatic and biostratigraphic implications. *Bull. de la Soc. Geolog. de France* 170:697–706.

Beard, K. C. and Dawsons, M. R. 2001, Early Wasatchian mammals from the Gulf Coastal Plain of Mississippi: Biostratigraphic and paleobiogeographic implications, in: Gunnell, G. F., ed., *Eocene Biodiversity: Unusual Occurrences and Rarely Sampled Habitats*, Kluwer Academic/Plenum Publishers, New York, pp. 75–94.

Beard, K. C. and Wang, J. 2004, The eosimiid primates (Anthropoidea) of the Heti Formation, Yuanqu Bain, Shanxi and Henan Provinces, People's Republic of China. *J. Hum. Evol.* 46:401–432.

Beard, K. C., Qi, T., Dawson, M. R., Wang, B., and Li, C. 1994, A diverse new primate fauna from middle Eocene fissure-fillings in southeastern China. *Nature* 368:604–609.

Beard, K. C., Tong, Y., Dawson, M. R., Wang, J., and Huang, X. 1996, Earliest complete dentition of an anthropoid primate from the late middle Eocene of Shanxi Province, China. *Science* 272:82–85.

Begun, D. R. 2003, Planet of the apes. *Sci. Am.* 289:74–83.

Begun, D. R., Gulec E., and Geraads D. 2003, Dispersal patterns of Eurasian hominoids: Implications from Turkey. *Deinsea* 10:23–39.

Begun, D. R. and Nargolwalla, M. C. 2004, Late Miocene hominid biogeography: Some recent perspectives. *Evol. Anthropol.* 13:234–238.

Begun, D. R., Ward, C. V., and Rose, M. D. 1997, Events in hominoid evolution. In: Begun, D. R., Ward, C. V., and Rose, M. D., eds., *Function, Phylogeny, and Fossils: Miocene Hominoid Evolution and Adaptation*. Plenum Press, New York, pp. 389–415.

Behrensmeyer, A. K., Todd, N. E., Potts, R., and McBrinn, G. E. 1997) Late Pliocene faunal turnover in the Turkana Basin, Kenya and Ethiopia. *Science* 278:1589–1594.

Belmaker, M. 2002, First evidence of the presence of Theropithecus sp. in the Southern Levant. *Israeli J. Zool.* 48:165.

Benefit, B. R. and Pickford, M. 1986, Miocene fossil cercopithecoids from Kenya. *Am. J. Phy. Anthropol.* 69:441–464.

Bernor, R. L. 1983, Geochronology and zoogeographic relationships of Miocene Hominoidea, in: Ciochon, R. L. and Corruccini, R. S., eds., *New Interpretations of Ape and Human Ancestry*, Plenum Press, New York.

Brandon-Jones, D. 1996, The Asian Colobinae (Mammalia: Cercopithecidae) as indicators of Quaternary climatic change. *Biol J. Linn. Soc.* 59:327–350.

Bromage, T. G., Schrenk, F., and Juwayeyi, Y. M. 1995, Paleobiogeography of the Malawi Rift: Age and vertebrate paleontology of the Chiwondo Beds, northern Malawi. *J. Hum. Evol.* 28:37–57.

Chivers, D. J. 1977, The lesser apes, in: Ranier, H. S. H. o. M. and Bourne, G. H., eds., *Primate Conservation*, Academic Press, New York, pp. 539–598.

Ciochon, R. L. and Gunnell, G. F. 2002, Eocene primates from Myanmar: historical perspectives on the origin of Anthropoidea. *Evol. Anthropol.* 11:156–168.

Ciochon, R. L. and Gunnell, G. F. 2004, Eocene large-bodied primates of Myanmar and Thailand: morphological considerations and phylogenetic affinities, *Anthropoid Origins: New Visions*, Kluwer Academic/Plenum Publishers, New York, pp. 249–282.

Clyde, W. C., Kahn, I. H., and Gingerich, P. D. 2003, Stratigraphic response and mammalian dispersal during initial India-Asia collision: Evidence from the Ghazij Formation, Balochistan, Pakistan. *Geology* 31:1097–1100.

Colyn, M., Gautier-Hion, A., and Verheyen, W. 1991, A re-appraisal of palaeoenvironmental history in Central Africa: Evidence for a major fluvial refuge in the Zaire Basin. *J. Biogeogr.* 18:403–407.

Cote, S. M. 2004, Origins of the African hominoids: an assessment of the palaeobiogeographical evidence. *Comptes Rend. Paleoevol.* 3:323–340.

Dagosto, M. 1988, Implications of postcranial evidence for the origin of euprimates. *J. Hum. Evol.* 17:35–56.

Dashzeveg, D. and McKenna, M. C. 1977, Tarsioid primate from the early Tertiary of the Mongolian People's Republic. *Acta Palaeontol. Polonica* 22:119–137.

Delson, E. 1973, Fossil Colobine Monkeys of the Circum-Mediterranean Region and the Evolutionary History of the Cercopithecidae (Primates, Mammalia). Ph. D., Columbia University, New York.

Delson, E. 1975, Paleoecology and zoogeography of the Old World Monkeys, in: Tuttle, R. H., ed., *Primate Functional Morphology and Evolution*, Mouton, Paris, pp. 37–64.

Delson, E. 1980, Fossil macaques, phyletic relationships and a scenario of deployment, in: Lindberg, D. G., ed., *The Macaques: Studies in Ecology, Behavior, and Evolution*, Van Nostrand Reinhold, New York, pp. 10–30.

Delson, E. 1994, Evolutionary history of the colobine monkeys in paleoenvironmental perspective, in: Davies, A. G. and Oates, J. F., eds., *Colobine Monkeys: Their Ecology, Behaviour and Evolution*, Cambridge University Press, Cambridge, pp. 11–43.

deMenocal, P. 2004, African climate change and faunal evolution during the Pliocene-Pleistocene. *Earth Planet. Sci. Lett.* 220:3–24.

Dennell, R. 2003, Dispersal and colonization, long and short chronologies: how continuous is the Early Pleistocene record for hominids outside East Africa? *J. Hum. Evol.* 45:421–440.

Donoghue, M.J. and Moore, B.R. 2003, Toward an Integrative Historical Biogeography *Integr. Comp. Biol.* 43:261–270.

Dixon, E. J. 2001, Human colonization of the Americas: timing, technology and process. *Quat. Sci. Rev.* 20:277–299.

Eeley, H. A. C. and Foley, R. A. 1999, Species richness, species range size and ecological specialization among African primates: Geographical patterns and conservation implications. *Biodivers. Conserv.* 8:1033–1056.

Eeley, H. A. C. and Lawes, M. J. 1999, Large-scale patterns of species richness and species range size in anthropoid primates, in: Fleagle, J. G., Janson, C. H., and Reed, K. E., eds., *Primate Communities*, Cambridge University Press, Cambridge, pp. 191–219.

Eronen, J. T. and Rook, L. 2004, The Mio-Pliocene European primate fossil record: Dynamics and habitat tracking. *J. Hum. Evol.* 47:323–341.

Eudey, A. A. 1980, Pleistocene glacial phenomena and the evolution of Asian macaques, in: Lindberg, D. G., ed., *Macaques: Studies in Ecology, Behavior, and Evolution*, Van Nostrand Reinhold, New York, pp. 52–83.

Fleagle, J. G. 1999, *Primate Adaptation and Evolution*. Academic Press, San Diego.

Fleagle, J.G. and Kay, R.F. 1985, The Paleobiology of Catarrhines, in: Delson, E., ed., *Ancestors: The Hard Evidence*, Alan R. Liss, New York, pp.23–36.

Fleagle, J. G., Kay, R. F., and Anthony, M. R. L. 1997, Fossil New World monkeys, in: Kay, R. F., Madden, R. F., Cifelli, R. L., and Flynn, J. J., eds., *Vertebrate Paleontology in the Neotropics: The Miocene Fauna of La Venta*. Smithsonian Institution Press, Washington, pp. 473–495.

Fleagle, J.G. and Reed, K.E. 1999, Phylogenetic and Temporal Perspectives on Primate Ecology, in: Fleagle, J. G., Janson, C. H., and Reed, K. E., eds., *Primate Communities*, Cambridge University Press, New York, pp. 92–115.

Fleagle, J. G. and Tejedor, M. F. 2002, Early platyrrhines of southern South America, in: Hartwig, W. C., ed., *The Primate Fossil Record*, Cambridge University Press, New York, pp. 161–173.

Frost, S. R. (In press), African Pliocene and Pleistocene cercopithecid evolution and global climatic change, in: Bobe, R., Alemseged, Z., and Behrensmeyer, A. K., eds., *Hominin Environments in the East African Pliocene: An Assessment of the Faunal Evidence*.

Ganzhorn, J. U. 1998, Nested patterns of species composition and their implications for lemur biology in Madagascar. *Folia Primatol.* 69:332–341.

Gazin, C. L. 1958, A review of the Middle and Upper Eocene primates of North America. *Smithsonian Misc. Collect.* 136:1–112.

Gebo, D. L., Dagosto, M., Beard, K. C., and Qi, T. 2001, Middle Eocene primate tarsals from China: implications for haplorhine evolution. *Am. J. Phys. Anthropol.* 116:83–107.

Gebo, D. L., Dagosto, M., Beard, K. C., Qi, T., and Wang, J. 2000, The oldest known anthropoid postcranial fossils and the early evolution of higher primates. *Nature* 404:276–278.

Gibert, J., Ribot, F., Gibert, L., Leakey, M., Arribas,A., and Martinez, B. 1995, Presence of the cercopithecid genus Theropithecus in Cueva Victoria (Murcia, Spain). *J. Hum. Evol.* 28:487–493.

Gingerich, P. D. 1990, African dawn for primates. *Nature* 346:411.

Gingerich, P. D., Dashzeveg, D., and Russell, D. E. 1991, Dentition and systematic relationships of Altanius orlovi (Mammalia, Primates) from the early Eocene of Mongolia. *Geobios* 24:637–646.

Gingerich, P. D. and Ting, S. 2004, Paleocene-Eocene boundary and faunal change in relation to climate. *J. Vertebr. Paleontol.* 24:64A.

Ginsberg, L. and Mein, P. 1986, Tarsius thailandica nov. sp., Tarsiidae (Primates, Mammalia) fossile d'Asie. *Comptes Rendus de l'Acad. des Sci. (Paris)* 304:1213–1215.

Godfrey, L. R. and Jungers, W. L. 2002, Quaternary fossil lemurs, in: Hartwig, W. C., ed., *The Primate Fossil Record*, Cambridge University Press, New York, pp. 97–122.

Godfrey, L. R. and Jungers, W. L. 2003, The extinct sloth lemurs of Madagascar. *Evol. Anthropol.* 12:252–263.

Godinot, M. 1994, Early North African primates and their significance for the origin of Simiiformes (= Anthropoidea), in: Fleagle, J. G., and Kay, R. F., eds., *Anthropoid Origins*, Plenum Press, New York, pp. 235–295.

Gunnell, G. F. 2002, Notharctine primates (Adapiformes) from the early to middle Eocene (Wasatchian-Bridgerian) of Wyoming: Transitional species and the origins of *Notharctus* and *Smilodectes. J. Hum. Evol.* 43:353–380.

Gunnell, G. F. and Miller, E. R. 2001, Origin of Anthropoidea: Dental evidence and recognition of early anthropoids in the fossil record, with comments on the Asian anthropoid radiation. *Am. J. Phys. Anthropol.* 114:177–191.

Gupta, V. L. and Sahni, A. 1981, *Theropithecus delsoni*, a new cercopithecine species from the Upper Siwaliks of India. *Bull. Indian Geol. Assoc.* 14.

Hamilton, A. C. 1988, Guenon evolution and forest history, in: Gautier-Hion, A., Bourliere, F., Gautier, J. P., and Kingdon, J., eds., *A Primate Radiation: Evolutionary Biology of the African Guenons*, Cambridge University Press, Cambridge, pp. 13–34.

Harcourt, A. H. 2000, Latitude and latitudinal extent: a global analysis of the Rapoport effect in a tropical mammalian taxon: Primates. *Biogeogr.* 27:1169–1182.

Harrison, T. 2002, Late Oligocene to middle Miocene catarrhines from Afro-Arabia, in: Hartwig, W. C., ed., *The Primate Fossil Record*, Cambridge University Press, New York, pp. 311–338.

Harrison, T. (In press) The zoogeographic and phylogenetic relationships of early catarrhine primates in Asia. *Anthropol. Sci.*

Harrison, T. and Gu, Y. M. 1999, Taxonomy and phylogenetic relationships of early Miocene catarrhines from Sihong, China. *J. Hum. Evol.* 37:225–277.

Harrison, T., Krigbaum, J., and Manser, J. 2005, Primate biogeography and ecology on the Sunda Shelf islands: a paleontological and zooarchaeological perspective, in: Lehman, S. and Fleagle, J. G., eds., *Primate Biogeography*, Kluwer Scientific Press.

Hartenberger, J. L. and Marandat, B. 1992, A new genus and species of early Eocene primate from North Africa. *Hum. Evol.* 7:9–16.

Hartwig, W. C. 1994, Patterns, puzzles, and perspectives on platyrrhine origins, in: Corruccini, R. S. and Ciochon, R. L., eds., *Integrative Paths to the Past: Paleoanthropological Advances in Honor of F. Clark Howell*, Prentice-Hall, New York, pp. 69–93.

Hartwig, W. C. and Meldrum, D. J. 2002, Miocene platyrrhines of the northern Neotropics, in: Hartwig, W. C., ed., *The Primate Fossil Record*, Cambridge University Press, New York, pp. 175–188.

Heesy, C. P., Samonds, K., and Stevens, N. 2005, Biogeographic origins of primate higher taxa, in: Lehman, S. and Fleagle, J. G., eds., *Primate Biogeography*, Kluwer Scientific Press.

Hickey, L. J., West, R. M., Dawson, M. R., and Choi, D. K. 1983, Arctic terrestrial biota: Paleomagnetic with mid-northern latitudes during the late Cretaceous and early Tertiary. *Science* 221:1153–1156.

Hoffecker, J. F. 2004, *A Prehistory of the North: Human Settlement of the Higher Latitudes*, Rutgers University Press, New Brunswick, NJ.

Hoffstetter, R. and Lavocat, R. 1970, Decouverie dans le Deseadien de Bolivie de genres pentalophodenies appuyant les affinities africaines des Rongeurs Caviornorphes. *Comptes Rendus de l'Acad. des Sci. (Paris)* 271:172–175.

Hooker, J. J. 1992, British mammalian paleocommunities across the Eocene-Oligocene transition and their environmental implications, in: Prethero, D. R. and Berggren, W. A., eds., *Eocene-Oligocene Climatic and Biotic Evolution*, Princeton University Press, Princeton, pp. 172–175.

Hooker, J. J. 2000, Paleogene mammals: Crises and ecological change, in: Culver, S. J. and Rawson, P. F., eds., *Biotic Response to Global Change: The Last 145 Million Years*, Cambridge University Press, New York, pp. 333–349.

Hooker J. J. 2004, A new omomyid primate from the UK Early Eocene: Its phylogenetic and palaeobiogeographic implications. *J. Vertebr. Paleontol.* 24:72A.

Hooker, J. J., Collinson M. E., and Sille N. P. 2004, Eocene-Oligocene mammalian faunal turnover in the Hampshire Basin, UK: Calibration to the global time scale and the major cooling event. *J. Geol. Soc., Lond.* 161:161–172.

Hooker, J. J. and Dashzeveg, D. 2003, Evidence for direct mammalian faunal interchange between Europe and Asia near the Paleocene-Eocene boundary. *Geol. Soc. Am. Sp. Paper* 369:479–500.

Hooker, J. J., Russell, D. E., and Phelizon, A. 1999, A new family of Plesiadapiformes (Mammalia) from the Old World lower Paleogene. *Palaeontologica* 42:377–407.

Horovitz, I. and MacPhee, R. D. E. 1999, The Quaternary Cuban platyrrhine *Paralouatta varonai* and the origin of Antillean monkeys. *J. Hum. Evol.* 36:33–68.

Houle, A. 1999, The origin of platyrrhines: an evaluation of the Antarctic scenario and the floating island model. *Am. J. Phy. Anthropol.* 109:541–559.

Hsu, K. J., Ryan, W. B. F., and Cita M. B. 1973, Late Miocene desiccation of the Mediterranean. *Nature* 242:240–244.

Iturralde-Vinent, M. A. and MacPhee, R. D. E. 1999, Paleogeography of the Caribeean region: Implications for Cenozoic biogeography. *Bull. Am. Mus. Nat. Hist.* 238:1–95.

Jacobs, L. L. 1981, Miocene lorisid primates from the Pakistan Siwaliks. *Nature* 289:585–587.

Jungers, W. L., Godfrey, L. R., Simons, E. L., Wunderlich, R. E., Richmond, B. G., and Chatrath, P. S. 2002, Ecomorphology and behavior of giant extinct lemurs from Madagascar, in: Plavcan, J. M., Kay, R. F., Jungers, W. L., and van Schaik, C. P., eds., *Reconstructing Behavior in the Primate Fossil Record*, Kluwer Academic/Plenum Publishers, New York, pp. 371–411.

Kay, R. F., Ross, C., and Williams, B. A. 1997, Anthropoid origins. *Science* 275:797–803.

Kay, R. F., Williams, B. A., Ross, C. F., Takai, M., and Shigehara, N. 2004, Anthropoid origins: A phylogenetic analysis, in: Ross, C. F. and Kay, R. F., eds., *Anthropoid Origins: New Visions.* Kluwer Academic/Plenum Publishers, New York, pp. 91–135.

Krause, D. W. 2001, Fossil molar from a Madagascan marsupial—The discovery of a tiny tooth from the Late Cretaceous period has sizeable implications. *Nature* 412:497–498.

Krause, D. W., Hartman, J. H., and Wells, N. A. 1997, Late Cretaceous vertebrates from Madagascar: Implications for biotic change in deep time, in: Goodman, S. M., and Patterson, B. D., eds., *Natural Change and Human Impact in Madagascar*, Smithsonian Institution Press, Washington, DC, pp. 3–43.

Krause, D. W. and Maas, M. C. 1990, The biogeographic origins of Late Paleocene-Early Eocene mammalian immigrants to the Western Interior of North America. *Geol. Soc. Am. Sp. Paper* 243:71–105.

Kumar, K., Hamrick, M. W., and Thewissen, J. G. M. 2002, Middle Eocene prosimian primate from the Subathu Group of Kalakot, northwestern Himalaya, India. *Curr. Sci.* 83:1255–1259.

Lahr, M. M. and Foley, R. A. 1994, Multiple dispersals and modern human origins. *Evol. Anthropol.* 3:48–60.

Leakey, M. G., Feibel, C. S., Bernor, R. L., M. H. J., Cerling, T. E., Stewart, K. M., Storrs, G. W., Walker, A., Werdelin, L., and Winkler, A. J. 1996, Lothagam: A record of faunal change in the Late Miocene of East Africa. *J. Vertebr. Paleontol.* 16:556–570.

Leakey, M. G., Teaford, M. F., and Ward, D. V. 2003, Cercopithecidae from Lothagam, in: Leakey, M. G. and Harris, J. M., eds., *Lothagam: The Dawn of Humanity in Eastern Africa*, Columbia University Press, New York, pp. 201–248.

Lehman, S. M. 2004, Distribution and diversity of primates in Guyana: A biogeographic analysis. *Int. J. Primatol.* 25:73–95.

MacLatchy, L., Gebo, D. L., Kityo, R., and Pilbeam, D. 2000, Postcranial functional morphology of *Morotopithecus bishopi*, with implications for the evolution of modern ape locomotion. *J. Hum. Evol.* 39:159–183.

MacPhee, R. D. E. and Horovitz, I. 2002, Extinct Quaternary platyrrhines of the Great Antilles and Brazil, in: Hartwig, W. C., ed., *The Primate Fossil Record*, Cambridge University Press, New York, pp. 189–200.

MacPhee, R. D. E., Horovitz, I., Aredondo, O., and Jeminez-Vasquez, O. 1995, A new genus for the extinct Hispaniolan monkey *Saimiri bernensis* (Rimoli, 1977), with notes on its systematic position. *Am. Mus. Novitates* 3134:1–21.

MacPhee, R. D. E. and Jacobs, L. L. 1986, *Nycticeboides simpsoni* and the morphology, adaptations, and relationship of Miocene Siwalik Lorisidae. *Contributions to Geology, Univeristy of Wyoming, Special Paper* 3:131–161.

Marivaux, L., Welcomme, J.-L., Antoine, P.-O., Metais, G., Baloch, I. M., Benammi, M., Chaimanee, Y., Ducrocq, S., and Jaeger, J.-J. 2001, A fossil lemur from the Oligocene of Pakistan. *Science* 294:587–591.

Marshall, M. V. and Sugardjito, J. 1986, Gibbon systematics, in: Swindler, D. R. and Erwin, J., eds., *Comparative Primate Biology, Vol. 1: Systematics, Evolution, and Anatomy*, Alan R. Liss, New York, pp. 137–185.

McCall, R. A. 1997, Implications of recent geological investigations of the Mozambique Channel for the mammalian colonization of Madagascar. *Proc. R. Soc. Lond., Ser. B* 264:663–665.

McKee, J. K. 1996, Faunal turnover patterns in the Pliocene and Pleistocene of southern Africa. *South Afr. J. Sci.* 92:111–113.

McKee, J. K. 2001, Faunal turnover rates and mammalian biodiversity of the late Pliocene and Pleistocene of eastern Africa. *Paleobiology* 27:500–511.

McKenna, M. C. 1973, Sweepstakes, filters, corridors, Noah's arks, and beached Viking funeral ships in palaeogeography, in: Tarling, D. H. and Runcorn, S. K., eds., *Implications of Continental Drift to the Earth Sciences*, Academic Press, London, pp. 295–308.

McKenna, M. C. 1975, Fossil mammals and Early Eocene North Atlantic land continuity. *Ann. Missouri Bot. Gardens* 62:335–353.

McKenna, M. C. 1980, Eocene paleolatitude, climate and mammals of Ellesmere Island. *Palaeogeogr., Palaeoclimatol., Palaeoecol.* 30:349 362.

Miller, E. R., Gunnell, G. F., and Martin, R. D., 2005, Deep time and the search for anthropoid origins. *Yrbk Phys. Anthropol.* 48:60–95.

Ni, X., Wang, Y., Hu, Y., and Li, C. 2004, A euprimate skull from the early Eocene of China. *Nature* 427:65–68.

Oates, J. F. 1988, The distribution of Cercopithecus monkeys in West African forests, in: Gautier-Hion, A., Bourliere, F., Gautier, J. P., and Kingdon, J., eds., *A Primate Radiation: Evolutionary Biology of the Guenons*, Cambridge University Press, Cambridge, pp. 79–103.

Pan, Y. R. and Jablonski, N. G. 1987, The age and geographical distribution of fossil cercopithecids in China. *Hum. Evol.* 2:59–69.

Pickford, M. 1993, Climate change, biogeography, and *Theropithecus*, in: Jablonski, N. G., ed., *Theropithecus: The Rise and Fall of a Primate Genus*, Cambridge University Press, Cambridge, pp. 227–243.

Prothero, D. R. and Berggren, W. A., eds. 1992, *Eocene-Oligocene Climatic and Biotic Evolution*, Princeton University Press, Princeton.

Qi, T. and Beard, K. C. 1998, Late Eocene sivaladapid primate from Guangxi Zhuang autonomous region, People's Republic of China. *J. Hum. Evol.* 35:211–220.

Rae, T. C. 1993, Phylogenetic Analyses of Proconsulid Facial Morphology. Ph. D., State University of New York at Stony Brook, Stony Brook.

Rasmussen, D. T. 1994, The different meanings of a tarsioid-anthropoid clade and a new model of anthropoid origins, in: Fleagle, J. G. and Kay, R. F., eds., *Anthropoid Origins*, Plenum Press, New York, pp. 335–360.

Rasmussen, D. T., Conroy, G. C., and Simons, E. L. 1998, Tarsier-like locomotor specializations in the Oligocene primate *Afrotarsius*. *Proc. Nat. Acad. Sci. USA* 95:14848–14850.

Rasmussen, D. T. and Nekaris, K. A. 1998, The evolutionary history of lorisiform primates. *Folia Primatol.* 69:250–285.

Reed, K. E. and Fleagle, J. G. 1995, Geographic and climatic control of primate diversity. *Proc. Nat. Acad. Sci. USA* 92:7874–7876.

Rogl, F. 1999, Circum-Mediterranean Miocene paleogeography, in: Rossner, G. E. and Heissig, K., eds., *The Miocene Land Mammals of Europe: The Continental European Miocene*, Dr. Friedrich Pfeil, Munich, pp. 39–48.

Rook, L., Martinez-Navarro B., and Clark Howell F. 2004, Occurrence of *Theropithecus* sp. in the Late Villafranchian of Southern Italy and implication for Early Pleistocene "out of Africa" dispersals. *J. Hum. Evol.* 47:247–277.

Rose, K. D. 1995, The earliest primates. *Evol. Anthropol.* 3:159–173.

Ross, C. F. 1994, The craniofacial evidence for anthropoid tarsier relationships, in: Fleagle, J. G. and Kay, R. F., eds., *Anthropoid Origins*, Plenum Press, New York, pp. 469–548.

Ross, C. F. and Kay, R. F., eds. 2004, *Anthropoid Origins: New Visions*. Kluwer Academic/Plenum Publishers, New York.

Ross, C. F., Williams, B. A., and Kay, R. F. 1998, Phylogenetic analysis of anthropoid and tarsier relationships. *J. Hum. Evol.* 35:221–306.

Rossie, J. B. and Seiffert, E. R. 2005, Continental paleobiogeography as phylogenetic evidence, in: Lehman, S. and Fleagle, J. G., eds., *Primate Biogeography*, Kluwer Scientific Press.

Schwartz, J. H. 1992, Phylogenetic relationships of African and Asian lorisids, in: Matano, S., Tuttle, R. H., Ishida, H. and Goodman, M., eds., *Topics in Primatology, vol. 3, Evolutionary Biology, Reproductive Endocrinology, and Virology*, University of Tokyo Press, Tokyo, pp. 65–81.

Scott, W. B. 1913, *A History of Land Mammals in the Western Hemisphere*. The MacMillan Company, New York.

Seiffert, E. R., Simons, E. L., and Attia, Y. 2003, Fossil evidence for an ancient divergence of lorises and galagos. *Nature* 422: 421–424.

Seiffert, E. R., Simons, E. L., and Simons, C. V. M. 2004, Phylogenetic, biogeographic, and adaptive implications of new fossil evidence bearing on crown anthropoid origins and early stem catarrhine evolution, in: Ross, C. F. and Kay, R. F., eds., *Anthropoid Origins: New Visions*, Kluwer Academic/Plenum Publishers, New York, pp. 157–181.

Seiffert, E. R., Simons, E. L., Clyde, W. C., Rossie, J. B., Attia, Y., Bown, T. M., Chatrath, P., and Mathison, M. E. 2005, Basal anthropoids from Egypt and the antiquity of Africa's higher primate radiation. *Science* 310:300–304.

Sige, B., Jaeger, J. J., Sudre, J., and Vianey-Liaud, M. 1990, *Altiatlasius koulchii* n. gen. et sp., primate omomyide du Paleocene superieur du Maroc, et les origines des euprimates. *Palaeontographica Abteilung A* 214:31–56.

Simons, E. L. 1961, The dentition of *Ourayia*: Its bearing on relationships of omomyid primates. *Postilla* 54:1–20.

Simons, E. L. 1997, Discovery of the smallest Fayum Egyptian primates (Anchomomyini, Adapidae). *Proc. Nat. Acad. Sci. USA* 94:180–184.

Simons, E. L. 2003, The fossil record of tarsier evolution, in: Wright, P. C., Simons, E. L. and Gursky, S., eds., *Tarsiers: Past, Present and Future*, Rutgers University Press, New Brunswick, NJ.

Simons, E. L. and Bown, T. M. 1985, *Afrotarsius chatrathi*, first tarsiiform primate (? Tarsiidae) from Africa. *Nature* 313:475–477.

Simons, E. L., Rasmussen, D. T., and Gingerich, P. D. 1995, New cercamoniine adapid from Fayum, Egypt. *J. Hum. Evol.* 29:577–589.

Stanley, S. M. 1996, *Children of the Ice Age*, WH Freedman and Company, New York.

Stehli, F. and Webb, S. D., eds. 1985, *The Great American Biotic Interchange*. Plenum Press, New York.

Stewart, M. B. and Disotell, T. R. 1998, Primate evolution- in and out of Africa. *Curr. Biol.* 8:R582–R588.

Strait, D. S. and Wood, B. A. 1999, Early hominid biogeography. *Proc. Nat. Acad. Sci. USA* 96:9196–9200.

Stringer, C. 2000, Human evolution: How an African primate became global, in: Culver, S. J. and Rawson, P. F., eds., *Biotic Response to Global Change: The Last 145 Million Years*, Cambridge University Press, Cambridge, pp. 379–390.

Szalay, F. S. and Delson, E. 1979, *Evolutionary History of the Primates*, Academic Press, New York.

Takai, M., Anaya, F., Shigehara, N., and Segotuchi, T. 2000, New fossil materials of the earliest New World monkey, *Branisella boliviana*, and the problem of platyrrhine origins. *Am. J. Phys. Anthropol.* 111:263–281.

Takai, M. and Shigehara, N. 2004, The Pondaung primates, enigmatic "possible anthropoids" from the latest Middle Eocene, Central Myanmar, in: Ross, C. F. and Kay, R. F., eds., *Anthropoid Origins. New Visions*, Kluwer Academic/Plenum Publishers, New York, pp. 283–321.

Turner, A. and Wood, B. A. 1993, Taxonomic and geographic diversity in robust australopithecines and other Plio-Pleistocene larger mammals. *J. Hum. Evol.* 24:147–168.

Vrba, E. S. 1988, Late Pliocene climatic events and hominid evolution, in: Grine, F. E., ed., *Evolutionary History of the "Robust" Australopithecines*, Aldine de Gruyter, New York, pp. 405–426.

Vrba, E. S. 1992, Mammals as a key to evolutionary theory. *J. Mammal.* 73:1–28.

Webb, C.O., Ackerly, D.D., McPeek, M.A., and Donoghue, M.J. 2003, Phylogenies and Community Ecology. *Ann. Rev. Ecol. Syst.* 33:475–505.

Wesselman, H. B. 1985, Fossil micromammals as indicators of climatic change about 2.4 Myr ago in the Omo valley, Ethiopia. *South Afr. J. Sci.* 81:260–261.

Wheeler, B. 2003, Community ecology of the middle Miocene primates of La Venta, Colombia: The relationship between divergence time and ecological diversity. *Am. J. Phys. Anthropol. Suppl.* 16:223.

Whybrow, P. J. and Andrews, P. 2000, Response of Old World terrestrial vertebrate biotas to Neogene climate change, in: Culver, S. J. and Rawson, P. F., eds., *Biotic Response to Global Change: The Last 145 Million Years*, Cambridge University Press, New York, pp. 350–366.

Wing, S. L., Harrington, G. J., Smith, F. A., Bloch, J. I., Boyer, D. M., and Freeman, K. H. 2005, Transient floral change and rapid global warming at the Paleocene-Eocene boundary. *Science* 310:993–996.

Yoder, A. 1997, Back to the future: A synthesis of strepsirrhine systematics. *Evol. Anthropol.* 6:11–22.

Yoder, A., Cartmill, M., Ruvolo, M., Smith, K., and Vilgalys, R. 1996, Ancient single origin for Malagasy primates. *Proc. Nat. Acad. Sci. USA* 93:5122–5126.

Yoder, A. and Yang, Z. 2004, Divergence dates for Malagsy lemurs estimated from multiple gene loci: geological and evolutionary context. *Mol. Ecol.* 13:757–773.

Yoder, A. D., Burns, M. M., Zehr, S., Delefosse, T., Veron, G., Goodman, S. M., and Flynn, J. J. 2003, Single origin of Malagasy Carnivora from an African ancestor. *Nature* 421:734–737.

Zachos, J., Pagani, M., Sloan, L., Thomas, E., and Billups, K. 2001, Trends, rhythms, and aberrations in global climate 65 Ma to present. *Science* 292:686–693.

CHAPTER FOURTEEN

Biogeographic Origins of Primate Higher Taxa

Christopher P. Heesy, Nancy J. Stevens, and Karen E. Samonds

ABSTRACT

Cladistic character reconstruction has become an increasingly popular method used to infer areas of origin in biogeographic studies. However, no study to date has assessed the role that fossils play in center-of-origin reconstructions for the order Primates. Fossils preserve more information about the 'where' and the 'when' key extinct groups were present than would be apparent in analyses that focused solely on extant taxa. This paper examines the sensitivity of cladistic character reconstruction to ingroup and outgroup tree topologies when critical fossil taxa are included in the cladistic analysis of Primates. Specifically, reconstruction sensitivity is examined at the basal primate, strepsirrhine, haplorhine and anthropoid nodes to outgroup choice. Results demonstrate that biogeographic reconstructions are extremely sensitive to outgroup choice and internal tree topology and suggest caution in interpretations of areas of origin from phylogenies that do not include fossil taxa.

Key Words: Outgroup, topology, character evolution, area of origin, fossils

Christopher P. Heesy • Department of Anatomy New York College of Osteopathic Medicine, New York Institute of Technology, Old Westbury, New York 11568   Nancy J, Stevens • Department of Biomedical Sciences, 228 Irvine Hall, College of Osteopathic Medicine, Ohio University, Athens, OH 45701   Karen E. Samonds • Doctoral Program in Anatomical Sciences, Department of Anatomical Sciences, Health Sciences Center, T-8, Stony Brook University, Stony Brook, New York 11794-8081

Primate Biogeography, edited by Shawn M. Lehman and John G. Fleagle.
Springer, New York, 2006.

INTRODUCTION

On which continent did primates originate? Relationships between historical events and biogeographic patterns have long been of interest to natural historians (e.g., Wallace, 1876; Perrier de la Bathie 1936; Paulian, 1961; Simpson, 1965). A number of methodologies exist to examine the biogeographical history of a given taxonomic group (reviewed in Crisci *et al.*, 2003). For example, dispersal approaches emphasize the importance of the movement of organisms through space and time, considering the dispersal abilities of individual taxa to result in the present distribution patterns (Myers and Giller, 1988). In contrast, proponents of vicariance biogeography assert that biogeographic patterns result primarily when habitats and their resident biota are split by the emergence of barriers. Evolution in these now-separate biotas occurs in isolation via allopatric speciation and drift results in differing distribution patterns at different places. When vicariance patterns of many groups of unrelated taxa conform with one another, it may be inferred that abiotic processes have intervened to separate habitats (Pielou, 1979; Myers and Giller, 1988).

It is likely that both dispersal and vicariance mechanisms contribute to the biogeographic patterns observed today, and it is difficult to unravel their individual roles in the evolution of a taxonomic group. For this reason, several recent studies of vertebrate distributions have relied on phylogenetic vicariance biogeography approaches (e.g., Raxworthy and Nussbaum, 1994, 1996; Raxworthy *et al.*, 1998, 2002). In these studies, an understanding of taxonomic relationships among groups precedes the understanding of biogeographical patterns, and endemic taxa are the "derived characters" that allow one to reconstruct biogeographical history (Myers and Giller, 1988).

Such approaches are convenient in that they can utilize existing phylogenies to examine the biogeographic history of a group. In addition, cladistic data can be used to infer the center of origin of a group (Bremer, 1992, 1995; Crisci *et al.*, 2003). Using this approach, the areas inhabited by the group are optimized onto the tree using maximum parsimony. It can be inferred from the optimization analysis that the more primitive members of the group are found closer to the center of origin for that group (Hennig, 1966; Bremer, 1992, 1995; Crisci *et al.*, 2003). Yet no study to date has assessed the role that fossils play in reconstructions of center of origin. Moreover, the effects of differing tree topologies and outgroup taxa upon the robusticity of biogeographic inferences are not well understood.

Most recent work on eutherian supraordinal biogeography is based on molecular phylogenies (e.g., Springer *et al.*, 1997; Murphy *et al.*, 2001a). For our focus on the area of origin of primates, a molecular phylogeny seems a particularly inadequate starting point. This becomes clear when comparing the distributions of extant and fossil primate taxa. Extant nonhuman primates are found in Madagascar, southern Asia, sub-Saharan Africa, and South America (Fleagle, 1999). However, many Eocene- and Oligocene-age primate fossils are known from broader distributions in Asia and Africa, as well as North America, Europe, and continental India (Fleagle, 1999; see also Marivaux *et al.*, 2001). For the purposes of phylogenetic reconstruction, it has been argued that fossils preserve characters more closely approximating the ancestral condition, in addition to features entirely absent in extant taxa (Gauthier *et al.*, 1988; Donoghue *et al.*, 1989). The same could also be suggested for biogeographic reconstruction; fossils preserve more about "where" and "when" those primates existed than would a simple consideration of extant primates alone. This point is further illustrated in Table 1. When continent of origin is optimized onto various morphologically and molecularly based phylogenies of eutherians, the majority of the molecular phylogenies imply an Asian origin of primates as well as strepsirrhines, haplorhines, and anthropoids. However, two fossil-based morphological phylogenies reconstruct a North American origin, which is not implied by any of the molecular phylogenies. This difference emphasizes the importance of including fossil taxa and character states in biogeographic reconstructions.

The most comprehensive work to date on the biogeographic origin of primates using fossil data is by Beard (1998; see also Beard, this volume), who reconstructed an Asian origin for primates and at least 12 other placental groups. The congruence of the analyses is surprising. Of considerable interest is the topology of the primate tree used in Beard's analysis. The topology is based in part upon his Primatomorpha hypothesis (Beard, 1993), in which Dermoptera + Plesiadapiformes (broadly considered) and Primates are sister taxa. These relationships have been called into question based primarily on otic and postcranial evidence (Bloch and Silcox, 2001; Silcox, 2002). Other prominent studies of primate phylogeny are also incongruent with Beard's tree (e.g., Shoshani *et al.*, 1996; Kay *et al.*, 1997; Ross *et al.*, 1998). This lack of consensus erodes confidence not only in Beard's biogeographic reconstruction, but also in any such reconstruction. In other words, are area of origin reconstructions overly sensitive to tree topology? As is shown in Table 1, many but not all tree topologies suggest an Asian origin for the Order Primates. However, none

Table 1. Continent of Origin for Major Primate Nodes, by Advocated Phylogeny

Study	Primates node	Strepsirrhini node	Haplorhini node	Anthropoidea node	Type of data
Arnason *et al.*, 2002	Asia	Asia	—	Asia	molecular-various
Beard, 1998	**Asia**	**Asia**	**Asia**	**equivocal**	**morphological-MP**
Bloch and Boyer, 2002	**Asia**	—	—	—	**morphological-MP**
Eizirik *et al.*, 2001	Laurasia	—	—	—	molecular-MP,ML
Gunnell *et al.*, 2002	**North America**	—	—	**Africa**	**morphological-MP**
Kay *et al.*, 1992	**Asia**	—	—	—	**morphological-MP**
Liu *et al.*, 2001	Asia	—	—	—	matrix representation
Madsen *et al.*, 2001	Africa	—	—	—	molecular-ML
Marivaux *et al.*, 2001	**North America**	**equivocal (Asia, Africa)**	—	—	**morphological-MP**
Murphy *et al.*, 2001a	Asia	—	Asia	Asia	molecular-MP,ML,D
Murphy *et al.*, 2001b	Asia	—	—	—	molecular-Bayesian
Ni *et al.*, 2004	**North America**	**North America**	**Asia**	**Asia/Africa**[1]	**morphological-MP**
Norejko, 1999	Asia	—	—	—	morphological-MP
Novacek, 1992	**Asia**	—	—	—	**morphological-T**
Novacek and Wyss, 1986	Asia	—	—	—	morphological-MP
Purvis, 1995a; - and Webster, 1999	—	Asia	Asia	Asia	matrix representation
Ross *et al.*, 1999 (preferred)	**Asia**	**Asia**	**Asia**	**equivocal (Asia, Africa)**	**morphological-MP**
Shoshani *et al.*, 1996	**Asia**	**Asia**	**Asia**	**Asia**	**morphological and molecular-MP**

Studies in bold indicate those that incorporate data on fossil taxa.

MP = maximum parsimony, ML = maximum likelihood, D = distance, T = taxonomic

[1] The area of origin for anthropoids reconstructed using the Ni et al. (2004) topology depends on whether *Eosimias* is considered an anthropoid. If so, then anthropoids originated in Asia, if not, then anthropoids originated in Africa.

of these studies represents both living and fossil diversity. Inclusion of fossil taxa can have dramatic impact on biogeographic reconstructions (Stewart and Disotell, 1998). It is also true that we are uncertain of the outgroup for the primate order.

This study employs a dense representative phylogeny to examine the robusticity of reconstructions of primate biogeographic area of origin. It systematically evaluates the biogeographic origin of the Order Primates, as well as its major subgroups (e.g., Strepsirrhini, Haplorhini, and Anthropoidea). The purpose is to examine the sensitivity of character reconstruction at the basal primate, strepsirrhine, haplorhine, and anthropoid nodes to outgroup choice. In order to do this, we first generate a composite cladistic phylogeny of extant and fossil primate taxa using published data sets. We then evaluate the biogeographic implications of this phylogeny using multiple assumption sets of outgroup taxa. Finally, we discuss the relative support for the various reconstructions of the basal primate and anthropoid nodes.

MATERIALS AND METHODS

Matrix Representation Using Parsimony

Cladistic biogeography relies upon robust phylogenies. To date, no study has generated a cladistically based phylogeny of all major primate clades, including both extant and extinct taxa resolved to the generic or species level. A phylogeny encompassing fossil taxa is desirable in the reconstruction of trait evolution because fossils often preserve characters in states that more closely approximate the ancestral condition or that are entirely absent in extant taxa (Gauthier et al., 1988). In addition, a phylogeny that includes modern and fossil taxa offers the opportunity to analyze biogeographic distributions through time.

For this reason a composite phylogeny of extant and extinct primates was generated for this analysis using a cladistically based method, matrix representation using parsimony (Baum, 1992; Ragan, 1992; Purvis, 1995a,b). Following this method, a matrix was constructed by recoding source cladistic, phenetic (e.g., UPGMA), and taxonomic studies and scoring each taxon in a clade with "1," each taxon in the sister clade with "0," and all others with "?" (Purvis, 1995a,b; see also Sanderson et al., 1998). These data are hereafter known as "matrix elements" (Bininda-Emonds and Bryant, 1998), because they code for node/clade membership and do not directly represent phylogenetic character information. Subjecting the matrix elements to parsimony analysis produces trees that are the most parsimonious representations of the hierarchical information derived from the source analyses (Baum, 1992; Ragan, 1992; Purvis 1995a,b; Bininda-Emonds and Bryant, 1998). Trees are rooted by scoring an all-"0" outgroup (Ragan, 1992; Purvis, 1995b). For this analysis, phylogenetic sources of data incorporating both extant and extinct taxa were used in order to generate a composite tree of living and fossil primates. The following studies were used as sources of phylogenetic information: Fleagle and Kay (1987), Beard et al. (1991, 1994), Jungers et al. (1991), Begun (1995), Purvis (1995a), Rose (1995a), Begun and Kordos (1997), Benefit and McCrossin (1997), Kay et al. (1997), Horovitz and Meyer (1997), Jaeger et al. (1998), Harris and Disotell (1998), Kay et al. (1998), Ross et al. (1998), Purvis and Webster (1999), Horovitz et al. (1998), Fleagle (1999), Horovitz (1999), Horovitz and MacPhee (1999), Gebo et al. (2000), Ross (2000 (summary of analysis in press)), Seiffert et al. (2000). In the cases of *Eosimias*, *Archaeolemur*, *Palaeo-propithecus*, and *Megaladapis*, each of which was present in some analyses as resolved to the generic level, the generic monophyly was assumed and species were manually inserted as sister taxa.

The data matrix of 226 matrix elements for 165 extant and extinct primate taxa was subjected to maximum parsimony analysis in PAUP 3.0s+1 (Swofford and Begle, 1993) with the following parameters: Branch and Bound search algorithm using the furthest addition sequence, unordered matrix elements, uninformative matrix elements ignored, and collapse option enabled. In addition, the analysis was conducted without weighting or partitioning the matrix elements.

Character Mapping of Biogeographic Data

The areas of biogeographic origin were reconstructed for primate higher taxa in MacClade 4.0 by optimizing geographic area (continent) onto each tree using maximum parsimony, which reconstructs the most parsimonious sequence of changes to produce the observed character state distribution (Maddison and Maddison, 1992, 2000). This method of optimization has been successfully applied to biogeographic analyses where continents or subcontinents were the minimum geographic unit coded as a trait (Beard, 1998; Strait and Wood, 1999; Murray, 2001). Character and taxon coding are described in the Appendix (sections 1 and 2). Continental distribution was treated as an unordered, multistate character. No constraints on dispersal were applied. In other words, taxa in this analysis could theoretically disperse from Asia to South America. For illustrative purposes, major clades, such as the Lemuriformes, Platyrrhini, and Catarrhini, were condensed when the reconstructed node value was unequivocal. The maximum parsimony option in MacClade yields the set of equally most parsimonious solutions to the optimization of a trait for a given phylogeny. Nodes and internodes for which multiple solutions are possible are reconstructed as equivocal. This set of equally most parsimonious solutions includes optimizations that favor parallelisms (accelerated transformations, or ACCTRAN) and reversals (decelerated transformations, or DELTRAN) as well as all other parsimonious solutions. ACCTRAN and DELTRAN are specific models of character optimization and do not necessarily demonstrate the most appropriate solution to the evolution of the trait of interest because they may not apply to all characters simultaneously.

Putative outgroup taxa include Plesiadapiformes (e.g., Bloch and Boyer, 2002), Scandentia (e.g., Jacobs, 1980), and Dermoptera (e.g., Beard, 1993). The effects of outgroup choice on the biogeographic reconstruction of major primate nodes were explored by varying outgroup combinations. Outgroup variations were coded for major continents from which fossil and living primates are known, those being Africa, Asia, North America, and Europe. Equivocal node reconstructions were considered unresolvable based on the current data.

RESULTS

Composite Phylogeny of Primates

The maximum parsimony analysis yielded 29 equally most parsimonious composite trees of 235 steps. The summary of the strict consensus composite tree with Lemuroidea, Lorisoidea, Ceboidea, Cercopithecinae, Colobinae, and Hominoidea compressed is shown in Figure 1. The complete strict consensus summary file is presented in the Appendix (section 3). The Rescaled Consistency Index is 0.95 (CI = 0.96, RI = 0.99), and is a measure of the congruence of source trees rather than of matrix element homoplasy (Bininda-Emonds and Bryant, 1998). These values are high because composite trees contain far fewer homoplastic and uninformative matrix elements than the characters used in the original phylogenetic analyses.

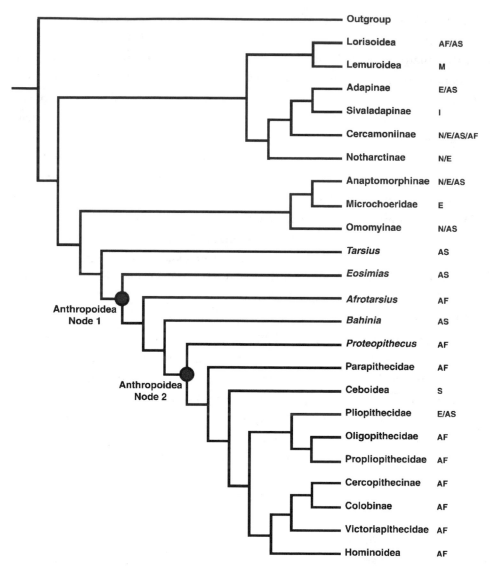

Figure 1. Summary composite tree of Primates with Lorisoidea, Lemuroidea, Ceboidea, Cercopithecinae Colobinae, and Hominoidea compressed. Continents to which taxa are endemic are indicated to the right. Abbreviations are: N—North America, E—Europe, As—Asia, Af—Africa, S—South America, M—Madagascar, and I—India. The definition of Anthropoidea is controversial. In this study, we reconstruct area of origin based on two node definitions of anthropoids: Anthropoidea Node 1 is defined by considering *Eosimias* and *Bahinia* as undisputed anthropoids. Anthropoidea Node 2 does not include *Eosimias* and *Bahinia* in Anthropoidea. The node is defined as *Proteopithecus* + all later anthropoids.

The monophyly of Strepsirrhini and Haplorhini are supported (Figure 1). Adapoidea is the sister group to all other strepsirrhines, and Omomyoidea is the sister group to all other haplorhines. *Eosimias, Afrotarsius chatrathi*, and *Bahinia pondaungensis* are "basal" anthropoids, or are sister taxa to a [*Proteopithecus* [Parapithecidae [Catarrhini + Platyrrhini]]] clade. The [Pliopithecidae [Oligopithecidae + Propliopithecidae]] clade is the sister group to all other catarrhines (Figure 1).

Results of Character Mapping of Biogeographic Data

Basal Primates

Reconstruction of the continental area of origin for all primates using various outgroup assumption sets yields seven unequivocal solutions (Table 2). These outgroup assignments also have varying effects on the reconstruction of strepsirrhine areas of origin (haplorhine origins are discussed with anthropoids, see below). A European origin of primates is supported by assigning either Europe or Europe + North America character states to the outgroups. A European origin of primates also reconstructs the Adapiformes + Strepsirrhini node as arising in Europe. An African origin of primates is supported when either Africa or Africa + North America character states are assigned to the outgroups. An African outgroup also reconstructs Adapiformes + Strepsirrhini as African. Three outgroup character state assignments lead to a reconstructed Asian origin: Asia (Scandentia + Dermoptera), Asia ((Scandentia + Siwaliks tupaiid) + Dermoptera), and Asia + North America. In these three cases, the strepsirrhine-based nodes are equivocal.

Anthropoidea

Assigning varying outgroup character states to the entire primate tree is largely irrelevant for node reconstructions of haplorhines and anthropoids (Table 2). The majority of haplorhine nodes are reconstructed as Asian in origin, with the minority exceptions for those that generate equivocal reconstructions. The primary anthropoid node reconstructions are completely consistent and unequivocal. If *Eosimias* and *Bahinia* are considered undisputed anthropoids, then anthropoids originated in Asia. If, however, the basal primate node is defined by *Proteopithecus* + all later anthropoids, then anthropoids originated in Africa.

Table 2. Continent of Origin for Major Primate Nodes Using Maximum Parsimony Character Reconstruction

Outgroup assumption set	Primates node	Strepsirrhini node	Adapiformes +Strepsirrhini	Haplorhini node	Omomyiformes+ Haplorhini	Anthropoidea node 1	Anthropoidea node 2
Asia (Scan+Derm)	Asia	equivocal	equivocal	Asia	Asia	Asia	Africa
Asia ((Scan+fossil)+Der)	Asia	equivocal	equivocal	Asia	Asia	Asia	Africa
Africa	Africa	Africa	Africa	Asia	Asia	Asia	Africa
North America	equivocal	equivocal	equivocal	equivocal	equivocal	Asia	Africa
Europe	Europe	equivocal	Europe	equivocal	equivocal	Asia	Africa
Asia+Africa	equivocal	equivocal	equivocal	Asia	Asia	Asia	Africa
Asia+North America	Asia	equivocal	equivocal	Asia	Asia	Asia	Africa
Asia+Europe	equivocal	equivocal	equivocal	Asia	equivocal	Asia	Africa
Africa+North America	Africa	Africa	Africa	Asia	Asia	Asia	Africa
Africa+Europe	equivocal	equivocal	equivocal	Asia	equivocal	Asia	Africa
North America+Europe	Europe	equivocal	Europe	Asia	equivocal	Asia	Africa

Anthropoidea Node 1 = includes *Eosimias* and *Bahinia* as undisputed anthropoids. **Anthropoidea Node 2** = does not include *Eosimias* and *Bahinia*. The node is defined as *Proteopithecus* + all later anthropoids.

DISCUSSION

Based on the results of this study, primates could be suggested to have orig-inated in Africa, Asia, or Europe. These reconstructions based on theoretical outgroup choices are not unwarranted by fossil evidence. An African origin of primates is consistent with the suggestion that *Altiatlasisus* is a basal primate (Gingerich, 1990; Sige *et al.*, 1990). The phylogenetic position and significance of *Altiatlasius* is, however, highly debatable with some researchers suggesting omomyid affinities (e.g., Sige *et al.*, 1990), and others plesiadapiform affinities (Hooker *et al.*, 1999). Similarly, an Asian origin of primates is consistent with the suggestion that *Altanius* from Mongolia is a basal primate (Dashzeveg and McKenna, 1977; Gingerich *et al.*, 1991). However, the affinities of *Altanius* are also debated, quite possibly because it shares similarities with both omomyids and adapids, and plesiadapiformes (Rose and Krause, 1984; Gingerich *et al.*, 1991; reviewed in Gunnell and Rose, 2002; Ni *et al.*, 2004).

The earliest undisputed Adapiformes and Omomyiformes appeared nearly simultaneously in Europe and North America (Rose and Fleagle, 1981; Rose *et al.*, 1994; Rose, 1995b;). This distribution seemingly would provide poten-tial resolution to the question of origin. However, it is important to consider but one example that was very tentatively suggested by Covert (2002) that either the Asian *Altanius* or the African *Altiatlasius* represent the stem group from which Omomyiformes and Adapiformes originated. In such cases, the ge-ographic distribution of later prosimians is irrelevant to reconstructing the area of origin of earlier groups.

The largely unequivocal results for anthropoid origins illustrate both the strengths and major weaknesses of cladistic biogeography using character re-construction methodology. The hypothesis that basal anthropoids originated in Asia is based primarily on the disputed phylogenetic position of *Eosimias* (e.g., Kay *et al.*, 1997; Ross *et al.*, 1998; see also Beard, 2002). As with any fossil taxon of this importance, the affinities of *Eosimias* have been disputed from the beginning (Culotta, 1992; Godinot, 1994; Simons and Rasmussen, 1994; Simons, 1995a). The *Eosimias* problem illustrates the weight that one taxon with an unusual character state can have on subsequent reconstructions. The vast majority of fossils and data on early anthropoid evolution come from North Africa (Simons, 1995b). If in the future the position of *Eosimias* should be re-solved differently, then reconstruction of the area of origin using an approach like that of this study will likely suggest an African origin of anthropoids.

Character reconstruction using a parsimony algorithm (e.g., Heesy and Ross, 2001) is a second stage process that is entirely dependent on topological resolution of the tree. If ambiguity or error is present in a base tree, then the resulting character reconstruction data are probably not robust. We would also argue that cladistic biogeography using the character reconstruction method is especially sensitive to missing data. Just as fossils may contain important features for the purposes of phylogenetic reconstruction and comparative analysis that are not found in extant taxa (Gauthier *et al.*, 1988; Donoghue *et al.*, 1989), our results suggest that fossils also represent geographic distribution data not necessarily found among extant taxa. These results call into question all biogeographic hypotheses based solely on molecular phylogenies (e.g., Springer *et al.*, 1997). However, at present, our results suggest that biogeographic reconstruction using character reconstruction when simultaneously considering both fossil and living taxa can only add potential sources for areas of origin, not discriminate among those that have been suggested for primate higher taxa.

CONCLUSIONS AND SUMMARY

Primates are among the best-documented taxa in the mammalian fossil record. As such, they provide a useful test case for understanding effects of different phylogenetic interpretations upon biogeographic reconstructions. This study has used multiple competing phylogenies, including a new comprehensive composite tree incorporating fossil taxa to evaluate the area of origin of primate higher taxa. It has examined the robusticity of biogeographic inferences, based on the sensitivity of such reconstructions to tree topology. Results demonstrate that biogeographic reconstructions are extremely sensitive to outgroup choice and internal tree topology and suggest caution in interpretations of primate and anthropoid areas of origin from phylogenies that do not include fossil taxa. Moreover, it has been shown that even a single taxon can have a powerful effect upon area of origin interpretations. Perhaps not surprisingly, it is only with greater phylogenetic resolution that a clearer understanding of the biogeographic origins of primate higher taxa will emerge.

ACKNOWLEDGEMENTS

We would like to thank John Fleagle and Shawn Lehman for organizing this volume, for inviting us to participate, and for their seemingly inexhaustible patience in waiting for our contribution. In addition, we thank Patrick O'Connor,

Margaret Hall, John Fleagle, John Hunter, and LeaAnn Jolley for helpful comments. A preliminary version of this work was presented at the Society of Vertebrate Paleontology Sixtieth Annual Meeting (Stevens and Heesy, 2000).

APPENDIX

Character Description and States.

One character, continental distribution of taxa, was coded as: North America = 1, Europe – 2, Asia = 3, Africa = 4, South America = 5, Madagascar = 6, and India = 7. Note that some taxa were coded as a multistate because representatives are found on multiple continents.

Taxon Coding

Tupaiinae 3; Ptilocercinae 3; *Lemur* 6; *Hapalemur* 6; *Eulemur* 6; *Varecia variegata* 6; *Varecia v. rubra* 6; *Lepilemur* 6; *Avahi* 6; *Indri* 6; *Propithecus verreauxi* 6; *Propithecus v. coquereli* 6; *Microcebus* 6; *Mirza* 6; *Cheirogaleus* 6; *Allocebus* 6; *Phaner* 6; *Daubentonia* 6; *Galago* 4; *Otolemur* 4; *Euoticus* 4; *Galagoides* 4; *Arctocebus* 4, *Loris* 3, *Nycticebus* 3, *Perodicticus* 3, *Tarsius* 3; *Trachypithecus* 3; *Presbytis* 3; *Semnopithecus* 3/7; *Nasalis* 3; *Simias* 3; *Rhinopithecus* 3; *Pygathrix* 3; *Piliocolobus* 4; *Colobus* 4/7; *Papio* 4; *Theropithecus* 4; *Lophocebus* 4; *Cercocebus* 4; *Macaca* 2/3; *Mandrillus* 4; *Cercopithecus* 1; *Chlorocebus* 4, *Miopithecus* 4; Hominoidea 3/4; *Callithrix* 5; *Cebuella* 5; *Saguinus* 5; *Callimico* 5; *Leontopithecus* 5; *Saimiri* 5; *Cebus* 5; *Aotus* 5; *Tremacebus* 5; *Callicebus* 5; *Pithecia* 5; *Cacajao* 5; *Chiropotes* 5; *Alouatta* 5; *Brachyteles* 5; *Lagothrix* 5; *Ateles* 5; Notharctinae 1/2; Cercamoniinae 1/2/3/4; Adapinae 2/3; Sivaladapinae (*Sivaladapis/ Indraloris*) 7; Microchoeridae 2; Anaptomorphinae 1/2/3; Omomyinae 1/3; *Eosimias* 3; *Afrotarsius* 4; *Proteopithecus* 4; Parapithecidae 4; Oligopithecidae 4; Pliopithecidae 2/3; Propliopithecidae 4; Victoriapithecidae 4; *Komba* 4; *Mioeuoticus* 4; *Progalago* 4; *Bahinia* 3; *Mohanamico* 5; *Lagonimico* 5; *Patasola* 5; *Carlocebus* 5; *Cebupithecia* 5; *Nuciruptor* 5; *Paralouatta* 5; *Antillothrix* 5; *Proteropithecia* 5; *Stirtonia* 5; *Archaeolemur* 6; *Hadropithecus* 6; *Palaeopropithecus* 6; *Megaladapis* 6.

Summary Tree File Generated in this Study.

((Ptilocercinae, Tupaiinae), (((Notharctinae, (Cercamoniinae, (Adapinae, Sivaladapinae))), (((((*Galago*, *Galagoides*), *Otolemur*, *Euoticus*), *Komba*),

((((*Arctocebus*, *Loris*), *Nycticebus*), *Perodicticus*), *Mioeuoticus*, *Progalago*)),
(*Daubentonia*, (((((*Microcebus*, *Mirza*), *Cheirogaleus*), (*Allocebus*, *Phaner*)),
(((((*Lemur*, *Hapalemur*), *Eulemur*), (*Varecia_v._rubra*, *Varecia_variegata*)),
((*Lepilemur*, *Megaladapis*), (((*Avahi*, (*Indri*, (*Propithecus_verreauxi*, *Pro-
pithecus_v._coquereli*))), *Palaeopropithecus*), (*Archaeolemur*, *Hadropithe-
cus*))))))))), ((((Microchoeridae, Anaptomorphinae), Omomyinae), (((*Tarsius*,
Tarsius_eocaenus), *Xanthorhysis*, *Afrotarsius*), (*Eosimias*, (*Bahinia*, (*Proteop-
ithecus*, (Parapithecidae, (((((((*Callithrix*, *Cebuella*), *Saguinus*, ((*Callimico*,
Patasola), *Carlocebus*), *Leontopithecus*, *Mohanamico*, *Lagonimico*), (*Saimiri*,
Cebus)), (*Tremacebus*, *Aotus*)), (((*Callicebus*, (*Paralouatta*, *Antillothrix*)),
(((((*Pithecia*, (*Cacajao*, *Chiropotes*)), *Cebupithecia*), *Proteropithecia*), *Nu-
ciruptor*)), ((*Alouatta*, *Stirtonia*), (*Brachyteles*, (*Lagothrix*, *Ateles*))))),
(((Oligopithecidae, Propliopithecidae), Pliopithecidae), ((Victoriapithecidae,
(((((((*Trachypithecus*, *Presbytis*), *Semnopithecus*), (*Nasalis*, *Simias*)), (*Rhino-
pithecus*, *Pygathrix*)), (*Piliocolobus*, *Colobus*)), (((((*Papio*, *Theropithecus*),
Lophocebus), (*Cercocebus*, *Mandrillus*)), *Macaca*), ((*Cercopithecus*, *Chlorocebus*),
Miopithecus)))), Hominoidea)))))))))))

REFERENCES

Arnason, U., Adegoke, J. A., Bodin, K., Born, E. W., Esa, Y. B., Gullberg, A., Nilsson, M., Short, R. V., Xu, X., and Janke, A. 2002, Mammalian mitogenomic relationships and the root of the eutherian tree. *Proc. Nat. Acad. Sci. USA* 99:8151–8156.

Baum, B. R. 1992, Combining trees as a way of combining data sets for phylogenetic inference, and the desirability of combining gene trees. *Taxon* 41:3–10.

Beard, K. C. 1993, Phylogenetic systematics of the Primatomorpha, with special reference to Dermoptera, in: Szalay, F. S., Novacek, M. J., and McKenna, M. C., eds., *Mammal Phylogeny: Placentals,* Springer-Verlag, New York, pp. 129–150.

Beard, K. C. 1998, East of Eden: Asia as an important center of taxonomic origination in mammalian evolution. *Bull. Carnegie Mus. Nat. Hist.* 34:5–39.

Beard, K. C. 2002, Basal anthropoids, in: Hartwig, W. C., ed., *The Primate Fossil Record.*, Cambridge University Press, New York, pp. 133–149.

Beard, K. C., Krishtalka, L., and Stucky, R. K. 1991, First skulls of the Early Eocene primate *Shoshonius cooperi* and the anthropoid-tarsier dichotomy. *Nature* 349:64–67.

Beard, K. C., Qi, T., Dawson, M. R., Wang, B., and Li, C. 1994, A diverse new primate fauna from middle Eocene fissure-fillings in southeastern China. *Nature* 368:604–609.

Begun, D. R. 1995, Late Miocene European orang-utans, gorillas, humans, or none of the above? *J. hum. Evol.* 29:169–180.

Begun, D. R. and Kordos, L. 1997, Phyletic affinities and functional convergence in *Dryopithecus* and other Miocene and living hominids, in: Begun, D. R., Ward, C. V., and Rose, M. D., eds., *Function, Phylogeny and Fossils. Miocene Hominoid Evolution and Adaptations*, Plenum Press, New York, pp. 291–316.

Benefit, B. R. and McCrossin, M. L. 1997, Earliest known Old World monkey skull. *Nature* 388:368–371.

Bininda-Emonds, O. R. P. and Bryant, H. N. 1998, Properties of matrix representation with parsimony analyses. *Syst. Biol.* 47:497–508.

Bloch, J. I. and Boyer, D. M. 2002, Grasping primate origins. *Science* 298:1606–1610.

Bloch, J. I. and Silcox, M. 2001, New basicrania of Paleocene-Eocene *Ignacius*: Re-evaluation of the plesiadapiform dermopteran link. *Am. J. phys Anthropol.* 116:184–198.

Bremer, K. 1992, Ancestral areas: A cladistic reinterpretation of the center of origin concept. *Syst. Biol.* 41:436–445.

Bremer, K. 1995, Ancestral areas: Optimization and probability. *Syst. Biol.* 44:255–259.

Covert, H. H. 2002, The earliest fossil primates and the evolution of prosimians: Introduction, in: Hartwig, W. C., ed., *The Primate Fossil Record.* Cambridge University Press, New York, pp. 13–20.

Crisci, J. V., Katinas, L., and Posadas, P. 2003, *Historical Biogeography: An Introduction*, Harvard University Press, Cambridge.

Culotta, E. 1992, A new take on anthropoid origins. *Science* 256:1516–1517.

Dashzeveg, D. and McKenna, M. C. 1977, Tarsioid primate from the early tertiary of the Mongolian people's republic. *Acta Palaeontol. Pol.* 22(2):119–137.

Donoghue, M. J., Doyle, J. A., Gauthier, J., Kluge, A. G., and Rowe, T. 1989, The importance of fossils in phylogeny reconstruction. *Ann. Rev. Ecol. Syst.* 20:431–460.

Eizirik, E., Murphy, W. J., and O'Brien, S.J. 2001, Molecular dating and biogeography of the early placental mammal radiation. *J. Heredity* 92:212–219.

Fleagle, J. G. 1999, *Primate Adaptation and Evolution*, 2nd edn., Academic Press, San Diego.

Fleagle, J. G. and Kay, R. F. 1987, The phyletic position of the Parapithecidae. *J. hum. Evol.* 16:483–532.

Gauthier, J., Kluge, A. G., and Rowe, T. 1988, Amniote phylogeny and the importance of fossils. *Cladistics* 4:105–209.

Gebo, D. L., Dagosto, M., Beard, K. C., Qi, T., and Wang, J. 2000, The oldest known anthropoid postcranial fossils and the early evolution of higher primates. *Nature* 404:276–278.

Gingerich, P. D. 1990, African dawn for primates. *Nature* 346:411.

Gingerich, P. D., Dashzeveg, D., and Russell, D. E. 1991, Dentition and systematic relationships of *Altanius orlovi* (Mammalia, Primates) from the Early Eocene of Mongolia. *Geobios* 24:637–646.

Godinot, M. 1994, Early North African Primates and their significance for the origin of Simiiformes (=Anthropoidea), in: Fleagle, J. G. and Kay, R. F., eds., *Anthropoid Origins*, Plenum Press, New York.

Gunnell, G. F., Ciochon, R. L., Gingerich, P. D., and Holroyd, P. A. 2002, New assessment of *Pondaungia* and *Amphipithecus* (Primates) from the Late Middle Eocene of Myanmar, with a comment on 'Amphipithecidae'. *Contrib. Mus. Paleont. Univ. Mich.* 30:337–372.

Gunnell, G. F. and Rose, K. D. 2002, Tarsiiformes: Evolutionary history and adaptation, in: Hartwig, W. C., ed., *The Primate Fossil Record*, Cambridge University Press, New York, pp. 45–82.

Harris, E. E. and Disotell, T. R. 1998, Nuclear gene trees and the phylogenetic relationships of the mangabeys (Primates: Papionini). *Mol. Biol. Evol.* 15:892–900.

Heesy, C. P. and Ross, C. F. 2001, Evolution of activity patterns and chromatic vision in primates: Morphometrics, genetics and cladistics. *J. Hum. Evol.* 40:111–149.

Hennig, W. 1966, *Phylogenetic Systematics*, University of Illinois Press, Urbana.

Hooker, J. J., Russell, D. E., and Phelizon, A. 1999, A new family of Plesiadapiformes (Mammalia) from the Old World lower Paleogene. *Palaeontologia* 42:377–407.

Horovitz, I. 1999, A phylogenetic study of living and fossil platyrrhines. *Am. Mus. Novit.* 3269:1–40.

Horovitz, I. and MacPhee, R. D. E. 1999, The quarternary Cuban platyrrhine *Paralouatta varonai* and the origin of Antillean monkeys. *J. hum. Evol.* 36:33–68.

Horovitz, I. and Meyer, A. 1997, Evolutionary trends in the ecology of new world monkeys inferred from a combined phylogenetic analysis of nuclear, mitochondrial, and morphological data, in: Givnish, T. C. and Sytsma, K. J., eds, *Molecular Evolution and Adaptive Radiation*, Cambridge University Press, Cambridge, pp. 189–224

Horovitz, I., Zardoya, R., and Meyer, A. 1998, Platyrrhine systematics: A simultaneous analysis of molecular and morphological data. *Am. J. Phys. Anthropol.* 106:261–281.

Jacobs, L. L. 1980, Siwalik fossil tree shrews, in: Luckett, W. P., ed., *Comparative Biology and Evolutionary Relationships of Tree Shrews*, Plenum Press, New York, pp. 205–216.

Jaeger, J.-J., Soe, A. N., Aung, A. K., Benammi, M., Chaimanee, Y., Tun, T., Thein, T., and Ducrocq, S. 1998, New Myanmar middle Eocene anthropoids: An asian origin for catarrhines? *C R Acad. Sci. (Paris)* 321:953–959.

Jungers, W. L., Godfrey, L. R., Simons, E. L., Chatrath, P. S., and Rakotosamimanana, B. 1991, Phylogenetic and functional affinities of *Babakotia* (Primates), a fossil lemur from northern Madagascar. *Proc. Natl. Acad. Sci. USA* 88:9082–9086.

Kay, R. F., Johnson, D., and Meldrum, D. J. 1998, A new pitheciin primate from the middle Miocene of Argentina. *Am. J. Primatol.* 45:317–336.

Kay, R. F., Ross, C., and Williams, B. A. 1997, Anthropoid origins. *Science* 275:797–804.

Kay, R. F., Thewissen, J. G. M., and Yoder, A. D. 1992, Cranial anatomy of *Ignacius graybullianus* and the affinities of the Plesiadapiformes. *Am. J. Phys. Anthropol.* 89:477–498.

Liu, F. -G. R., Miyamoto, M. M., Freire, N. P., Ong, P. Q., Tennant, M. R., Young, T. S., and Gugel, K. F. 2001, Molecular and morphological supertrees for eutherian (placental) mammals. *Science* 291:1786–1789.

Maddison, W. P. and Maddison, D. R. 1992, *MacClade Version 3. Analysis of Phylogeny and Character Evolution*, Sinauer Associates, Sunderland, MA.

Maddison, D. R. and Maddison, W. P. 2000, *MacClade 4: Analysis of Phylogeny and Character Evolution*. Version 4.0. Sinauer Associates, Sunderland, MA.

Madsen, O., Scally, M., Douady, C. J., Kao, D. J., DeBry, R. W., Atkins, R., Amrine, H. M., Stanhope, M. J., de Jong, W. W., and Springer, M. S. 2001, Parallel adaptive radiations in two major clades of placental mammals. *Nature* 409:610–614.

Marivaux, L., Welcomme, J.-L., Antoine, P.-O., Metais, G., Baloch, I. M., Benammi, M., Chaimanee, Y., Ducrocq, S., and Jaeger, J.-J. 2001, A fossil lemur from the Oligocene of Pakistan. *Science* 294:587–591.

Murphy, W. J., Eizirik, E., Johnson, W. E., Zhang, Y. P., Ryder, O. A., and O'Brien, S. J. 2001a, Molecular phylogenetics and the origins of placental mammals. *Nature* 409:614–618.

Murphy, W. J., Eizirik, E., O'Brien, S. J., Madsen, O., Scally, M., Douady, C. J., Teeling, E., Ryder, O. A., Stanhope, M. J., de Jong, W. W., and Springer, M. S. 2001b, Resolution of the early placental mammal radiation using bayesian phylogenetics. *Science* 294:2348–2351

Murray, A. M. 2001, The fossil record and biogeography of the Cichlidae (Actinopterygii: Labroidei). *Biol. J. Linn. Soc.* 74:517–532.

Myers, A. A. and Giller, P. S. (eds.). 1988, *Analytical Biogeography. An Integrated Approach to the Study of Animal and Plant Distributions*. Chapman and Hall, London.

Ni, X., Wang, Y., Hu, Y., and Li, C. 2004, A euprimate skull from the early Eocene of China. *Nature* 427:65–68.

Norejko, J. 1999, *Comparative Myology of Archontan Mammals*, MS. Thesis, State University of New York at Stony Brook.

Novacek, M. J. 1992, Mammalian phylogeny: Shaking the tree. *Nature* 356:121–125.

Novacek, M. J. and Wyss, A. R. 1986, Higher-level relationships of the recent eutherian orders: Morphological evidence. *Cladistics* 2:257–287.

Paulian, R. 1961, *La zoographie de Madagascar et des iles voisines*. L'Institut de Recherche Scientifique Tananarive-Tsimbazaza. Antananarivo.

Pielou, E. C. 1979, *Biogeography*, John Wiley and Sons, Somerset, NJ.

Perrier de la Bathie, H. 1936, *Biogeographie des plantes de Madagascar*, Societe Editions, Geographiques, Maritimes et Coloniales, Paris.

Purvis, A. 1995a, A composite estimate of primate phylogeny. *Phil. Trans. R. Soc. Lond.* 348:405–421.

Purvis, A. 1995b, A modification to Baum and Ragan's method for combining phylogenetic trees. *Syst. Biol.* 44:251–255.

Purvis, A., and Webster, A. J. 1999, Phylogenetically independent comparisons and primate phylogeny, in: Lee, P.C., ed., *Comparative Primate Socioecology*, Cambridge University Press, New York, pp. 44–70.

Ragan, M. A. 1992, Phylogenetic inference based on matrix representation of trees. *Mol. Phylogenet. Evol.* 1:53–58.

Raxworthy, C. J., Andreone, F., Nussbaum, R. A., Rabibisoa, N., and Randriamahazo, H. 1998, Amphibians and reptiles of the Anjanaharibe-Sud Massif, Madagascar: Elevational distribution and regional endemicity. *Fieldiana Zool.* 90: 79–92.

Raxworthy, C. J., Forstner, M. R. J., and Nussbaum, R.A. 2002, Chameleon radiation by oceanic dispersal. *Nature* 415: 784–787.

Raxworthy, C. J. and Nussbaum, R. A. 1994, A rainforest survey of amphibians, reptiles and small mammals at Montagne D'Ambre, Madagascar. *Biol. Conserv.* 69(1): 65–73.

Raxworthy, C. J. and Nussbaum, R. A. 1996, Amphibians and reptiles of the Réserve Naturelle Intégrale d'Andringitra, Madagascar: A study of elevational distribution and regional endemicity. *Fieldiana Zool.* 85: 158–170.

Rose, K. D. 1995a, Anterior dentition and relationships of the early Eocene omomyids *Arapahovius advena* and *Teilhardina demissa*, sp. nov. *J. hum. Evol.* 28:231–244.

Rose, K. D. 1995b, The earliest primates. *Evol. Anthropol.* 3:159–173.

Rose, K. D. and Fleagle, J. G. 1981, The fossil history of nonhuman primates in the Americas, in: Coimbra-Filho, A. F. and Mittermeier, R. A., eds., *Ecology and Behavior of Neotropical Primates*, Academia Brasileira de Ciencias, Rio de Janeiro, pp. 111–167.

Rose, K. D., Godinot, M., and Bown, T. M. 1994, The early radiation of Euprimates and the initial diversification of Omomyidae, in: Fleagle, J. G. and Kay, R. F., eds., *Anthropoid Origins*. Plenum Press, New York, pp. 1–28.

Rose, K. D. and Krause, D. W. 1984, Affinities of the primate *Altanius* from the Early Tertiary of Mongolia. *J. Mammalogy* 65:721–726.

Ross, C. F. 2000, Into the light: The origin of Anthropoidea. *Ann. Rev. Anthropol.* 29:147–194.

Ross, C., Williams, B., and Kay, R. F. 1998, Phylogenetic analysis of anthropoid relationships. *J. hum. Evol.* 35:221–306.

Sanderson, M. J., Purvis, A., and Henze, C. 1998, Phylogenetic supertrees: assembling the trees of life. *TREE* 13:105–109.

Seiffert, E. R., Simons, E. L., and Fleagle, J. G. 2000, Anthropoid humeri from the late Eocene of Egypt. *Proc. Nat. Acad. Sci. USA* 97:10062–10067.

Shoshani, J., Groves, C. P., Simons, E. L., and Gunnell, G. F. 1996, Primate phylogeny: Morphological vs molecular results. *Mol. Phylo. Evol.* 5:102–154.

Sigé, B., Jaeger, J.-J., Sudre, J., and Vianey-Liaud, M. 1990, *Altiatlasius koulchii* n. gen. et sp., primate omomyide du Paleocene superieur du Maroc, et lles origines des euprimates. *Palaeontographica A* 214:31–56.

Silcox, M. 2002, The phylogeny and taxonomy of plesiadapiformes. *Am. J. phys. Anthropol.* Suppl. 34: 141–142.

Simons, E. L. 1995a, Skulls and anterior teeth of *Catopithecus* (Primates: Anthropoidea) from the Eocene and anthropoid origins. *Science* 268:1885–1888.

Simons, E. 1995b, Egyptian Oligocene primates: a review. *Yrbk. Phys. Anthropol.* 38:199–238.

Simons, E. L. and Rasmussen, T. 1994, A whole new world of ancestors: Eocene anthropoideans from Africa. *Evol. Anthropol.* 3:128–139.

Simpson, G. G. 1965, *The Geography of Evolution*, Capricorn Books, New York.

Springer, M. S., Cleven, G. C., Madsen, O., de Jong, W. W., Waddell, V. G., Amrine, H. M., and Stanhope, M. J. 1997, Endemic African mammals shake the phylogenetic tree. *Nature* 388:61–64.

Stevens, N. J. and Heesy, C. P. 2000, Biogeographic origins of primate higher taxa. *J. Vert. Paleont.* 20, Supplement to Number 3: 71A.

Stewart, C.-B., and Disotell, T. R. 1998, Primate evolution—in and out of Africa. *Curr. Biol.* 8:R582–R588.

Strait, D. S. and Wood, B. A. 1999, Early hominid biogeography. *Proc. Nat. Acad. Sci. USA* 96:9196–9200.

Swofford, D. L. and Begle, D. P. 1993, *PAUP. Phylogenetic Analysis Using Parsimony*, Laboratory of Molecular Systematics, Smithsonian Institution, Washington, DC.

Swofford, D. L. and Maddison, W. P. 1992, Parsimony, character-state reconstructions, and evolutionary inferences, in: Mayden, R.L., ed., *Systematics, Historical Ecology, and North American Freshwater Fishes*, Stanford University Press, Stanford, CA, pp. 186–223.

Wallace, A. R. 1876, *The Geographical Distribution of Animals*, MacMillan, London.

CHAPTER FIFTEEN

Mammalian Biogeography and Anthropoid Origins

K. Christopher Beard

ABSTRACT

The continuity of phylogenetic descent requires that sister taxa originate in the same place and at the same time. Resolving phylogenetic relationships can therefore aid in reconstructing remote paleobiogeographic events. The order Primates is hierarchically nested within an exclusively Asian branch of the mammalian family tree, suggesting that Primates originated in Asia. Likewise, an Asian origin for Anthropoidea is supported by the geographic distributions of its sister group (Tarsiiformes) and various stem anthropoid taxa (Eosimiidae and Amphipithecidae). Although an African origin for Primates and/or Anthropoidea has been advocated repeatedly in the past, potential sister groups for either Primates or Anthropoidea are conspicuously lacking from the living and fossil biotas of that continent. The dispersal history of Malagasy lemurs and South American platyrrhines demonstrates that primate dispersal into new terrains often sparks adaptive radiation and morphological innovation. The precocious (Paleocene) dispersal of basal anthropoids from Asia to Africa may have instigated an adaptive radiation that yielded the modern anthropoid bauplan.

Key Words: Africa, China, Primates, Eosimiids, Laurasiatheria, Euanchonta, Colonizaton

K. Christopher Beard • Section of Vertebrate Paleontology, Carnegie Museum of Natural History, 4400 Forbes Avenue, Pittsburgh, PA 15213

Primate Biogeography, edited by Shawn M. Lehman and John G. Fleagle.
Springer, New York, 2006.

INTRODUCTION

Paleontologists have long believed that Africa played a pivotal role during the course of primate evolution (e.g., McKenna, 1967; Walker, 1972; Gingerich, 1986, 1990; Sigé *et al.*, 1990). Though the African origin of such familiar primate groups as Catarrhini and Hominidae is widely accepted, pinpointing the continent on which earlier and more basal primate clades arose has proven to be far more contentious. For example, recent attempts to determine the geographic roots of Anthropoidea—living and fossil catarrhines and platyrrhines ("crown anthropoids") as well as the extinct species that are more closely related to this clade than is *Tarsius* ("stem anthropoids")—embrace a wide range of controversies. These include debates about the phylogenetic affinities of putative basal anthropoid taxa, divergent opinions on how the taxon Anthropoidea should be defined, disagreements regarding the relevance and reliability of the fossil record versus neontological data sets (especially long sequences of nucleotides), as well as the methods followed to reconstruct historical biogeography at such deep phylogenetic nodes. Regardless of these ongoing disputes, many paleoprimatologists have looked to Africa as the most promising locus of anthropoid origins (Hoffstetter, 1977; Fleagle and Kay, 1987; Holroyd and Maas, 1994; Ciochon and Gunnell, 2002). Among the most important factors contributing to this Afrocentric perspective has been the sequence whereby early fossil anthropoids have been recovered. For decades, late Eocene and early Oligocene strata in the Fayum region of Egypt yielded the only uncontested Paleogene fossils for reconstructing early anthropoid evolution (Simons, 1995). This limited occurrence in space and time led many authors to conclude that anthropoids originated in Africa sometime near the Eocene-Oligocene boundary (e.g., Rasmussen and Simons, 1992).

The discovery of substantially older, though highly fragmentary, anthropoid fossils in middle Eocene strata in Algeria overturned the chronology of this classical hypothesis of anthropoid origins (Godinot and Mahboubi, 1992, 1994; Godinot, 1994). At the same time, these early Algerian anthropoids merely corroborated the prevailing notion that anthropoids originated in Africa. The precocious record of African anthropoids, along with paleogeographic and paleobiogeographic evidence for significant endemism among African Paleogene mammals as a whole, continues to persuade some workers that Africa is the most probable ancestral homeland for anthropoids (Holroyd and Maas, 1994; Ciochon and Gunnell, 2002, 2004).

Shortly after the middle Eocene anthropoids from Algeria were described, fossils pertaining to a previously unknown group of stem anthropoids— designated as eosimiids—began to be unearthed in Asia (Beard et al., 1994, 1996; Jaeger et al., 1999; Beard and Wang, 2004). The discovery of *Eosimias* and closely related forms at middle Eocene sites in central and eastern China, along with the discovery of *Bahinia* at slightly younger sites in Myanmar, has revived the possibility that anthropoids originated in Asia—an old idea that was previously thought to be discredited. Paleontological support for an Asian origin of anthropoids was originally founded on *Pondaungia* and *Amphipithecus* from Myanmar, both of which belong to a second group of Asian Eocene primates known as amphipithecids (Pilgrim, 1927; Colbert, 1937; Ba Maw et al., 1979; Ciochon et al., 1985). Although anatomical evidence bearing on amphipithecids has improved markedly in recent years, their anthropoid affinities continue to be debated (Ciochon and Holroyd, 1994; Jaeger et al., 1998, 2004; Chaimanee et al., 2000; Ciochon et al., 2001; Beard, 2002; Gunnell et al., 2002; Ciochon and Gunnell, 2002, 2004; Marivaux et al., 2003; Kay et al., 2004; Takai and Shigehara, 2004). As a result of these latest discoveries in Asia, the fossil record now offers an ambiguous signal regarding the birthplace of the anthropoid clade.

Recently, substantial progress has been achieved in resolving the molecular systematics of placental mammals and charting the fossil record of early Cenozoic mammals in Africa. Advances in both of these areas provide a fresh means of evaluating the geographic component of anthropoid origins. After reviewing some of the recent developments in these fields, I will summarize how the new data clarify the biogeography of anthropoid origins. I will conclude by exploring some possible links between the dispersal history of early anthropoids and the evolution of their diagnostic suite of morphological synapomorphies.

MAJOR FEATURES OF PLACENTAL MAMMAL PHYLOGENY

Although many details of placental mammal phylogeny remain to be resolved, phylogenetic analyses of long sequences of nucleotides routinely support four broad associations of taxa (Springer et al., 1997, 2004; Madsen et al., 2001; Murphy et al., 2001a, 2001b). These groups include Xenarthra (primarily South American sloths, anteaters, and armadillos), Afrotheria (elephants, sirenians, hyracoids, elephant shrews, aardvarks, tenrecs, and golden moles), Laurasiatheria (perissodactyls, artiodactyls, cetaceans, carnivorans, pangolins, bats, hedgehogs, shrews, moles, and solenodons), and Euarchontoglires (primates, tree shrews, flying lemurs, rodents, and lagomorphs). Many, though certainly not all, of these groupings are consistent with evidence from paleontology and comparative anatomy (e.g., Domning et al., 1986; Beard, 1993; Gingerich et al., 2001; Meng et al., 2003).

Even a cursory examination of the taxa comprising each of the four major groups of placental mammals reveals a strong biogeographic imprint on group composition. Xenarthrans, for example, can confidently be regarded as a South American clade of placental mammals, both because of the group's extant diversity there and its early appearance in the South American fossil record (Rose and Emry, 1993). Likewise, both phylogenetic and biostratigraphic evidence supports an African origin for afrotheres and a Laurasian birthplace for laurasiatheres, just as their names would suggest (Springer et al., 1997, 2004; Beard, 1998a; also see below). Euarchontoglires, the group that includes the order Primates and its nearest relatives, also appears to carry a strong biogeographic signal.

From a purely biogeographic perspective, it is clear that primates evolved from an exclusively Asian branch of the mammalian family tree. Primates are the most diverse and successful living members of the clade Euarchonta, which also includes tree shrews (order Scandentia) and flying lemurs (order Dermoptera). Both scandentians and dermopterans are currently restricted to southern and southeastern Asia. The fossil record of tree shrews and flying lemurs is meager, but the only fossils that can be attributed with confidence to these groups likewise come from Asia (Chopra et al., 1979; Jacobs, 1980; Tong, 1988; Ducrocq et al., 1992). We can safely conclude, then, that at least two of the three ordinal-level members of Euarchonta originated in Asia. This information alone does not allow us to infer that Primates also arose in Asia, because there is no consensus regarding the phylogenetic relationships among the three major euarchontan clades. Parsimony would suggest that Primates originated in Asia if the sister group of Primates could be shown to be either Scandentia or Dermoptera. On the other hand, if the sister group of Primates turns out to be a Scandentia + Dermoptera clade, broader phylogenetic context would be required to ascertain the most parsimonious birthplace for Primates. Molecular phylogenetic studies indicate that the sister group of Euarchonta is Glires (rodents and lagomorphs) (Madsen et al., 2001; Murphy et al., 2001a, 2001b). In contrast to tree shrews and flying lemurs, Glires eventually achieved a widespread distribution. However, the fossil record of such basal Glires as Heomys, Tribosphenomys, and Mimotona in Asia is sufficiently dense and ancient to indicate that this group too must have originated on that continent (Meng *et al.*, 1994, 2003; Dawson, 2003).

Following the methodology outlined by Beard (1998a), one can generate a "phylogenetically derived biogeographic reconstruction" for Primates and their relatives simply by optimizing the biogeographic distributions of terminal taxa onto internal nodes of the cladogram for Euarchontoglires (Figure 1). This

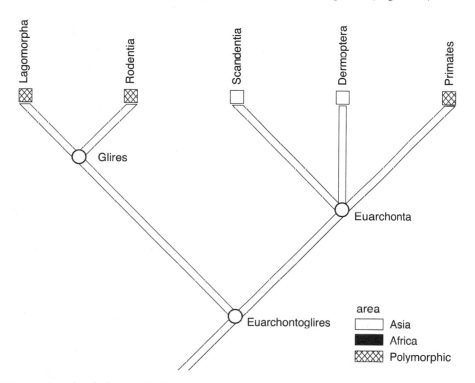

Figure 1. A phylogenetically derived biogeographic reconstruction for the five ordinal-level crown groups of Euarchontoglires and some key fossil taxa. The geographic ranges of Scandentia and Dermoptera are restricted to Asia, while those for Rodentia, Lagomorpha, and Primates include Asia (actually Laurasia) and Africa. Relatively recent range extensions for Rodentia, Lagomorpha, and Primates into South America and/or Australia are ignored here. Likewise, the European and North American records of living and fossil rodents, lagomorphs, and primates are excluded from this analysis, because the fossil record reveals that each of these groups dispersed to North America and Europe during the early Cenozoic, after having originated elsewhere (e.g., Beard, 1998a; Beard and Dawson, 1999). *A posteriori* optimization of the geographic character onto interior nodes of the cladogram indicates that the last common ancestors of Euarchontoglires, Glires, and Euarchonta are most parsimoniously interpreted as having lived in Asia, with subsequent (and independent) dispersal of rodents, lagomorphs, and primates from Asia to Africa.

procedure unambiguously supports an Asian origin for Primates. The geographic distributions of certain key fossil taxa corroborate this finding. Plesiadapiforms, which are often regarded as "archaic primates," are well documented and diverse on the three northern continents (Gingerich, 1976; Szalay and Delson, 1979; Gunnell, 1989; Beard and Wang, 1995), but undoubted plesiadapiforms have never been found in Africa [contrary to Tabuce *et al.* (2004), the morphologically aberrant Azibiidae from the Eocene of Algeria are not regarded as plesiadapiforms here]. Likewise, *Altanius*, which is commonly cited as a stem euprimate (that is, a sister group of the Strepsirhine + Haplorhini clade) is only known to occur in Mongolia (Gingerich *et al.*, 1991).

Despite the overwhelming phylogenetic evidence that Primates originated in Asia, the oldest euprimate currently documented in the fossil record hails from Africa. The source of this apparent conflict between phylogeny and the fossil record, late Paleocene *Altiatlasius* from Morocco, will be discussed at length later in this chapter. At present, it is sufficient to note that what little is known about the morphology of *Altiatlasius* indicates that it lies nested deeply within Primates. Evaluated within the broader context of mammalian phylogeny and primate biogeography, *Altiatlasius* signifies surprisingly early dispersal rather than any deeper phylogenetic history of Primates on the African continent.

THE FOSSIL RECORD AND AFRICAN BIOGEOGRAPHY: GARDEN OF EDEN OR MELTING POT?

One way to assess the conflicting biogeographic signals yielded by phylogeny and the fossil record is by comparing them with the broader pattern of mammalian evolution in Africa. In other words, if we momentarily ignore the debate over Africa's role as a potential cradle for Primates and/or Anthropoidea, how important was Africa as a biogeographic source for other living and extinct mammalian clades? Did Africa function as a constant wellspring of mammalian diversity throughout the late Mesozoic and Cenozoic? Or did the modern African mammal fauna develop by a process of accretion, through the stepwise addition of immigrant taxa to a core fauna dominated by endemic African forms?

Paleontologists and molecular systematists agree that a significant fraction of living placental mammal clades originated in Africa. For example, paleontologists have long advocated a common African ancestry for elephants (Proboscidea), dugongs and manatees (Sirenia), and hyraxes (Hyracoidea)

(Simpson, 1945; McKenna, 1975). Recent advances in molecular systematics have shown that this central group of African endemic mammals can now be extended to include elephant shrews (Macroscelidea), aardvarks (Tubulidentata), golden moles (Chrysochloridae), and tenrecs (Tenrecidae), yielding the anatomically heterogeneous assemblage known as Afrotheria (Springer *et al.*, 1997; Madsen *et al.*, 2001; Murphy *et al.*, 2001a, b). Although the monophyly of Afrotheria would likely never have been suspected on the basis of morphological criteria alone, the African fossil record corroborates an African origin for most, if not all, afrotherian taxa. Africa has yielded the world's only fossil elephant shrews, golden moles, and tenrecs (Butler and Hopwood, 1957; Tabuce *et al.*, 2001). By wide margins, the earliest representatives of Proboscidea, Hyracoidea, and Tubulidentata are also confined to Africa (Patterson, 1975; Court and Hartenberger, 1992; Court and Mahboubi, 1993; Gheerbrant *et al.*, 2002). Presumably because of their aquatic lifestyle, sirenians were able to disperse widely during the early phases of their evolutionary history (Domning, 2001). Nevertheless, their phylogenetic position as close relatives of proboscideans and hyracoids is uncontested, suggesting that they too originated in Africa.

Besides afrotheres and several extinct taxa that are difficult to place on the mammalian family tree (e.g., Palaeoryctidae), the early Cenozoic fossil record of Africa is remarkable for lacking several groups of mammals that are otherwise widespread and abundant. Apparently, these taxa (various groups of laurasiatheres and euarchontoglireans) originated elsewhere and invaded Africa multiple times during the Cenozoic. Primates, first documented by late Paleocene *Altiatlasius* from Morocco, were among the first of these exotic mammals to disperse to Africa successfully. Rodents -by far the most diverse and ecologically successful group of mammals alive today—invaded Africa at least twice (and more likely, three or more times) during the Eocene (Jaeger *et al.*, 1985; Marivaux *et al.*, 2002; Dawson *et al.*, 2003). Primitive zegdoumyid rodents (which have been cited as potential relatives of anomalurids, but which are more plausibly linked with glirids) first appear in early-middle Eocene strata of Tunisia and Algeria (Vianey-Liaud *et al.*, 1994). As noted previously, rodents are known to have originated in Asia, so this first appearance of rodents in Africa serves as an important biogeographic datum. It either reflects dispersal directly from Asia to Africa or indirectly from Asia to Africa via Europe. Subsequent episodes of rodent dispersal to Africa emit less ambiguous biogeographic signals. Undoubted anomalurids appear alongside the earliest African phiomyids at the late-middle Eocene site of Bir el Ater in northeastern Algeria (Jaeger *et al.*,

1985). Both anomalurids and phiomyids seemingly dispersed directly from Asia to Africa (Marivaux *et al.*, 2002; Dawson *et al.*, 2003). The earliest African artiodactyls—the hippo-like anthracotheres—appear alongside anomalurid and phiomyid rodents in the late-middle Eocene, thus providing a further example of the successful invasion of Africa by an Asian group of mammals (Ducrocq, 1997). By the time the first anthracotheres show up in Africa, a wide diversity of artiodactyl groups is established on the northern continents. Given our current understanding of Tethyan paleogeography, each of the preceding groups of mammalian immigrants to Africa—primates, rodents, and anthracotheres—must have arrived there via sweepstakes dispersal from the north.

A major episode of African faunal turnover occurred near the Oligocene-Miocene boundary, when a land bridge became established that connected Africa with Eurasia for the first time (Jolivet and Faccenna, 2000; Kappelman *et al.*, 2003). Utilizing this direct overland route, numerous additional groups of Laurasian mammals—including perissodactyls (rhinos and chalicotheres), artiodactyls other than anthracotheres (pigs, giraffes, and tragulids), carnivorans (cats, viverrids, and amphicyonids), true lipotyphlans (hedgehogs and shrews), and lagomorphs (pikas)—invaded Africa during the early Miocene. Other ecologically important groups of extant African mammals (including bovids, equids, canids, and leporids) dispersed to Africa later in the Miocene. Like most of the Laurasian mammals that preceded them, the earliest records of these taxa on the northern continents significantly antedate their first occurrences in Africa.

The preceding overview of the African fossil record demonstrates that what we currently know about African paleontology agrees with results from molecular systematics regarding the endemism of living afrotheres. The flip side of this coin also holds. That is, those groups of mammals currently residing in Africa that are not afrotheres all seem to have originated elsewhere, notably Asia. Moreover, in terms of their diversity and abundance, these exotic mammalian groups—including primates, rodents, lagomorphs, lipotyphlans, carnivorans, artiodactyls, and perissodactyls—have come to dominate modern African ecosystems. Since their heyday in the early Cenozoic, afrotheres have diminished in abundance and diversity, almost certainly as a result of competition and/or predation at the hands of these northern invaders. The modern African mammal fauna therefore evolved as successive groups of Laurasian mammals insinuated themselves into native African ecosystems harboring fewer and fewer endemic afrotheres. Primates are but one of the Laurasian mammal groups that attained high taxonomic diversity and ecological prominence after dispersing into this African melting pot.

PRIMATE DISPERSAL AND ADAPTIVE RADIATION

There are many examples in evolutionary biology whereby dispersal fosters an adaptive radiation among the organisms that succeed in colonizing a new terrain. Among primates, this pattern is best exemplified by the colonization of Madagascar by ancestral lemurs and the colonization of South America by early platyrrhine monkeys. Both of these evolutionary radiations were instigated by "sweepstakes" dispersal of an ancestral stock of primates into an ecologically appropriate region harboring few, if any, potential mammalian competitors. Given that primates were among the first Laurasian mammals to invade Africa in the early Cenozoic, we can only assume that the initial colonization of Africa by primates such as *Altiatlasius* would have triggered its own adaptive radiation. To appreciate the potential significance of this poorly documented radiation of early Cenozoic primates in Africa, let us first review briefly what happened when primates colonized two other isolated landmasses, the island of Madagascar and the island continent of South America.

Because the fossil record of Madagascar is so inadequate, little is known about the early colonization of that island by ancestral lemurs. Geographic proximity to Africa, where the sister group of Malagasy lemurs still survives in the form of lorises and galagos, suggests that early lemurs probably rafted across the Mozambique Channel. Once these early lemurs arrived on Madagascar, they spawned a broad, monophyletic radiation that significantly expanded the envelope of primate ecomorphospace (Tattersall, 1982; Yoder *et al.*, 1996). Various lemur taxa developed novel anatomical structures, allowing them to exploit unique ecological niches. *Daubentonia*, for example, combines a vaguely rodent-like dentition, enlarged external ears, and highly elongated, claw-bearing manual third digits to achieve its ecological convergence upon woodpeckers (Cartmill, 1974). *Palaeopropithecus* and its close relatives evolved strongly curved phalanges and other postcranial autapomorphies as part of a sloth-like adaptation for suspensory folivory (Jungers *et al.*, 1997). *Archaeolemur* and *Hadropithecus* share a bilophodont molar pattern and various postcranial features with terrestrial cercopithecoid monkeys (Godfrey *et al.*, 1997). Other groups of Malagasy lemurs retain suites of anatomical traits in common with early Cenozoic primates. Living lemurids, for example, share numerous features in common with Eocene notharctids, while the postcranial adaptations of Eocene omomyids are often compared with those of modern cheirogaleids (Gregory, 1920; Gebo, 1988; Dagosto *et al.*, 1999). Considered as a whole, the breadth of the Malagasy lemur radiation is astonishing, particularly in light of the fact that it occurred within the confines of an island encompassing roughly 2% of the area subsumed by Africa.

Compared with the lemuriform example, the adaptive radiation that followed the colonization of South America by early platyrrhine monkeys was relatively modest. Nevertheless, it produced the only primates to be equipped with a prehensile tail (atelines), a clade of specialized seed predators with unique dental adaptations (the pitheciines), a small-bodied clade bearing claws rather than nails on their digits (the callitrichines), and the only nocturnal anthropoid (*Aotus*). Despite the morphological and ecological innovations forged by platyrrhines, the large disparity in area between South America and Madagascar suggests that the platyrrhine radiation might well have been more expansive than it actually was. A potential explanation lies in the antiquity of the lemuriform and platyrrhine radiations (Fleagle and Reed, 2004). In the absence of direct information from the fossil record, Yoder and Yang (2004) interpret molecular phylogenetic data as indicating that lemurs have been radiating in Madagascar since the Paleocene, some 62 Ma. This is more than twice the age of the earliest known South American monkey (late Oligocene *Branisella boliviana*, roughly 25 Ma).

Regardless of their relative breadths, the lemuriform and platyrrhine radiations suffice to illustrate the basic concept that primate dispersal into a new territory often sparks an adaptive radiation. Beyond simply generating additional taxonomic diversity, these adaptive radiations also yield novel anatomical features and unique ecological strategies.

EARLY ANTHROPOIDS FROM AFRICA AND ASIA

By the middle Eocene, early anthropoids show a remarkably broad geographic distribution, ranging from western Algeria to eastern China (Figure 2; Beard, 2002). However, even by this early date, important anatomical features distinguish African and Asian anthropoids. All of the African anthropoids described until now from this interval—including *Algeripithecus*, *Tabelia*, and *Biretia*—are documented solely by isolated teeth. Despite this meager fossil record, the phylogenetic position of these animals is uncontroversial because they bear

Figure 2. Map of Africa and Eurasia, showing the wide geographic distribution of Paleogene anthropoids.

such typically anthropoid features as bunodont upper and lower molars, large upper molar hypocones, and the loss or reduction of lower molar paraconids (Bonis *et al.*, 1988; Godinot and Mahboubi, 1992, 1994; Godinot, 1994). In contrast, Asian middle Eocene anthropoids—typified by *Eosimias*—retain numerous primitive features, including the absence of upper molar hypocones and the presence of large, cuspidate paraconids on all lower molars. Accordingly, although Asian eosimiids are documented by anatomically superior specimens, their anthropoid status remains controversial (Gunnell and Miller, 2001; Ciochon and Gunnell, 2002; Schwartz, 2003; Simons, 2003). Nevertheless, comprehensive phylogenetic analyses indicate that eosimiids are basal anthropoids (Kay *et al.*, 2004), a conclusion that is upheld by multiple derived characters in the eosimiid dentition, lower face, and ankle region (Beard *et al.*, 1994, 1996; Jaeger *et al.*, 1999; Gebo *et al.*, 2000, 2001; Beard and Wang, 2004).

The anatomical disparity between African anthropoids such as *Algeripithecus* and Asian anthropoids such as *Eosimias* suggests that cladogenesis between the two groups occurred substantially before the middle Eocene. At the same time, three lines of evidence indicate that the earliest anthropoids arose in Asia, rather than Africa. The first of these consists of the extremely basal phylogenetic position of *Eosimias* and its Asian relatives on the anthropoid family tree (Beard *et al.*, 1994, 1996; Gebo *et al.*, 2000, 2001; Beard and Wang, 2004; Kay *et al.*, 2004). If anthropoids arose some place other than Asia, it is difficult to explain why Asian eosimiids consistently show up near the base of anthropoid phylogenies (e.g., Kay *et al.*, 2004; Seiffert *et al.*, 2004). Indeed, this predicament is exacerbated by a second factor supporting an Asian origin for anthropoids—the fact that *Tarsius* and its fossil relatives are restricted to Asia (or to Laurasia, if North American omomyids and European microchoerids are regarded as tarsiiforms) (Beard, 1998b). A wide variety of anatomical, paleontological, and molecular evidence indicates that tarsiers and their extinct relatives are the sister group of anthropoids (e.g., Martin, 1990; Kay *et al.*, 2004; Schmitz and Zischler, 2004). By definition, sister taxa originate at the same time and in the same place (Beard, 1998a). The restricted geographic range of living and fossil tarsiers therefore severely constrains the realm of possible locations where anthropoids may have originated. The final line of support for an Asian origin for anthropoids comes from paleontological and molecular evidence suggesting that the initial diversification of primates into strepsirhines, tarsiiforms, and anthropoids occurred very rapidly (e.g., Beard and MacPhee, 1994; Yoder, 2003; Eizirik *et al.*, 2004). Given the overwhelming evidence that primates as a whole originated in Asia, rapid cladogenesis of the order into its three major subdivisions would have left little time for intercontinental dispersal to intervene and complicate an otherwise simple biogeographic pattern.

The very early dichotomy between anthropoids and tarsiiforms that is implied by available molecular and paleontological data explains how African and Asian anthropoids were able to diverge so widely by the middle Eocene. Further corroboration for the antiquity of the anthropoid clade comes from phylogenetic analysis of the earliest known euprimate.

PHYLOGENETIC AND BIOGEOGRAPHIC
SIGNIFICANCE OF *ALTIATLASIUS*

For no other reason than its age, *Altiatlasius* from the late Paleocene of Morocco figures prominently in any discussion of the phylogeny and biogeography of early primates. Given the significance of *Altiatlasius*, it is unfortunate that we know so little about the anatomy of this creature. *Altiatlasius* is documented by approximately ten isolated teeth and one lower jaw fragment bearing the germ of an unerupted M_2 (Sigé *et al.*, 1990). Not surprisingly, this meager anatomical record has led to conflicting phylogenetic reconstructions. *Altiatlasius* was originally described as an "omomyid," but the phylogenetic scheme adopted by Sigé *et al.* (1990, Figure 1) placed *Altiatlasius* as the sister group of Simiiformes or Anthropoidea. Subsequent workers have generally supported and elaborated upon this viewpoint (Godinot, 1994; Beard, 1998b; Seiffert *et al.*, 2004). Alternatively, Hooker *et al.* (1999) interpreted *Altiatlasius* as a plesiadapiform on the basis of upper molar characters, including an elongated and buccally oriented postmetacrista and a large parastyle, that they regarded as being more primitive than those of any known euprimate. However, Beard and Wang (2004) have recently demonstrated that these same features occur in *Eosimias*, thereby enhancing the likelihood that *Altiatlasius* is a basal anthropoid.

The phylogenetic position of *Altiatlasius* is reexamined here in light of the new information regarding upper molar anatomy in *Eosimias* published by Beard and Wang (2004). An obvious impediment to this analysis is the problem of missing data for *Altiatlasius*. Indeed, although Sigé *et al.* (1990) ascribed an isolated lower premolar to *Altiatlasius*, this attribution is not accepted here because the tooth differs fundamentally from those of other early anthropoids, omomyids, and adapiforms. Similar reservations were expressed by Rose *et al.* (1994, p. 12). We therefore remain ignorant of such basic aspects of the dentition of *Altiatlasius* as the dental formula, the size and orientation of the lower incisors, and the anatomy of all upper and lower tooth crowns anterior to the molars. Despite these obstacles, the upper and lower molar anatomy of *Altiatlasius* seemingly emits a strong phylogenetic signal. A branch-and-bound search of 25 dental characters distributed across 11 taxa (Appendices 1 and 2) using PAUP 4.0b10 (Swofford, 2002) yielded a single most parsimonious tree, which is illustrated in Figure 3.

Perhaps the most intriguing result of the cladistic analysis performed here is the support it offers for a close phylogenetic tie between *Altiatlasius* and *Eosimias*. Given the dental characters under consideration, these taxa are most parsimoniously interpreted as sister groups. Regardless of whether this putative *Altiatlasius* + *Eosimias* clade withstands further scrutiny, both taxa appear to be stem anthropoids, lying outside a clade including crown anthropoids such as *Saimiri* and primitive Fayum anthropoids such as *Arsinoea* and *Proteopithecus*.

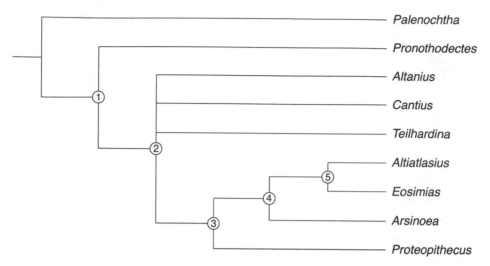

Figure 3. The most parsimonious tree recovered from branch-and-bound search in PAUP 4.0b10 (Swofford, 2002) of the character-taxon matrix given in Appendix 2. With the exception of Character 12, all characters were treated as unordered (see Appendix 1). Tree length = 45; consistency index = 0.689.

Anatomical support for regarding *Altiatlasius* as a basal anthropoid comes from its peculiar upper and lower molar structure, which differs markedly from the pattern common to other early euprimates (Figure 4). Several authors have emphasized the remarkable similarity in the dentitions of basal adapiforms and omomyids (Godinot, 1978; Rose and Bown, 1991; Rose *et al.*, 1994). A very different dental pattern characterizes *Altiatlasius, Eosimias,* and other basal anthropoids. In contrast to those of early adapiforms and omomyids, the upper molars of *Eosimias* and *Altiatlasius* bear a complete lingual cingulum, an enlarged parastyle, a well-developed and buccally oriented postmetacrista that terminates in a weak metastyle, and a paracone and metacone that are situated internally on the crown (away from the labial margin). Additionally, the upper molars of *Eosimias* and *Altiatlasius* lack any trace of the postprotocingulum, a structure that is present in basal adapiforms and omomyids. The lower molars of *Eosimias* and *Altiatlasius* differ from those of basal adapiforms and omomyids in having protoconids that are taller and more voluminous than their corresponding metaconids, entoconids that are shifted mesially to lie near the base of the postvallid, and hypoconulids that project distally beyond the remainder of the postcristid. Finally, the paraconid and metaconid cusps on the lower molars of *Eosimias* and *Altiatlasius* do not become increasingly connate from front

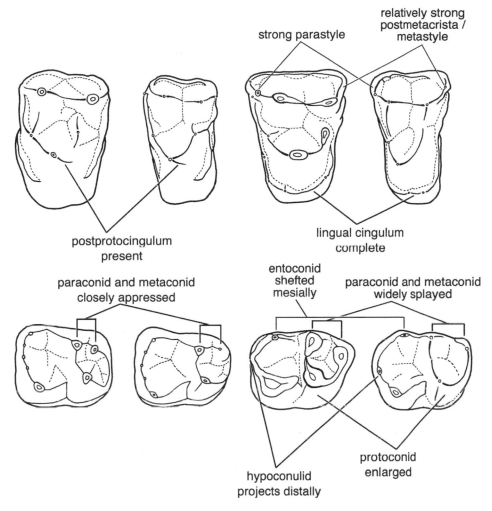

Figure 4. Schematic drawings of upper and lower second molars in some early euprimates. From left to right, the taxa depicted are as follows: the early Eocene adapiform *Cantius*, the early Eocene omomyid *Teilhardina*, the late Paleocene stem anthropoid *Altiatlasius*, and the middle Eocene stem anthropoid *Eosimias*. Various phylogenetically significant dental features are highlighted.

to back across the molar series as they do in basal adapiforms and omomyids. Given how paltry our current knowledge of *Altiatlasius* is, this taxon shares an extraordinary number of features in common with *Eosimias*. Although the polarity of dental characters is not always easy to establish among basal euprimates, a significant fraction of the features held in common by *Eosimias* and *Altiatlasius* are likely to be derived.

If the phylogenetic analysis performed here is accurate (or even roughly so), *Altiatlasius* cannot be interpreted as a basal euprimate, despite its age. Its nested phylogenetic position within the euprimate radiation, along with the absence of potential primate sister taxa in Africa, indicates that *Altiatlasius* dispersed to Africa from elsewhere, with the most obvious option being Asia.

BIOGEOGRAPHY AND ANTHROPOID ORIGINS: THE SHORT FUSE EXPLODES IN AFRICA

Over the years, scientific attempts to illuminate anthropoid origins have produced several hypotheses that bear upon the temporal, phylogenetic, biogeographic, and adaptive contexts of this important macroevolutionary transformation. Prior to the mid-1990s, most researchers assumed that anthropoids originated relatively late in the Paleogene, despite lingering disagreements over the phylogenetic position of anthropoids with respect to other living and fossil primates (e.g., Gingerich, 1980; Delson and Rosenberger, 1980; Rasmussen and Simons, 1992; Rasmussen, 1994). Two main factors appear to have contributed to the notion that anthropoids originated sometime near the Eocene-Oligocene boundary. The first of these was a dearth of anthropoid fossils dating significantly before the end of the Eocene. The second was a persistent—yet typically unacknowledged—influence from the *Scala naturae* positing that, because early anthropoids were "more advanced" than their prosimian relatives, they must have taken longer to evolve. Regardless of the conflicting phylogenetic reconstructions of different researchers, we can conveniently refer to all hypotheses that advocate such late Paleogene dates as "long fuse" versions of anthropoid origins (Figure 5). Given the dramatic climatic and biotic events that transpired near the Eocene-Oligocene boundary, these "long fuse" versions of anthropoid origins naturally set the stage for broader attempts to decipher both the biogeographic and adaptive contexts of anthropoid origins (e.g., Cachel, 1979; Rasmussen and Simons, 1992; Rasmussen, 1994; Holroyd and Maas, 1994).

Recent discoveries of anthropoid fossils in North Africa and Asia dating to the earlier part of the Paleogene allow us to reject these traditional "long fuse" versions of anthropoid origins, for the simple reason that their chronology has now been falsified. Biogeographic and adaptive hypotheses regarding anthropoid origins that are contingent upon such a long fuse can likewise be rejected. Unless one accepts an inordinately early date for primate origins, we are left with

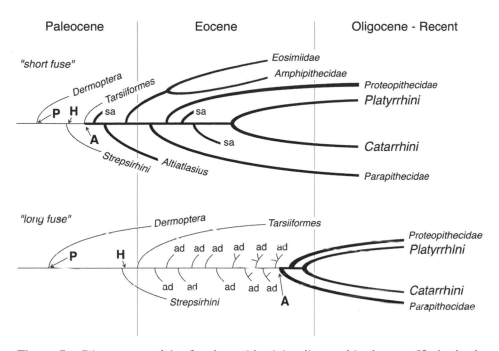

Figure 5. Divergent models of anthropoid origins discussed in the text. Under both models, the order Primates (abbreviated as "P") originated in the Paleocene, when the sister group of Primates (depicted here as Dermoptera) branched away from the earliest members of the primate clade. Likewise, both models accept the monophyly of Haplorhini (abbreviated as "H"), which originated sometime later in the Paleocene, coinciding with cladogenesis between ancestral strepsirhines and ancestral haplorhines. The models disagree on the timing of anthropoid origins, the definition of Anthropoidea (abbreviated as "A"), and the phylogenetic position of crown anthropoids and their closest relatives. Under the short fuse model, Anthropoidea originates in the late Paleocene, when the lineages culminating in modern tarsiiforms and anthropoids split. Extinct taxa (including *Altiatlasius*, Eosimiidae, Amphipithecidae, Parapithecidae, and Proteopithecidae) that are more closely related to crown anthropoids than to tarsiiforms are, by definition, stem anthropoids (abbreviated as "sa"). Under the long fuse model, Anthropoidea originates in the late Eocene, when stem anthropoids are thought to have crossed an arbitrary grade-level boundary separating them from their nearest "prosimian" relatives. In the version of the long fuse model shown here, anthropoids are depicted as being the descendants of adapiforms (abbreviated as "ad"). Other versions of the long fuse would regard anthropoids as having descended from omomyids or some other haplorhine group, yet the chronology of anthropoid origins would remain largely unaffected.

"short fuse" models of anthropoid origins, whereby relatively little time elapsed between the origin of the order Primates and the establishment of its three major clades, Strepsirhine, Tarsiiformes, and Anthropoidea (Figure 5). Although certain proponents of the molecular clock do indeed advocate surprisingly early dates for primate origins (e.g., Eizirik *et al.*, 2004), even these workers typically agree with the "short fuse" prediction that the origin of anthropoids occurred soon after the first primates evolved.

As we have seen, basic chronological frameworks such as the "short fuse" model advocated here can impact biogeographic and adaptive hypotheses regarding anthropoid origins as well. Indeed, with the demise of adaptive hypotheses that sought to link anthropoid origins with the shifting global climate and environments near the Eocene-Oligocene boundary, there is currently no compelling explanation for how and why anthropoids achieved their modern ecological dominance. Given this lack of understanding, recent advances in several of the areas reviewed above—including enhanced resolution of mammalian phylogeny, improvements in our knowledge of the early Cenozoic record of Africa, and theoretical progress in understanding the interplay between phylogeny and biogeography—suggest the following alternative hypothesis.

Basal anthropoids originated in Asia, as did the earliest primates before them. Molecular and paleontological estimates of the timing of the dichotomy between anthropoids and tarsiiforms overwhelmingly support the "short fuse" model, suggesting that the most recent common ancestor of these taxa lived no later than the late Paleocene (Beard and MacPhee, 1994; Beard, 1998b; Meireles *et al.*, 2003; Eizirik *et al.*, 2004). Soon after the anthropoid clade was established in Asia, one or more lineages dispersed to Africa, where basal anthropoids are documented by *Altiatlasius* by the late Paleocene. These basal anthropoids were among the first Laurasian mammals (and the first primates) to succeed in colonizing the ancestral homeland of the afrotheres. There, they encountered minimal ecological competition and experienced a rapid and expansive evolutionary radiation that remains poorly documented in the African fossil record. Many, perhaps even most, of the diagnostic morphological synapomorphies that distinguish modern anthropoids from tarsiiforms and strepsirhines evolved during this early and explosive radiation of African anthropoids. As a result, some ten million years after their initial colonization of Africa, middle Eocene anthropoids on that continent were already equipped with most of the features that distinguish modern anthropoids from their living and fossil "prosimian" relatives. In contrast, contemporary Asian anthropoids such as *Eosimias* never experienced the opportunity to radiate in splendid isolation in Africa. They retained numerous primitive features in common with *Altiatlasius* as a result.

How believable is this account of anthropoid origins? Like most freshly minted hypotheses, it requires further testing before it can be widely accepted. Nevertheless, the "short fuse" model of anthropoid origins is far more consistent with current knowledge of mammalian and primate phylogeny, the fossil record, and the complicated ways that phylogeny and biogeography influence one another than are any of its competitors. For example, even if we have yet to reach any consensus on identifying the sister group of primates, we know that primates are not afrotheres. Furthermore, even if we admit that the early Cenozoic fossil record of Africa remains inadequately sampled, we must acknowledge that most ordinal-level taxa of living placental mammals originated in Laurasia (more specifically, Asia). The probability that both primates and anthropoids originated in Asia is therefore extremely high, even if it is not yet proven. Likewise, while it is obvious that better specimens of *Altiatlasius* are necessary to render any final judgement on its phylogenetic position, the dental similarities shared by it and *Eosimias* are remarkable, suggesting a close phylogenetic relationship. Few would contest the fact that these similarities far outweigh any resemblances between *Altiatlasius* and other basal euprimates. Accordingly, until anatomical evidence emerges that would suggest otherwise, we must assume that *Altiatlasius* marks the initial colonization of Africa by primates, rather than any deeper phylogenetic history of primates there. Judging by the examples provided by Malagasy lemurs and South American platyrrhines, the dispersal of early primates such as *Altiatlasius* to an isolated, tropical land mass populated by a largely endemic mammalian fauna must have triggered a tremendous adaptive radiation. One product of this radiation may well have been the modern anthropoid bauplan, having been molded from one sharing much in common with primitive haplorhines. If so, chance and historical contingency, rather than long-term adaptive trends and the *Scala naturae*, account for this major macroevolutionary transformation among primates.

ACKNOWLEDGEMENTS

I thank John Fleagle for providing the opportunity to contribute to this volume. An earlier version of the manuscript was improved by comments made by Erik Seiffert, John Fleagle, and two anonymous reviewers. Thanks also to the numerous colleagues with whom I have worked in the field in Asia in recent years, including Yaowalak Chaimance, Daniel L. Gebo, Jean-Jacques Jaeger, Bernard Marandat, Laurent Marivaux, Ni Xijun, and Wang Yuanqing. Mark Klingler skillfully prepared the figures. This research has been supported by National Science Foundation grant BCS 0309800.

APPENDIX 1

Character descriptions

1. i1 orientation: procumbent (0); or relatively vertical (1).
2. i1 size: i1 < i2 (0); i1 = i2 (1); or i1 > i2, or clear evidence of i1 hypertrophy if i2 absent (2).
3. p1: present (0); or absent (1).
4. p2 root configuration: double-rooted (0); or single-rooted (1).
5. p2 lingual cingulid: absent or incomplete (0); or complete (1).
6. p2 size: p2 significantly smaller than p3 (0); p2 equal to or larger than p3 (1).
7. p3 exodaenodonty: absent (0); or present (1).
8. p3 lingual cingulid: absent or incomplete (0); or complete (1).
9. p3 mesial root location: directly mesial to distal root (0); mesiobuccal to distal root (1); or fused with distal root (2).
10. p4 exodaenodonty: absent (0); or present (1).
11. p4 mesial root location: directly mesial to distal root (0); mesiobuccal to distal root (1); or fused with distal root (2).
12. p4 metaconid: absent (0); present, located inferior and distal to protoconid (1); or present, located higher on crown and lingual to protoconid (2) (*ordered*).
13. m1-2 protoconid size: equal in height to metaconid (0); or taller and more voluminous than metaconid (1).
14. m1-2 protoconid location: closely appressed with metaconid (0); or widely spaced buccolingually from metaconid (1).
15. m1-2 entoconid location: at distolingual corner of talonid (0); or shifted mesially near the base of the postvallid (1).
16. m1-2 hypoconulid development: indistinct (0); or prominent (1).
17. m1-2 hypoconulid location: central, being incorporated within postcristid (0); distal, projecting beyond postcristid (1); or lingual, being "twinned" with entoconid (2).

18. m2-3 paraconid: more closely connate with metaconid than is the case on
    m1 (0); similar in disposition relative to metaconid as on m1 (1); or absent
    or highly reduced (2).

19. m3 hypoconulid: projects distally beyond level of hypoconid and ento-
    conid, forming a "talonid heel" (0); or evinces weak distal projection, fail-
    ing to form a "talonid heel" (1).

20. P3-4 protocone location: fully lingual (0); or mesial and buccal, away from
    lingual margin of crown (1).

21. M1-2 lingual cingulum: absent or incomplete (0); or complete (1).

22. M1-2 postprotocingulum: absent (0); or present (1).

23. M1-2 parastyle: absent or only weakly developed (0); or distinct (1).

24. M1-2 conules: present (0); or absent or greatly reduced (1).

25. M1-2 postmetacrista and metastyle: absent or only weakly developed (0);
    or present, strongly developed (1).

Taxon-character matrix used in parsimony analysis

Palenochtha	0211?	0??00	00000	00100	01100
Pronothodectes	02110	01001	00000	00000	01000
Altanius	0100?	?1001	01000	01000	01000
Cantius	00000	00100	01000	00000	01000
Teilhardina	02011	00000	01000	00000	01000
Altiatlasius	?????	?????	??111	111??	10101
Eosimias	10111	01111	11111	11111	10111
Arsinoea	10111	11111	11111	1201?	?????
Proteopithecus	10111	11111	12000	12211	10010
Saimiri	10111	11121	22111	00211	10010
Tarsius	02111	01101	01100	00001	10000

REFERENCES

Ba Maw, Ciochon R. L., and Savage D. E. 1979, Late Eocene of Burma yields earliest
    anthropoid primate, *Pondaungia cotteri. Nature* 282, 65–67.

Beard, K. C. 1993, Phylogenetic systematics of the Primatomorpha, with special refer-
    ence to Dermoptera, in: Szalay, F. S., Novacek, M. J., McKenna, M. C., eds., *Mammal
    Phylogeny: Placentals.* Springer-Verlag, New York, pp. 129–150.

Beard, K. C. 1998a, East of Eden: Asia as an important center of taxonomic origination
    in mammalian evolution. *Bull. Carnegie Mus. Nat. Hist.* 34, 5–39.

Beard, K. C. 1998b, A new genus of Tarsiidae (Mammalia: Primates) from the middle
    Eocene of Shanxi Province, China, with notes on the historical biogeography of
    tarsiers. *Bull. Carnegie Mus. Nat. Hist.* 34, 260–277.

Beard, K. C. 2002, Basal anthropoids, in: Hartwig, W. C., ed., *The Primate Fossil Record.*
    Cambridge University Press, Cambridge, pp. 133–149.

Beard, K. C. and Dawson, M. R. 1999, Intercontinental dispersal of Holarctic land
    mammals near the Paleocene/Eocene boundary: Paleogeographic, paleoclimatic and
    biostratigraphic implications. *Bull. Soc. géol. France* 170, 697–706.

Beard, K. C. and MacPhee, R.D.E. 1994, Cranial anatomy of *Shoshonius* and the an-
    tiquity of Anthropoidea, in: Fleagle, J. G. and Kay, R. F., eds., *Anthropoid Origins.*
    Plenum Press, New York, pp. 55–97.

Beard, K. C. and Wang, J. 1995, The first Asian plesiadapoids (Mammalia: Primato-
    morpha). *Ann. Carnegie Mus.* 64, 1–33.

Beard, K. C. and Wang, J. 2004, The eosimiid primates (Anthropoidea) of the Heti
    Formation, Yuanqu Basin, Shanxi and Henan Provinces, People's Republic of China.
    J. Hum. Evol. 46, 401–432.

Beard, K. C., Qi, T., Dawson, M. R., Wang, B., and Li, C. 1994, A diverse new primate
    fauna from middle Eocene fissure-fillings in southeastern China. *Nature* 368, 604–
    609.

Beard, K. C., Tong, Y., Dawson, M. R., Wang, J., and Huang, X. 1996, Earliest complete
    dentition of an anthropoid primate from the late middle Eocene of Shanxi Province,
    China. *Science* 272, 82–85.

Bonis, L. de, Jaeger, J.-J., Coiffat, B., and Coiffat, P.-E. 1988, Découverte du plus
    ancien primate catarrhinien connu dans l'Éocène supérieur d'Afrique du Nord. *C. R.*
    Acad. Sci., Paris (Sér. II) 306, 929–934.

Butler, P. M. and Hopwood, A. T. 1957, Insectivora and Chiroptera from the Miocene
    rocks of Kenya colony. *Fossil Mammals of Africa* 13, 1–35.

Cachel, S. 1979, A paleoecological model for the origin of higher primates. *J. Hum.*
    Evol. 8, 351–359.

Cartmill, M. 1974, *Daubentonia, Dactylopsila,* woodpeckers and klinorhynchy, in: Mar-
    tin, R. D., Doyle, G. A., and Walker, A. C., eds., *Prosimian Biology.* Duckworth,
    London, pp. 655–670.

Chaimanee, Y., Tin Thein, Ducrocq, S., Aung Naing Soe, Benammi, M., Than Tun,
    Thit Lwin, San Wai, and Jaeger, J.-J. 2000, A lower jaw of *Pondaungia cotteri* from
    the late middle Eocene Pondaung Formation (Myanmar) confirms its anthropoid
    status. *Proc. Nat. Acad. Sci.* 97, 4102–4105.

Chopram S. R. K., Kaul, S., and Vasishat, R. N. 1979, Miocene tree shrews from the
    Indian Sivaliks. *Nature* 281, 213–214.

Ciochon, R. L. and Gunnell, G. F. 2002, Chronology of primate discoveries in Myan-
    mar: influences on the anthropoid origins debate. *Yearb. phys. Anthrop.* 45, 2–35.

Ciochon, R. L. and Gunnell, G. F. 2004, Eocene large-bodied primates of Myanmar and Thailand: morphological considerations and phylogenetic affinities, in: Ross, C. F., Kay, R. F., eds., *Anthropoid Origins: New Visions*. Kluwer Academic/Plenum Publishers, New York, pp. 249–282.

Ciochon, R. L. and Holroyd, P. A. 1994, The Asian origin of Anthropoidea revisited, in: Fleagle, J. G., Kay, R. F., eds., *Anthropoid Origins*. Plenum Press, New York, pp. 143–162.

Ciochon, R. L., Savage, D. E., Thaw Tint, and Ba Maw 1985, Anthropoid origins in Asia? New discovery of *Amphipithecus* from the Eocene of Burma. *Science* 229, 756–759.

Ciochon, R. L., Gingerich, P. D., Gunnell, G. F., and Simons, E. L. 2001, Primate postcrania from the late middle Eocene of Myanmar. *Proc. Nat. Acad. Sci.* 98, 7672–7677.

Colbert, E. H. 1937, A new primate from the upper Eocene Pondaung Formation of Burma. *Am. Mus. Novit.* 951, 1–18.

Court, N. and Hartenberger, J.-L. 1992, A new species of the hyracoid mammal *Titanohyrax* from the Eocene of Tunisia. *Palaeontology* 35, 309–317.

Court, N. and Mahboubi, M. 1993, Reassessment of lower Eocene *Seggeurius amourensis*: aspects of primitive dental morphology in the mammalian order Hyracoidea. *J. Paleontol.* 67, 889–893.

Dagosto, M., Gebo, D. L., and Beard, K. C. 1999, Revision of the Wind River faunas, early Eocene of central Wyoming. Part 14. Postcranium of *Shoshonius cooperi* (Mammalia: Primates). *Ann. Carnegie Mus.* 68, 175–211.

Dawson, M. R. 2003, Paleogene rodents of Eurasia. *Deinsea* 10, 97–126.

Dawson, M. R., Tsubamoto, T., Takai, M., Egi, N., Soe Thura Tun, and Chit Sein 2003, Rodents of the family Anomaluridae (Mammalia) from southeast Asia (middle Eocene, Pondaung Formation, Myanmar). *Ann. Carnegie Mus.* 72, 203–213.

Delson, E. and Rosenberger, A. L. 1980, Phyletic perspectives on platyrrhine origins and anthropoid relationships, in: Ciochon, R. L. and Chiarelli, A. B., eds., *Evolutionary Biology of the New World Monkeys and Continental Drift*. Plenum Press, New York, pp. 445–458.

Domning, D. P. 2001, The earliest known fully quadrupedal sirenian. *Nature* 413, 625–627.

Domning, D. P., Ray, C. E., and McKenna, M. C. 1986, Two new Oligocene desmostylians and a discussion of tethytherian systematics. *Smithson. Contr. Paleobiol.* 59, 1–56.

Ducrocq, S. 1997, The anthracotheriid genus *Bothriogenys* (Mammalia, Artiodactyla) in Africa and Asia during the Paleogene: Phylogenetical and paleobiogeographical relationships. *Stuttgarter Beiträge zur Naturkunde (Ser. B)* 250, 1–44.

Ducrocq, S., Buffetaut, E., Buffetaut-Tong, H., Jaeger, J.-J., Jongkanjanasoontorn, Y., and Suteethorn, V. 1992, First fossil flying lemur: A dermopteran from the late Eocene of Thailand. *Palaeontology* 35, 373–380.

Eizirik, E., Murphy, W. J., Springer, M. S., and O'Brien, S. J. 2004, Molecular phylogeny and dating of early primate divergences, in: Ross, C. F. and Kay, R. F., eds., *Anthropoid Origins: New Visions.* Kluwer Academic/Plenum Publishers, New York, pp. 45–64.

Fleagle, J. G. and Kay R. F. 1987, The phyletic position of the Parapithecidae. *J. Hum. Evol.* 17, 483–532.

Fleagle, J. G. and Reed, K. E. 2004, The evolution of primate ecology: patterns of geography and phylogeny, in: Anapol, F., German, R. Z., and Jablonski, N. G., eds., *Shaping Primate Evolution.* Cambridge University Press, Cambridge, pp. 353–367.

Gebo, D. L. 1988, Foot morphology and locomotor adaptation in Eocene primates. *Folia Primatol.* 50, 3–41.

Gebo, D. L., Dagosto, M., Beard, K. C., Qi, T., and Wang, J. 2000, The oldest known anthropoid postcranial fossils and the early evolution of higher primates. *Nature* 404, 276–278.

Gebo, D. L., Dagosto, M., Beard, K. C., and Qi, T. 2001, Middle Eocene primate tarsals from China: Implications for haplorhine evolution. *Am. J. phys. Anthropol.* 116, 83–107.

Gheerbrant, E., Sudre, J., Capetta, H., Iarochène, M., Amaghzaz, M., and Bouya, B. 2002, A new large mammal from the Ypresian of Morocco: Evidence of surprising diversity of early proboscideans. *Acta Palaeontol. Pol.* 47, 493–506.

Gingerich, P. D. 1976, Cranial anatomy and evolution of early Tertiary Plesiadapidae (Mammalia, Primates). *Univ. Michigan Pap. Paleontol.* 15, 1–141.

Gingerich, P. D. 1980, Eocene Adapidae, paleobiogeography, and the origin of South American Platyrrhini, in: Ciochon, R. L. and Chiarelli, A. B. eds., *Evolutionary Biology of the New World Monkeys and Continental Drift.* Plenum Press, New York, pp. 123–138.

Gingerich, P. D. 1986, Early Eocene *Cantius torresi*—oldest primate of modern aspect from North America. *Nature* 320, 319–321.

Gingerich, P. D. 1990, African dawn for primates. *Nature* 346, 411.

Gingerich, P. D., Dashzeveg, D., and Russell, D. E. 1991, Dentition and systematic relationships of *Altanius orlovi* (Mammalia, Primates) from the early Eocene of Mongolia. *Geobios* 24, 637–646.

Gingerich, P. D., ul Haq, M., Zalmout, I. S., Khan, I. H., and Malkani, M. S. 2001, Origin of whales from early artiodactyls: hands and feet of Eocene Protocetidae from Pakistan. *Science* 293, 2239–2242.

Godfrey, L. R., Jungers, W. L., Wunderlich, R. E., and Richmond, B. G. 1997, Reappraisal of the postcranium of *Hadropithecus* (Primates, Indroidea). *Am. J. phys. Anthropol.* 103, 529–556.

Godinot, M. 1978, Un nouvel Adapidé (primate) de l'Éocène inférieur de Provence. *C. R. Acad. Sci., Paris (Sér. D)* 286, 1869–1872.

Godinot, M. 1994, Early North African primates and their significance for the origin of Simiiformes (= Anthropoidea), in: Fleagle, J. G. and Kay, R. F., eds., *Anthropoid Origins.* Plenum Press, New York, pp. 235–295.

Godinot, M. and Mahboubi, M. 1992, Earliest known simian primate found in Algeria. *Nature* 357, 324–326.

Godinot, M. and Mahboubi, M. 1994, Les petits primates simiiformes de Glib Zegdou (Éocène inférieur à moyen d'Algérie). *C. R. Acad. Sci., Paris (Sér. II)* 319, 357–364.

Gregory, W. K. 1920, On the structure and relations of *Notharctus*, an American Eocene primate. *Mem. Am. Mus. Nat. Hist.* (n.s.) 3, 49–243.

Gunnell, G. F. 1989, Evolutionary history of Microsyopoidea (Mammalia, ?Primates) and the relationship between Plesiadapiformes and Primates. *Univ. Michigan Pap. Paleontol.* 27, 1–157.

Gunnell, G. F. and Miller, E. R. 2001, Origin of Anthropoidea: dental evidence and recognition of early anthropoids in the fossil record, with comments on the Asian anthropoid radiation. *Am. J. phys. Anthropol.* 114, 177–191.

Gunnell, G. F., Ciochon, R. L., Gingerich, P. D., and Holroyd P. A. 2002, New assessment of *Pondaungia* and *Amphipithecus* (Primates) from the late middle Eocene of Myanmar, with a comment on 'Amphipithecidae.' *Contrib. Mus. Paleont. Univ. Michigan* 30, 337–372.

Hoffstetter, R. 1977, Phylogénie des Primates: confrontation des résultats obtenus par les diverses voies d'approche du problème. *Bull. Mém. Soc. Anthrop. Paris (Sér. 13)* 4, 327–346.

Holroyd, P. A. and Maas, M. C. 1994, Paleogeography, paleobiogeography, and anthropoid origins, in: Fleagle, J. G. and Kay, R. F., eds., *Anthropoid Origins.* Plenum Press, New York, pp. 297–334.

Hooker, J. J., Russell, D. E., and Phélizon, A. 1999, A new family of Plesiadapiformes (Mammalia) from the Old World lower Paleogene. *Palaeontology* 42, 377–407.

Jacobs, L. L. 1980, Siwalik fossil tree shrews, in: Luckett, W. P., ed., *Comparative Biology and Evolutionary Relationships of Tree Shrews.* Plenum Press, New York, pp. 205–216.

Jaeger, J.-J., Denys, C., and Coiffat, B. 1985, New Phiomorpha and Anomaluridae from the late Eocene of north-west Africa: Phylogenetic implications, in: Luckett, W. P., Hartenberger, J.-L., eds., *Evolutionary Relationships Among Rodents: A Multidisciplinary Analysis.* Plenum Press, New York, pp. 567–588.

Jaeger, J.-J., Aung Naing Soe, Aye Ko Aung, Benammi, M., Chaimanee, Y., Ducrocq, R.-M., Than Tun, Tin Thein, and Ducrocq, S. 1998, New Myanmar middle Eocene anthropoids. An Asian origin for catarrhines? *C. R. Acad. Sci., Paris (Sciences de la vie)* 321, 953–959.

Jaeger, J.-J., Tin Thein, Benammi, M., Chaimanee, Y., Aung Naing Soe, Thit Lwin, Than Tun, San Wai, and Ducrocq, S. 1999, A new primate from the middle Eocene of Myanmar and the Asian early origin of anthropoids. *Science* 286, 528–530.

Jaeger, J.-J., Chaimanee, Y., Tafforeau, P., Ducrocq, S., Aung Naing Soe, Marivaux, L., Sudre, J., Soe Thura Tun, Wanna Htoon, and Marandat, B. 2004, Systematics and paleobiology of the anthropoid primate *Pondaungia* from the late middle Eocene of Myanmar. *C. R. Palevol.* 3, 243–255.

Jolivet, L. and Faccenna, C. 2000, Mediterranean extension and the Africa-Eurasia collision. *Tectonics* 19, 1095–1106.

Jungers, W. L., Godfrey, L. R., Simons, E. L., and Chatrath, P. S. 1997, Phalangeal curvature and positional behavior in extinct sloth lemurs (Primates, Palaeopropithecidae). *Proc. Natl. Acad. Sci.* 94, 11998–12001.

Kappelman, J., Rasmussen, D. T., Sanders, W. J., Feseha, M., Bown, T., Copeland, P., Crabaugh, J., Fleagle, J., Glantz, M., Gordon, A., Jacobs, B., Maga, M., Muldoon, K., Pan, A., Pyne, L., Richmond, B., Ryan, T., Seiffert, E. R., Sen, S., Todd, L., Wiemann, M. C., and Winkler, A. 2003, Oligocene mammals from Ethiopia and faunal exchange between Afro-Arabia and Eurasia. *Nature* 426, 549–552.

Kay, R. F., Williams, B. A., Ross, C. F., Takai, M., and Shigehara, N. 2004, Anthropoid origins: A phylogenetic analysis, in: Ross, C. F. and Kay, R. F., eds., *Anthropoid Origins: New Visions.* Kluwer Academic/Plenum Publishers, New York, pp. 91–135.

Madsen, O., Scally, M., Douady, C. J., Kao, D. J., DeBry, R. W., Adkins, R., Amrine, H. M., Stanhope, M. J., de Jong, W. W., and Springer, M. S. 2001, Parallel adaptive radiations in two clades of placental mammals. *Nature* 409, 610–614.

Marivaux, L., Vianey-Liaud, M., Welcomme, J.-L., and Jaeger, J.-J. 2002, The role of Asia in the origin and diversification of hystricognathous rodents. *Zool. Scri.* 31, 225–239.

Marivaux, L., Chaimanee, Y., Ducrocq, S., Marandat, B., Sudre, J., Aung Naing Soe, Soe Thura Tun, Wanna Htoon, and Jaeger, J.-J. 2003, The anthropoid status of a primate from the late middle Eocene Pondaung Formation (central Myanmar): Tarsal evidence. *Proc. Natl. Acad. Sci.* 100, 13173–13178.

Martin, R. D. 1990, *Primate Origins and Evolution: A Phylogenetic Reconstruction.* Chapman and Hall, London.

McKenna, M. C. 1967, Classification, range, and deployment of the prosimian primates, in: Problèmes Actuels de Paléontologie (Évolution des Vertebrés). *Éditions du CNRS, Paris (Colloques Internationaux du Centre National de la Recherche Scientifique,* no. 163), pp. 603–610.

McKenna, M. C. 1975, Toward a phylogenetic classification of the Mammalia, in: Luckett, W. P. and Szalay, F. S., eds., *Phylogeny of the Primates: A Multidisciplinary Approach*. Plenum Press, New York, pp. 21–46.

Meireles, C. M., Czelusniak, J., Page, S. L., Wildman, D. E., Goodman, M. 2003, Phylogenetic position of tarsiers within the order Primates: Evidence from γ-globin DNA sequences, in: Wright, P. C., Simons, E. L., and Gursky, S., eds., *Tarsiers: Past, Present, and Future*. Rutgers University Press, New Brunswick, pp. 145–160.

Meng, J., Wyss, A. R., Dawson, M. R., and Zhai, R. 1994, Primitive fossil rodent from Inner Mongolia and its implications for mammalian phylogeny. *Nature* 370, 134–136.

Meng, J., Hu, Y., and Li, C. 2003, The osteology of *Rhombomylus* (Mammalia, Glires): Implications for phylogeny and evolution of Glires. *Bull. Am. Mus. Nat. Hist.* 275, 1–247.

Murphy, W. J., Eizirik, E., Johnson, W. E., Zhang, Y. P., Ryder, O. A., and O'Brien, S. J. 2001a, Molecular phylogenetics and the origins of placental mammals. *Nature* 409, 614–618.

Murphy, W. J., Eizirik, E., O'Brien, S. J., Madsen, O., Scally, M., Douady, C. J., Teeling, E., Ryder, O. A., Stanhope, M. J., de Jong, W. W., and Springer, M. S. 2001b, Resolution of the early placental mammal radiation using Bayesian phylogenetics. *Science* 294, 2348–2351.

Patterson, B. 1975, The fossil aardvarks (Mammalia: Tubulidentata). *Bull. Mus. Comp. Zool.* 147, 185–237.

Pilgrim, G. E. 1927, A *Sivapithecus* palate and other primate fossils from India. *Memoirs of the Geological Survey of India, Palaeontologia Indica*, n.s. 14, 1–26.

Rasmussen, D. T. 1994, The different meanings of a tarsioid-anthropoid clade and a new model of anthropoid origin, in: Fleagle, J. G. and Kay, R. F., Eds., *Anthropoid Origins*. Plenum Press, New York, pp. 335–360.

Rasmussen, D. T. and Simons, E. L. 1992, Paleobiology of the oligopithecines, the earliest known anthropoid primates. *Int. J. Primatol.* 13, 477–508.

Rose, K. D. and Bown, T. M. 1991, Additional fossil evidence on the differentiation of the earliest euprimates. *Proc. Natl. Acad. Sci.* 88, 98–101.

Rose, K. D. and Emry, R. J. 1993, Relationships of Xenarthra, Pholidota, and fossil "edentates": The morphological evidence, in: Szalay, F. S., Novacek, M. J., and McKenna, M. C., eds., *Mammal Phylogeny: Placentals*. Springer-Verlag, New York, pp. 81–102.

Rose, K. D., Godinot, M., and Bown, T. M. 1994, The early radiation of euprimates and the initial diversification of Omomyidae, in: Fleagle, J. G. and Kay, R. F., eds., Anthropoid Origins. Plenum Press, New York, pp. 1–28.

Schmitz, J. and Zischler, H. 2004, Molecular cladistic markers and the infraordinal phylogenetic relationships of primates, in: Ross, C. F. and Kay, R. F., eds.,

Anthropoid Origins: New Visions. Kluwer Academic/Plenum Publishers, New York, pp. 65–77.

Schwartz, J. H. 2003, How close are the similarities between *Tarsius* and other primates? in: Wright, P. C., Simons, E. L., and Gursky, S., eds., *Tarsiers: Past, Present, and Future.* Rutgers University Press, New Brunswick, pp. 50–96.

Seiffert, E. R., Simons, E. L., and Simons, C. V. M. 2004, Phylogenetic, biogeographic, and adaptive implications of new fossil evidence bearing on crown anthropoid origins and early stem catarrhine evolution, in: Ross, C. F. and Kay, R. F., eds., *Anthropoid Origins: New Visions.* Kluwer Academic/Plenum Publishers, New York, pp. 157–181.

Sigé, B., Jaeger, J.-J., Sudre, J., and Vianey-Liaud, M. 1990, *Altiatlasius koulchii* n. gen. et sp., primate omomyidé du Paléocène supérieur du Maroc, et les origines des euprimates. *Palaeontographica* (Abt. A) 214, 31–56.

Simons, E. L. 1995, Egyptian Oligocene primates: A review. *Yearb. phys. Anthrop.* 38, 199–238.

Simons, E. L. 2003, The fossil record of tarsier evolution, in: Wright, P. C., Simons, E. L., and Gursky, S., eds., *Tarsiers: Past, Present, and Future.* Rutgers University Press, New Brunswick, pp. 9–34.

Simpson, G. G. 1945, The principles of classification and a classification of mammals. *Bull. Am. Mus. Nat. Hist.* 85, 1–350.

Springer, M. S., Cleven, G. C., Madsen, O., de Jong, W. W., Waddell, V. G., Amrine, H. M., and Stanhope, M. J. 1997, Endemic African mammals shake the phylogenetic tree. *Nature* 388, 61–64.

Springer, M. S., Stanhope, M. J., Madsen, O., and de Jong, W. W. 2004, Molecules consolidate the placental mammal tree. *Trends Ecol. Evol.* 19, 430–438.

Swofford, D. L. 2002, *PAUP: Phylogenetic Analysis Using Parsimony, Version 4.0b10* (Altivec). Sinaeur Associates, Sunderland, MA.

Szalay, F. S., and Delson, E. 1979, *Evolutionary History of the Primates.* Academic Press, New York.

Tabuce, R., Coiffat, B., Coiffat, P.-E., Mahboubi, M., and Jaeger, J.-J. 2001, A new genus of Macroscelidea (Mammalia) from the Eocene of Algeria: A possible origin for elephant-shrews. *J. Vert. Paleontol.* 21, 535–546.

Tabuce, R., Mahboubi, M., Tafforeau, P., and Sudre, J. 2004, Discovery of a highly-specialized plesiadapiform primate in the early-middle Eocene of northwestern Africa. *J. Hum. Evol.* 47, 305–321.

Takai, M. and Shigehara, N. 2004, The Pondaung primates, enigmatic "possible anthropoids" from the latest middle Eocene, central Myanmar, in: Ross, C. F. and Kay, R. F.; eds., *Anthropoid Origins: New Visions.* Kluwer Academic/Plenum Publishers, New York, pp. 283–321.

Tattersall, I. 1982, *The Primates of Madagascar.* Columbia University Press, New York.

Tong, Y. 1988, Fossil tree shrews from the Eocene Hetaoyuan Formation of Xichuan, Henan. *Vertebrata PalAsiatica* 26, 214–220.

Vianey-Liaud, M., Jaeger, J.-J., Hartenberger, J.-L., and Mahboubi, M. 1994, Les rongeurs de l'Eocene d'Afrique nord-occidentale [Glib Zegdou (Algerie) et Chambi (Tunisie)] et l'origine des Anomaluridae. *Palaeovertebrata* 23, 93–118.

Walker, A. 1972, The dissemination and segregation of early primates in relation to continental configuration, in: Bishop, W. W. and Miller, J. A., eds., *Calibration of Hominoid Evolution: Recent Advances in Isoptopic and Other Dating Methods to the Origin of Man.* Scottish Academic Press, Edinburgh, pp. 195–218.

Yoder, A. D. 2003, The phylogenetic position of genus *Tarsius*: Whose side are you on? in: Wright, P. C., Simons, E. L., and Gursky, S., eds., *Tarsiers: Past, Present, and Future.* Rutgers University Press, New Brunswick, pp. 161–175.

Yoder, A. D. and Yang, Z. 2004, Divergence dates for Malagasy lemurs estimated from multiple gene loci: Geological and evolutionary context. *Mol. Ecol.* 13, 757–773.

Yoder, A. D., Cartmill, M., Ruvolo, M., Smith, K., and Vilgalys, R. 1996, Ancient single origin for Malagasy primates. *Proc. Natl. Acad. Sci.* 93, 5122–5126.

CHAPTER SIXTEEN

Continental Paleobiogeography as Phylogenetic Evidence

James B. Rossie and Erik R. Seiffert

ABSTRACT

Morphological convergence between members of clades isolated on different landmasses can mislead phylogenetic analyses, and can imply intercontinental dispersals that are unlikely given reconstructed paleogeography. It is argued here that paleobiogeographic data, like chronostratigraphic data, are relevant to the process of inferring phylogenetic relationships. In order to allow the dynamic history of continental paleogeography to influence phylogenetic analysis, a "chronobiogeographic character" is developed here, and implemented in two phylogenetic case studies from the primate fossil record. The chronobiogeographic character allows for a tradeoff between morphological character debt and "chronobiogeographic debt", and provides a more explicit test of phylogenetic hypotheses that imply complex (and perhaps unlikely) biogeographic histories.

Key Words: biogeography, phylogenetics, paleobiogeography, primates

James B. Rossie • Department of Anthropology, Stony Brook University, Stony Brook, NY 11794-4364, U.S.A.   **Erik R. Seiffert** • Department of Earth Sciences and Museum of Natural History, University of Oxford, Parks Road, Oxford, OX1 3PR, United Kingdom.

Primate Biogeography, edited by Shawn M. Lehman and John G. Fleagle.
Springer, New York, 2006.

INTRODUCTION

Primates are remarkable among non-aquatic, non-volant placental mammals in that every major extant primate clade is, at some point in its evolutionary history, likely derived from an ancestor that successfully migrated across a substantial body of water to colonize a distant landmass. Indeed, two such dispersals that were responsible for much of extant primate diversity—the lemuriform strepsirrhine and platyrrhine anthropoid radiations—might not have been possible had it not been for even earlier hypothesized dispersals of stem anthropoids and stem strepsirrhines across the Tethys Sea to Afro-Arabia from northern continents (e.g., Beard, 1998; Kay *et al.*, 2004; Seiffert *et al.*, 2004). It is impossible to reconstruct what would have happened to platyrrhines and lemurs had these random twigs along their stem lineages not successfully crossed the Atlantic Ocean and Mozambique Channel, respectively, but we do know that the other members of their stem lineages went extinct. Along with the similarly itinerant rodents, we extant primates certainly owe much of our existence to Paleogene paleogeography and improbable sea voyages.

Despite primates' and rodents' relative success at overwater dispersal, a number of molecular phylogenetic analyses that have been carried out over the last decade have nevertheless provided strong support for the hypothesis that successful dispersal events such as these are probably exceedingly rare, and that the evolutionary dynamics of colonization can produce monophyletic radiations that are otherwise not easily detectable using morphological data alone (either due to morphological evolutionary inertia or rampant morphological convergence with distantly related taxa on other landmasses). Molecular approaches have, for instance, demonstrated that the Malagasy strepsirrhines are unambiguously monophyletic (Porter *et al.*, 1995, 1997; Yoder *et al.*, 1996; Yoder, 1997; Roos *et al.*, 2004), despite previous claims to the contrary (Szalay and Katz, 1973; Cartmill, 1975; Schwartz and Tattersall, 1985), and have also detected a number of other monophyletic mammalian radiations, such as the immigrant carnivorans (Yoder *et al.*, 2003) and tenrecs (Olson and Goodman, 2003) that now coexist with lemurs in Madagascar. Each of these Malagasy radiations were apparently allowed for by similarly unlikely colonization events. At higher levels in placental mammalian phylogeny, supraordinal clades such as the endemic Afro-Arabian superorder Afrotheria (containing such morphologically diverse taxa as elephants, hyraxes, sea cows, elephant-shrews, aardvarks, golden moles, and tenrecs) have only recently been recognized using retroposons as well as nuclear and mitochondrial DNA sequences (Murphy *et al.*, 2001b; Nikaido *et al.*, 2003). It is not our intention to suggest that molecular evidence will always solve such biogeographic problems correctly. Indeed, its inapplicability to almost all extinct taxa limits its ability to address many questions, but the evolutionary phenomena revealed by these molecular data do suggest that there may be other options for extracting more information from the fossil record than is currently appreciated.

There has long been an interest in employing phylogenies derived from morphological or molecular data to infer historical biogeographic patterns and the geographic causality of phylogenetic events (Platnick and Nelson, 1978; Nelson and Platnick, 1981; Grande, 1985; Humphries and Parenti, 1986; Brooks, 1990; Morrone and Carpenter, 1994; Ronquist, 1997; Hunn and Upchurch, 2001), and this topic has garnered even greater interest now that molecular studies have provided such strong evidence for congruence between phylogenetic and historical biogeographic patterns. But might this phenomenon indicate that paleobiogeography should actually contribute to the recovery of phylogeny, rather than simply being inferred from it? After all, in the case of afrotherian monophyly, which is now overwhelmingly supported by genomic data but potentially little, or no, morphological data (but see Robinson and Seiffert, 2004; Seiffert, 2003), one might reasonably conclude that the past distribution of terrestrial afrotherians in Afro-Arabia actually provides a signal of greater phylogenetic valence than that of their morphology. The same could arguably be said of Malagasy lemurs as well, which were long considered by many to be paraphyletic or polyphyletic based strictly on morphological data (Szalay and Katz, 1973; Cartmill, 1975; Schwartz and Tattersall, 1985). This question should thus be of great interest to paleontologists, for in the absence of genetic data for extinct taxa, any additional means for improving estimates of relationships among such fossil organisms—and even features not encoded in the genome, like stratigraphy (Fisher, 1992, 1994) or biogeography (Seiffert *et al.*, 2003; Seiffert *et al.*, 2004)—should be welcomed. At present, however, investigators have no recourse but to either dismiss or accept cladograms that imply multiple crossings of vast seaways (e.g., Ducrocq, 1999; Ross, 2000), because, unlike stratigraphy—which is now occasionally incorporated into phylogenetic analyses of extinct taxa (e.g., Thewissen, 1992; Polly, 1997; Bodenbender and Fisher, 2001; Bloch *et al.*, 2001; Muldoon and Gunnell, 2002; Finarelli and Clyde, 2004)—it is not yet clear how paleontologists might best gauge the relative importance of morphology versus paleobiogeography in phylogenetic analyses of fossil taxa.

In this paper we discuss the possible benefits of, and problems associated with, adding the historical continental distribution of living and extinct taxa, in the form of a "chronobiogeographic character," to the mix of morphological and molecular evidence currently being employed in phylogenetic analyses of the primate radiation. We use two case studies—one derived from a previous phylogenetic analysis of Paleogene primates (Seiffert *et al.*, 2004), and another from a phylogenetic analysis of Neogene catarrhines (Rossie and MacLatchy, 2004)—to illustrate the utility of the method. Both examples include fossil and living species because it is our contention that analysis of biogeography (whether preceding or following phylogenetic analysis) must be approached diachronically, or in "time slices" (Hunn and Upchurch, 2001).

AIM AND JUSTIFICATION

Our aim is to develop a method by which paleobiogeographic data can be incorporated directly into the process of phylogenetic analysis, as Fisher (1992) has already done with temporal data in creating the stratocladistic method. We take this "hard" approach (*sensu* Hunn and Upchurch, 2001) because alternative "soft" approaches to cladistic biogeography—that is, those that deny a direct role for paleobiogeography in phylogenetic reconstruction, and instead generally only attempt to determine the geographic causality of phylogenetic patterns—do not allow researchers to satisfactorily *test* those hypotheses that may explain how paleogeography has influenced organismal phylogeny and paleobiogeography. Moreover, the mapping of a geographical character onto the cladogram of a group in order to infer its biogeographical history (e.g., Stewart and Disotell, 1998) gives no consideration to how improbable the implied dispersal events might be.

As a time/space character has, to the best of our knowledge, never previously been employed in a phylogenetic analysis of living and extinct taxa, it is perhaps incumbent upon us to demonstrate that the chronobiogeographic character addresses a real and problematic shortcoming of previous methods. The weakness of available methodology, as we see it, is that there is currently no satisfactory way to allow paleobiogeographic evidence to influence a phylogenetic analysis, or to provide a more rigorous test of biogeographically unparsimonious phylogenetic hypotheses. But are there legitimate reasons for wanting to do so? As noted above, there is ample evidence that endemic clades may sometimes defy recognition when only morphology is considered, and that some cladograms imply patterns of paleo-dispersal that are difficult to believe. These problems may provide sufficient motivation, but there must also be some reason to believe that paleobiogeographic data have some relevance to phylogenetic history— that such data provide evidence bearing on phylogenetic relationships. Should this be the case, it would also be reassuring to know whether the addition of paleobiogeographic data could improve the accuracy of phylogenetic analyses. Paleobiogeographic data represent a combination of biogeographic and chronostratigraphic information, and the method outlined below essentially adds a biogeographic element to stratocladistic methodology (see Fisher 1992, 1994). The bearing of chronostratigraphic data on phylogenetic relationships, and the success of stratocladistics in increasing phylogenetic accuracy and resolution has received ample discussion (Thewissen, 1992; Fisher, 1994; Wagner, 1995, 2001; Fox *et al.*, 1999; Alroy, 2001), so we will focus the following discussion on biogeographic data.

Relevance of Biogeography

The evolution of a clade is a process—one that imprints certain patterns on the natural world. Systematists study these patterns in an effort to achieve a description of the process that produced them. This description is the evolutionary history of the clade. The most important pattern employed in this pursuit is the distribution of morphological and/or molecular characters among known extant and fossil species. Insofar as the divergence of lineages involves the acquisition of new characters (or states), each recognizable clade can be diagnosed by at least one character—a synapomorphy (Eldredge and Cracraft, 1980; Wiley, 1981). Each character is a synapomorphy at some level of phylogenetic inclusiveness (Stevens, 1984), and the distribution of characters among species reflects the nested hierarchy of phylogenetic relationships within the clade. That hierarchy can be inferred (through the logical process of abduction, see Sober, 1988) by arranging the taxa into a hierarchy (usually depicted as a cladogram) that minimizes the number of times that each character must arise independently in order to explain its taxonomic distribution—that is, the number of times that its presence is not a reflection of the true nested hierarchy of relationships. Such instances require *ad hoc* explanation (homoplasy), and favoring the hypothesis that minimizes the number of *ad hoc* explanations is standard logical and scientific practice—in systematics, it is commonly known as the parsimony criterion (Wiley, 1981; Farris, 1983; Sober, 1988). This 'most parsimonious' hypothesis of relationships is the one that best fits the observed data (Kluge and Farris, 1969; Sober, 1988; Farris, 1983).

Evolution produces temporal and geographic patterns as well (Eldredge and Cracraft, 1980; Cracraft, 1981). Because each divergence of lineages (speciation) occurred at a single point in time, no members of the clades that the diverging lineages might give rise to can occur in the fossil record at a point in time earlier than the divergence. For this reason, we expect some consilience between the relative temporal order of taxa in the record and the relative order of nodes in a cladistic hypothesis of their relationships, and this expectation is often met (Gauthier et al., 1988; Norell and Novacek, 1992; Huelsenbeck, 1994; Siddall, 1998). Still, most phylogenetic hypotheses require some *ad hoc* explanations of non-preservation in the fossil record—'ghost lineages' in our sense of the term (see also Fox et al., 1999). We would not argue that the fossil record is so complete that ghost lineages are unlikely any more than we would argue that morphological homoplasy is unlikely. Fortunately, the application of the parsimony criterion to both sorts of *ad hoc* explanations (as in stratocladistics) does not rely on such arguments (see Farris, 1983; Fox et al., 1999).

Likewise, speciation occurs in either a single place or two directly adjacent places (Mayr, 1942, 1963; Bush, 1975), and it is no surprise that Darwin (1859) saw in the geographical distribution of species evidence for their shared common ancestry. However, the biogeographic pattern produced by the evolution of a clade is more convoluted than the temporal pattern. If a lineage diverges into two somewhere (for instance, Europe), the descendants of these diverging lineages could occur anywhere—provided that they can get there. They can also reticulate: they can disperse to Afro-Arabia and spawn descendents, one of which returns to Europe. This is why there is no inherent order or polarity to purely biogeographic data. However, certain logical necessities must still hold. Most importantly, terrestrial species cannot have a sister species in a geographical region separated from it by a third intervening region unless one of the two species (or an unknown intermediate taxon) once existed in the intervening region. This is not much of a logical constraint, but it does require *ad hoc* explanations for such absences, and these should not be postulated beyond necessity. Our method infers hypotheses of evolutionary history by minimizing the number of *ad hoc* explanations needed to explain the patterns of temporal, geographic, and character distribution produced by evolution. Such hypotheses will have greater explanatory power than even stratocladistic solutions, and will be highly testable.

The notion of "dispersal" may give the impression that use of biogeographic patterns in a phylogenetic analysis would constitute employing process assumptions. We make no more assumptions about the process of dispersal than does any investigator employing an ordered morphological character. We merely assert that a change in geographic 'state' must pass through an intermediate state, and we do so with better justification. As will be discussed below, we differentially weight character state changes in our biogeographic character, because we believe that some changes are less likely than others. While such assumptions might be objectionable to some, it has become commonplace in molecular systematics to make assumptions about the relative likelihood of different state changes (transition/transversion), and space precludes even a terse summary of the assumptions that are involved in routine morphological character coding (see Gift and Stevens, 1997; Wiens, 2001), but in both cases the assumptions are made at a lower level than that of phylogenetic inference. As long as assumptions about the relative difficulty of certain types of dispersal are not being tested by the parsimony analysis, they are acceptable background knowledge.

Such lower level assumptions appear to be acceptable as long as they have a good justification, as in the case of transition and transversion, and at present we see no possible justification for believing that a mammalian dispersal across hundreds of miles of open ocean is more likely than dispersal across a chain of islands. More importantly, what we assume is not the probability of different types of dispersal, but the *relative* probability (see below). A similar approach has been suggested for stratocladistics as well, in which one weights each chronostratigraphic unit according to how well-sampled the ingroup taxa are (Fisher, 1992). Again, this is logically equivalent to the assumption routinely made in using an ordered character (Pogue and Mickevich, 1990; Wiens, 2001).

METHODS

Although it has rarely been done, paleobiogeographic data could be added to a phylogenetic analysis in several ways. For example, in their phylogenetic analysis of primates, Seiffert *et al.* (2003) included a simple unconstrained (unordered) biogeographic character with states composed of the major landmasses on which the living and extinct study taxa were found. Such a character generates one

step of parsimony debt (we will refer to this as "biogeographic debt") for any
implied dispersal between landmasses. The circumstances under which a lineage might change its geographical distribution vary considerably, however,
and, among other things, an unconstrained biogeographic character threatens
to recover a sequence of dispersal events in the history of any given clade that
makes little sense given the relative positions of major landmasses through time.
Incorporating these differences into a weighting scheme for biogeographic debt
is complicated, but an improvement was provided by Seiffert *et al.* (2004), who
used likely dispersal routes between major landmasses to create a step-matrix for
the biogeographic character that more appropriately penalized those phylogenetic hypotheses that incurred considerable biogeographic debt. Unfortunately
this approach is only of clear use if all ingroup taxa are derived from a specific
time period during which continental configurations were relatively stable, but
phylogenetic analyses of major primate clades are rarely so restricted in their
temporal sampling. If, for instance, extant African and Eurasian taxa are included
in an analysis for purposes of broader morphological or molecular sampling, it
is obvious that a "dispersal" between present-day Afro-Arabia and Asia is not
nearly as unlikely as a dispersal from Afro-Arabia to Asia during the early Paleogene, when these areas were separated by a vast Tethys Sea (e.g., Holroyd
and Maas, 1994; Scotese, 2002). The likelihood of such events in the evolutionary history of a clade can only be addressed if temporal data are incorporated. This need to view biogeography in "time-slices" has been recognized for
some time (Grande, 1985), but has only recently become popular (Hunn and
Upchurch, 2001; Upchurch *et al.*, 2002). For our purposes, the biogeographic
character should have a temporal component because the likelihood of dispersal, and the possibility of vicariance, both depend on knowledge of the nature
(or lack) of the connection between two or more landmasses *at the time of the
hypothesized event*. Accordingly, temporal and biogeographical data are incorporated into a single character in the "chronobiogeographic" approach developed
here.

The Chronobiogeographic Character

The temporal portion of the chronobiogeographic character is essentially that
of the stratocladistic character, constructed using the step-matrix approach described by Fisher (1992), although the matrix must be expanded so that there
is a column and row for each stratum and landmass (time/place) within which

an ingroup taxon has been sampled (see Figure 1). This portion of chrono-biogeographic debt penalizes those phylogenetic hypotheses that require the presence of ghost lineages; that is, undocumented "stratum-crossings" in the usual stratocladistic sense (Fisher, 1992; Fox *et al.*, 1999; Bloch *et al.*, 2001). As with the original stratocladistic character, parsimony debt is only accrued when ghost lineages are identified in any given stratigraphic unit, and not when a lineage crosses from one unit into an adjacent one. This step of the chrono-biogeographic analyses requires a chronostratigraphic framework similar to that used in stratocladistics.

As in stratocladistics, the ability to correlate strata across all landmasses limits the extent to which the chronostratigraphic framework can be subdivided. A more pernicious issue concerns the long stretches of time, most apparent in the Paleogene analysis below, between fossil and extant taxa. For instance, inclusion of several chronostratigraphic intervals between that analysis' penultimate chronostratigraphic unit (Late Oligocene) and Recent may be desirable in order to penalize certain hypotheses that would thereby incur substantial stratigraphic debt through that interval, but stratocladistics countermands the use of chronostratigraphic units for which no ingroup taxa are sampled—the rationale being that if no primates could have been found at those times and places, then there is no basis for penalizing a hypothesis that identifies a ghost lineage there (e.g., Bloch *et al.*, 2001). A related problem concerns the use of additional chronostratigraphic units within which many extinct primate taxa have been sampled that are not clearly relevant to the hypotheses being tested by the phylogenetic analysis. For instance, the case study of Paleogene primates discussed below was designed to test competing hypotheses relevant to the anthropoid origins debate, but for purposes of broader morphological character sampling and outgroup comparison, respectively, this analysis also samples crown platyrrhine and lorisiform strepsirrhine clades that have a Miocene fossil record. Ideally, in order to justify the use of, for instance, an early Miocene chronostratigraphic unit, all of the known Miocene platyrrhines and lorisiforms should also be sampled. The same can be said of the broad Eocene omomyiform and adapiform radiations, for which so many taxa are now known, but only a few of which are sampled herein. These sampling issues are simply beyond the scope of this analysis; such problems will be remedied to some extent in future analyses (Seiffert *et al.*, in prep.), but for the time being we suspect that they are, fortunately, not particularly relevant to the primary hypotheses being tested.

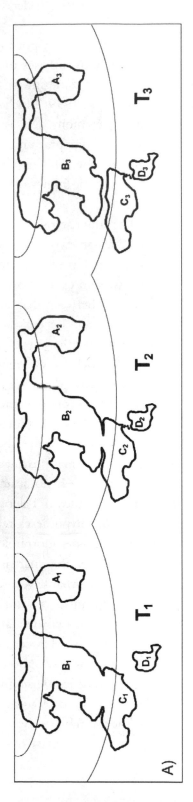

A)

B)

T_1 T_2 T_3

Figure 1. (a) Fragmentation of four hypothetical continents at three different time intervals—T_1 (oldest interval) to T_3 (youngest)—illustrating the five different paleogeographic arrangements that are of interest for weighting dispersal events in the chronobiogeographic character's step-matrix. Note that during T_1, landmasses A, B, and C are all broadly connected, but A and B are broadly connected only at high latitude; dispersals at lower latitudes (between B and C) are weighted zero step for interval T_1, but dispersal from A to B or from B to A is weighted one step throughout T_1–T_3 due to the constraining influence of climate on high-latitude dispersal. D is far away from C, with a substantial body of water and no evidence for island chains—dispersal between C and D thus weighted four steps during T_1. During interval T_2, landmasses C and D have converged, and are connected by an ephemeral island chain that could facilitate dispersal across the water barrier separating the two; dispersal from C to D, or from D to C, is accordingly weighted two steps during this interval and in the next. During interval T_3, C has broken away from B; the two landmasses are still closely situated but there is no evidence for island chains that would facilitate overwater dispersal, and as such dispersal between these two landmasses is weighted three steps during T_3. (b) Step-matrix constraining dispersals between the landmasses in (a). Note that each interval/landmass (time/place) has its own column and row, but that only certain dispersals are scored—i.e., those from older times/places to younger times/places, and between continents during the same interval. Dispersals that require a lineage to move backward in time are not allowed [and assigned irreversible or infinity (i)]. The total value in each cell represents chronobiogeographic debt, which is a combination of stratigraphic debt and biogeographic debt, with the biogeographic debt portion of the equation explained within parentheses next to the total value for biogeographic debt. As an example, moving from D_1 to A_3 (ninth column, fourth row) would incur one step of stratigraphic debt (for leaving a ghost lineage in T_2), but because the dispersal could have happened at any point between T_1 and T_3, the simplest biogeographic solution is found during T_2, when a dispersal from D to C would cost two steps, a dispersal from C to B would cost no steps, and a dispersal from B to A would cost one step (due to the high-latitude connection). Thus the cell contains $1 + 3(2 + 0 + 1) = 4$, i.e., (stratigraphic debt = 1) + (biogeographic debt = 3; because a dispersal from D to C would cost two steps, a dispersal from C_2 to $B_2 = 2 +$ a dispersal from B_2 to $A_2 = 1$) = (chronobiogeographic debt = 4).

The biogeographical portion of the character penalizes those phylogenetic hypotheses that require either (a) dispersal across high-latitude land connections during time periods when climate would not have favored such dispersal, (b) dispersal across some type of water barrier, at any latitude, or (c) potentially both, in cases where a lineage is inferred to have moved through one or more intermediate landmasses. A zero step dispersal is equivalent to a range expansion or shift that involves no geographical barrier. This weighting scheme suits the scope of our analyses, but researchers can employ finer biogeographical divisions, provided that they can define them and justify their relative likelihood. Assessment of the likelihood of dispersal from one time/place to another requires some knowledge of the land connections present during each temporal interval (see Appendix 1). Individual dispersal events were weighted as follows:

0: landmasses broadly connected at low latitude, uni- or bidirectional primate dispersal not constrained by climate
1: landmasses broadly connected at high latitude, and/or uni- or bidirectional primate dispersal constrained by climate
2: filtered uni- or bidirectional primate dispersal between isolated but closely situated landmasses likely allowed for by ephemeral island chains
3: filtered uni- or bidirectional primate dispersal between isolated landmasses not obviously facilitated by ephemeral island chains, but landmasses nevertheless closely situated
4: landmasses widely separated, extensive overwater dispersal necessary for primate dispersal

The biogeographic debt for each cell is determined by identifying the least costly opportunity for dispersal (or vicariance) that has occurred over the course of the time interval (however long) separating the two states of interest. For example, the cell describing a state change from the upper-early Miocene of Africa to the late Miocene of Eurasia is scored by identifying the paleogeographic arrangement(s) most favorable for dispersal (i.e., that or those which would require the least amount of biogeographic debt) for the circumscribed time intervals (upper-early Miocene, middle Miocene, and late Miocene); this is because the intercontinental dispersal could have occurred at any point between the two times/places of interest. In this case, there would be no biogeographic debt, because the crossing could have occurred during the Langhian Regression in the middle Miocene (Bernor and Tobien, 1990; Andrews et al., 1996) through range expansion. Accordingly, a phylogenetic hypothesis requiring this state change is only penalized one step—specifically, one unit of stratigraphic debt, for leaving a ghost lineage during the middle Miocene. Similarly, when two or more dispersal routes between adjacent landmasses are possible for a state change (e.g., dispersal from Europe to South America via either Afro-Arabia or North America), that sequence of intercontinental dispersals that incurs the least amount of biogeographic debt is used to weight the biogeographic component of the chronobiogeographic character. This minimizes the number of *ad hoc* hypotheses required to explain the paleobiogeographic data, in keeping with the parsimony principle (Farris, 1983).

The chronobiogeographic character is irreversible with respect to time, but not geography. That is, the upper right hand half of the step-matrix contains the various character states that have been determined for the chronobiogeographic character, while the lower left hand corner is composed primarily of "infinity" (irreversible) states (see Figure 1) so that it is impossible for a lineage to disperse backwards in time. The complete asymmetry of the step-matrix is violated only in order to allow bidirectional state changes between continents during the same temporal interval (e.g., from the early Eocene of Afro-Arabia to the early Eocene of Europe, and *vice versa*). The character cannot be ordered because there is no polarity to biogeographic data (see Hunn and Upchurch, 2001); the ordering required by temporal data is present in the character in the form of the stratigraphic debt, which increases by counting intervening strata, but ignoring biogeography. There is no consequence to which of the geographic regions within the oldest chronostratigraphic unit is coded as state "0," since only the chronostratigraphic data have polarity.

Figure 2 illustrates total debt calculation for a simple example of four taxa scored for two morphological characters and the chronobiogeographic character from Figure 1. The four taxa (W–Z) are found on different landmasses (A–D, connected as in Figure 1) at different times (T_1–T_3, with T_1 being oldest). Figure 2a shows the cladogram representing the most parsimonious interpretation of the distribution of morphological characters (left), and the most parsimonious chronobiogeographic hypothesis that is congruent with its topology (right). The basal taxon W is found on landmass A during time T_1, so all phylogenies require a dispersal across the high-latitude connection between A and B, which costs one step. There is no cost for dispersing between landmasses B and C until time T_3. As in stratocladistics, lineages passing from one time interval to the next incur no parsimony debt. In order for taxon X to be the sister

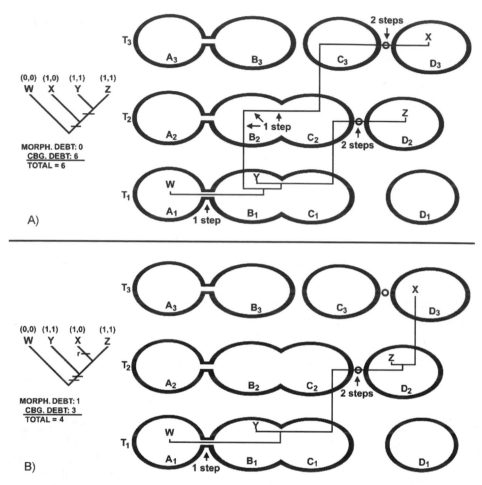

Figure 2. Examples illustrating the calculation of character and chronobiogeographic parsimony debt for two competing hypotheses using the step matrix from Figure 1. Schematized land masses A–D are arranged as in Figure 1. Taxa W, X, Y, and Z are found in A_1, D_3, B_1, and D_2, respectively. The cladistic hypotheses differ in their interpretation of the morphological state '0' in taxon X. Chronobiogeographic analysis finds it more parsimonious to consider this state a reversal in taxon X, while cladistic analysis of the two morphological characters alone interprets it as a symplesiomorphy. Note that as in Fox et al. (1999) morphological ("morph") debt is the number of homoplasies (annotated as 'r' for the reversal in 2b), not the total number of state changes (annotated as hash marks).

of a (Y,Z) clade, the X lineage must have a ghost lineage on the B–C landmass during time T_2 (one step) and must cross an island chain from landmass C to D during either time T_2 (not shown) or T_3 (two steps). This hypothesis incurs six steps of chronobiogeographic debt: one stratigraphic, and five geographic. The cladogram in figure 2b (left) is a less parsimonious morphological solution that interprets the morphological state '0' in taxon X as a reversal rather than as a symplesiomorphy. This generates one step of morphological debt, but the most parsimonious chronobiogeographic hypothesis (right) that fits this cladogram requires only three steps—all geographic. Taxa X and Z now share a common ancestral lineage that crossed from landmass B–C to D during time T_2. Note that no additional reduction in chronobiogeographic debt would be achieved by placing any of the taxa in an ancestral position. Note also that stratigraphic debt alone would make the two phylogenies equally parsimonious.

The chronobiogeographic character cannot be analyzed alongside morphological data in the popular program PAUP (Swofford, 1998) because the step-matrix associated with the character is inevitably internally inconsistent or "violates the triangle inequality"—this means that, for example, a change from state 1 to state 3 may be determined by the user to require one step, but elsewhere in the step-matrix individual changes from state 1 to state 2, and from state 2 to state 3, may themselves cost no steps, implying that a change from state 1 to state 2 to state 3 should theoretically require no extra steps (see Maddison and Maddison (1992) for discussion of why temporal data can justify acceptance of this "violation"). When PAUP processes a data file, the program automatically adjusts user defined step matrices to eliminate all such inconsistencies, and unfortunately in so doing changes its intended internal dynamics. Due to this problem, chronobiogeographic analyses must proceed much as does stratocladistic analysis, in which the most parsimonious clado-grams (MPCs) recovered from a parsimony analysis of morphological data are imported into MacClade and compared for length differences following ad-dition of the chronobiogeographic character to the morphological character matrix. The most thorough search procedure involves calculation of a "debt ceiling" (see Fisher 1992), within which the most parsimonious tree(s) must exist, and subsequent examination of all trees shorter than that length; however this is often impractical due to the sheer number of trees that may be recov-ered under the debt ceiling (often tens of thousands). In such cases the next best option is to employ manual branch swapping within MacClade (see Bloch *et al.*, 2001) in combination with the "search above" tool, in order to deter-mine whether trees shorter than the one congruent with the most parsimonious cladogram(s) derived from parsimony analysis of morphological data alone can be recovered by trading off morphological debt for stratigraphic debt.

In a departure from stratocladistic analysis, which conservatively places an-cestral nodes in the youngest possible stratum (see Maddison and Maddison,

1992), the chronobiogeographic character will place an ancestral node in the next (older) stratum if doing so decreases parsimony debt (and is thus more conservative). This is made possible by two factors—the internal inconsistency of the character, and the fact that the step-matrix algorithm used by MacClade does not limit ancestral state reconstructions to states found in terminal taxa above a given node. The program can therefore insert intermediate states at internal nodes if doing so reduces parsimony debt. Ultimately, this procedure should recover the most parsimonious paleobiogeographical scenario that is consistent with the combined evidence, even if this means hypothesizing the presence of an ancestor in a time and place where none has been found.

It should be noted that the behavior of the chronobiogeographic character can only be investigated through case studies that include fossil taxa (see below), or through experiments using simulated phylogenies (as done by Fox et al., 1999). It cannot be tested on extant taxa with well-corroborated molecular phylogenies because these lack a temporal element. However, we suspect that the character's effect on phylogenetic accuracy and precision will be similar to that of stratocladistics (Fox et al., 1999) for two reasons. The first is obvious: the chronobiogeographic character is half chronostratigraphic. More importantly, if our understanding of why chronostratigraphic data constitute phylogenetically meaningful information is correct, then the same should be true of biogeographic data. Both data types represent patterns in the natural world that are the direct result of evolutionary history.

Most objections to the use of temporal or geographical data imply that the imperfect nature of the fossil record will lead them to have a corrupting effect on phylogenetic accuracy (e.g., Smith, 2000). However temporal order and geographic position of fossil taxa cannot provide false information, they can at worst only provide misleading information under certain circumstances of missing data. Since the same is true of morphological characters (Felsenstein, 1978), this alone can hardly undermine their use. Moreover, since the three types of data will not necessarily be misleading simultaneously, their combined use may actually help to minimize the impact of missing data within each class of information (see Kim, 1993, for a discussion of such benefits).

To illustrate the utility of the chronobiogeographic character for different types of problems in primate, and particularly anthropoid primate, phylogeny, we provide two case studies below—the first derived from a previously published analysis of Paleogene primates by Seiffert et al. (2004), and the second derived from an analysis of Neogene catarrhines more recently undertaken by J.B.R.

As the analysis of Paleogene primates has already been discussed extensively in Seiffert *et al.* (2004), most of the commentary that follows is concerned primarily with illustrating the impact of the chronobiogeographic character on the phylogenetic results from that dataset. The analysis of the Neogene catarrhine radiation is discussed at greater length and in more detail as this study has, at the time that this manuscript is being submitted, only been briefly mentioned in an abstract (Rossie and MacLatchy, 2004).

Case Study 1: Paleogene Primate Phylogeny and the Biogeography of Anthropoid Origins

Seiffert *et al.* (2004) provided a phylogenetic analysis of 31 Paleogene and 10 extant anthropoid and non-anthropoid primates, employing 273 morphological characters, one nuclear gene (IRBP), and one mitochondrial gene (cytochrome b), with the intention of testing competing phylogenetic hypotheses bearing on the issues of stem and crown anthropoid origins. As mentioned previously, this analysis employed a "constrained" biogeographic character associated with a step-matrix whose internal dynamics were derived from assessments of the likely dispersal routes between major landmasses. Seiffert *et al.* (2004) did not figure a consensus tree derived from an analysis that excluded the biogeographic character, but such a tree is provided here (Figure 3A), based on parsimony analysis of morphological and molecular data alone, in order to illustrate the increased phylogenetic resolution provided by adding a single biogeographic character treated either as unconstrained (Figure 3B), or as constrained by a step-matrix (Figure 3C). As can be seen, there is very little resolution in the strict consensus of the 109 MPCs derived from the analysis of morphological and molecular data alone (all results described henceforth are derived from heuristic searches in PAUP 4.0b10 that employed random addition sequence and 1000 replicates). However, examination of individual MPCs reveals that this irresolution is due in large part to various poorly known "wild-card" taxa; early Paleocene *Purgatorius* from North America, middle-late middle Eocene Asian eosimiids, later Eocene Asian amphipithecids, and the African prosimians *Djebelemur* and "*Anchomomys*" *milleri* all find a number of equally parsimonious positions among various living and extinct primate clades. Adding an unconstrained biogeographic character (Figure 3B) increases resolution only slightly; for instance, a (parapithecid, arsinoeid (proteopithecid, crown anthropoid)) clade is recovered in all 23 MPCs, but other robust clades,

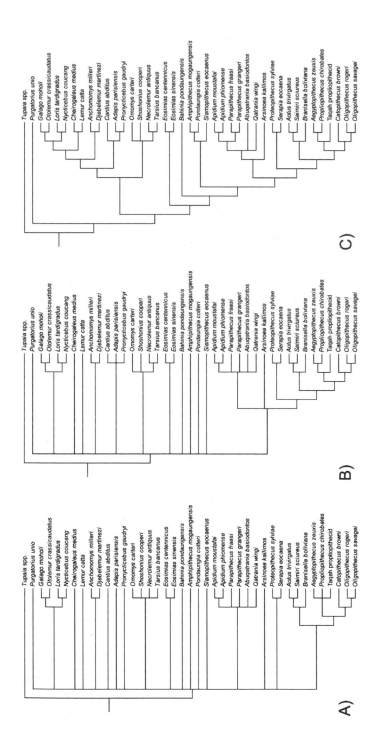

Figure 3. Strict consensus trees (cladograms) derived from analysis of matrix described in Case Study 1 (Paleogene primates). (A) Strict consensus of 109 MPCs derived from analysis of molecular and morphological data alone, no biogeographic character included. Tree length (TL) (measured in MacClade) = 2983, consistency index (CI) = 0.54, retention index (RI) = 0.52, rescaled consistency index (RCI) = 0.28. (B) Strict consensus of 23 MPCs derived from analysis of molecular and morphological data and an unconstrained biogeographic character. TL = 2993, CI = 0.54, RI = 0.52, RCI = 0.28. (C) Strict consensus of four MPTs recovered following addition of the chronobiogeographic character. Tree length = 3023; MacClade will not calculate CI, RI, or RCI for user-defined characters such as the chronobiogeographic character.

Table 1. Chronostratigraphic states for Paleogene primate analysis

Chronobiogeographic state	Taxa
0) Danian of North America	*Purgatorius unio*
1) Ypresian of North America	*Shoshonius cooperi, Cantius abditus*
2) Ypresian of Afro-Arabia	*Djebelemur martinezi*
3) Lutetian of North America	*Omomys carteri*
4) Lutetian of Asia	*Eosimias sinensis*
5) Bartonian of Asia	*Amphipithecus mogaungensis, Bahinia pondaungensis, Eosimias centennicus, Pondaungia cotteri*
6) Priabonian of Afro-Arabia	*Abuqatrania basiodontos, Anchomomys milleri, Arsinoea kallimos, Catopithecus browni, Oligopithecus savagei, Proteopithecus sylviae, Qatrania wingi, Serapia eocaena*
7) Priabonian of Europe	*Adapis parisiensis, Necrolemur antiquus, Pronycticebus gaudryi*
8) Priabonian of Asia	*Siamopithecus eocaenus*
9) Rupelian of Afro-Arabia	*Aegyptopithecus zeuxis, Apidium moustafai, Apidium phiomense, Parapithecus grangeri, Propliopithecus chirobates*
10) Chattian of South America	*Branisella boliviana*
11) Recent of Asia	*Loris tardigradus, Nycticebus coucang, Tarsius bancanus, Tupaia* spp.
12) Recent of South America	*Aotus trivirgatus, Saimiri sciureus*
13) Recent of Madagascar	*Cheirogaleus medius, Lemur catta*
14) Recent of Afro-Arabia	*Galago moholi, Otolemur crassicaudatus*
?) Controversial or unknown age	*Oligopithecus rogeri, Parapithecus fraasi,* Taqah propliopithecid

such as crown Strepsirrhini, are not recovered until the biogeographic character is constrained by likely dispersal routes encoded in a step-matrix (see Seiffert *et al.*, 2004, Figure 1). However, even the analysis with the constrained biogeographic character is influenced by the recovery of some relationships that seem unlikely given temporal and biogeographic considerations—for example, in three of the 12 MPCs the primitive early Paleocene plesiadapiform *Purgatorius* forms a clade with middle Eocene *Eosimias*.

Complete resolution of such temporal and biogeographic peculiarities is only found following addition of the chronobiogeographic character (Figure 3c; Table 1, Appendix 2). For purposes of chronobiogeographic analysis, taxa were conservatively scored according to the geologic age in which they are found (Danian in the case of *Purgatorius*; Ypresian, Lutetian, Bartonian, or Priabonian for Eocene taxa; Rupelian for early Oligocene Fayum anthropoids [in this analysis those taxa from Fayum Quarries G, I, M, and V], and Chattian for *Branisella*). The "debt ceiling" approach was found to be impractical as an exhaustive search method, but extensive manual branch swapping within MacClade revealed that chronobiogeographic analysis resolves the tree almost completely, with the sole polytomies derived from uncertainty regarding the relative positions of *Omomys* and *Shoshonius* among tarsiiforms and *Abuqatrania* and *Qatrania* among parapithecids. From a biogeographic standpoint, it is of great interest that the most parsimonious topologies: (1) convincingly recover *Djebelemur* and "*Anchomomys*" *milleri* as "advanced" stem strepsirrhines that are more closely related to crown Strepsirrhini than are the "classic" adapids from northern continents, and accordingly support an Afro-Arabian origin for crown Strepsirrhini, and (2) recover the Eocene Asian taxa *Eosimias*, *Bahinia*, and Amphipithecidae—which otherwise appeared as problematic "wild-card" taxa in the previous analyses—as a paraphyletic group of stem anthropoids that support an Asian origin for that clade. Importantly, amphipithecids can *not* be placed more parsimoniously with either derived early Oligocene propliopithecid catarrhines (Chaimanee *et al.*, 1997; Ducrocq, 1999; Kay *et al.*, 2004; Tabuce and Marivaux, 2005), nor with older, or contemporaneous, adapids (Ciochon and Gunnell, 2002; Gunnell *et al.*, 2002; Kay *et al.*, 2004), and eosimiids cannot be placed with tarsiiforms (Rasmussen, 2002), following addition of the chronobiogeographic character. As is now widely accepted, platyrrhines are found to be deeply nested within an otherwise Afro-Arabian anthropoid clade (Fleagle and Kay, 1987; Kay *et al.*, 2004; Seiffert *et al.*, 2004), but *Proteopithecus* cannot be more parsimoniously placed as an African stem or crown platyrrhine (Takai *et al.*, 2000).

As with all phylogenetic analyses that must, at some point, limit their taxon or character sampling, this analysis has some such shortcomings that will be taken into account and improved upon in future studies (Seiffert *et al.*, in prep.). A few criticisms have already been raised by Ross and Kay (2004), who were primarily concerned by the fact that the analysis recovered Prosimii (i.e., a *Tarsius*-Strepsirrhini clade)—now well supported by nucleotide sequences (Murphy *et al.*, 2001a; Eizirik *et al.*, 2004)—rather than Haplorhini (a *Tarsius*-Anthropoidea clade), which is both the most parsimonious morphological hypothesis (e.g., Kay *et al.*, 2004) and is also supported by a number of short interspersed nuclear elements (SINEs) (Schmitz and Zischler, 2004). These authors suggested that the results might be different were the analysis to: (1) include additional data from plesiadapiform and extant dermopteran outgroups, and (2) sample more widely from the corpus of DNA sequence data now available for members of the major extant primate clades (Murphy *et al.*, 2001a; Eizirik *et al.*, 2004). These points are well taken; additional sampling of living and extinct euarchontans, and members of the larger Euarchontoglires clade, would certainly provide greater confidence in these phylogenetic results were they to hold, and additional sampling of molecular data is similarly preferable, although we suspect that sampling of a molecular data set such as that of Eizirik *et al.* (2004), which strongly supports Prosimii, is unlikely to recover Haplorhini given the same morphological data set. Other improvements to the analysis could derive from sampling of various taxa on certain key landmasses for the primate and anthropoid origins debates, such as Asia and Africa—for instance, late Paleocene African *Altiatlasius* and the basal early Eocene Asian taxa *Altanius* and *Teilhardina asiatica* could influence interpretation of the geographic origin of major primate clades.

Case Study 2: Neogene Catarrhines

The character matrix (Appendix 4) for Neogene catarrhines was developed in order to determine the phylogenetic affinities of a newly discovered primitive catarrhine (Rossie and MacLatchy, 2004), and will appear in its full form elsewhere. Here it is presented in an abbreviated form, without the new taxon. The study focuses on the early diversification of major catarrhine clades such as cercopithecoids, pliopithecoids, nyanzapithecines, and hominoids. Accordingly, taxa that would be relevant to investigation of questions such as the position of the Eurasian Miocene apes (e.g., *Sivapithecus*, *Dryopithecus*) were not included. The matrix of 183 morphological characters scored across 34 taxa was based on published descriptions (e.g., refs in Begun, 2002; Harrison, 2002), and on study of original fossil material in the National Museum of Kenya and the Uganda National Museum. The chronostratigraphic subdivisions were based on an unpublished synthesis of East African chronostratigraphy (the details of which are available from J.B.R. upon request), and subsequent correlation with the European MN system. Three of the ingroup taxa—*Victoriapithecus*, *Dendropithecus macinnesi*, and *Limnopithecus legetet*—could be assigned polymorphic codings for the chronobiogeographic character because their temporal ranges span multiple adjacent strata. However, because the "make ancestral" function is inoperable in the chronobiogeographic analysis, the first appearance datum (FAD) of the taxa is all that concerns us, so they are coded for the older state. *Victoriapithecus* was considered to have a FAD in the early Miocene on the basis of the cercopithecoid molar from Napak (Pilbeam and Walker, 1968), even though this is probably a different species from the middle Miocene *Victoriapithecus macinnesi*.

Parsimony analysis of the morphological data set yielded 12 MPCs with a tree length of 809. The strict consensus of these MPCs is shown in figure 4A. The structure of the cladogram accords well with current views (e.g., Harrison, 2002) in that it features well-supported clades such as Plio-pithecoidea, Nyanzapithecinae, Cercopithecoidea (including *Victoriapithecus*), and crown Hominoidea, as well as monophyly of Neogene catarrhines with respect to *Aegyptopithecus* and *Catopithecus*. However, relationships between these clades, and among most of the early and middle Miocene taxa are left unresolved.

Addition of the chronobiogeographic character (Appendix 3, scored as character 184 in Appendix 4) reduced the set of 12 MPCs to two most parsimonious *trees* (MPTs) with tree lengths of 813. Such an increase in resolution is not surprising since it is a common result of adding chronostratigraphic data (Fox *et al.*, 1999). As in the first case study, the "debt ceiling" approach proved impractical in that there were thousands of morphological cladograms of 813 steps or less, but extensive manual branch-swapping and use of the "search above" tool revealed no more parsimonious solutions than these two. The strict consensus of the two trees (Figure 4b) exhibits several improvements in resolution. Although the cercopithecoids are still not precisely placed, there is now a polychotomous stem hominoid group consisting of, at a minimum, *Kalepithecus, Dendropithecus, Limnopithecus legetet*, the four *Proconsul* species, and the Nyanzapithecinae (with *Simiolus* at its base). *Equatorius* and the *Afropithecus-Morotopithecus* clade appear as successive sister taxa of crown hominoids.

Detailed comparison of this more precise tree with recent phylogenetic hypotheses (Pilbeam, 1996; Begun *et al.*, 1997; Harrison, 2002) is complicated by fundamental disagreements among current investigators. The principal disagreement concerns the position of cercopithecoids. Of the known African Miocene 'apes' either some (Rose *et al.*, 1992), none (Andrews, 1978), or all but *Morotopithecus* (Harrison, 2002) are viewed as stem catarrhines rather than hominoids. This seems like a major issue, but it boils down to the position of one clade—Cercopithecoidea. Cercopithecoids are rarely included in computer-driven cladistic analyses of Miocene catarrhines, even those which purport to establish stem hominoid positions for some taxa (e.g., Begun *et al.*, 1997). Their inclusion here at least offers the possibility of differentiating stem catarrhines from stem hominoids, and does so in some obvious cases (*Catopithecus* and *Aegyptopithecus* at one end, and *Equatorius, Afropithecus*, and *Morotopithecus* on the other). Still, it remains unclear exactly where

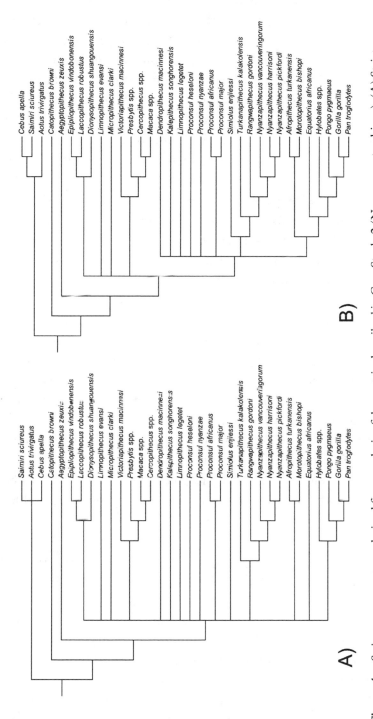

Figure 4. Strict consensus trees derived from analysis of the matrix described in Case Study 2 (Neogene catarrhines). (A) Strict consensus of 12 MPCs derived from an analysis of morphological data alone. $TL = 809$, $CI = 0.33$, $RI = 0.61$, $RCI = 0.20$. (B) Strict consensus of two MPTs following addition of the chronobiogeographic character. $TL = 813$.

along this spectrum cercopithecoids fit in, and the present results should not be taken as concrete evidence for the hominoid status of, for example, *Dendropithecus*. Our results support a more traditional placement of most Miocene 'apes' as stem hominoids, including all but two of the small-bodied apes. While we actually agree with the arguments of Rose *et al.* (1992) and Harrison (1982; 1987; 1988; 2002), who view all small-bodied 'apes' as stem catarrhines, our results highlight what a complicated issue the precise position of cercopithecoids is. The predominance of autapomorphic character states in cercopithecoids certainly contributes to their ambiguous position. Indeed, the number of crown catarrhine synapomorphies that can be scored in fossil species is far too low to compensate for other, plausibly primitive, similarities between taxa like *Kalepithecus* and crown hominoids. Beyond this issue, there are a few other oddities worthy of brief discussion.

Harrison's (2002) Dendropithecoidea (*Simiolus*, *Micropithecus*, and *Dendropithecus*) fail to unite in this analysis for several possible reasons. This trio of genera suffers from "crossing synapomorphies" and missing data. While *Micropithecus* and *Simiolus* share many dental characters, *Dendropithecus* resembles these taxa primarily in canine and p3 morphology, and differs considerably in postcanine morphology. *Simiolus* and *Dendropithecus* share many features of the humerus, but this element remains unknown for *Micropithecus clarki* (Harrison, 1982; Rose *et al.*, 1992). The net effect of this conflict appears to be sufficient to disperse these taxa throughout the tree. Dendropithecoidea may indeed be a real clade (*Dendropithecus* is in the same clade as *Simiolus* in both MPTs, but in different places), but as Harrison (2002) noted, at least their postcranial similarities may be symplesiomorphies. *Simiolus* is placed as a basal nyanzapithecine for the same reasons as in the study by Singleton (2000)—principally its mesiodistally expanded molars.

A final caveat concerns the *Afropithecus-Morotopithecus* clade. Although some have argued for their specific synonymy (Pickford, 2002), these two genera differ dramatically in the adaptations revealed by their postcranial remains (Ward et al., 1993; MacLatchy, 2004). *Afropithecus* closely resembles *Proconsul* (Rose, 1997), whereas *Morotopithecus* provides the earliest evidence for crown-hominoid-like orthograde posture and a mobile shoulder joint (MacLatchy, 2004). Unfortunately, the two taxa are known from virtually no overlapping postcranial elements, so that the missing data for each taxon can be reconstructed (by PAUP) as the state present in the other. The sister-taxon relationship implied here results from the presence of some dental similarities, combined with PAUP's unfortunate indifference toward functionally nonsensical chimeras. This problem is not insurmountable if more apes are included in the analysis, as has been shown elsewhere (Young and MacLatchy, 2004).

Our primary interest here is in the effect of the chronobiogeographic character on the cladistic results. While addition of the chronobiogeographic character was sufficient to improve resolution, it only increased tree length by four steps in the two MPTs, and six steps in the other ten trees. This is obviously due to the fact that Old World paleogeography from the late Oligocene-Recent did not constrain dispersal as much as at it did during the Paleogene (e.g., Kappelman *et al.*, 2003). Most of the lineages that cross between Eurasia and Afro-Arabia could easily have done so in the middle or late Miocene, when no biogeographical debt would have been accrued. For example, the chronobiogeographic history of the hominoid lineage including *Equatorius* and crown apes involves some stratigraphic debt but no biogeographical debt, because all of the implied range expansions or dispersals occurred in the middle or late Miocene.

Another reason for the mild effect of the chronobiogeographic character in this analysis is that most of the ambiguously placed taxa in the analysis of morphological characters alone are from the same time and place—the middle-early Miocene of Africa. Moreover, two of the clades that were well supported by morphology alone already have members in the middle-early Miocene of Africa, so the chronobiogeographic character does not encourage attraction between these clades and the other contemporaneous species. This being the case, one might wonder why the introduction of the chronobiogeographic character leads to, for example, a more precise placement of *Simiolus*. The massive polytomies produced by morphological homoplasy are a grid-lock of equally parsimonious scenarios of character change among the study taxa. Introduction of stratigraphic data, or chronobiogeographic data, provides a basis for preferring one reconstruction of a character's pattern of change over another, otherwise equally parsimonious, scenario (Gingerich, 1976; Bretsky, 1979). By acting as a "tie-breaker" for just one character, the chronobiogeographic character can affect the position of several taxa, even if some of the taxa have the same chronobiogeographic character state (as in the *Simiolus* case). The analysis of Neogene catarrhines, in which the chronobiogeographic character has a mild effect, demonstrates just how small a 'nudge' is needed to achieve this. In this regard chronobiogeographic (like stratocladistic) analysis provides another type of solution to a problem raised by Fleagle (1997, p. 1), who posed the following question: "given the presence of rampant homoplasy in most data sets and the limits of parsimony, what more can we do to identify the correct phylogeny among many alternatives?"

Tracing the chronobiogeographic character on the two MPTs in order to reconstruct ancestral nodes reveals logical results. Most clades originate in the middle-early Miocene of Afro-Arabia or earlier. The pliopithecoid stem lineage resides either in the late-early Miocene of Afro-Arabia or Eurasia, or in the middle-early Miocene of Afro-Arabia. The common ancestor of pliopithecoids and more advanced catarrhines is reconstructed as having lived in either the middle-early Miocene or Oligocene of Afro-Arabia. The *Equatorius*-crown hominoid clade is reconstructed as having originated in either the late-early Miocene of Afro-Arabia or the middle Miocene of Afro-Arabia or Eurasia. The reconstructed biogeographical history of the cercopithecoid taxa provides an example of how MacClade exploits the triangle inequality of the chronobiogeographic character. Because we included only the early-to-middle Miocene *Victoriapithecus* and three living cercopithecoid genera, it might appear that three steps of stratigraphic debt (state 2 to 9) should be required along the branch between *Victoriapithecus* and the Eurasian pair, *Macaca* and *Presbytis*. However, by reconstructing this branch as an intermediate state—either middle Miocene of Eurasia or Afro-Arabia, or as late Miocene of Eurasia—it reduces stratigraphic debt to two steps.

DISCUSSION AND FUTURE DIRECTIONS

The foregoing examples clearly demonstrate that the chronobiogeographic character can provide a considerable improvement in phylogenetic resolution over parsimony analysis of morphological and molecular data alone. An important next step is to determine whether such increased resolution does indeed represent an improvement in phylogenetic accuracy, as is the case with stratocladistics (Fox *et al.*, 1999). Those who prefer to restrict phylogenetic analysis to intrinsic properties of organisms (morphological and molecular data), or who prefer "soft" approaches to the use of temporal or paleogeographic data in phylogenetics, might argue that such "resolution" is simply an illusion of improved phylogenetic accuracy that derives from ill-advised inclusion of confounding extra-biological data in phylogenetic analysis. However, even those workers who use such soft approaches do so because they recognize that phylogenetic patterns observable in the fossil record are influenced by time and paleogeography. In our opinion, inclusion of chronobiogeographic data directly into phylogenetic analysis represents an improvement for these very reasons, and, just as important, further reduces the need for additional *ad hoc* explanations derived from extra-biological considerations such as time or paleogeography.

There are various reasons to suspect that, given appropriate implementation, chronobiogeographic analysis almost certainly does increase phylogenetic accuracy. For instance, as with stratocladistic analysis (Fox *et al.*, 1999), chronobiogeographic analysis brings phylogenetic analysis of living and extinct taxa closer to the realm of likelihood, and further away from the relatively blind approach of traditional maximum parsimony, which is incapable of detecting the constraining temporal component of cladogenesis. The simulation study undertaken by Fox *et al.* (1999) has already demonstrated that inclusion of the temporal portion of our chronobiogeographic character (i.e., the stratigraphic character) improves phylogenetic accuracy over parsimony analysis of morphological character data alone, and it is only logical that consideration of paleogeography would similarly restrict the range of phylogenetic options to those that make most sense given the totality of data that systematists must integrate to infer phylogenetic patterns. In this sense chronobiogeographic analysis represents the next logical step in the move towards analyzing "total evidence" in phylogenetics, while also allowing for the inclusion of a model-based phylogenetic method derived from assessments of stratigraphic succession and the likelihood of dispersal events through space and time. Unfortunately the likelihood of certain types of dispersals cannot be calculated in the same way, or at least with the same ease, that one can calculate, for instance, a transition/transversion ratio for molecular phylogenetic analysis, but such constraints on our phylogenetic inferences (however simple) nevertheless represent what we consider to be an important methodological step forward in paleobiology. Of course whatever subjectivity may be associated with creating a chronobiogeographic step-matrix does not differentiate this type of character from most subjectively delineated morphological characters employed in phylogenetic analyses of living and extinct taxa (see Wiens, 2001).

The development of chronobiogeographic analysis also provides the first major methodological step that we are aware of towards overcoming the sorts of confounding phylogenetic signals that are so often associated with colonization of, and subsequent adaptive radiation on, isolated landmasses. It has long been recognized that clades radiating in isolation tend to evolve morphological adaptations that resemble those of distantly related, but functionally analogous, taxa on other landmasses, but the remarkable extent to which these trends have concealed phylogenetic signal has only recently come to be exposed by molecular phylogenetic methods (Madsen *et al.*, 2001; Springer *et al.*, 2004). Chronobiogeographic analysis finally provides a sound means for penalizing those phylogenetic hypotheses that preferentially accrue biogeographic debt over morphological homoplasy, while still controlling for changes in paleo-geography and the possibility of vicariance or geo-dispersal (*sensu* Lieberman and Eldredge, 1996). An important test of the utility of the chronobiogeo-graphic character will come from analyses of geographically widespread clades such as Placentalia, within which morphological phylogenetic signal suggests extensive biogeographic homoplasy, but molecular phylogenetic signal instead suggests that geographically isolated endemic clades have evolved detailed mor-phological homoplasies with distantly related taxa on other landmasses (e.g., Madsen *et al.* 2001).

Another important use of the chronobiogeographic character relates to the common practice of reconstructing the biogeographic history of a clade. Several times in the two case studies above, the chronobiogeographic character equivo-cates (in MacClade) over the location and/or time of a reconstructed ancestor or internal branch; for example, some of the Neogene lineages may have in-habited either the middle Miocene of Afro-Arabia or Eurasia. This is perfectly sensible, because the two states are separated by zero steps—they are essentially a single state, and could, in fact, be scored that way in the step-matrix for the character. In other words, in terms of paleobiogeography, the middle Miocene of Afro-Arabia and Eurasia are one and the same. As we have argued above, hypothesized phylogenies should not be penalized for requiring dispersals be-tween these well-connected landmasses during this time. However, the logical complement to this has not been appreciated in recent years. For instance, some authors have claimed that the stem lineage of the (*Gorilla*, (*Pan*, *Homo*)) clade originated in Eurasia because this minimizes the number of 'dispersals' re-quired to account for the distribution and relationships of fossil and living apes (e.g., Stewart and Disotell, 1998). However, these dispersals are likely to have occurred during the middle or late Miocene, in which case it is difficult to see why their number should fall under the purview of parsimony. If tectonic and eustatic events had rendered Afro-Arabia and Eurasia a continuous landmass during that time, then these "dispersals" were what Lieberman and Eldredge (1996) term "geo-dispersals." The difference between this sort of event and the traditional concept of a dispersal is that here, the likelihood of the event having occurred has nothing to do with the number of ingroup taxa involved—the paleogeographical connection is a matter of record. Accordingly, phylogenies that imply several exploitations of a known geo-dispersal are no less parsimo-nious than those that imply fewer. The reason for this is simple—these dispersals require no additional *ad hoc* explanations. The explanation for all of them is the connected landmass, and this is not *ad hoc*. Whether the middle Miocene saw several ape lineages, or a single founding lineage spread into Eurasia cannot be determined by counting the number of "dispersals" on a cladogram if these were geo-dispersals that accrue no parsimony debt. Using the chronobiogeographic character in such cases would reveal the true ambiguity of the clade's biogeographic history, and failure to employ such an approach will surely yield biogeographic precision at the price of inaccuracy.

The utility of the chronobiogeographic character could be expanded in several ways, and we hope that certain improvements can be made to this nascent approach. We would prefer to be able to limit ancestral node reconstructions to only the states found in terminal taxa above the node in question. This would permit a slightly less charitable approach by not allowing ancestral nodes to be pushed down into older strata in which no plausible ancestors are known. On a related topic, it would be desirable to implement the "make ancestral" tool of the stratocladistic character (Maddison and Maddison, 1992) so that ancestral taxa themselves could attest to intermediate chronobiogeographic states. The ability to make taxa ancestral could also potentially lead to the discovery of MPTs with different topologies than those found with the present method. However, this can only happen if taxa span multiple chronostratigraphic units, and possess few or no autapomorphies (if the number of autapomorphies exceeds the number of strata occupied by a taxon, it can never be made ancestral). Lastly, the violation of the triangle inequality principle seems to be the only obstacle to using the chronobiogeographic character in PAUP. If this could be remedied, it would greatly improve the method by permitting a complete search of all possible chronobiogeographically augmented trees—thus obviating the "debt ceiling" approach that is so often impossible in MacClade.

ACKNOWLEDGMENTS

We would like to thank John Fleagle and Shawn Lehman for inviting us to contribute to this volume. We also thank Jonathan Bloch, David Fox, David Strait, and two anonymous reviewers for helpful comments on the manuscript.

APPENDIX 1

Weightings for biogeographic debt are based on paleogeographic reconstructions described and synthesized in various sources (e.g., Smith et al., 1994;

Holroyd and Maas, 1994; Rögl, 1999; Scotese, 2002). These reconstructions are based on a variety of evidence, both biotic and abiotic. Brief justifications and notes on weightings are given below. Discussion of terrestrial fauna is included in some cases to illustrate the fit or lack thereof between the paleogeographical evidence and faunal evidence, but weighting is based on the paleogeographic evidence, even when the two are in disagreement (e.g., 'Dispersals between Afro-Arabia and Asia' below).

1. *Direct dispersal between Afro-Arabia and North America, Asia and South America, South America and Madagascar, Asia and Madagascar, North America and Madagascar, Europe and South America, and Europe and Madagascar.* With the exception of direct dispersal between Asia and Madagascar, none of these dispersal possibilities have been seriously entertained by biologists as plausible given the sheer overwater distance that an organism would need to travel in order to successfully accomplish such a colonization. For this reason these options are weighted as "infinity" to limit movement to more likely dispersal routes.

2. *Direct dispersals between Afro-Arabia and Madagascar and between Afro-Arabia and South America.* Both of these options are weighted four steps throughout the Cenozoic given the strong paleogeographical evidence that such movements would have required extensive overwater travel. Unlike the dispersal scenarios in the foregoing example, however, there is strong evidence that mammalian lineages have indeed dispersed from Afro-Arabia to Madagascar, and from Afro-Arabia to South America. It should be noted that McCall (1997) has argued that an island chain connecting Afro-Arabia and Madagascar from the middle Eocene through the early Miocene would have facilitated mammalian dispersal from Afro-Arabia to Madagascar, but divergence dates calculated for various crown clades of Malagasy mammals do not match up well with this scenario (Yoder *et al.*, 2003), and furthermore the geophysical data on which the landbridge hypothesis are based are not compelling (Rogers *et al.*, 2000; Krause *et al.*, 2003).

3. *Dispersals between Afro-Arabia and Asia.* All intervals for dispersal between Afro-Arabia and Asia are weighted four steps until the early Miocene because the two were consistently separated by the Tethys Seaway. Despite paleogeographic reconstructions that provide no hint of a physical connection between these landmasses during the Eocene, there is clear evidence for multiple mammalian dispersals between Asia and Afro-Arabia during the Paleogene (entirely, or almost entirely, unidirectional, from Asia

to Afro-Arabia) in the form of hystricognathous and anomaluroid rodents (Jaeger *et al.*, 1985), anthracotheriid artiodactyls (Ducrocq, 1997), possibly creodonts (e.g., Gheerbrant, 1995), marsupials, as well as anthropoid (Beard, this volume), tarsiiform (Simons and Bown, 1985), and adapiform primates. Most of these dispersals appear to have occurred by the early or middle Eocene, and evidence for continued geographic isolation until the latest Oligocene is provided by 27–28 Ma deposits in Ethiopia (Kappelman *et al.*, 2003) that preserve no evidence for other, previously undocumented, Laurasian clades having dispersed into Afro-Arabia by that time. The fossil record of various non-sirenian paenungulate taxa outside of Afro-Arabia during the Paleogene (Radulesco *et al.*, 1976; Domning *et al.*, 1986; Maas *et al.*, 1998; Thewissen *et al.*, 2000) is complex, and it is unclear to what extent these dispersals, presumably out of Afro-Arabia, were facilitated by an aquatic or semi-aquatic lifestyle (Court, 1993, 1994).

Later in the Cenozoic, extensive faunal exchange was not permitted between Afro-Arabia and Eurasia until the latest Oligocene or early Miocene, when the Afro-Arabian plate collided with Eurasia (Bernor *et al.*, 1988). Because of the complex interplay of environmental (belts of sub-parathethyan woodland) and geographical (ever-changing paratethys seaway) factors, this is best considered a filtered "one step" connection. This holds until the middle Miocene (~16 Ma) when the Langhian Regression created a broad "zero step" connection by virtue of at least two major sea level lowstands (Haq *et al.*, 1987; Bernor, 1983; Bernor and Tobien, 1990; Andrews *et al.*, 1996). The late Miocene and Recent are weighted zero step as well because of the extensive faunal interchange that resumed in the Turolian, coinciding with a major sea level lowering at around 10 Ma (Bernor *et al.*, 1988; Bernor, 1983).

4. *Dispersals between Afro-Arabia and Europe.* Paleogeographic reconstructions indicate that northwest Africa was, at various times during the Paleogene, very closely situated to Europe (<100 km) via the Iberian Peninsula, and Holroyd and Maas (1994) considered primate dispersal from Europe to Afro-Arabia to be more likely than dispersal from Asia based on paleogeographic grounds alone. However there is no strong evidence for island chains connecting the two landmasses at any time during the Paleogene, and with the possible exception of adapisoriculids (Gheerbrant, 1995), adapiform primates, zegdoumyid rodents (Vianey-Liaud *et al.*, 1994), the creodont *Apterodon,* and marsupials, there is little evidence for mammalian faunal exchange between Europe and Afro-Arabia during the

Paleogene. Based on proximity, however, we weighted dispersals between these landmasses as three steps throughout the first case study (within which there are no Oligocene-Recent European taxa sampled); note that in the second case study Europe and Asia are treated together as Eurasia, given the broad geographic connection that had been established by that time.

5. *Dispersals between Asia and North America.* Paleogeographic evidence, along with evidence for extensive faunal exchange, between Asia and North America (primarily from Asia into North America) during the Paleogene (Beard, 1998; Beard and Dawson, 1999) leads us to weight dispersals between North America and Asia one step throughout the Paleogene, as climate appears to have been a limiting factor on dispersal even during the late Paleocene and early Eocene (Beard, 2002). Climate was obviously increasingly a limiting factor on dispersal as temperature decreased through the Cenozoic (Zachos *et al.*, 2001), but there is nevertheless little evidence for filtering of dispersals between Eurasia and North America into the early Miocene (e.g., Graham, 1999; Tedford *et al.*, 2004). The Bering Strait opened as long ago as 7.4 Ma (Marinkovitch and Gladenkov, 1998), and the route is weighted two steps for the late Miocene to Recent.

6. *Dispersals between North America and Europe.* Europe and North America were connected briefly during the early Paleogene, facilitating the movement of mammalian taxa between these continents (e.g., Beard and Dawson, 1999; Hooker and Dashzeveg 2003), but rifting in the North Atlantic had eliminated this connection by the early Eocene (e.g., Ritchie and Hitchen, 1996). Therefore for the early Eocene this connection is weighted one step; thereafter it is weighted four.

7. *Dispersals between Europe and Asia.* Faunal exchange between Europe and Asia during the Paleogene was limited by the West Siberian Sea (Iakovleva *et al.*, 2001), although a few examples of mammalian faunal exchange near the Paleocene-Eocene boundary have recently been identified by Hooker and Dashzeveg (2003). Because of this geographical barrier, dispersals between Asia and Europe are weighted three steps until the Oligocene, when dispersal between Europe and Asia once again became possible due to land connections.

8. *Dispersals between North America and South America.* Dispersal between these landmasses is weighted four steps throughout the Paleogene, as there is no evidence for a connection between these landmasses during the Paleogene or Miocene. The Panamanian land bridge did not form until the early Pliocene; thereafter, this low latitude connection is weighted zero step.

APPENDIX 2. Step-matrix for Case Study 1. Individual cells contain stratigraphic debt + biogeographic debt = total chronobiogeographic debt.

	Danian NA	Ypresian NA	Ypresian Afr	Lutetian NA	Lutetian Asia	Bartonian Asia	Priabonian Afr	Priabonian Eur	Priabonian Asia	Rupelian Afr	Chattian SA	Recent Asia	Recent SA	Recent Mad	Recent Afr
Danian NA	0	0+0=0	0+4=4	1+0=1	1+1=2	2+1=3	3+4=7	3+1=4	3+1=4	4+4=8	5+4=9	6+1=7	6+0=6	6+4=10	6+1=7
Ypresian NA	inf	0	0+4=4	0+0=0	0+1=1	1+1=2	2+4=6	2+1=3	2+1=3	3+4=7	4+4=8	5+1=6	5+0=5	5+4=9	5+1=6
Ypresian Afr	inf	0+4=4	0	0+4=4	0+4=4	1+4=5	2+0=2	2+3=5	2+4=6	3+0=3	4+4=8	5+0=5	5+1=6	5+4=9	5+0=5
Lutetian NA	inf	inf	inf	0	0+1=1	0+1=1	1+5=6	1+4=5	1+1=2	2+5=7	3+4=7	4+1=5	4+0=4	4+4=8	4+1=5
Lutetian Asia	inf	inf	inf	0+1=1	0	0+0=0	1+4=5	1+3=4	1+0=1	2+4=6	3+5=8	4+0=4	4+1=5	4+3=7	4+0=4
Bartonian Asia	inf	inf	inf	inf	inf	0	0+4=4	0+3=3	0+0=0	1+4=5	2+5=7	3+0=3	3+1=4	3+3=6	3+0=3
Priabonian Afr	inf	inf	inf	inf	inf	inf	0	0+3=3	0+4=4	0+0=0	1+4=5	2+0=2	2+1=3	2+4=6	2+0=2
Priabonian Eur	inf	inf	inf	inf	inf	inf	0+3=3	0	0+3=3	0+3=3	1+7=8	2+0=2	2+1=3	2+4=6	2+0=2
Priabonian Asia	inf	inf	inf	inf	inf	inf	0+4=4	0+3=3	0	0+4=4	1+5=6	2+0=2	2+1=3	2+4=6	2+0=2
Rupelian Afr	inf	inf	inf	inf	inf	inf	inf	inf	inf	0	0+4=4	1+0=1	1+1=2	1+4=5	1+0=1
Chattian SA	inf	inf	inf	inf	inf	inf	inf	inf	inf	inf	0	0+1=1	0+0=0	0+5=5	0+1=1
Recent Asia	inf	inf	inf	inf	inf	inf	inf	inf	inf	inf	inf	0	0+1=1	inf	0+0=0
Recent SA	inf	inf	inf	inf	inf	inf	inf	inf	inf	inf	inf	0+1=1	0	0+5=5	0+0=0
Recent Mad	inf	inf	inf	inf	inf	inf	inf	inf	inf	inf	inf	inf	0+5=5	0	0+4=4
Recent Afr	inf	inf	inf	inf	inf	inf	inf	inf	inf	inf	inf	0+0=0	0+1=1	0+4=4	0

APPENDIX 3. Step-matrix for Case Study 2. Contents of cells as in Appendix 2. See Appendix 4 for taxon codings.

	L.Eocene Afr	E.Oligo. Afr	M.E. Mio. Afr	L.E. Mio. Afr	L.E. Mio. Eurasia	M. Mio. Afr	M. Mio. Eurasia	L. Mio. Eurasia	Recent Afr	Recent Eurasia	Recent SA
L.Eo. Afr	0	0	1	2	2+1=3	3	3+0=3	4+0=4	5	5+0=5	5+4=9
E.Oligo. Afr	inf	0	0	1+0=1	1+1=2	2+0=2	2+0=2	3+0=3	4+0=4	4+0=4	4+4=8
M.E. Mio. Afr	inf	inf	0	0	0+1=1	1+0=1	1+0=1	2+0=2	3+0=3	3+0=3	3+4=7
L.E. Mio. Afr	inf	inf	inf	0	0+1=1	0+0=0	0+0=0	1+0=1	2+0=2	2+0=2	2+4=6
L.E. Mio. Eursia	inf	inf	inf	1	0	0+0=0	0+0=0	1+0=1	2+0=2	2+0=2	2+4=6
M. Mio. Afr	inf	inf	inf	inf	inf	0	0+0=0	0+0=0	1+0=1	1+0=1	1+4=5
M. Mio. Eur	inf	inf	inf	inf	inf	0	0	0+0=0	1+0=1	1+0=1	1+4=5
L. Mio. Eurasia	inf	inf	inf	inf	inf	inf	inf	0	0+0=0	0+0=0	0+4=4
Recent Afr	inf	inf	inf	inf	inf	inf	inf	inf	0	0+0=0	0+4=4
Recent Eurasia	inf	inf	inf	inf	inf	inf	inf	inf	0	0	0+4=4
Recent SA	inf	inf	inf	inf	inf	inf	inf	inf	4	4	0

APPENDIX 4

Characters and matrix for Neogene catarrhine analysis.

1. *Incisor crown waisting*: (0) absent; (1) present.
2. *I2 cingulum*: (0) present; (1) absent.
3. *Upper I1 shape*: (0) labiolingually thin; (1) labiolingually thick.
4. *Upper I2 size*: (0) much smaller than I1; (1) slightly smaller than I1.
5. *Upper canine bilateral compression*: (0) very compressed; (1) compressed; (2) less compressed.
6. *P2s*: (0) present; (1) absent.
7. *P3 crown shape*: (0) triangular; (1) sub-ovoid; (2) very ovoid.
8. *P4 crown shape*: (0) triangular; (1) sub-ovoid; (2) ovoid.
9. *Upper premolar cusp heteromorphy*: (0) marked; (1) reduced; (2) subequal.
10. *Upper premolar cusp volume*: (0) buccolingually narrow; (1) inflated, not crowded; (2) inflated and crowded.
11. *P3 distal margin*: (0) convex; (1) angled; (2) straight.
12. *P3 MD length*: (0) very broad; (1) broad; (2) narrow; (3) very narrow.
13. *P4 MD length*: (0) very broad; (1) broad; (2) narrow; (3) very narrow.
14. *P3 transverse crest*: (0) linking cusps; (1) cusps separated by groove.
15. *P3 buccal cingulum*: (0) absent; (1) present.
16. *P3 buccal face*: (0) triangular; (1) diamond-shaped; (2) mesiobuccal expansion.
17. *P3 lingual face*: (0) sloping; (1) vertical.
18. *P3 lingual cingulum*: (0) inflated; (1) present; (2) absent.
19. *P4 lingual cingulum*: (0) inflated; (1) present; (2) absent.
20. *Molar cusp form*: (0) high & sharp; (1) low & rounded; (2) high & inflated; (3) high & very inflated.
21. *M1 crown shape*: (0) very broad; (1) broad; (2) narrow; (3) very narrow.
22. *M2 crown shape*: (0) very broad; (1) broad; (2) narrow; (3) very narrow.
23. *M1 metacone & hypocone size*: (0) similar to mesial cusps; (1) smaller than mesial cusps.
24. *M2 cusp at mesiolingual corner of crown*: (0) absent ; (1) present.
25. *M1-M2 waisting*: (0) absent; (1) present.
26. *M1-M2 crown tapering*: (0) none; (1) distal tapering; (2) strong tapering.
27. *M1-M3 lingual cingulum*: (0) strong; (1) reduced; (2) none.
28. *M2-M3 buccal cingulum*: (0) present; (1) absent.

29. *M1-M2 mesial shelf*: (**0**) none; (**1**) narrow cingulum; (**2**) pronounced cingulum; (**3**) shelf.

30. *M1-M3 size sequence*: (**0**) M1 < M2 < M3; (**1**) M1 < M3 < M2; (**2**) M1~M3 < M2; (**3**) M3 < M1 < M2; (**4**) M3 < M2 < M1.

31. *M3 distal moiety*: (**0**) reduced; (**1**) unreduced.

32. *Crista obliqua*: (**0**) present; (**1**) absent.

33. *Metacone Size*: (**0**) < protocone; (**1**) > / = protocone.

34. *Protocone position*: (**0**) aligned with paracone; (**1**) distal to paracone & crown margin.

35. *Protoconule size*: (**0**) large; (**1**) small; (**2**) absent.

36. *Hypocone-metacone crista*: (**0**) discontinuous; (**1**) continuous; (**2**) true hypoloph.

37. *Hypocone position*: (**0**) in line with metacone; (**1**) distal to metacone; (**2**) distobuccal; (**3**) more lingual than protocone.

38. *Prehypocrista position*: (**0**) meets base of protocone; (**1**) meets crista obliqua.

39. *Trigon size*: (**0**) broad; (**1**) narrow; (**2**) very narrow & crescentic.

40. *Talon size*: (**0**) absent; (**1**) small; (**2**) large relative to trigon.

41. *Lower incisor lingual enamel*: (**0**) present; (**1**) absent.

42. *Lower i1 shape*: (**0**) narrow & tall; (**1**) broader & shorter.

43. *Lower i2 shape*: (**0**) narrow & tall; (**1**) broader & shorter.

44. *Lower i2 lingual cingulum*: (**0**) absent; (**1**) present.

45. *i2 distal margin*: (**0**) angled; (**1**) straight.

46. *p3 mesiolingual beak*: (**0**) absent; (**1**) present.

47. *p3 metaconids*: (**0**) present; (**1**) absent.

48. *p3 sectoriality*: (**0**) non-sectorial; (**1**) poor; (**2**) moderate; (**3**) moderate to strong; (**4**) strong.

49. *p4 length*: (**0**) very narrow; (**1**) narrow; (**2**) broad; (**3**) very broad.

50. *p4 buccal flare*: (**0**) slight; (**1**) strong.

51. *p4 mesial fovea height*: (**0**) slightly above distal basin; (**1**) much higher than distal basin.

52. *Pliopithecine triangle*: (**0**) none; (**1**) distal arm; (**2**) mesial & distal arms.

53. *Protoconid-metaconid*: (**0**) oblique; (**1**) transverse.

54. *m1-2 paraconids*: (**0**) present; (**1**) absent.

55. *m1 crown shape*: (**0**) very narrow; (**1**) long, narrow; (**2**) broad; (**3**) very broad.

56. *m2 crown shape*: (**0**) Very narrow; (**1**) long, narrow; (**2**) broad; (**3**) very broad.

57. *m1 crown tapering*: (**0**) tapering anteriorly; (**1**) no tapering.

58. *m1-m3 buccal cingulum*: (**0**) well-developed; (**1**) only in notches; (**2**) none.

59. *m1-2 hypoconulid*: (**0**) absent; (**1**) appressed to entoconid; (**2**) median; (**3**) buccal; (**4**) twinned with hypoconid.

60. *m1 cristid obliqua*: (**0**) oblique; (**1**) slightly oblique; (**2**) straight to protoconid.

61. *m1-m2 crown outline*: (**0**) rectangular; (**1**) slightly ovoid; (**2**) ovoid.

62. *m1-m2 hypoconulids*: (**0**) present; (**1**) absent.

63. *m2 cristid obliqua*: (**0**) straight; (**1**) slightly oblique; (**2**) oblique.

64. *m1-m2 mesial lophid*: (**0**) absent; (**1**) present.

65. *m1-m3 lingual notch depth*: (**0**) shallow; (**1**) deep opening into talonid.

66. *m1-m3 mesial fovea*: (**0**) broad; (**1**) narrow.

67. *m1-m3 distal fovea*: (**0**) wide; (**1**) narrow; (**2**) narrow and distally projecting.

68. *m3 shape*: (**0**) relatively untapered; (**1**) tapers distally; (**2**) tapers strongly, entoconid indistinct.

69. *m3 hypoconulid & entoconid linked by crest*: (**0**) yes; (**1**) no.

70. *m3 hypoconid & entoconid linked by crest*: (**0**) no; (**1**) yes; (**2**) true distal lophid.

71. *Inferior transverse torus*: (**0**) present; (**1**) absent.

72. *Superior transverse torus position*: (**0**) at 1/3 height of symphysis; (**1**) mid-symphysis height.

73. *Mandibular condyle height*: (**0**) lower than coronoid; (**1**) equal or higher.

74. *Lumbar centra*: (**0**) dorsoventrally compressed and small; (**1**) expanded; (**2**) very expanded.

75. *Transverse process inclination*: (**0**) ventrally inclined; (**1**) neutral; (**2**) dorsally and caudally inclined.

76. *Transverse processes*: (**0**) on body; (**1**) on pedicle.

77. *Accessory processes*: (**0**) present; (**1**) absent or reduced.

78. *Vertebral body keeling*: (**0**) present; (**1**) absent.

79. *Spinous process inclination*: (**0**) cranially inclined; (**1**) caudally inclined.

80. *Lumbar centra spooling*: (**0**) present; (**1**) absent or reduced.

81. *Lumbar centrum hollowing*: (**0**) present; (**1**) absent or reduced.

82. *Lumbar #*: (**0**) 7; (**1**) 6; (**2**) 5; (**3**) 4.

83. *Sacral #*: (0) 3; (1) 4; (2) 5; (3) 6.

84. *Ischial spine*: (0) distal to acetabulum; (1) level with acetabulum.

85. *Acetabulum*: (0) expanded w/raised lip; (1) cranially expanded lunate surface; (2) symetrical lunate surface.

86. *Sacrum shape*: (0) broad; (1) narrow.

87. *Scapula shape*: (0) vertebral border < caudal; (1) long vertebral border.

88. *Glenoid curvature*: (0) more curved craniocaudally than dorsoventrally; (1) uniformly and moderately curved.

89. *Glenoid orientation*: (0) faces laterally; (1) faces cranially.

90. *Scapular spine*: (0) oblique to axial border; (1) low angle to axial border.

91. *Scapula position*: (0) lateral; (1) dorsal on thorax.

92. *Sternum*: (0) long narrow; (1) short broad.

93. *Medial trochlear keel*: (0) anterior to epicondyle; (1) projects distal to epicondyle.

94. *Lateral trochlear keel*: (0) weak; (1) moderate; (2) strong.

95. *Medial epicondyle orientation*: (0) very retroflexed; (1) slight retroflex; (2) medial.

96. *Humeral head orientation*: (0) <15 degrees; (1) >15 degrees.

97. *Dorsal epitrochlear fossa*: (0) present; (1) absent.

98. *Olecranon fossa*: (0) shallow; (1) deep, sharp laterally.

99. *Entepicondylar foramen*: (0) present; (1) absent.

100. *Olecranon fossa articular surface*: (0) no extension into fossa; (1) extends into fossa laterally.

101. *Trochlear breadth*: (0) < capitular; (1) > capitular.

102. *Distal humerus anteroposterior thickness*: (0) very thin; (1) intermediate; (2) deep.

103. *Zona conoidea*: (0) shallow, broad; (1) deep, narrow.

104. *Capitulum*: (0) no distal expansion of surface; (1) distal expansion of articular surface.

105. *Trochlear proximal border*: (0) straight; (1) V-shaped for coronoid beak.

106. *Humeral shaft curvature*: (0) straight; (1) retroflected; (2) strongly retroflected.

107. *Deltoid insertion*: (0) on proximal half of shaft; (1) on distal half.

108. *Humeral head shape*: (0) oblong, extending between tuberosities; (1) less oblong; (2) spherical w/tuberosities anterior.

109. *Tuberosity sizes*: (0) lesser <60% of greater; (1) 60–80%.

110. *Bicipital groove*: (0) broad; (1) intermediate; (2) narrow.

111. *Humeral head height*: (**0**) lower than tuberosities; (**1**) equal or slightly higher; (**2**) much higher.
112. *Humeral/femoral head size*: (**0**) humeral head not > femoral; (**1**) humeral head > femoral.
113. *Ulnar olecranon process size (% of sigmoid notch length)*: (**0**) very short < 50; (**1**) short 51–80; (**2**) long 80–100; (**3**) very long >100.
114. *Olecranon orientation*: (**0**) proximal; (**1**) posterior.
115. *Sigmoid notch shape*: (**0**) long, narrow; (**1**) low, wide.
116. *Sigmoid median keel*: (**0**) none; (**1**) strong keel.
117. *Ulna shaft cross-section*: (**0**) ML < 66% of AP; (**1**) ML > 66% of AP.
118. *Ulna shaft bowing*: (**0**) convex dorsally; (**1**) concave or straight dorsally.
119. *Distal ulna shape*: (**0**) ML narrow; (**1**) ML broad.
120. *Styloid-triquetral articulation*: (**0**) large styloid and articular surface; (**1**) small styloid and articular surface.
121. *Distal radioulnar articulation*: (**0**) proximodistally narrow; (**1**) proximodistally expanded.
122. *Distal ulnar facet on radius*: (**0**) faces medially; (**1**) faces proximomedially.
123. *Radial notch position*: (**0**) anterior to lateral third of shaft; (**1**) anterior to middle of shaft.
124. *Radial notch orientation*: (**0**) anterolateral; (**1**) lateral.
125. *Radial head outline*: (**0**) oval w/flat posterolateral area; (**1**) smaller flat area; (**2**) round.
126. *Radial head lateral lip*: (**0**) large; (**1**) present; (**2**) absent.
127. *Radial notch shape*: (**0**) single oval facet; (**1**) two faces @ 90 degree angle; (**2**) two separated facets.
128. *Dorsal ridge on radius*: (**0**) absent; (**1**) present.
129. *Radial styloid process*: (**0**) prominent; (**1**) reduced.
130. *Radial lunate articular surface*: (**0**) AP narrow; (**1**) venrtally expanded.
131. *Os centrale*: (**0**) free; (**1**) fused.
132. *Capitate distal surface*: (**0**) concave; (**1**) biconcave.
133. *Trapezium-MC1 joint*: (**0**) sellar; (**1**) modified hinge; (**2**) non-sellar.
134. *Metacarpal dorsal ridges*: (**0**) absent; (**1**) present.
135. *MC1 prox articulation*: (**0**) no lateral extension; (**1**) lateral extension.
136. *Metacarpal II medial facet*: (**0**) undivided; (**1**) divided by ligament pit.
137. *Hamate distolateral edge*: (**0**) uninterrupted; (**1**) interrupted by ligament pit.
138. *Scaphoid beak*: (**0**) absent; (**1**) present.

139. *Hamate hamulus*: (**0**) not projecting; (**1**) distally projecting.
140. *Capitate & hamate shape*: (**0**) long, narrow; (**1**) proximodistally short.
141. *Distal metacarpal shape*: (**0**) widest palmerly; (**1**) quadrilateral.
142. *Tibial condylar facets*: (**0**) symmetrical; (**1**) medial larger; (**2**) lateral larger.
143. *Distal fibula*: (**0**) robust, flares laterally; (**1**) reduced, less lateral flare.
144. *Femoral head shape*: (**0**) spherical and separate from neck; (**1**) blends w/neck superoposteriorly.
145. *Head height*: (**0**) below trochanter; (**1**) at or above trochanter.
146. *Femoral neck tubercle*: (**0**) present; (**1**) absent.
147. *Gluteal tuberosity*: (**0**) absent; (**1**) ridge-like; (**2**) distinct.
148. *Femoral condyles*: (**0**) nearly symmetrical; (**1**) medial larger.
149. *Patellar groove*: (**0**) narrow, deep; (**1**) broad, shallow; (**2**) wider than shaft.
150. *Talar head*: (**0**) width = height; (**1**) width > height.
151. *Fibular facet of talus*: (**0**) vertical; (**1**) laterally projecting.
152. *Curve of medial trochlear margin (talus)*: (**0**) gently curved; (**1**) strongly curved.
153. *Ectal facet curve*: (**0**) gently concave; (**1**) sharply concave.
154. *Ectal facet shape*: (**0**) broadest anterolaterally; (**1**) broadest proximomedially.
155. *Anterior calcaneal length*: (**0**) > one third total length; (**1**) one third; (**2**) < one third.
156. *Posterior calcaneoastragular joint*: (**0**) faces medially; (**1**) faces dorsally.
157. *Sustentacular facet*: (**0**) undivided; (**1**) divided.
158. *Calcaneocuboid joint*: (**0**) round facet w/deep pit; (**1**) smaller pit.
159. *Heel tubercle*: (**0**) present; (**1**) absent.
160. *Metatarsal II lateral facet*: (**0**) divided; (**1**) single.
161. *Entocuneiform-MT I facet*: (**0**) convex w/medial extension; (**1**) less convex w/ slight extension.
162. *Postglenoid foramen*: (**0**) large; (**1**) reduced; (**2**) absent.
163. *Clivus orientation*: (**0**) protruding; (**1**) more vertical.
164. *Ectotympanic*: (**0**) annular; (**1**) semi-tubular; (**2**) tubular.
165. *Pyriform inferior border*: (**0**) narrow V-shaped; (**1**) wider & more rounded; (**2**) flat, horizontal.
166. *Supraorbital region*: (**0**) costae; (**1**) tori or arches; (**2**) strong inflated tori.
167. *Facial profile*: (**0**) prosthion, rhinion, glabella in line; (**1**) stepped.
168. *Infraorbital surface of maxilla*: (**0**) slopes posteroinferiorly; (**1**) slopes anteroinferiorly.

169. *Snout length*: (0) protruding; (1) orthognathic.
170. *Palate shape*: (0) narrow anteriorly; (1) parallel tooth rows; (2) bowed laterally.
171. *Frontal trigon*: (0) present; (1) absent.
172. *Premaxilla-nasal contact*: (0) excluded or minimal; (1) mid-nasal; (2) extensive.
173. *Orbit inferior margin*: (0) flush with cheek; (1) sharp protruding; (2) bar-like protruding.
174. *Interorbital region*: (0) broad; (1) narrow.
175. *Orbit position*: (0) high; (1) low.
176. *Nasal length*: (0) long; (1) short.
177. *Nasal bridge*: (0) flat; (1) rounded horizontally.
178. *Nasal shape*: (0) narrow superiorly; (1) broad superiorly.
179. *Frontal sinus*: (0) present; (1) absent.
180. *Premaxilla ascending wing*: (0) broad; (1) narrow.
181. *Buccal pouches*: (0) absent; (1) present.
182. *Horizontal palatine process*: (0) broad; (1) narrow.
183. *Atrioturbinal ridge*: (0) present; (1) absent.
184. *Chronobiogeographic character*: (0) late Eocene Africa; (1) early Oligocene Africa; (2) middle-early Miocene Africa; (3) late-early Miocene Africa (4) late-early Miocene Eurasia; (5) middle Miocene Africa; (6) middle Miocene Eurasia; (7) late Miocene Eurasia; (8) Recent Africa; (9) Recent Eurasia; (10) Recent South America.

Aotus trivirgatus 0 0 1 0 2 0 0 0 0 0 0 1 0 0 0 0 1 1 0 1 1 1 0 0 1 1 1 0 4 0
0 0 0 1 0 3 1 1 2 0 1 0 1 0 0 0 0 3 0 1 0 1 1 2 2 1 2 0 0 0 1 1 0 0 0 - 1 - 1 1 0
0 0 0 0 0 0 0 0 0 0 0 1 0 0 0 0 0 0 0 0 1 0 0 0 0 0 0 0 0 0 0 1 0 1 0 0 1 1
2 0 0 0 0 0 0 0 0 0 1 1 0 0 0 0 0 0 0 0 1 0 0 ? 0 0 0 0 0 2 0 1 1 0 2 0 1 0 0 0
0 1 0 0 0 1 1 ? 0 1 1 0 1 1 1 0 1 2 1 1 1 1 1 0 1 1 1 0 0 0 0 **10**

Saimiri sciureus 0 0 0 0 2 0 0 0 0 0 0 0 0 0 0 0 - 0 0 0 0 0 1 0 0 1 0 1 0 4 0 0
0 0 1 0 3 0 1 2 0 0 0 1 0 0 0 0 3 0 1 0 0 1 3 3 1 2 0 0 0 1 2 0 0 0 - 1 - 1 1 0 0
0 0 0 0 0 0 0 0 0 0 0 1 0 0 0 0 0 0 0 0 1 0 0 0 0 / 1 0 1 0 0 0 0 1 0 0 1 1 1 1
1 0 0 0 0 0 0 0 0 0 1 1 0 1 0 0 0 0 0 0 1 0 0 0 0 0 0 0 2 0 0 0 0 2 0 0 0 0 0
0 0 0 0 0 1 1 0 0 0 0 0 2 1 1 1 1 0 1 1 0 1 0 1 1 0 1 0 0 0 0 0 **10**

Cebus apella 0 0 1 1 2 0 0 0 0 0 0 0 0 1 0 0 0 1 1 1 0 0 0 0 0 1 1 1 0 4 0 0 0 0
1 1 3 1 0 1 0 1 1 1 0 0 0 0 3 0 0 0 0 1 3 3 1 2 0 0 0 1 1 0 0 0 - 1 - 1 0 1 0 0 0
0 0 0 0 0 0 1 0 1 2 0 0 0 0 0 0 0 0 0 0 0 0 0 0 0 1 0 0 0 1 0 0 1 1 1 1 1 0 0

0 0 0 0 0 0 0 1 1 0 1 0 0 0 0 0 0 1 0 0 0 0 0 0 0 0 1 0 0 1 0 1 0 2 0 1 0 0 0 0
0 0 0 1 0 0 0 0 0 1 1 1 1 0 2 1 1 0 1 0 0 1 0 0 1 0 0 0 **1 0**

Catopithecus browni 0 0 0 1 2 1 0 0 0 0 2 3 1 - 1 0 - 1 1 0 0 0 1 1 0 0 0 0 1 3
0 0 0 1 1 - 3 - 0 1 0 0 0 ? 0 0 1 1 0 0 1 0 0 0 1 1 0 2 1 0 1 0 2 0 1 0 1 0 1 0 1
0 0 ? ? ? ? ? ? ? ? ? ? ? ? ? ? ? ? ? ? ? 0 0 1 ? 0 0 0 0 1 0 0 ? ? 1 0 ? ? ? ? ? ? ? ? ? ? ?
? 0 1 1 2 ? ? 1 1 1 0 0 ? ? ? ? ? ? ? 1 0 0 1 0 ? 0
0 0 0 1 0 0 1 0 1 1 ? 0 ? 0 ? **0**

Aegyptopithecus zeuxis 0 0 0 0 2 1 0 2 1 0 1 1 1 1 1 0 - 0 0 1 0 0 0 0 0 0 0 1
1 0 0 1 0 1 0 3 0 0 1 0 0 0 0 0 0 1 1 3 0 0 0 0 1 2 3 1 0 2 2 0 0 0 0 0 1 1 1 0
0 0 1 0 ? ? ? ? ? ? ? ? ? ? ? ? ? ? ? ? ? ? ? 0 ? 0 0 1 0 0 0 0 0 1 0 0 0 0 1 0 0 1 0 1 ? 3 0
0 0 ? 0 ? 0 ? ? 0 1 ? ? 0 0 ? ? ? ? 0 ? ? ? ? ? ? ? ? 0 1 0 2 0 2 1 1 1 0 0 0 ? 0 0 ? 1
? 1 0 0 0 0 0 1 0 0 0 1 0 0 0 0 1 1 0 0 ? 0 0 **1**

Dionysopithecus shuangouensis 1 ? 0 ? 1 1 ? 1 1 0 ? ? 1 ? ? ? ? ? 1 1 1 1 1 0 0 0 0
0 2 2 0 0 0 0 1 0/1 0 0 1 1 0 ? 0 1 0 0 0/1 1 1 0 ? 1/2 0 0 2 1 0 0 2 1 0 0 2
0 0 1 1 1 0 0 ?
? ?
? ? ? ? ? ? ? ? ? ? ? ? ? **4**

Epipliopithecus vindobonensis 1 0 1 0 1 1 0 2 1 0 0 1 1 0 1 0 0 2 1 0 0 0 0 0 0
0 0 0 1 2 0 0 1 0 0 0 0 1 1 0 0 0 0 0 0 1 1 2 0 0 1 0 1 2 1 0 0 4 0 0 0 0 2 0 0
1 2 0 0 1 1 1 0 0 0 0 0 0 0 0 1 0 ? ? 1 ? 0 0 ? ? 0 0 0 1 0 0 0 0 0 0 0 1 0 0 0 0
0 1 1 0 1 ? 2 0 0 0 0 0 0 0 0 0 0 0 0 0 0 0 ? ? 1 0 0 0 0 ? ? ? 0 2 ? 0 1 0 0 1
2 1 1 0 0 0 0 1 1 1 0 0 0 ? 0 1 0 1 1 0 1 0 1 ? 2 0 1 1 ? 1 0 1 ? 0 ? **6**

Laccopithecus robustus 1 0 1 1 2 1 1 2 1 1 0 1 1 0 1 0 0 2 1 0 1 1 0 0 0 0 1 0 1
2 0 0 0 0 1 0 0 0 0 1 0 0 0 1 0 0 0 1 1 0 0 0 0 1 2 1 0 1 3 0 1 0 2 0 0 1 2 0 0
0 ?
? 0 ? 0 1 1 0 1 ? 1 1
2 0 1 1 ? ? ? 1 ? 0 ? **7**

Victoriapithecus macinnesi 0 0 0 0 1 1 0 0 1 0 0 3 2 1 0 2 0 2 2 1 2 3 1 0 1 1
1 1 0 2 0 0 1 0 2 0 2 1 - 1 0 1 1 1 0 0 0 4 1 0 0 0 1 1 2 2 1 2 2 - - 0 - 1 0 1 1
1 0 2 0 ? 0 ? ? ? ? ? ? ? ? ? 0 2 ? ? ? ? ? 0 ? 1 0 0 0 0 1 1 1 0 2 0 0 0 2 0 0 1 0 0 ?
3 1 0 0 0 ? 0 0 0 ? 1 0 0 ? 2 ? ? ? ? ? 0 0 0 ? ? 1 ? ? 0 ? 1 1 0 ? 0 ? 0 0 0 0 1 ? 0 ?
1 ? 0 ? 1 ? 0 2 0 0 0 1 0 2 0 ? 0 1 0 0 1 0 1 1 ? 0 1 **2**

Cercopithecus sp. 0 0 1 0 0 1 0 1 1 2 0 3 3 0 0 0/2 1 2 2 0 3 3 1 0 1 1 0 1 0 3
0 1 0 0 2 2 0 0 - 0 1 1 1 0 0 0 1 4 0 0 0 0 1 1 0 1 1 2 - - - 1 - 1 0 1 1 1 - 2 0 1
0 1 0 0 0 0 0 0 0 0 0 2 0 0 0 0 0 0 0 1 0 0 0 0 1 1 1 0 2 0 0 0 1 0 0 0 1 0 0

3 1 0 0 1 1 0 0 0 0 1 0 0 1 2 0 0 0 0 0 0 0 0 0 0 1 0 0 0 0 1 1 0 0 0 0 0 0 0 0
1 1 1 ? 1 1 1 0 1 2 0 2 0 1 1 1 0 2 1 2 0 1 0 0 1 0 1 1 1 0 1 8

Macaca spp. 0 0 0 1 0 1 0 2 1 0 0 3 3 0 0 2 1 2 2 2 3 3 1 0 1 1 0 1 0 ? 1 1 1 0
2 2 0 0 - 0 1 1 1 0 0 0 1 4 2 0 0 0 1 1 0 1 1 2 - - - 1 - 1 0 1 1 1 - 2 0 1 0 1 0 0
0 0 0 0 0 0 ? 0 2 0 0 0 0 0 0 0 1 0 0 0 1 1 1 1 2 0 0 0 2 0 0 1 1 0 0 3 0 0 0
0 0 0 0 0 0 1 0 0 0 2 0 0 0 0 0 0 0 0 0 0 1 0 0 0 0 1 1 0 0 0 0 0 0 0 0 1 1 0 1
1 1 1 0 1 1 0 2 0 1 1 1 0 2 1 2 0 1 0 0 1 0 1 1 1 0 1 9

Presbytis spp. 0 0 0 0 0 1 0 0 0 0 0 3 2 0 0 2 1 2 2 0 3 3 1 0 1 1 0 1 0 2 1 1 1
0 2 2 0 0 - 0 0 1 1 0 0 0 0 / 1 4 2 0 1 0 1 1 1 2 1 2 - - - 1 - 1 1 0 0 1 - 2 0 1 1
1 0 0 0 0 0 0 0 0 0 0 2 0 0 0 0 0 0 0 1 0 0 0 0 1 1 1 0 2 0 0 0 1 0 ? 1 0 0 0 1
0 0 0 0 1 0 0 0 0 1 0 0 0 1 0 1 0 0 0 0 0 0 1 0 1 0 0 0 1 1 1 0 1 0 0 0 0 0 0 0
1 2 ? 1 1 1 0 1 2 0 2 0 1 1 1 1 2 1 2 0 0 1 1 1 0 1 1 0 0 1 9

Proconsul major 0 ? 0 0 1 1 0 2 1 0 0 1 1 0 0 0 - 2 1 0 1 1 0 0 0 0 0 0 2 1 1 0
1 0 0 0 1 0 0 1 0 1 1 0 0 0 1 2 3 1 0 0 1 1 2 2 1 0 2 2 0 0 0 0 0 0 0 0 1 1 0 1 0
? 1 1 ?
? 0 ? ? ? ? ? ? ? ? ? ? ? ?
? ? ? ? ? 2

Proconsul africanus 0 ? 0 0 1 1 0 2 0 0 2 1 1 0 1 0 1 2 2 0 1 1 0 0 0 0 0 0 2 1
0 0 0 0 0 0 1 0 0 1 0 1 1 0 ? 0 1 2 3 1 0 0 1 1 2 2 1 1 2 2 1 0 0 0 0 0 0 0 2 0 0
1 0 ?
? ?
? ? ? ? ? ? ? 2

Nyanzapithecus harrisoni ? ? ? ? ? 1 ? ? ? ? ? ? ? ? ? ? ? ? 3 3 3 0 1 1 2 0 1 3 1 0
0 0 1 0 0 1 0 2 2 ? ? ? ? ? 0 1 2 0 0 1 0 1 1 0 0 1 0 3 2 2 0 0 0 1 1 1 1 1 1 ? 0 ?
? ?
? ?
? ? ? ? 5

Nyanzapithecus pickfordi 0 0 1 1 0 1 2 2 2 0 3 3 1 1 0 1 2 0 3 3 3 1 1 1 2 1
1 3 0 1 0 0 1 1 0 2 1 2 2 0 1 1 1 0 ? 1 ? 0 0 1 0 ? 1 0 0 1 1 2 2 ? 0 0 0 1 1 1 1
1 1 ? 0 ? ? ? ? ? ? ? ? ? ? ?
? 0 ? 1 ? ? ? ? ? ? ? ?
? ? ? ? ? ? ? ? ? 5

Equatorius africanus 0 0 0 0 1 1 1 2 1 0 1 1 1 0 2 1 2 1 / 2 1 2 3 0 0 0 0 1 1
1 1 1 0 1 0 1 0 3 0 0 1 0 0 0 1 0 1 1 1 3 0 0 0 1 1 1 2 1 2 3 2 1 0 0 0 0 0 0 1 1
0 0 0 1 ? ? ? ? ? 0 ? ? ? ? ? ? ? 0 ? ? 1 ? 1 0 ? 1 0 ? 1 1 1 1 1 ? ? 1 1 0 1 1 0 0 1 1

1 1 ? 1 1 0 0 0 0 0 1 ? ? 0 1 ? 1 0 ? 0 1 ? 1 1 1 ? 1 0 ? 0 0 1 0 0 ? 2 1 1 ? ? ? ? ?
? 1 ? 1 ? 5

Afropithecus turkanensis 0 0 1 0 2 1 1 ? 1 1 0 0 0 0 0 ? 0 0 0 1 1 2 0 0 0 1 1 ?
1 0 1 0 1 1 ? ? 1 ? 1 1 0 0 0 1 0 0 1 1 3 0 0 0 1 1 ? 2 1 1 2 2 1 0 0 0 0 0 1 1 0
0 0 0 ? 1 1 ? ? ? 1 ? ? ? ? ? ? ? ? ? ? ? ? ? 0 ? ? ?
? ? ? ? ? ? ? ? ? 0 0 0 0 1 1 ? ? 0 0 ? 0 ? ? ? ? ? ? 1 1 ? 0 ? ? ? 0 ? ? ? ? 0 ? 1 0 0
1 0 1 0 0 0 0 0 0 1 0 1 ? 0 1 3

Nyanzapithecus vancouveringorum ? ? ? ? ? 1 2 2 2 2 0 2 3 1 1 ? ? 0 0 2 3 3 0 0
0 1 1 2 0 1 0 0 1 1 0 1 0 1 2 ? ? ? ? ? ? ? ? 3 0 1 0 1 1 0 0 1 1 3 2 2 0 0 0 1 1
1 1 ? 0 1 1 ? 0 ? ? ? ? ? ? ? ? ? ? 0 1 1 1 1 ? ? ? ? ?
? ?
? ? ? ? ? ? ? ? ? ? ? ? ? 3

Simiolus enjiessi 0 ? ? ? 0 1 0 0 0 0 2 1 1 0 0 1 - 1 1 0 1 3 0 0 0 0 0 1 2 0 1 0 1
1 1 1 3 0 1 2 0 0 0 0 0 0 1 3 1 0 0 0 1 1 0 0 1 1 2 2 2 0 0 0 1 1 1 1 0 0 0 0 ? ?
? ? ? ? ? ? ? ? ? ? ? ? ? ? ? ? ? 0 0 1 ? 0 0 1 0 1 0 0 0 0 0 0 ? ? ? ? ? ? ? ? ? ? ? ? ? ? ?
? ? ? ? ? ? ? ? ? 0 ? ? ? ? ? ? ? ? 0 1 1 ? ? ? 1 0 ? 0 ? ? ? ? ? ? ? ? 0 ? ? ? ? 1 1 ? ? ? 0
? 0 ? ? ? ? 1 ? ? ? 3

Micropithecus clarki 0 0 0 0 2 1 0 2 0 0 2 1 1 0 0 0 0 2 1 0 2 2 0 0 0 0 0 0 0 1 3
0 0 1 1 1 0 / 1 3 0 1 1 0 0 0 1 1 0 1 3 2 0 0 0 0 1 1 2 1 1 2 2 2 0 1 0 0 0 1 2 0
0 0 1 ? 0 ? ? ? ? ? ? ? ? ? ? ? ? ? ?
? 1 ? 1 ? ? 0 1 1 ? ?
1 ? 1 ? ? ? ? 1 ? ? 1 2

Kalepithecus songhorensis 0 0 0 0 1 1 0 0 0 0 0 2 2 0 0 1 1 2 1 1 1 1 0 0 0 0 0
0 2 ? ? 0 1 0 0 0 3 0 0 1 0 0 0 1 0 0 1 1 2 1 0 0 0 1 2 3 1 0 3 2 1 0 0 0 0 1 0 1
0 / 1 0 0 / 1 0 ?
? ? 0 0 0 ? ? ? ? ? ? ? ? ? 0 ? ? ? ? ? ? 0 ? ? ? ? ? ? ? ? 1 1 1 1 0 0 ? ? ? ? ? ? ? 0 ? 1
? ? ? ? ? ? ? ? ? ? ? ? ? 1 ? ? 0 2

Limnopithecus legetet 0 0 0 0 1 1 0 1 1 0 0 1 1 0 1 0 1 0 0 2 1 0 1 2 0 0 0 0 0 0 2
1 0 0 0 1 1 3 0 0 1 0 1 0 1 0 0 0 / 1 1 2 0 1 0 1 1 1 2 1 0 3 2 0 0 0 0 0 0 0 0 1
0 1 1 0 ?
1 0 ? ? 0 1 ? 0 0 0 ? ? ? ? ? ? ? ? ? ? ? ? ? 0 1 ? 0 ? ? 1 1 0 0 ? 1 0 0 0 0 ? ? ? ? ? ?
? ? ? ? ? ? ? ? ? ? ? ? ? 2

Limnopithecus evansi 0 ? 0 ? 1 1 0 1 1 0 0 1 1 1 1 - 1 1 1 0 1 0 0 0 1 0 0 2 1
0 0 0 0 1 0 3 0 0 1 0 0 1 1 0 0 1 2 1 0 0 0 1 1 1 2 1 0 2 2 1 0 0 0 0 1 0 1 0 0
0 0 ? 0 0 ? ? ? ?

? ? 0 0 ? ? ? ? ? ? ? ? ? ? ? ? ? ? ? ? ? ? ? 0 1 0 1 ? ? ? 1 0 0 0 ? ? ? ? ? ? ? 0 ? 0 ? ? ? ? ?
? ? ? 1 ? ? ? ? 1 ? 0 0 2

Dendropithecus macinnesi 0 0 0 0 1 0 0 0 0 1 1 1 0 0 1 1 2 1 0 0 1 0 0 0 0 0
0 2 1 0 0 0 0 1 1 0 0 0 1 0 1 1 1 0 0 1 3 2 1 0 0 1 1 1 1 1 0 3 2 0 0 0 0 0 1 0
1 0 0 0 0 ? ? ? ? ? ? ? ? ? ? ? ? ? ? 0 ? ? ? ? ? ? ? 0 0 1 ? 0 0 1 0 1 0 0 0 0 0 0 ? ? 0 ? ? 3
0 1 0 1 ? ? ? ? ? 0 0 0 1 0 0 ? 0 ? ? ? ? ? ? ? ? ? 0 ? ? ? 0 ? ? ? ? ? 1 1 1 0 0 0 0 0 0 0
? ? ? 1 ? 0 ? ? ? ? 0 ? ? ? ? ? ? ? ? ? 0 0 2

Rangwapithecus gordoni 0 ? 0 1 0 1 1 2 1 0 0 1 2 1 1 0 - 0 0 0 3 3 1 0 0 1 0 0
2 0 1 0 ? 1 0 0 1 0 1 2 0 0 0 1 0 0 1 2 1 0 0 0 1 1 0 0 1 0 3 2 0 0 0 0 1 0 0 1
0 0 1 0 ?
? ? ? ? 0 0 ? ? ? ? ? ? ? ? ? ? ? ? ? ? ? 0 1 1 0 ? ? 1 1 0 0 ? ? 1 0 0 ? ? ? ? 1 ? 1 ? ? ?
? 0 ? ? ? ? ? ? ? ? ? ? 0 ? 2

Turkanapithecus kalakolensis ? ? ? ? 1 1 1 2 1 2 0 2 2 1 0 0 - 1 0 ? 2 3 1 1 0 1 0
0 3 1 1 0 0 1 1 0 2 0 2 1 ? ? ? ? ? ? 1 ? ? ? ? 0 1 1 ? 0 1 1 3 ? 1 0 0 0 1 1 1 1 ? ? 1
1 0 ? 1 0 1 0 1 1 ? ? 1
0 1 0 1 1 0 0 0 1 ? ? ? ? ? ? ? ? ? ? ? ? ? 0 1 1 1 1 1 1 1 1 ? ? ? ? ? ? ? ? ? 0 1 1
1 0 ? 1 0 0 0 0 0 0 1 0 ? ? 1 3

Proconsul nyanzae 0 0 0 0 1 1 0 0 0 0 0 1 1 0 0 1 1 2 1 1 1 2 0 0 0 0 0 0 0 2 1 1
0 1 1 0 1 3 0 0 1 0 0 0 ? 0 0 1 2 3 0 0 0 1 1 2 2 1 0 3 2 0 0 0 0 0 0 0 0 1 0 0 1
1 ? 0 0 0 0 0 0 0 0 ? ? 0 ? 1 ? ? ? ? ? ? ? ? ? ? ? 1 1 1 ? 1 1 1 0 ? ? ? ? ? ? ? ? ? ? ? ?
? ? ? ? ? ? ? ? ? ? 0 0 ? ? 1 ? 0 ? 0 ? ? ? 0 0 1 0 1 0 1 1 1 0 1 0 ? 0 0 0 1 ? 0 ? 0 ?
0 ? ? 1 0 1 ? 0 ? 0 0 0 0 1 ? 1 ? 0 ? 3

Proconsul heseloni 0 0 0 0 1 1 0 0 1 0 0 1 1 0 0 1 1 2 2 1 1 1 0 0 0 0 0 1 1 1 0
0 0 0 0 0 3 0 0 1 0 0 0 ? 0 0 1 2 3 1 0 0 1 1 2 1 1 0 3 2 0 0 0 0 0 0 0 0 1 1 0 0 0
1 0 0 0 0 0 0 0 0 0 ? ? ? ? 1 0 0 0 0 0 1 0 1 1 0 1 1 1 1 1 1 1 1 0 1 0 1 ? 0 ? ? ?
1 0 0 1 1 1 0 0 0 0 1 1 1 0 0 0 1 0 0 0 0 1 0 0 1 0 0 0 ? 0 ? ? ? 0 0 1 1 1 0 0 1
0 1 0 0 1 0 0 2 0 2 0 1 1 0 0 1 0 0 0 0 0 0 1 0 1 ? 0 1 3

Gorilla gorilla 0 1 1 1 0 1 0 1 1 0 0 2 2 0 0 2 1 1 1 0 2 3 1 0 1 1 1 1 1 3 1 0 0
1 0 0 3 0 1 2 0 0 1 1 0 1 1 1 3 0 1 0 1 1 1 2 1 0 3 2 0 0 0 0 1 0 0 1 0 0 0 1 1
2 2 1 1 1 1 1 1 3 0/2 1 0 1 1 1 1 1 1 0 2 1 1 1 1 1 1 1 1 1 1 1 0 1 2 0 2 1 1
0 0 1 1 1 0 1 1 1 0 0 1 1 2 0 1 1 1 1 1 0 1 1 1 1 0 1 1 1 1 0 0 1 1 0 1 2 1 1 1
0 0 2 1 0 1 0 1 0 2 1 2 2 2 1 1 0 1 1 0 0 0 0 0 0 0 0 1 0 1 1 8

Pan troglodytes 0 1 1 1 1 1 1 1 0 0 2 2 0 0 2 1 2 2 0 2 2 1 0 1 1 1 1 1 2 1 0
0 1 1 0 3 0 1 2 0 1 1 1 0 1 1 1 3 0 0 0 1 1 2 2 1 2 3 2 0 0 0 0 0 0 0 0 1 1 0 0 1
1 2 2 1 1 1 1 1 1 3 0/2 1 0 1 1 1 1 1 1 0 0 2 2 1 1 1 1 1 1 1 1 1 1 1 0 1 2 0 2 1

1 0 1 1 1 1 0 1 1 1 1 0 1 1 2 0 1 1 1 1 1 0 1 1 1 1 0 1 1 1 1 0 0 1 1 0 1 2 1 1
1 0 0 2 1 0 0 0 0 0 2 0 2 2 2 1 1 0 1 1 0 0 0 0 1 0 0 0 1 0 1 1 **8**

Pongo pygmaeus 0 1 1 0 1 1 0 1 2 1 0 3 2 1 0 2 1 2 2 1 2 3 1 0 0 0 2 1 0 2 1 0
0 0/1 0 0 1 0 1 1 0 1 1 1 0 1 1 1 3 1 0 0 1 1 2 3 1 1 3 2 0 0 0 0 0 0 0 1 0 0 0
1 1 2 2 1 1 1 1 1 1 3 2 1 0 1 1 1 1 1 1 0 2 2 1 1 1 1 1 1 1 1 1 0 1 2 0 2 1
1 0 0/1 1 1 1 0 1 1 1 1 0 1 2 2 0 0 1 1 0 1 0 0 1 1 1 0 0 1 1 1 0 0 1 1 0 1 2 1
1 1 0 0 2 0 0 0 0 0 2 0 2 2 0 1 1 0 1 0 0 0 1 0 1 0 0 1 1 0 1 1 **9**

Hylobates spp. 0 0 1 1 0 1 0 0 0 0 0 3 3 0 0 2 1 2 1 0 2 3 1 0 0 0 1 1 1 3 1 0 0
0 0 0 3 0 1 1 0 1 1 1 0 0 1 2 1 0 1 0 1 1 1 2 1 2 3 2 0 0 0 0 1 0 1 1 1 0 0 1 1
2 1 0 0/1 1 1 1 1 2 1/2 1 0 1 1 1 1 1 1 0 0 2 2 1 1 1 1 1 1 1 1 1 1 0 0 2 0 2
1 1 0 0 1 1 1 0 1 1 1 0 0 1 2 2 0 0 0 1 0 1 2 0 - 1 ? 0 0 1 1 0 0 0 1 1 0 0 2 0 1
1 0 0 1 0 0 0 1 ? 0 2 1 2 2 1 1 0 1 0 1 0 2 0 1 1 0 1 1 1 0 0 1 **9**

Morotopithecus bishopi ? 0 1 0 2 1 0 1 1 1 0 1 0 0 1 2 - 1 1 1 2 3 1 0 0 0 0 0 2
1 0 0 1 0 0 1 1 ? 0 1 ? ? ? ? ? ? 1 ? ? ? ? 0 1 1 2 ? 1 1 2 2 0 0 0 0 0 1 1 ? ? ? ? ? ? 2
2 1 1 1 1 1 1 ? ? ? ? ? ? 1 ?
? ? ? ? ? ? ? ? ? ? ? ? ? ? ? ? ? 0 1 0 ? ? 2 ? ? ? ? ? ? ? ? ? ? ? ? 0 ? 1 ? 0 1 0 1 ? 0 0 0
0 0 0 0 0 1 ? 1 1 2 **2**

REFERENCES

Alroy, J. 2001, Stratigraphy in phylogeny reconstruction—reply to Smith (2000). *J. Paleontol.* 76:587–589.

Andrews, P. 1978, A revision of the Miocene Hominoidea of East Africa. *Bull. British Museum (Natural History), Geology Series* 30:85–224.

Andrews, P. J., Harrison, T., Delson, E., Bernor, R. L., and Martin, L. 1996, Distribution and biochronology of European and Southwest Asian Miocene catarrhines, in: Bernor, R. L., Fahlbusch, V., and Mittmann, H., eds., *The Evolution of Western Eurasian Neogene Mammalian Faunas*. Columbia University Press, New York, pp. 168–207.

Beard, K. C. 1998, East of Eden: Asia as an important biogeographic center of taxonomic origination in mammalian evolution. *Bull. Carnegie Museum Nat. Hist.* 34:5–39.

Beard, K. C. 2002, East of Eden at the Paleocene/Eocene boundary. *Science* 295:2028–2029.

Begun, D. R. 2002, The Pliopithecoidea, in: Hartwig, W. C., ed., *The Primate Fossil Record*. Cambridge University Press, Cambridge, pp. 221–240.

Begun, D. R., Ward, C. V., and Rose, M. D. 1997, Events in hominoid evolution, in: Begun, D. R., Ward, C. V., and Rose, M. D., eds., *Function, Phylogeny, and Fossils: Miocene Hominoid Evolution and Adaptations*. Plenum, New York, pp. 389–415.

Bernor, R. L. 1983, Geochronology and zoogeographic relationships of Miocene Hominoidea, in: Ciochon, R. L., and Corruccini, R. S., eds., *New Interpretations of Ape and Human Ancestry*. Plenum Press, New York, pp. 21–64.

Bernor, R. L., Flynn, L. J., Harrison, T., Hussain, S. T., and Kelley, J. 1988, *Dionysopithecus* from southern Pakistan and the biochronology and biogeography of early Eurasian catarrhines. *J. Hum. Evol.* 17:339–358.

Bernor, R. L. and Tobien, H. 1990, The mammalian geochronology and biogeography of Pasalar (middle Miocene, Turkey). *J. Hum. Evol.* 19:551–568.

Bloch, J. I., Fisher, D. C., Rose K. D., and Gingerich, P. D. 2001, Stratocladistic analysis of Paleocene Carpolestidae (Mammalia, Plesiadapiformes) with description of a new late Tiffanian genus. *J. Vert. Paleontol.* 21:119–131.

Bodenbender, B. E., and Fisher, D. C. 2001, Stratocladistic analysis of blastoid phylogeny. *J. Paleontol.* 75:351–369.

Bretsky, S. S. 1979, Recognition of ancestor-descendant relationships in invertebrate paleontology, in: Cracraft, J. and Eldredge, N., eds., *Phylogenetic Analysis and Paleontology*. Columbia University Press, New York, pp. 113–164.

Brooks, D. R. 1990, Parsimony analysis in historical biogeography and coevolution: Methodological and theoretical update. *Syst. Zool.* 39.14 30.

Bush, G. L. 1975, Modes of animal speciation. *Ann. Rev. Ecol. Sys.* 6:339–364.

Cartmill, M. 1975, Strepsirhine basicranial structures and the affinities of the Cheirogaleidae, in: Luckett, W. P. and Szalay, F. S., eds., *Phylogeny of the Primates: A Multidisciplinary Approach*. Plenum Press, New York, pp. 313–356.

Chaimanee, Y., Suteethorn, V., Jaeger, J. J., and Ducrocq, S. 1997, A new late Eocene anthropoid primate from Thailand. *Nature* 385:429–431.

Ciochon, R. L. and Gunnell, G. F. 2002, Eocene primates from Myanmar: Historical perspectives on the origin of Anthropoidea. *Evol. Anthropol.* 11:156–168.

Court, N. 1993, Morphology and functional anatomy of the postcranial skeleton in *Arsinoitherium* (Mammalia, Embrithopoda). *Palaeontographica, Abt. A* 226:125–169.

Court, N. 1994, Limb posture and gait in *Numidotherium koholense*, a primitive proboscidean from the Eocene of Algeria. *Zool. J. Linn. Soc.* 111:297–338.

Cracraft, J. 1981, Pattern and process in paleobiology: The role of cladistic analysis in systematic paleontology. *Paleobiology* 7:456–468.

Darwin, C. 1859, *The Origin of Species*. Murray, London.

Domning, D., Ray, C. E., and McKenna, M. C. 1986, Two new Oligocene desmostylians and a discussion of tethytherian systematics. *Smithson. Contrib. Paleobiol.* 59:1–56.

Ducrocq, S. 1997, The anthracotheriid genus *Bothriogenys* (Mammalia, Artiodactyla) in Africa and Asia during the Paleogene: Phylogenetical and paleobiogeographical relationships. *Stuttgarter Beitr. Naturk., Ser. B* 250:1–44.

Ducrocq, S. 1999, *Siamopithecus eocaenus*, a late Eocene anthropoid primate from Thailand: Its contribution to the evolution of anthropoids in Southeast Asia. *J. Hum. Evol.* 36:613–635.

Eizirik, E., Murphy, W. J., Springer, M. S., and O'Brien, S. J. 2004, Molecular phylogeny and dating of early primate divergences, in: Ross, C. F. and Kay, R. F., eds., *Anthropoid Origins: New Visions*. Kluwer Academic Press, New York, pp. 45–64.

Eldredge, N. and Cracraft, J. 1980, *Phylogenetic Patterns and the Evolutionary Process*. Columbia University Press, New York.

Farris, J. S. 1983, The logical basis of phylogenetic analysis, in: Platnick, N. and Funk, V. A., eds., *Advances in Cladistics*, Vol. 2. Columbia University Press, New York.

Felsenstein, J. 1978, Cases in which parsimony or compatibility methods will be positively misleading. *Sys. Zool.* 27:401–410.

Finarelli, J. A. and Clyde, W. C. 2004, Reassessing hominoid phylogeny: Evaluating congruence in the morphological and temporal data. *Paleobiology* 30:614–651.

Fisher, D. C. 1992, Stratigraphic parsimony, in: Maddison, W. P. and Maddison, D. R., eds., *MacClade: Analysis of Phylogeny and Character Evolution*. Sinauer Associates, Sunderland, pp. 124–129.

Fisher, D. C. 1994, Stratocladistics: Morphological and temporal patterns and their relation to phylogenetic process, in: Grande, L. and Rieppel, O., eds., *Interpreting the Hierarchy of Nature—From Systematic Patterns to Evolutionary Process Theories*. Academic Press, Orlando, FL, pp. 133–171.

Fleagle, J. G. 1997, Beyond parsimony. *Evol. Anthropol.* 6:1.

Fleagle, J. G. and Kay, R. F. 1987, The phyletic position of the Parapithecidae. *J. Hum. Evol.* 16:483–532.

Fox, D. L., Fisher, D. C., and Leighton, L. R. 1999, Reconstructing phylogeny with and without temporal data. *Science* 284:1816–1819.

Gauthier, J., Kluge, A. G., and Rowe, T. 1988, Amniote phylogeny and the importance of fossils. *Cladistics* 4:105–209.

Gheerbrant, E. 1995, Les mammifères paléocenes du Bassin d'Ouarzazate (Maroc). III. Adapisoriculidae et autres mammifères (Carnivora, ?Creodonta, Condylarthra, ?Ungulata et *incertae sedis*). *Palaeontographica, Abt. A* 237:39–132.

Gift, N. and Stevens, P. F. 1997, Vageries in the delimitation of character states in quantitative variation—an experimental study. *Syst. Biol.* 46:112–125.

Gingerich, P. D. 1976, Cranial anatomy and evolution of early Tertiary Plesiadapidae (Mammalia: Primates). *University of Michigan, Museum of Paleontology, Papers on Paleontology* 15:1–141.

Graham, A. 1999, *Late Cretaceous and Cenozoic history of North American vegetation north of Mexico*. Oxford University Press, New York.

Grande, L. 1985, The use of paleontology in systematics and biogeography, and a time control refinement for historical biogeography. *Paleobiology* 11:234–243.

Gunnell, G. F., Ciochon, R. L., Gingerich, P. D., and Holroyd, P. A. 2002, New assessment of *Pondaungia* and *Amphipithecus* (Primates) from the late middle Eocene of Myanmar, with a comment on "Amphipithecidae". *Contributions from the Museum of Paleontology, University of Michigan* 30:337–372.

Harrison, T. 1982, Small-bodied Apes from the Miocene of East Africa. Ph.D., University College London, London.

Harrison, T. 1987, The phylogenetic relationships of the early catarrhine primates: a review of the current evidence. *J. Hum. Evol.* 16:41–80.

Harrison, T. 1988, A taxonomic revision of the small catarrhine primates from the early Miocene of East Africa. *Folia Primatol.* 50:59–108.

Harrison, T. 2002, Late Oligocene to middle Miocene catarrhines from Afro-Arabia, in: Hartwig, W. C., ed., *The Primate Fossil Record*. Cambridge University Press, Cambridge pp. 311–338.

Holroyd, P. A. and Maas, M. C. 1994, Paleogeography, paleobiogeography, and anthropoid origins, in: Fleagle, J. G. and Kay, R. F., eds., *Anthropoid Origins*. Plenum Press, New York, pp. 297–334.

Hooker, J. J. and Dashzeveg, D. 2003, Evidence for direct mammalian faunal interchange between Europe and Asia near the Paleocene-Eocene boundary, in: Wing, S. L., Gingerich, P. D., Schmitz, B., and Thomas, E., eds., *Causes and Consequences of Globally Warm Climates in the Early Paleogene. Geological Society of America Special Paper 369*. Geological Society of America, Boulder, Colorado, pp. 479–500.

Huelsenbeck, J. 1994, Comparing the stratigraphic record to estimates of phylogeny. *Paleobiology* 20:470–483.

Humphries, C. J. and Parenti, L. R. 1986, *Cladistic Biogeography*. Oxford University Press, Oxford.

Hunn, C. A. and Upchurch, P. 2001, The importance of time/space in diagnosing the causality of phylogenetic events: Towards a "chronobiogeographical" paradigm? *Syst. Biol.* 50:391–407.

Iakovleva, A. I., Brinkhuis, H., and Cavagnetto, C. 2001, Late Paleocene—early Eocene dinoflagellate cysts from the Turgay Strait, Kazakhstan: Correlations across ancient seaways. *Palaeogeogr. Palaeoclimatol. Palaeoecol.* 172:243–268.

Jaeger, J.-J., Denys, C., and Coiffait, B. 1985, New Phiomorpha and Anomaluridae from the late Eocene of north-west Africa: Phylogenetic implications, in: Luckett, W. P. and Hartenberger, J.-L., eds., *Evolutionary Relationships among Rodents—A Multidisciplinary Analysis*. Plenum Press, New York, pp. 567–588.

Kappelman, J., Rasmussen, D. T., Sanders, W. J., Feseha, M., Bown, T., Copeland, P., Crabaugh, J., Fleagle, J., Glantz, M., Gordon, A., Jacobs, B., Maga, M., Muldoon, K., Pan, A., Pyne, L., Richmond, B., Ryan, T., Seiffert, E. R., Sen, S., Todd, L., Wiemann, M. C., and Winkler, A. 2003, Oligocene mammals from Ethiopia and faunal exchange between Afro-Arabia and Eurasia. *Nature* 426:549–552.

Kay, R. F., Williams, B. A., Ross, C. F., Takai, M., and Shigehara, N. 2004, Anthropoid origins: A phylogenetic analysis, in: Ross, C. F. and Kay, R. F., eds., *Anthropoid Origins: New Visions*. Kluwer, New York, pp. 91–135.

Kim, J. 1993, Improving the accuracy of phylogenetic estimation by combining different methods. *System. Biol.* 42:331–340.

Kluge, A. G. and Farris, J. S. 1969, Qualitative phyletics and the evolution of anurans. *System. Zool.* 18:1–32.

Krause, D. W., Evans, S. E., and Gao, K.-Q. 2003, First definitive record of Mesozoic lizards from Madagascar. *J. Vert. Paleontol.* 23:842–856.

Lieberman, B. S. and Eldredge, N. 1996, Trilobite biogeography in the Middle Devonian: Geological processes and analytical methods. *Paleobiology* 22:66–79.

Maas, M. C., Thewissen, J. G. M., and Kappelman, J. 1998, *Hypsamasia seni* (Mammalia, Embrithopoda) and other mammals from the Eocene Kartal Formation of Turkey. *Bull. Carn. Mus. Nat. Hist.* 34:286–297.

MacLatchy, L. 2004, The oldest ape. *Evol. Anthropol.* 13:90–103.

Maddison, W. P. and Maddison, D. R. 1992, *MacClade: Analysis of Phylogeny and Character Evolution, Version 3.0*. Sinauer Associates, Inc, Sunderland.

Madsen, O., Scally, M., Douady, C. J., Kao, D. J., DeBry, R. W., Adkins, R., Amrine, H. M., Stanhope, M. J., de Jong, W. W., and Springer, M. S. 2001, Parallel adaptive radiations in two major clades of placental mammals. *Nature* 409:610–614.

Marinkovitch, L. and Gladenkov, A. Y. 1998, An early opening of the Bering Strait. *Nature* 397:149–151.

Mayr, E. 1942, *Systematics and the Origin of Species*. Columbia University Press, New York.

Mayr, E. 1963, *Animal Species and Evolution*. Harvard University Press, Cambridge.

McCall, R. A. 1997, Implications of recent geological investigations of the Mozambique Channel for the mammalian colonization of Madagascar. *Proc. R. Soc. Lond. B* 264:663–665.

Morrone, J. J. and Carpenter, J. M. 1994, In search of a method for cladistic biogeography: An empirical comparison of component analysis, Brooks parsimony analysis, and three-area statements. *Cladistics* 10:99–153.

Muldoon, K. M. and Gunnell, G. F. 2002, Omomyid primates (Tarsiiformes) from the Early Middle Eocene at South Pass, Greater Green River Basin, Wyoming. *J. Hum. Evol.* 43:479–511.

Murphy, W. J., Eizirik, E., Johnson, W. E., Zhang, Y. P., Ryder, O. A., and O'Brien, S. J. 2001a, Molecular phylogenetics and the origins of placental mammals. *Nature* 409:614–618.

Murphy, W. J., Eizirik, E., O'Brien, S. J., Madsen, O., Scally, M., Douady, C. J., Teeling, E., Ryder, O. A., Stanhope, M. J., de Jong, W. W., and Springer, M. S. 2001b, Resolution of the early placental mammal radiation using Bayesian phylogenetics. *Science* 294:2348–2351.

Nelson, G. and Platnick, N. I. 1981, *Systematics and Biogeography: Cladistics and Vicariance.* Columbia University Press, New York.

Nikaido, M., Nishihara, H., Hukumoto, Y., and Okada, N. 2003, Ancient SINEs from African endemic mammals. *Mol. Biol. Evol.* 20:522–527.

Norell, M. A. and Novacek, M. J. 1992, Congruence between superpositional and phylogenetic patterns: comparing cladistic patterns with fossil records. *Cladistics* 8:319–337.

Olson, L. E. and Goodman, S. M. 2003, Phylogeny and biogeography of tenrecs, in: Goodman, S. M. and Benstead, J. P., eds., *The Natural History of Madagascar.* The University of Chicago Press, Chicago, pp. 1235–1242.

Pickford, M. L. 2002, New reconstruction of the Moroto hominoid snout and a re-assessment of its affinities to *Afropithecus turkanensis. Hum. Evol.* 17:1–19.

Pilbeam, D. 1996, Genetic and morphological records of the Hominoidea and hominid origins: A synthesis. *Mol. Phylogen. Evol.* 5:155–168.

Pilbeam, D. R. and Walker, A. C. 1968, Fossil monkeys from the Miocene of Napak, North-east Uganda. *Nature* 220:657–660.

Platnick, N. I. and Nelson, G. 1978, A method of analysis for historical biogeography. *Syst. Zool.* 27:1–16.

Pogue, M. G. and Mickevich, M. 1990, Character definitions and character state delineation: The bête noire of phylogenetic inference. *Cladistics* 6:319–361.

Polly, P. D. 1997, Ancestry and species definition in paleontology: A stratocladistic analysis of Paleocene-Eocene Viverravidae (Mammalia, Carnivora) from Wyoming. *Contributions from the Museum of Paleontology, University of Michigan* 30:1–53.

Porter, C. A., Page, S. L., Czelusniak, J., Schneider, H., Schneider, M. P. C., Sampaio, I., and Goodman, M. 1997, Phylogeny and evolution of selected primates as determined by sequences of the â-globin locus and 5' flanking regions. *Int. J. Primatol.* 18:261–295.

Porter, C. A., Sampaio, I., Schneider, H., Schneider, M. P. C., Czelusniak, J., and Goodman, M. 1995, Evidence on primate phylogeny from e-globin gene sequences and flanking regions. *J. Mol. Evol.* 40:30–55.

Radulesco, C., Iliesco, G., and Iliesco, M. 1976, Decouverte d'un Embrithopode nouveau (Mammalia) dans la Paléogène de la dépression de Hateg (Roumanie) et considération générales sur la géologie de la région. *N. Jb. Geol. Paläon. Monat.* 11:690–698.

Rasmussen, D. T. 2002, Early catarrhines of the African Eocene and Oligocene, in: Hartwig, W. C., ed., *The Primate Fossil Record.* Cambridge University Press, Cambridge, pp. 203–220.

Ritchie, J. D. and Hitchen, K. 1996, Early Paleogene offshore igneous activity to the northwest of the UK and its relationship to the North Atlantic Igneous Province. *Geolog. Soc. Lond. Spec. Publ.* 101:63–78.

Robinson, T. J. and Seiffert, E. R. 2004, Afrotherian origins and interrelationships: New views and future prospects. *Curr. Top. Dev. Biol.* 63:37–60.

Rogers, R. R., Hartman, J. H., and Krause, D. W. 2000, Stratigraphic analysis of Upper Cretaceous rocks in the Mahajanga Basin, northwestern Madagascar: implications for ancient and modern faunas. *J. Geol.* 108:275–301.

Rögl, F. 1999, Oligocene and Miocene palaeogeography and stratigraphy of the circum-Mediterranean region, in: Whybrow, P. J. and Hill, A., eds., *Fossil Vertebrates of Arabia.* Yale University Press, New Haven, pp. 485–500.

Ronquist, F. 1997, Dispersal-vicariance analysis: A new approach to the quantification of historical biogeography. *Syst. Biol.* 46:195–203.

Roos, C., Schmitz, J., and Zischler, H. 2004, Primate jumping genes elucidate strepsirrhine phylogeny. *Proc. Nat. Acad. Sci. USA.* 101:10650–10654.

Rose, M. D. 1997, Functional and phylogenetic features of the forelimb in Miocene hominoids, in: Begun, D. R., Ward, C. V., and Rose, M. D., eds., *Function, Phylogeny, and Fossils: Miocene Hominoid Evolution and Adaptations.* Plenum Press, New York, pp. 79–100.

Rose, M. D., Leakey, M. G., Leakey, R. E. F., and Walker, A. C. 1992, Postcranial specimens of *Simiolus enjiessi* and other primitive catarrhines from the early Miocene of Lake Turkana, Kenya. *J. Hum. Evol.* 22:171–237.

Ross, C. F. 2000, Into the light: The origin of Anthropoidea. *Ann. Rev. Anthropol.* 29:147–194.

Ross, C. F. and Kay, R. F. 2004, Anthropoid origins: Retrospective and prospective. in: Ross, C. F. and Kay, R. F., eds., *Anthropoid Origins: New Visions.* New York, Kluwer Academic Press, pp. 701–737.

Rossie, J. B. and MacLatchy, L. 2004, A new species of stem catarrhine from the early Miocene of Uganda. *Am. J. Phys. Anthropol. (Suppl. 38)* 123:170.

Schmitz, J. and Zischler, H. 2004, Molecular cladistic markers and the infraordinal phylogenetic relationships of primates, in: Ross, C. F. and Kay, R. F., eds., *Anthropoid Origins: New Visions.* Kluwer Academic Press, New York, pp. 65–77.

Schwartz, J. H. and Tattersall, I. 1985, Evolutionary relationships of living lemurs and lorises (Mammalia, Primates) and their potential affinities with European Eocene Adapidae. *Anthropol. Papers Am. Museum Nat. Hist.* 60:1–100.

Scotese, C. R. 2002, PALEOMAP reconstructions. Department of Geology, University of Texas at Arlington, Arlington.

Seiffert, E. R. 2003, A Phylogenetic Analysis of Living and Extinct Afrotherian Placentals. Ph.D., Duke University, Durham, North Carolina.

Seiffert, E. R., Simons, E. L., and Attia, Y. 2003, Fossil evidence for an ancient divergence of lorises and galagos. *Nature* 422:421–424.

Seiffert, E. R., Simons, E. L., and Simons, C. V. M. 2004, Phylogenetic, biogeographic, and adaptive implications of new fossil evidence bearing on crown anthropoid origins and early stem catarrhine evolution, in: Ross, C. F. and Kay, R. F., eds., *Anthropoid Origins: New Visions*. Kluwer Academic Press, New York, pp. 157–181.

Siddall, M. F. 1998, Stratigraphic fit to phylogenies: A proposed solution. *Cladistics* 14:201–208.

Simons, E. L. and Bown, T. M. 1985, *Afrotarsius chatrathi*, first tarsiiform primate (?Tarsiidae) from Africa. *Nature* 313:475–477.

Singleton, M. 2000, The phylogenetic affinities of *Otavipithecus namibiensis*. *J. Hum. Evol.* 38:537–573.

Smith, A. B. 2000, Stratigraphy in phylogeny reconstruction. *J. Paleontol.* 74:763–766.

Smith, A. G., Smith, D. G., and Funnell, B. M. 1994, *Atlas of Mesozoic and Cenozoic Coastlines*. Cambridge University Press, Cambridge.

Sober, E. 1988, *Reconstructing the Past: Parsimony, Evolution, and Inference*. MIT Press, Cambridge.

Springer, M. S., Stanhope, M. J., Madsen, O., and Jong, W. W. d. 2004, Molecules consolidate the placental mammal tree. *Trends Ecol. Evol.* 19:430–438.

Stevens, P. F. 1984, Homology and phylogeny: Morphology and systematics. *Syst. Bot.* 9:395–409.

Stewart, C.-B. and Disotell, T. R. 1998, Primate evolution—in and out of Africa. *Curr. Biol.* 8:R582–R588.

Swofford, D. L. 1998, *PAUP*. Phylogenetic Analysis Using Parsimony (*and Other Methods). Version 4*. Sinauer Associates, Sunderland.

Szalay, F. S. and Katz, C. C. 1973, Phylogeny of lemurs, lorises, and galagos. *Folia Primatol.* 19:88–103.

Tabuce, R. and Marivaux, L. 2005, Mammalian interchanges between Africa and Eurasia: an analysis of temporal constraints on plausible anthropoid dispersals during the Paleogene. *Anthropol. Sci.* 113:27–32.

Takai, M., Anaya, F., Shigehara, N., and Setoguchi, T. 2000, New fossil materials of the earliest New World monkey, *Branisella boliviana*, and the problem of platyrrhine origins. *Am. J. Phys. Anthropol.* 111:263–281.

Tedford, R. H., Storer, J. E., Albright, III L. B., Swisher III, C. C., and Barno, A. T. 2004, Mammalian biochronology of the Arikareean through Hemphillian interval (late Oligocene through early Pliocene epochs), in: Woodburne, M. O., ed., *Late Cretaceous and Cenozoic Mammals of North America*, pp. 169–231.

Thewissen, J. G. M. 1992, Temporal data in phylogenetic systematics: An example from the mammalian fossil record. *J. Paleontol.* 66:1–8.

Thewissen, J. G. M., Williams, E. M., and Hussain, S. T. 2000, Anthracobunidae and the relationships among Desmostylia, Sirenia, and Proboscidea. *J. Vert. Paleontol.* 20:73A.

Upchurch, P., Hunn, C. A., and Norman, D. B. 2002, An analysis of dinosaurian biogeography: Evidence for the existence of vicariance and dispersal patterns caused by geological events. *Proc. R. Soc. Lond. Series B* 269:613–621.

Vianey-Liaud, M., Jaeger, J.-J., Hartenberger, J.-L., and Mahboubi, M. 1994, Les rongeurs de l'Eocène d'Afrique nord-occidentale [Glib Zegdou (Algérie) et Chambi (Tunisie)] et l'origine des Anomaluridae. *Palaeovertebrata* 23:93–118.

Wagner, P. J. 1995, Stratigraphic tests of cladistic hypotheses. *Paleobiology* 21:153–178.

Wagner, P. J. 2001, Testing phylogenetic hypotheses with stratigraphy and morphology—a comment on Smith (2000). *J. Paleontol.* 76:590–593.

Ward, C. V., Walker, A. C., Teaford, M., and Odhiambo, I. 1993, Partial skeleton of *Proconsul nyanzae* from Mfangano Island, Kenya. *Am. J. Phys. Anthropol.* 90:77–111.

Wiens, J. J. 2001, Character analysis in morphological phylogenetics: problems and solutions. *Syst. Biol.* 50:689–699.

Wiley, E. O. 1981, *Phylogenetics. The Theory and Practice of Phylogenetic Systematics.* Wiley, New York.

Yoder, A. D. 1997, Back to the future: A synthesis of strepsirrhine systematics. *Evol. Anthropol.* 6:11–22.

Yoder, A. D., Burns, M. M., Zehr, S., Delefosse, T., Veron, G., Goodman, S. M., and Flynn, J. J. 2003, Single origin of Malagasy Carnivora from an African ancestor. *Nature* 421:734–737.

Yoder, A. D., Cartmill, M., Ruvolo, M., Smith, K., and Vilgalys, R. 1996, Ancient single origin for Malagasy primates. *Proc. Nat. Acad. Sci. USA.* 93:5122–5126.

Young, N. M. and MacLatchy, L. 2004, The phylogenetic position of *Morotopithecus*. *J. Hum. Evol.* 46:163–184.

Zachos, J., Pagani, M., Sloan, L., Thomas, E., and Billups, K. 2001, Trends, rhythms and aberrations in global climate 65 Ma to present. *Science* 292:686–693.

TAXONOMIC INDEX

SUBJECT INDEX

DATE DUE

GAYLORD   PRINTED IN U.S.A.

SCI QL 737 .P9 P72 2006

Primate biogeography